Elementary Statistics for Geographers

Second Edition

JAMES E. BURT
GERALD M. BARBER

THE GUILFORD PRESS
New York   London

To Moe and Jan,
Eric, Steven, Tom, Gordie, Mike, and Emily

© 1996 The Guilford Press
A Division of Guilford Publications, Inc.
72 Spring Street, New York, NY 10012

Marketed and distributed outside North America by Longman Group
Limited.

Printed in the United States of America

This book is printed on acid-free paper.

Last digit is print number: 9 8 7 6 5 4 3 2 1

Library of Congress Cataloging-in-Publication Data

Burt, James E.
    Elementary statistics for geographers. — 2nd ed. / by James E.
Burt, Gerald M. Barber.
      p. cm.
    Rev. ed. of: Elementary statistics for geographers / Gerald M.
Barber. c 1988.
    Includes bibliographical references and index.
    ISBN 0-89862-282-4   ISBN 0-89862-999-3 (pbk.)
    1. Geography—Statistical methods.   I. Burt, James E.
II. Barber, Gerald M.   Elementary statistics for geographers.
III. Title.
G70.3.B37   1996
910′.21—dc20   95-44912
                                     CIP

Preface

This edition is both a continuation and an extension of the first. Like the first edition, it rests on the premise that courses in statistical analysis must include fundamentals and theoretical underpinnings. Although we acknowledge that undergraduates often have a limited background in mathematics, we do not believe this justifies a cookbook approach to the subject, nor do we believe that mathematical rigor is beyond the capabilities of students who have yet to experience the joys of calculus. Simply put, our aim is to provide an algebra-based text which stresses basic principles. We believe this to be essential if students are to utilize statistical methods in their own research or if they are to evaluate critically the work of others who employ such methods.

For this reason, the book gives extensive treatment to certain topics commonly avoided or only briefly covered in many introductory statistical texts. For example, two chapters are devoted to probability theory and random variables. Many of the principles of statistical inference are based on probability, and it is foolish to suppose that geographers can somehow understand the logic of statistical inference without it. Likewise, the discussion of sampling distributions is more detailed than what is found in many texts, and the same is true for many other topics.

This edition extends the former version by including new material on exploratory data analysis, time series, and computer-intensive methods. Exploratory data analysis was covered briefly in the first edition; we have expanded the treatment here to a full chapter, as befits the increased role of EDA in geographical research. The inclusion of time series perhaps requires more explanation, given that the topic seldom appears in beginning texts. We see this as an oversight for several reasons. First and foremost, time series data are extremely common in all branches of geography; thus geographers need to be equipped with at least a few tools for using temporal data. Secondly, once a student has learned about linear regression, he or she is well-positioned to understand the basics of time series modeling. In other words, the essentials of time series can be acquired at little extra cost in terms of new material. Finally, we note that because time series are so common, geographers are likely to be operating on temporal data with or without formal exposure to the subject. We believe that even simple operations like running means should not be undertaken without appreciating the implications of the procedure. Because most students will not find time for a full course in time series, at least minimal coverage seems essential in an introductory text.

The coverage of time series is split over two chapters. The first treats fundamentals: basic terminology, why time series are special, and the mechanics of running means and other simple smoothing operators. In the later chapter we cover rudiments of time series models and take a deeper look at smoothing as a special case of filtering. With regard to the latter, our goal is twofold: (1) we try to show what running means and other common filters *do to* the data, as opposed to how they are *applied to* the data, and (2) we want to introduce the topic of designing a filter to accomplish a particular goal, as opposed to simply living with the effects of whatever filter is used.

The new chapter on computer-intensive methods is also unusual; indeed, we do not know of any other introductory geography text which covers them. This is surprising, given their widespread use these days (especially in physical geography), and given that almost no additional background is required to understand the techniques. The chapter includes a discussion of three popular resampling methods: bootstrapping, jackknifing, and cross-validation. It also discusses nonparametric methods for estimating probability density and point density.

Obviously, depth and breadth of coverage come at a price: this is a long book. Unless it is used in a year-long course, instructors will have to be selective with regard to what is assigned. With this in mind, we have attempted to make the chapters as self-contained as possible. Except for the introductory chapters on probability and sampling theory, a "pick and choose" approach will work well. For example, we know from experience that some instructors leave out the nonparametric chapter altogether with no downstream effects, whereas others skip different chapters and/or cover only certain subsections within chapters. Of course, some students will complain about a hopscotch approach. We find, however, that most students appreciate a book that covers more than what is treated in the course. Later, when confronted with unfamiliar methods in readings or on a research project, they can return to a book whose notational quirks they have already mastered, and can understand the new technique in the context of what was presented in the course. As we reflect on the contents of our bookshelves, it is precisely that kind of book that has proved most useful to us over the years. Though we do not presume to claim that this work will be of similarly lasting utility, we do offer it in the belief that it is better to cover too much than too little.

Many people deserve our thanks for their help in preparing this book. We are particularly grateful to students and teaching assistants at the Universities of Wisconsin and Victoria for telling us what worked and what didn't. Thanks also to Cort Willmott, Ed Aguado, and several anonymous reviewers for their comments on previous versions of the manuscript. We also appreciate the hard work by everyone at The Guilford Press involved wtih the project, especially our editor, Peter Wissoker. Most of all, we thank our families for so willingly accepting the cost of our preoccupation. To them we dedicate the book.

JAMES E. BURT
University of Wisconsin
GERALD M. BARBER
Kingston, Ontario

Contents

II. INFERENTIAL STATISTICS

1

Statistics and Geography

Hardly a day goes by without almost everyone being affected by the applications of statistics. In common usage, the word *statistics* refers to any collection of numerical data. *Vital statistics* such as birth rates and death rates, *economic indicators* such as unemployment rates and money supply figures, and *social statistics* such as juvenile delinquency rates are all statistics used in this sense of the word. The daily newspaper in any major city is likely to contain at least one or more stories that use numerical data. Statistical data of this type have been used for centuries by governments to improve their administration. In antiquity, statistics were compiled to determine the number of citizens from whom either taxes or military service could be required. The role of statistics expanded rapidly after the industrial revolution. Today, statistical data are collected by many individuals, government agencies, public and nonprofit organizations, and businesses. These data are organized in *information systems*, increasingly sophisticated computer-based frameworks for data collection, retrieval, and reporting.

The word *statistics* has another more specialized meaning. It is the *methodology* for collecting, presenting, and analyzing data. This methodology can be used as a basis for investigation in such diverse academic fields as education, physics and engineering, medicine, the biological sciences, and the social sciences including geography. Even traditionally nonquantitative disciplines in the humanities are finding increasing uses for statistical methodology. No matter what the discipline is that uses this methodology, statistical analysis begins with the collection of data. Analysis of this data are then undertaken for one of the following purposes:

1. To help *summarize* the findings of some inquiry, for example, a study of travel behavior of elderly or handicapped citizens or the estimation of timber reforestation requirements;
2. To obtain a better understanding of the phenomenon under study, primarily as an aid in generalization or *theory validation*, for example, to validate a theory of urban land rental patterns;
3. To make a *forecast* of some variable, for example, short-term interest rates, voter behavior, or real estate prices;

4. To *evaluate* the performance of some program, for example, a particular form of business subsidy, or an innovative medical or educational program or reform;
5. To help *select* a course of action among a number of possible alternatives, or to plan some system, for example, school locations.

That elements of statistical methodology can be used in such a variety of situations attests to its impressive versatility.

It is convenient to divide statistical methodology into two parts. *Descriptive statistics* deals with the organization and summary of data. The purpose of descriptive statistics is to replace what may be an extremely large set of numbers in some data set with a smaller number of summary measures. Whenever this replacement is made, there is inevitably some loss of information. It is impossible to retain *all* of the information in a data set using a smaller set of numbers. One of the principal goals of descriptive statistics is to minimize the effect of this information loss. Understanding which statistical measure should be used as a summary index in a particular case is another important goal of descriptive statistics. If we understand the derivation and use of descriptive statistics and are aware of its limitations, we can help avoid the propagation of misleading results. Much of the distrust of statistical methodology derives from its misuse in studies in which it has been inappropriately applied or interpreted. Just as the photographer can use a lens to distort a scene, so a statistician can distort the information in a data set through a poor choice of summary statistics. Understanding what descriptive statistics can tell us, *as well as what it cannot*, is a key concern of statistical analysis.

The second major part of statistical methodology is *inferential statistics*. In this form of statistical analysis, descriptive statistics is linked with probability theory so that an investigator can generalize the results of a study of a few individuals to some larger group. To clarify this process, it is necessary to introduce a few simple definitions. The set of persons, regions, areas, or objects in which a researcher has an interest is known as the *population* for the study.

DEFINITION: STATISTICAL POPULATION

A statistical population is the total set of elements (objects, persons, regions, neighborhoods, rivers, etc.) under examination in a particular study.

For instance, if a geographer is studying farm practices in a particular region, the relevant population consists of all farms in the region on a certain date or within a certain period of time. As a second example, the population for a study of voter behavior in a city would include all potential voters; these people are usually included in an eligible voters list.

In many instances, the statistical population under consideration is *finite*; that is, each element in the population can be listed. The eligible voters lists and the assessment rolls of a city or county are examples of finite populations. At other times, the population may be *hypothetical*. For example, a steel manufacturer wishing to test the quality of output may select a batch of 100 castings over a few weeks of production. The population under study is actually the *future* set of castings to be produced by the manufacturer using this equipment. Of course, this population does not exist and may have an infinitely large number of elements. Statistical analysis is relevant to both finite and hypothetical populations.

Usually, we are interested in one or more characteristics of the population.

DEFINITION: POPULATION CHARACTERISTIC

A population characteristic is any measurable attribute of an element in the population.

A fluvial geomorphologist studying stream flow in a certain watershed may be interested in a number of different measurable properties of these streams. Stream velocity, discharge, sediment load, and many other characteristic channel data may be collected during a field study. Because a population characteristic is likely to take on different values for different elements of the population, it is usually called a *variable*. The fact that the population characteristic does take on different values is what makes the process of statistical inference necessary. If a population characteristic does not vary within the population, it is of little interest to the investigator from an inferential point of view.

DEFINITION: VARIABLE

A variable is a population characteristic which takes on different values for the elements comprising the population.

Information about a population can be collected in two ways. The first is to determine the value of the variable(s) of interest for *each and every* element of the population. This is known as a population *census* or population *enumeration*. Clearly, it is a feasible alternative only for finite populations. It is extremely difficult, some would argue even impossible, for large populations. It is virtually impossible that a national decennial census of population in a large country will actually capture all of the persons in that population, but the errors can be kept to a minimum if the enumeration process is well designed.

DEFINITION: POPULATION CENSUS

A population census is a complete tabulation of a particular population characteristic for all elements in the population.

It is usually both necessary and desirable to determine the values of a variable for only a subset of the individuals in a population. This subset is called a *sample*.

DEFINITION: SAMPLE

A sample is a subset of the elements in the population and is used to make inferences about certain characteristics of the population as a whole.

For practical considerations, usually time and/or cost, it is convenient to sample rather than enumerate the entire population. Of course, sampling has one distinct disadvantage. Restricting our attention to a small proportion of the population makes it impossible to be as accurate about population characteristics as is possible with a complete census. The risk of making errors is increased.

DEFINITION: SAMPLING ERROR

Sampling error is the difference between the value of a population characteristic and the value of that characteristic inferred from a sample.

To illustrate sampling error, consider the population characteristic of the average selling price of homes in a given metropolitan area in a certain year. If each and every house is examined, it is found that the average selling price is $150,000. If only 25 homes per month are sampled, however, and the average selling price of the 300 homes in the sample (12 months × 25 homes) may be $120,000. All other things being equal, we could say that the difference of $150,000 − $120,000 = $30,000 is due to sampling error.

What do we mean by *all other things being equal*? Our error of $30,000 may be partly due to factors other than sampling. Perhaps the selling price for one home in the sample was incorrectly identified as $252,000, rather than $152,000. Many errors of this type occur in large data sets. Information obtained from personal interviews or questionnaires can contain factual errors from respondents owing to lack of recall, ignorance, or simply the respondent's desire to be less than candid.

DEFINITION: NONSAMPLING OR DATA ACQUISITION ERRORS

Errors that arise in the acquisition, recording, and editing of statistical data are called nonsampling or data acquisition errors.

In order that error, or the difference between the sample and the population, can be ascribed solely to sampling error, it is important to *minimize* nonsampling errors. Validation checks, careful editing, and instrument calibration are all methods used to reduce the possibility that nonsampling error is significantly increasing the total error, thereby distorting subsequent statistical inference.

The link between the sample and the population is probability theory. Inferences about the population are based on the information in the sample. The quality of these inferences depends on how well the sample reflects, or represents, the population. Unfortunately, short of a complete census of the population, there is no way of knowing how well a sample reflects the population. So, instead of attempting to select a *representative* sample, we select a *random* sample.

DEFINITION: RANDOM SAMPLE

A random sample is one in which every individual in the population has the same chance, or probability, of being included in the sample.

Basing our statistical inferences on random samples ensures unbiased findings. It is possible to obtain a highly unrepresentative random sample, but the chance of doing so is usually remote if the sample is large enough. In fact, because the sample has been randomly chosen, we can always determine the probability that the inferences made from the sample are misleading. This is why statisticians always make probabilistic judgments, never deterministic ones. The inferences are always qualified to the extent that random sampling error may lead to incorrect judgments.

The process of statistical inference is illustrated in Figure 1-1. Members, or units, of the population are selected in the process of sampling. Together, these units compose the sample. From this sample, inferences about the population are made. In short, sampling takes us from the population to a sample; statistical inference takes us from the sample back to the population.

The aim of statistical inference is to make statements about a population characteristic based on the information in a sample. There are two ways of making inferences. One type of inference is known as *estimation*.

DEFINITION: STATISTICAL ESTIMATION

Statistical estimation is the use of the information in a sample to estimate the value of an unknown population characteristic.

The use of political polls to estimate the proportion of voters in favor of a certain party or candidate is a well-known example of statistical estimation. *Estimates* are simply the statistician's best guess of the value of a population characteristic. From a *random sample* of voters, we try and guess what proportion of *all* voters will support a certain candidate.

Another way of making inferences about a population characteristic is *hypothesis testing*. In hypothesis testing, we hypothesize a value for some population characteristic and then determine the degree of support for this hypothesized value from the data in our random sample.

DEFINITION: HYPOTHESIS TESTING

Hypothesis testing is a procedure of statistical inference in which we decide whether the data in a sample support a hypothesis that specifies the value (or a range of values) of a certain population characteristic.

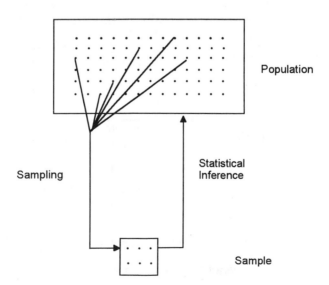

FIGURE 1-1. The process of statistical inference.

As an example, we may wish to use a political poll to find out whether some candidate holds an absolute majority of decided voters. Expressed in a statistical way, we wish to know whether the proportion of voters who intend to vote for the candidate exceeds a value of .50. We are not interested in the actual value of the population characteristic (the candidate's exact level of support), but in whether the candidate is likely to get a majority of votes. As you might guess, these two ways of making inferences are intimately related and differ more at the conceptual level than in practical application. The relation between them is so intimate that, for most purposes, either could be used to answer any problem. No matter which method is used, there are two fundamental elements of any statistical inference: the inference itself and our confidence in it.

   A useful synopsis of statistical analysis, including both descriptive and inferential techniques, is illustrated in Figure 1-2.

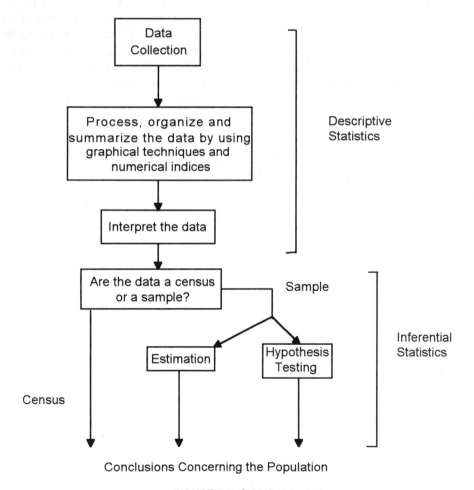

FIGURE 1-2. Statistical analysis.

1.1. Statistical Analysis and Geography

The application of statistical methods to problems in geography is relatively new. Only for about the last 30 years has statistics been an accepted part of the academic training of geographers. Nevertheless, there are earlier references to uses of descriptive statistics in literature cited by geographers. For example, several ninteenth-century researchers, including Carey (1858) and Ravenstein (1885), used statistical techniques in their studies of migration and other interactions. Elementary methods of descriptive techniques are seen commonly in the geographical literature of the early twentieth century. But for the most part, the three paradigms that dominated academic geography in the first half of the twentieth century—exploration, environmental determinism and possibilism, and regional geography—found few uses for statistical methods. Techniques for statistical inference were emerging at this time, but were not applied in the geographical literature.

Exploration

This paradigm is one of the earliest used in geography. Unexplored areas of the earth continued to hold the interest of geographers well into the current century. Explorations, funded by geographical societies such as the Royal Geographical Society (RGS) and the American Geographical Society (AGS), continued the tradition of geographers collecting, collating, and disseminating information about relatively obscure and unknown parts of the world. The research sponsored by these organizations helped lead to the establishment of academic departments of geography at several universities. But, given only a passing interest in generalization and an extreme concern for the unique, little of the data generated by this research was ever analyzed by conventional statistical techniques.

Environmental Determinism and Possibilism

Environmentalist determinists and possibilists focused on the role of the physical environment as a controlling variable in explaining the diversity of the human impact on the landscape. Geographers began to concentrate on how the physical environment controls human behavior, and some determinists went so far as to contend that virtually all aspects of human behavior were caused by environmental factors. Possibilists held a less extreme view, asserting that people are not totally passive agents of the environment, and had a long, and at times bitter, debate with determinists. Few geographers studied man–environment relations outside this paradigm; and very little attention was paid to statistical methodology.

Regional Geography

Proponents of regional geography reacted against the naive lawmaking attempts of the determinists and possibilists. Generalization of a different character was their goal. According to this paradigm, an *integration* or *synthesis* of the characteristics of areas or regions was to be undertaken by geographers. Ultimately, this would lead to a more or less complete knowledge of the areal differentiation of the world. Statistical methodology was limited to the systematic studies of population distribution, resources, industrial activity, and agricultural patterns. Emphasis was placed on the data collection and summary components of descriptive statistics. In fact, these systematic studies were seen as preliminary and subsidiary elements to the essential tasks of regional synthesis. The definitive work establishing this paradigm at the forefront of geographical research was Richard Hartshorne's *The Nature of Geography*, published in 1939.

Many of the contributions in this field discussed the problems of delimiting homogeneous regions. Each of the systematic specializations produced its own regionalizations. Together, these regions could be synthesized to produce a regional geography. A widely held view was that regional delimitation was a personal interpretation of the findings of many systematic studies. Despite the fact the map was considered one of the cornerstones of this approach, the analysis of maps using quantitative techniques was rarely undertaken. A notable exception was Weaver's (1954) multiattribute agricultural regionalization; however, this was not regarded as mainstream regional geography.

Beginning about 1950, the dominant approach to geographical research shifted away from regional geography and regionalism. To be sure, the transition took place over the succeeding two decades and did not proceed without substantial opposition. It was fueled by the increasing dissatisfaction with the regional approach and the gradual emergence of an acceptable alternative. Probably the first indication of what was to come was the rapid development of the systematic specialties of geography. The traditional systematic branches of physical, economic, historical and political soon were augmented by urban, marketing, resource management, recreation, transportation, population and social geography. These systematic specialties developed close links with related academic disciplines, historical geography with history, social geography with sociology, and so forth. Economic geographers in particular looked to the discipline of economics for modern research methods. Increased training in these so-called parent disciplines was suggested as an appropriate means of improving the quality of geographical scholarship. Throughout the 1950s and 1960s, the teaching of systematic specialties and research in these fields became much more important in university curricula. The historical subservience of the systematic fields to regional geography was reversed during this period.

The Scientific Method and Logical Positivism

The new paradigm that took root at this time focused on the use of the *scientific method*. This paradigm sought to exploit the power of the scientific method as a vehicle for establishing truly geographical laws and theories to explain spatial patterns. To some, geography was reduced to pure *spatial science*, though few held this rather extreme view. As it was applied in geography, the scientific method used the deductive approach to explanation favored by *positivist* philosophers.

The deductive approach is summarized in Figure 1-3. The researcher begins with a perception of some real world structure. A pattern, for example, the distance decay of some form of spatial interaction, leads the investigator to develop a model of the phenomenon from which a generalization or hypothesis can be formulated. An experiment or some other kind of test is used to see whether or not the model can be verified. Data are collected from the real world, and verification of the hypothesis or speculative law is undertaken. If the test proves successful, laws and then theories can be developed, heightening our understanding of the real world. If these tests prove successful in many different empirical applications,

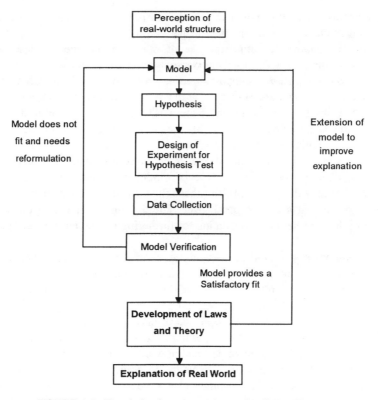

FIGURE 1-3. The deductive approach to scientific explanation.

then the *hypothesis* gradually comes to be accepted as a *law*. Ultimately, these laws are combined to form a *theory*. This approach obviously has many parallels with the methodology for statistics outlined in the introduction to this chapter.

The deduction-based scientific method began to be applied in virtually all fields of geography during the 1950s and 1960s. It remains particularly important in most branches of physical geography, as well as in urban, economic, and transportation geography. Part of the reason for this strength is the widespread use of the scientific method in the physical sciences and in the discipline of economics.

Quantification is essential to the application of the scientific method. Mathematics and statistics play central roles in the advancement of geographic knowledge using this approach. Because training in mathematics has not been viewed as essential by geographers, the statistical approach has been dominant and is now accepted as an important research tool by geographers. That is not to say that the methodology has been accepted throughout the discipline. Historical and cultural geographers shunned the new wave of quantitative, theoretical geography. Part of the reason for their skepticism was that early research using this paradigm tended to be long on quantification and short on theory. True positivists view quantification as only a means to an end—the development of theory through hypothesis testing. It cannot be said that this was clear to all those who practiced this approach to geographic generalization. Too often, research seemed to focus on those techniques that were available, not on the problem or issue at hand. The methods *themselves* are clearly insufficient to define the field of geography.

Many researchers advocating the use of the scientific method also defined the discipline of geography as *spatial science.* Human geography began to be defined in terms of *spatial* structures, *spatial* interaction, *spatial* processes, or *spatial* organization. Distance was seen as the key variable for geographers. Unfortunately, such a narrow view of the discipline seems to preclude much of the work undertaken by cultural and historical geographers. Physical geography, which had been brought back into geography with the onset of the quantitative revolution, was once again set apart from human geography. Reaction against geography as a spatial science occurred for several reasons. Chief among these was the disparity between the type of model promised by advocates of spatial science and what they delivered. Most of these theoretical models gave adequate descriptions of reality only at a very general level. The axioms upon which they were based seemed to provide a rather poor foundation for furthering the development of geographical theory.

By the mid-1960s, a field now known as *behavioral geography* was beginning to emerge. It was closely linked with psychology and drew many ideas from the rich body of existing psychological research. Proponents of this approach did not often disagree with the basic goals of logical positivism—the development of theory-based generalizations—only with how this task could be best accomplished. Behavioral geographers began to focus on individual spatial cognition and behavior, primarily from an inductive point of view. Rather than accept the unrealistic axioms of perfect knowledge and perfect rationality inherent in many models developed by this time, behavioral geographers felt that the use of more realistic assumptions about behavior might provide deeper insights into spatial structures and spatial behavior. Their inductive approach was seen as a way of providing the necessary input into a set of richer models based on the deductive mode. Statistical methodology has a clear role in this approach.

Contemporary Geography

In the last 15 to 20 years, the number of approaches manifest in the geographical literature has broadened considerably. Positivist approaches, based on the objectivity of scientific analysis of the world, now are part and parcel of mainstream geographical research. Two alternatives are becoming increasingly important and popular. First, there are approaches based on *humanistic philosophies,* Humanistic geographers take the view that people create *subjective* worlds in their own minds and that their behavior can be only understand using a method that can penetrate this subjectivity. By definition then, geographical laws developed from positivist precepts cannot have this capability. Humanistic geography appeals most strongly to historical, cultural social geographers, many of whom reacted strongly against the positive-based scientific method.

Structuralists reject both positivist and humanistic methods, arguing that explanations of observed spatial patterns cannot be made by a study of the pattern itself, but only by the establishment of theories to explain the development of the societal condition within which people must act. The structuralist alternative, exemplified by Marxism, emphasizes how human behavior is constrained by more general societal processes and can be understood only in those terms. For example, patterns of income segregation in contemporary cities can be understood only within the context of a class conflict between the bourgeoisie on one hand, and the proletariat, or workers, on the other. Understanding how power, and therefore resources, are allocated in a society is a prerequisite to a comprehension of its spatial organization.

What then is the role of statistics in contemporary geography? Why should we have a good understanding of the principles of statistical analysis? Certainly, statistics is an important component of the research methodology of virtually all systematic branches of geography. A substantial portion of the research in physical, urban, and economic geography employs increasingly sophisticated statistical analysis. Being able to properly evaluate the contributions of this research requires us to have a reasonable understanding of statistical methodologies.

Although many advocates of the more recent humanistic or Marxist approaches to geography reject outright the use of these methodologies because of their close affinity to the scientific method, there is actually an increasing use of statistical methods in empirical studies based within these paradigms. Many of the data collected from personal interviews, mail questionnaires, participant observation, and the like lend themselves to analysis by descriptive and sometimes inferential techniques. A basic understanding of statistical methodology may help to improve the empirical design of research exercises in studies using these paradigms.

For many geographers, the map is a fundamental building block of all research. Cartography is undergoing a period of rapid change in which computer-based methods continue to replace much conventional map compilation and production. Microcomputers linked to a set of powerful peripheral data storage and graphical devices are now essential tools for contemporary cartography. Maps are inherently mathematical and statistical objects and as such are one of the areas of geography where dramatic change will continue to take place for some time to come. This trend has forced many geographers to acquire better technical backgrounds in mathematics and computer science and has opened the door to the increased use of statistical and quantitative methods in cartography. Geographic Information Systems (GIS) are one manifestation of this phenomenon. Large sets of data are now stored, accessed, compiled, and subjected to various cartographic display techniques using video display terminals, plotting devices, and laser printers.

The analysis of the spatial pattern of a single map and the comparison of sets of interrelated maps are two cartographic problems for which statistical methodology has been an important source of ideas. Many of the fundamental problems of displaying data on maps have clear parallels to the problems of summarizing data through conventional descriptive statistics. An entire chapter of this text is devoted to descriptive spatial statistics and introduces many of these parallels.

Finally, statistical methods find numerous applications in applied geographical research. Retail location problems, transportation forecasting, and environmental impact assessment are three examples of applied fields in which statistical and quantitative techniques play a prominent role. Both private consulting firms and government planning agencies encounter problems in these areas on a day-to-day basis. It is impossible to underestimate the impact of the wide availability of microcomputers on the manner in which geographers can now collect, store and retrieve, analyze and display the data fundamental to their research. The methods employed by mathematical statisticians themselves have been fundamentally changed with the arrival and diffusion of this technology.

No course in statistics for geographers can afford to omit applied work with microcomputers in its curriculum.

In sum, statistical analysis is commonplace in contemporary geographical research and education, as it is in the other social, physical, and biological sciences. It is now being more thoughtfully and carefully applied than in the past and includes an ever widening array of specific techniques.

1.2. Use of Statistics in Geographical Research: Two Examples

To illustrate the versatility of statistical methodology and for a brief peek at how statistics is used in geographical research, let us now consider two detailed, yet conceptually simple, examples.

EXAMPLE 1-1. Analysis of Wind Speed and Direction   Statistical techniques and concepts are frequently applied in all the subfields of physical geography. Because of their close relations with the physical and life sciences, the systematic branches of physical geography, including geomorphology, hydrology, climatology and meteorology, and biogeography, felt the impact of the so-called quantitative revolution far earlier than the systematic branches of human geography. Descriptive statistics is used to investigate the morphology of physical systems as diverse as vegetative associations, drainage basins, and climatic types. The information generated in this research often finds practical use in applied studies of flood damage control, reforestation, avalanche prediction, wind power generation, and numerous other areas. The data sets used in these studies often contain thousands of records for different variables taken at close time intervals over long observation periods. For example, a series of temperature readings may consist of hourly recordings at a meteorological station taken each day for 20 or more years!! An obvious need exists for efficient graphical devices and summary descriptive measures for large data sets of this kind.

As an example, suppose a meteorologist has developed a data set containing measurements of wind speed and direction taken hourly at a series of stations for a 25-year period. Wind speeds are obtained from an anemometer mounted 33 ft (10 m) above the ground. Speed, measured in miles per hour, is taken as the average reading over a 1-min interval closest to the hour. Directions are measured on a continuous scale graduated in degrees true and to 16 points of the compass. In addition, the data set notes the frequency of *gusts* (rapid, brief increases in wind speed) and *calm* (the absence of appreciable air motion).

The task of making sense of this long series of wind records is made substantially easier by the use of many techniques from descriptive statistics. One of the simplest questions to be asked might be, Which of these recording stations has the highest wind speeds? We might even wish to classify the stations according to wind speeds as very windy, windy, relatively calm, and calm. A useful summary measure is the *average* wind speed by day, month, season, and year or for the entire 25-yr record of 219,000 observations (25 yr × 365 d × 24 h). Wind speeds vary systematically by hour of the day, month of the year, and at other regular frequencies. A summary table might be used to contrast the average wind speeds in two characteristic months of the year, based on the 25-yr observation period:

	Average Wind Speed, mi/h	
Station	January	July
A	8.4	5.4
B	2.2	6.5
C	7.1	7.8
D	25.2	18.0

From this table, it is easy to see that station D is the windiest, station B is the calmest and has a July maximum, station A has a January maximum, and station C has no marked monthly maximum. Other tables generated in this way might point out other phenomena that might require some sort of meteorological explanation. July and January might not be typical months at certain stations, for example, and an examination of the monthly means over a 12-month cycle might reveal other important patterns.

In addition to the typical wind speed likely to be observed at some location, the variability or persistence of wind speed may also be of interest. The variability of wind speed at station A is illustrated in Figure 1-4. This bar chart is a percentage summary of the wind speed readings for 1 yr and consists of 8760 (24 × 365) observations. Of these, 10% represent calm wind conditions. That is, 10 percent of the time we might expect calm wind conditions at station A. Wind speeds seem to remain mainly between 5 and 12 mi/h. Over 40% of the observations lie within these limits. Wind speed variations are an important consideration in studies of the dispersal of pollutants. Stations having dominantly calm conditions may be susceptible to pollutant concentrations. Locations with persistently high winds may be associated with rapid dispersal of pollutants. Of course, it is necessary to examine other variables such as temperature and precipitation patterns as well as topographic conditions before such conclusions could be reached.

Sometimes, the summary of wind speeds from such an extended series of measurements needs to be compiled in a particular fashion for a specific purpose. Consider the

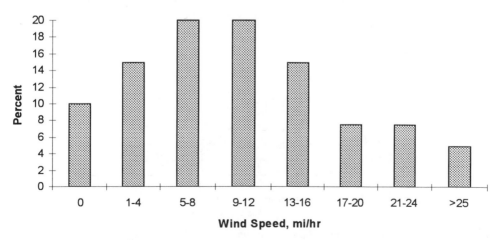

FIGURE 1-4. The variability of wind speed at station A.

problem of identifying potential sites for wind power generation. Figure 1-5 illustrates a form of *power duration curve*, a useful graphical device for estimating the viability of a site for wind power generation. The horizontal axis measures the number of hours of the year, from 0 to 8760. The vertical axis represents the observed wind speed. The power duration curve shows the number of hours per year (or in this case, the number of observations in the yearly wind speed series) for which the wind speed exceeds the value given on the vertical axis. Of particular importance in this curve is the percentage of time that the power generator cannot be in operation owing to either insufficient or excessive winds. If the wind speeds are below the *cut-in point*, there is insufficient wind to produce a significant amount of power. Such conditions prevail at this site for a total of $(8760 - b)$ h. At wind speeds above this value and up to a wind speed sufficient for maximum power generation from the wind turbines, power can be generated. Maximum power is generated at a certain wind speed; this is called the *rating capacity*. Wind speeds above the rating capacity provide no increase in power. However, at a critical value of wind speed known as the *furling point*, the wind is excessive, and the turbines must be shut down to avoid damage. This occurs for *a* hours of the year. The particular form of wind speed at a location can be used to estimate power production, since power production varies with the cube of the observed wind speed. In general, the larger the shaded area in a power duration curve, the greater the possibilities of power production. Ideally, a site with wind speeds that are constantly at the rating capacity is desirable. The power duration curve illustrated is clearly a descriptive graphical summary of wind speed, but is designed for a specific applied purpose. This contrasts with the conventional histogram, which is of more general application.

Summarizing the directional components of wind is another problem in descriptive statistics. Three different methods of portraying the directional components of wind are illustrated in Figure 1-6. In (a), the wind directions are summarized in a conventional

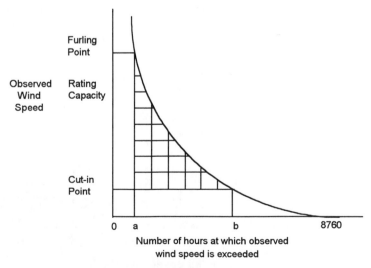

FIGURE 1-5. A power duration curve.

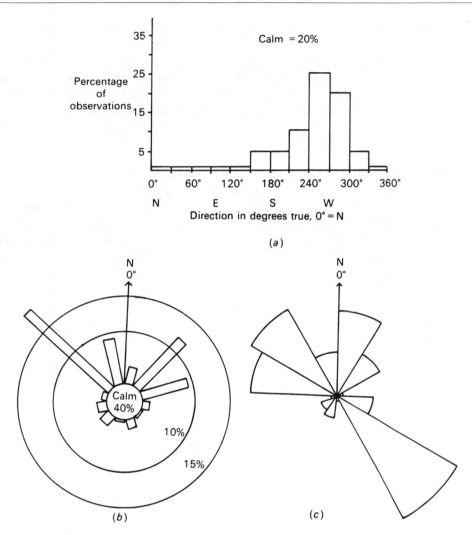

FIGURE 1-6. (a) Histogram of station A; (b) circular histogram of station D; (c) rose diagram of station C.

histogram. The percentage of observed wind directions in a yearly series is plotted on the vertical axis. The wind directions are aggregated to 30-degree sectors, although a 16-point compass aggregation of $22\frac{1}{2}$-degree sectors (N, NNE, NE, ENE, E, etc.) can also be used. For station A, the predominance of westerly winds is apparent from the figure. The *circular histogram* of Figure 1-6 (b) is an alternative, and possibly more easily interpretable, graphical portrayal of wind direction. In this format, the bars for the various directions are extended from a reference circle within which the percentage of calm winds is noted. The fact that winds at station D are dominantly northerly, especially from the northwest, is easy to see from this graph. A *rose diagram* is another easily interpretable summary graphic for directional data. A rose diagram for the observed winds at station C is illustrated in Figure 1-6 (c).

As this example clearly illustrates, the types of summary statistics and graphs useful for spatial data, such as an observed series of directions, often differ from the devices used to summarize ordinary variables such as wind speed. Part I of this test is devoted entirely to the methods of descriptive statistics. Conventional methods are discussed in Chapter 2, procedures for time series data in Chapter 3, and spatial data in Chapter 4. Relatively recent methods known as Exploratory Data Analysis are presented in Part IV, Chapter 16.

EXAMPLE 1-2. Residential Property Values   Geographers interested in cities often center their research on the theme of *urban spatial structure*. This term refers simultaneously to both the visible pattern of individual elements in the city and the underlying interrelationships that integrate these activities into a system. The study of the systematic variation of house prices and/or rents in cities is one example of research into urban spatial structure. Statistical methodology has been used extensively in studies of urban house prices in order to answer several key questions. First, what major explanatory variables account for the observed variations in individual house prices in cities? Second, what are the key spatial variables at work? Third, what is the relative importance of these spatial or locational variables vis-à-vis other factors? That is, are locational variables more or less important than characteristics of the house itself or of the neighborhood in which it is located? How do these results compare across cities?

The statistical approach to this topic often begins with the selection of a random sample from all the homes sold in some given period, for example, the previous 12 months. Usually, the period is selected so that there are no significant variations in interest rates or other key economic variables that may bias the selling price of homes during the study period. It would be foolish to have the period of our study cover a period in which interest rates moved between three and four percentage points. Also, houses sold through a court order or under unusual circumstances might be excluded from the sample on the grounds that they may be unrepresentative of the prevailing housing market in the city.

The most obvious variable to be collected is the actual selling price of the residential property but many other variables would also be included in the data set. As shown in Table 1-1, it is convenient to group these variables into three categories: (1) structural characteristics of the house, (2) neighborhood characteristics, and (3) location variables. *Structural* characteristics of the house include any measurement of the house or the lot upon which it is located. The size of the house in usable or developed square feet is one example of a structural variable. *Neighborhood characteristics* include measurable properties of the surrounding area such as the percentage in nonresidential land uses, crime rates, levels of pollution, and ethnic and racial breakdown of residents. Many empirical studies have found that houses located in neighborhoods with mixed land use or higher levels of crime and pollution have lower values than equivalent houses in exclusively residential districts. The *location* of the house within the urban area is another important factor affecting selling price. Historically, the principal variable used to measure this factor was the distance of the house from the central business district (CBD), but it is now more common to use *accessibility* measures that take into account the location of various activities throughout the city. There is considerable variability in the actual variables collected for studies of residential

TABLE 1-1

Factors Influencing House Price Variations in an Urban Area

Structural characteristics of the house	
Interior, sq ft	Number of bathrooms
Lot size, sq ft	Garage
Street frontage	Central heating
Number of rooms	Air conditioning
Number of bedrooms	Architectural significance

Neighborhood characteristics
Residential density
Percentage in nonresidential land use
Crime rate
Level of pollution
Presence of minorities

Location variables
Distance to central business district
Distance to major artery
Freeway access
Accessibility to employment, retail, and/or recreational facilities

property values. In some studies, well over 50 different variables are used, including such precise variables as the presence of thermopane windows or other energy saving devices, the existence or quality of landscaping, and the caliber of various building supply components. Comparing the results of many empirical studies is often hampered by the diversity of the variables included in the data sets upon which the analysis is based.

What can be done with these data? First, descriptive statistics are usually employed to summarize each of the variables gathered during data collection. What is the average selling price of houses in the sample? Real estate agents and housing market analysts often use changes in the average selling price to monitor the change in house prices in a city over time. Second, we might wish to examine the variability of house prices. How homogeneous are these selling prices about the average sale? A conventional histogram, such as the one used to describe wind speed variation in Figure 1-4, could be used for this purpose. Similar analyses would probably be undertaken for each variable collected so that we could have a good understanding of the averages and variability of these characteristics. As geographers, we might be especially interested in enumerating home sales by neighborhood, since this is a useful measure of neighborhood stability. The display of this data on a map would be a complementary technique to the descriptive statistics and graphs and would also help us understand the nature of the data included in our sample.

Third, we can examine the relationships *among* these variables. We might see, for example, whether the selling price of a house tends to be associated with distance from the central business district, the lot size, or the neighborhood crime rate. A scatter plot or scatter diagram can be used for this purpose. A scatter diagram relating house selling price and

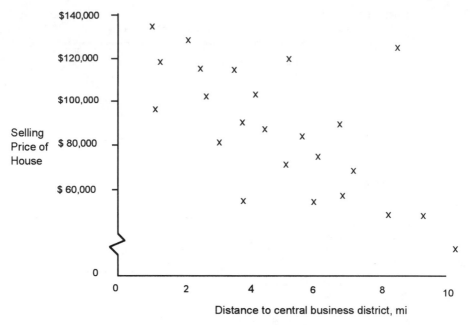

FIGURE 1-7. A scatter diagram.

distance of the house from the CBD of Figure 1-7 suggests a general decrease in price with distance, but several houses do not follow this trend. Methods for analyzing the relationship between two variables are discussed extensively in Part III of this textbook.

Fourth, studies of house price variations in cities are often concerned with identifying the *determinants* of house price variations. Unfortunately, no single theory of residential location and house price variation takes into account all of the variables listed in Table 1-1. Nevertheless, a number of empirical studies have found consistent effects for many of these variables as factors influencing residential selling prices in many North American and other Western cities. It is now usual to search for the complex interactions among these variables in a structured way. One of the more commonly used techniques is known as *hedonic price estimation*. In this form of estimation, we take the price of some aggregate product, such as a house, and divide it into implicit prices for the various attributes or components of the product. For a house, it might be argued that part of the total price paid for the house is to purchase square footage or the number of bedrooms, part is to purchase the locational attributes of the house, and another portion is paid for certain environmental variables such as low pollution levels. Although the consumer does not actually break down the components of the price, they are *implicit* in the bids consumers make and thus in the ultimate selling prices of the houses in the city.

In general, hedonic price estimation uses statistical techniques to determine the form of the function that links the price of the house to these attributes:

House selling price = *f*(structural characteristics, neighborhood
characteristics, location variables)

For example, suppose the results of the hedonic price estimation in some city indicate that the following equation describes house price variations:

$$\text{House selling price} = \$83,000 + \$15,000 \text{ (house size in ft}^2/1000)$$
$$- \$1,000 \text{ (concentration of CO, in ppm)}$$
$$- \$5,000 \text{ (distance to CBD in miles)}$$

By using this equation, it is possible to estimate the selling price of a 2000-ft^2 house in a neighborhood averaging 14 parts per million (ppm) of carbon monoxide and located 2 mi from the central business district as $83,000 + $15,000(2) − $1,000(14) − $5,000(2) = $89,000. The coefficients of this equation—$15,000, $1,000, and 5,000—can be thought of as the *implicit* prices that house buyers have used in making their bids for the houses in the sample. For example, an extra 1,000 ft^2 of space adds about $15,000 to the selling price of the average home.

Bivariate correlation and regression analysis is used to analyze the statistical relationship between two variables. These procedures are discussed in Part III of the text. When one variable (such as house selling price) is simultaneously related to several other variables (such as house size, carbon monoxide concentration, and distance to the CBD), the appropriate techniques are *multiple correlation and regression analysis*. These advanced techniques are briefly introduced in Chapter 13.

The two examples presented in this section help illustrate the variety of situations in geographical research in which statistical methodology is relevant. There are literally hundreds, if not thousands, of instances in geographical research in which the methods described in subsequent chapters of this text are useful. These applications span both physical and human geography. This range of application is further illustrated by examples for specific techniques presented throughout the remainder of this text.

1.3. Data

Although Figure 1-2 seems to suggest that statistical analysis begins with a data set, this is not strictly true. It is not unusual for a statistician to be consulted at the earliest stages of a research investigation. As the problem becomes clearly defined and questions of appropriate data emerge, the statistician can often give invaluable advice on sources of data, the methods used to collect them, and characteristics of the data themselves. A properly executed research design will yield data that can be used to answer the questions of concern in the study. The nature of the data used should never be overlooked. As a preliminary step, let us consider a few issues relating to the sources of data, the kinds of variables amenable to statistical analysis, and several characteristics of the data such as measurement scales, precision, and accuracy.

Sources of Data

Sometimes the data required for a particular study are already available, When these data are available in some form in various records kept by the institution or agency undertaking the study, the data are said to be from an *internal source*.

DEFINITION: INTERNAL DATA

Data available from existing records or files of an institution undertaking a study are data from an internal source.

For example, a meteorologist employed by a weather-forecasting service normally has many key variables, such as air pressure, temperature, and wind velocity, from a large array of computer files that are augmented hourly, daily, or at some predetermined frequency. Besides the ready availability of this data, the meteorologist has the added advantage of knowing a great deal about the instruments used to collect the data, the accuracy of the data, and possible errors. In-depth practical knowledge of many factors related to the methods of data collection is often invaluable in statistical analysis. For example, we may find that certain readings are always higher than we might expect. When we examine the source of the data, we might find that all the data were collected from a single instrument that was incorrectly calibrated.

When an *external data* source must be used, many important characteristics of the data may not be known.

DEFINITION: EXTERNAL DATA

Data obtained from an organization external to the institution undertaking the study are data from an external source.

Caution should always be exercised in the use of external data. Consider a set of urban populations extracted from a statistical digest summarizing population growth over a 50-yr period. Such a source may not record the exact areal definitions of the urban areas used as a basis for the figures. Moreover, these areal definitions may have changed considerably over the 50-yr study period, owing to annexations, amalgamations, and the like. Only a primary source such as the national census would record all the relevant information. Unless such characteristics are carefully recorded in the external source, users of the data may have the false impression that no anomalies exist in the data. It is not unusual to derive results in a statistical analysis that cannot be explained without a detailed knowledge of the data source. At times, statisticians are called upon to make comparisons among countries for which data collection procedures are different, data accuracy differs markedly, and even the variables themselves are defined differently. Imagine the difficulty of collecting data measuring the extent of poverty in all countries of the world!

Another useful distinction is between *primary* and *secondary* external data.

DEFINITION: PRIMARY DATA

Primary data are obtained from the organization or institution that originally collected the information.

DEFINITION: SECONDARY DATA

Secondary data are data obtained from a source other than the primary data source.

If you must use external data, always use the primary source. The difficulty with secondary sources is that they may contain data altered by recording or reediting of errors, selective data omission, rounding, aggregation, questionable merging of data sets from different sources, or various ad hoc procedures. For example, never use an encyclopedia to get a list of the 10 largest American cities, use the U.S. national census. It is surprising just how often research results are reported with erroneous conclusions—only because the authors were too lazy to use the primary data source or were unaware that it even existed.

Data Acquisition

When the data required for a study cannot be obtained from an existing source, they are usually collected during the course of the study. Data acquisition methods can be classified as either *experimental* or *nonexperimental*.

DEFINITION: EXPERIMENTAL METHOD OF DATA ACQUISITION

An experimental method of data acquisition is one in which some of the factors under consideration are controlled in order to isolate their effects on the variable or variables of interest.

Only in physical geography is this method of data collection prominent. Fluvial geomorphologists, for example, may use a flume to control such variables as stream velocity, discharge, bed characteristics, and gradient. Among the social sciences, the largest proportion of experimental data is collected in psychology.

DEFINITION: NONEXPERIMENTAL METHOD OF DATA ACQUISITION

A nonexperimental method of data collection or statistical survey is one in which no control is exercised over the factors that may affect the population characteristic of interest.

Studies in physical geography using field data fall into this category. Many carefully planned, detailed field studies have been seriously affected by instances of unforeseen aridity, excess moisture, insufficient snow pack, and the like. It is easy to see why experiments generally provide more useful data than surveys. This seeming advantage must be weighed against the obvious disadvantage of experiments in simulating real-world processes.

There are four common survey methods. Observation (or field study) requires monitoring of an ongoing activity and direct recording of data. This form of data collection avoids several of the more serious problems associated with other survey techniques, including incomplete data. However, faulty or incorrectly calibrated instruments can sometimes mitigate against the advantages of observational techniques. More than one field study has been prolonged due to instrument failure.

Two methods of data collection are often used to extract information from households or individuals: *personal interviews* and *telephone interviews*. In a personal interview, a trained interviewer asks a series of questions and records responses on a specially designed form. There are obvious advantages and disadvantages to this procedure. An alternative, and often cheaper, method of securing the data from a set of households is to send a *mail questionnaire*. This last method is often called self-enumeration because the respondent completes the questionnaire without assistance from the researcher. The disadvantages of this method include nonresponse, partial response, and low return rates for completed questionnaires. Factors affecting the quality of data from mail surveys include appropriate wording, proper question order, question types, layout, and design. For telephone and personal interviews, there is the added impact of the rapport developed between the interviewer and the subject. A useful typology of data sources is illustrated in Figure 1-8.

Characteristics of Data Sets

Statistical analysis cannot proceed until the available data have been assembled into a usable form.

DEFINITION: DATA SET

A data set is a collection of statistical information or values of the variables of interest in a study.

An example of a data set is shown in Table 1-2. In this example, the observational units are climatic stations. Five variables are contained within the data set. The information on every variable from one observational unit is often called an observation; it is also common to speak of the data value for a single variable as an observation because it is the observed value.

The variables in a data set can be divided into two types: quantitative and qualitative.

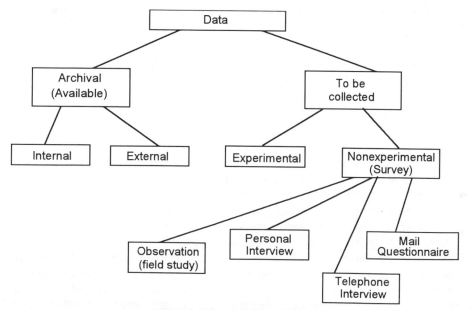

FIGURE 1-8. A typology of data sources.

TABLE 1-2
A Geographical Data Set

	Climatic station	Days per year with precipitation	Annual rainfall, cm	Mean January temperature, °C	Mean July temperature, °C	Coastal or inland
				Variables		
	A	114	71	6	16	C
	B	42	48	12	16	C
Observational	C	54	32	−4	21	I
units	D	32	28	−8	20	I
	E	41	129	16	18	C
	F	26	18	1	22	I
	G	3	8	24	29	I

DEFINITION: QUANTITATIVE VARIABLE

A quantitative variable is one in which the values or outcomes are expressed numerically.

Quantitative values can be obtained either by counting or by measurement. Discrete variables are those that can be obtained by counting. For example, the number of children in a family, the number of cars owned, the number of trips made in a day are all counting variables. The possible values of counting variables are the ordinary integers and zero: 0, 1, 2, ... Quantities such as rainfall, air pressure, or temperature are measured and can take on any continuous value depending on the accuracy of the measurement and recording instrument. Continuous variables are inherently different from discrete variables.

Qualitative variables are neither measured nor counted.

DEFINITION: QUALITATIVE VARIABLE

A qualitative variable is one for which the obtainable values are nonnumerical.

Qualitative variables are sometimes termed *categorical* variables because the observational units are placed into categories. Male/female, land-use type, occupation, and plant species are all examples of qualitative variables. These variables are defined by the set of classes into which an observation can be placed. In Table 1-2, for example, climatic stations are classified as either coastal (C) or inland (I).

Numerical values are sometimes assigned to qualitative variables. For example, the "yes" responses to a particular question in a survey may be assigned the value 1 and the "no" responses a value of 2. The variable gender may be identified by males = 1 and females = 0 (or vice-versa) . In both these examples, each category has been assigned an arbitrary numerical value. As we shall see, it is improper to perform most mathematical operations on qualitative variables expressed in this manner. Consider, for example, the operation of addition. Although this operation makes sense for a quantitative variable, it makes no sense to add the numerical values assigned to the variable gender.

Converting quantitative variables to qualitative variables is sometimes disadvantageous. In this case the possible numerical values of the quantitative variable are grouped into nonnumerical classes. For example, the quantitative variable called household income might be classified into three categories: low, medium, and high. Such a simplification can sometimes facilitate the interpretation of a large number of figures.

Besides being described as qualitative or quantitative, variables can also be classified according to the *scale of measurement* on which they are defined. This scale defines the amount of information the variable contains and what operations can be meaningfully undertaken and interpreted. The lowest scale of measurement is the *nominal scale*. Nominal scale variables are those *qualitative* variables that have no implicit ordering to their categories. Even though we sometimes assign numerical values to nominal variables, they have no meaning. Consider the variable in Table 1-2 that distinguishes coastal climatic stations (C) from inland ones (I). All that we can really do is distinguish between the two types of stations. In other words, we know that stations A, B, and E are coastal and therefore different from stations C, D, F, and G. Also, stations A, B, and E are all alike according to this variable. One way of summarizing a nominal variable is to count the number of observations in each category. This can be summarized in a bar graph or a simple table of the following form:

Category	Count or frequency	Proportion	Percentage
C	3	0.429	43
I	4	0.571	57
	7	1.000	100

The use of percentages or relative proportions to summarize such data is quite common. For example, we would say that 0.429, or 43%, of the climatic stations are coastal and 0.571, or 57%, are inland.

Proportional summaries are often extremely useful in comparing responses to similar questions from two or different surveys, or from different classes of respondents to the same question in a survey. For example, studies of migrants to cities of the Third World often include questions concerning the sources of information used by individual migrants in selecting their destination. A question of this type might be phrased in the following way: In reaching your decision to come here, you must have had some information about job possibilities, living conditions, income, etc. Which of the following gave you the *most* information? The question would then list a variety of potential sources of information: relatives, friends, newspapers, radio, or other. We could compare the proportional use of these information sources from different studies, or by gender or educational attainment of the respondent.

Table 1-3 summarizes the responses to this question and differentiates by gender of the respondent. Clearly, about two-thirds of the respondents cite family or friends as the dominant information source. However, men seem to rely more on family and less on friends in comparison to women. Also, men tend to use newspapers more frequently, and

TABLE 1-3
Percentage Distribution of Responses Concerning Primary Information Source about Migration Destinations Used by Migrants

	Age of respondent, yr	
Sources of information	14–21	22–64
Newspapers	13	7
Radio	3	2
Government labor office	2	3
Family members	40	28
Friends	27	41
School teacher	4	1
Career counselor	1	1
Other sources	10	17
Total	100	100

female migrants seem to use other sources more often. These results indicate the significant role played by kin and friends in the rural–urban migration process, but also point to potentially important differences in primary sources of information by gender. An interesting research question is whether these observed differences are truly important or not. If the respondents to the questionnaire were randomly selected, we may be able to use an inferential technique to determine whether this observed difference is due to random sampling or indicative of an important difference between males and females. Of course, we would also wish to compare the results of a number of similar studies from different cities in different countries before we would be able to confirm the importance of this hypothesis as a *general* rule. It may be limited to the current study.

If the categories of the *qualitative* variable can be put into order, then the scale of measurement of the variable is said to be *ordinal*. An example of an ordinal variable is the strength of opinion measured in responses to a question with the following categories:

Strongly agree	Agree	Neutral	Disagree	Strongly disagree
2	1	0	−1	−2
1	2	3	4	5
5	4	3	2	1

Three different numerical assignments are given, and each is consistent with an ordinal scale for this variable. In all cases, the stronger the agreement with the question the higher the value of the numerical assignment. In fact, any assignment can be used as long as the values assigned to the categories maintain the ordering implicit in the wordings attached to the categories. The assignment of values −200, −10, 270, 271, 9382 meets this criterion. Note that it doesn't matter which end of the scale is assigned the lowest values, nor does it matter if they are given negative values. All of the categories could be given negative values, all positive values, or a mixture of the two. Of course, the scales defined in the table above have the added advantage of simplicity.

The numerical difference between the values assigned to different categories has no meaning for an ordinal variable. We can neither subtract nor add the values of ordinal variables. Notice, however, that the numerical assignments used to define ordinal variables often use a constant difference or unit between successive categories. In the first scale defined above, each category differs from the next by a value of one. This does *not* mean that a respondent who checks the box for strongly agree is 2 units higher than a respondent who checks the box for neutral. We can only say that the first respondent agrees more with the question than the second respondent. Only statements about *order* can be made using the values assigned to the categories of an ordinal variable.

Quantitative or numerical variables, whether discrete or continuous, can be classified into two scales of measurement. Interval variables differ from ordinal variables in that the interval scale of measurement uses the concept of *unit distance*. The difference between any two numbers on this scale can always be expressed as some number of units. Both Fahrenheit and Celsius temperature scales are examples of interval-scale variables. Although it makes sense to compare differences of interval scale variables, it is not permissible to take ratios of the values. For example, we can say that 90°F is 45°F hotter than 45°F, but we cannot say that it is twice as hot. To see why, let us simply convert these temperatures to the Celsius scale: 90°F = 32°C and 45°F = 7°C. Note that 32°C is *not* twice as hot as 7°C.

At the highest level of measurement are ratio-scale variables. Any variable having the properties of an interval scale variable *as well as* a natural origin of zero is measured at the ratio scale of measurement. Distance measured in kilometers or miles, rainfall measured in centimeters or inches, and many other variables commonly studied by geographers are measured on the ratio scale. It is possible to compute ratios of such variables as well as to perform many other mathematical operations such as logarithms, powers, or roots. Because we can take ratios of distances, we can therefore say that a place 200 miles from us is twice as far as one 100 miles from us. Although there is a logical and theoretical distinction between ratio and interval variables, this distinction rarely comes into play in practice.

Of far greater interest is the specification of the level of measurement of a variable measured indirectly. For example, the scale of measurement of a variable constructed from the responses to a question or a set of questions in an interview may not be easily identified. This variable may be ordinal, interval, or even ratio. That is, respondents *may treat* the categories of scale with labels *strongly agree, agree, neutral, disagree, strongly disagree* as if they were part of an interval, not ordinal, scale. A significant amount of research in psychology has examined methods for deriving scales with interval properties from test questions that are, strictly speaking, only ordinal in character. Attitude scales and some measures of intelligence are two examples where interval properties are desirable. An investigator must sometimes decide what sorts of operations on the variables collected are meaningful, or whether the operations exceed the information contained in the variable. A summary of the scales of measurement, along with a list of permissible mathematical operations and examples, is given in Table 1-4.

1.4. Measurement Credibility

It goes without saying that the utility of any statistical analysis ultimately rests on the quality of the data used. If little confidence can be placed in the measurements, the results

TABLE 1-4
Levels of Measurement: A Summary

Level of measurement	Permissible mathematical operations[a]	Geographical examples
Nominal	A = B or A ≠ B, counting	Presence or absence of a road linking two cities or towns Land use types Gender
Ordinal	A < B or A > B or A = B	Preferences for different neighborhoods in rank order Ratings of shopping center attractiveness
Interval	Addition (A + B), subtraction (A − B), and multiplication (A × B)	Temperatures, °F or °C
Ratio	Take ratios A/B and compare Square roots Powers Logarithms Exponentiation	Distances (imperial or metric) Density (persons per unit area) Stream discharge Shopping center square footage Wheat or corn yield

[a]Permissible mathematical operations at each level of measurement include all those operations permissible for variables at lower levels of measurement.

will accrue little confidence, regardless of how sophisticated the analysis itself is. Central to the issue of measurement credibility are two general concepts, validity and accuracy. Validity is perhaps the more abstract of the two. Loosely speaking, it is the degree to which a variable measures what it is supposed to measure. In many physical science applications, validity will not be an issue. For example, the variable temperature is commonly used as a measure of thermal energy, or heat. The relation between heat and temperature is well known, thus there is not likely to be much debate about whether or not temperature measures an object's energy content. Furthermore, the principles of liquid thermometers are understood, thus it is clear that measurements taken by liquid thermometers do in fact reflect temperature. There are, however, instances in both physical and social sciences in which the situation is less clear, and measurement validity is questionable.

In the first case, it may be that the concept being studied is imperfectly defined. To pick just a few examples, we could mention intelligence, social status, drought, and ecosystem diversity. Everybody has a rough idea of what these concepts are, but precise definitions are elusive. To take another example, perhaps we want to know if air pollution control policies have improved air quality. Many air quality measures are available, including average pollutant level, maximum level, number of days above some threshold, and so forth. If the definition of "air quality" is vague, there will likely be questions about the validity of the variable chosen to measure air quality. A second, and perhaps less common, issue of validity arises when the concept is clear, but required data do not exist. For example, perhaps we have settled on a definition for "glaciation" as the land area covered by year-round ice. Direct measurements are not available for the period of study; thus we turn to proxy measures. We might have an accurate record of oxygen isotope changes, but do these changes track ice volume, or ice area, or precipitation, or some combination of precipitation and temperature? Unless we know that the measurements really do reflect the concept as defined, we must question their validity. An example from human geography is personal income. We might define it without ambiguity, but have a difficult time measuring it. Do we look at income tax returns, on which people have a strong incentive to under-report? Or do we conduct face-to-face interviews, in which people might overstate income?

Measurement credibility is also affected by measurement accuracy. We define accuracy as the correspondence between the value obtained from the instrument and the true value of the variable being measured. A thermometer is accurate if the measured value is close to the true temperature. (Note that this does not imply that temperature is a good measure of heat; thus accuracy need not imply validity.) As will be seen in Chapter 8, it is sometimes convenient to break accuracy down into predictable and unpredictable components. That is, we will write

$$\text{Total error} = \text{systematic error} + \text{random error}$$

The first component, systematic error, arises if the instrument consistently gives high or low values. Continuing with the temperature example, perhaps a thermometer has not been properly calibrated, so that it tends to read 2° above the true temperature. Figure 1-9(a) shows an extreme example, in which the measurements have only systematic error. Each of the five measurements is 2° above the corresponding true value. The observations are properly ordered, and are positioned correctly relative to each other, but all are wrong. They are wrong by the same amount; thus, the error is systematic from measurement to measurement. Systematic error is often referred to as *measurement bias*, a concept we will make more precise later in the text. For now, note that biased measurements are not necessarily useless. For example, if we were only after the difference between the observations, the bias in Figure 1-9(a) would be of no concern whatsoever.

Figure 1-9(b) shows another extreme, measurement errors that are completely random. Although the instrument is not very accurate, there is no tendency for measurements to be biased high or low. In calling the measurement error random, we must emphasize that this does not imply all errors are equally likely. Rather, the term means that we have no information about the individual errors. That is, the errors are unpredictable, or unsystematic. Random errors degrade accuracy by reducing the precision of the instrument. Formally, the precision of an instrument is the smallest difference that it can detect. If there is a large random component of error, differences between measurements cannot be attributed to changes in the variable, thus the device is imprecise. Real instruments, of course, are likely to contain both components of error, as shown in Figure 1-9(c). In the search for high overall accuracy, one must consider both sources of error. In particular, there are times when one will prefer a biased instrument over an unbiased alternative simply because the overall errors are smaller. In still other cases, the concern is almost exclusively with bias, so that random error components are very much less important.

When the results of a study based on empirical data are analyzed, a good first step is always to closely examine the operational definitions chosen for the variables. Are they suitable? Are they unambiguously defined? Are they the best data set that could have been used? Are they sufficiently precise? Are all variables unbiased? Could they be responsible for any misleading inferences? The need for caution in the use of data applies to all data sources, including those that are experimental, survey, and external or internal.

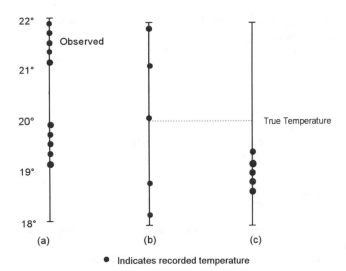

FIGURE 1-9. Differentiating between random and systematic error.

1.5. Summary

Statistical analysis includes methods used to collect, organize, present, and analyze data. Descriptive statistics refers to techniques used to describe data, either numerically or graphically. Inferential statistics includes methods used to make statements about a population characteristic on the basis of information from a sample. Statistical inference includes both hypothesis testing and estimation methods.

Within geography, most applications of statistical methodology are rather recent, having become a significant part of the research literature only after the "quantitative revolution" of the 1950s and 1960s. Statistical methodology is most commonly used by geographers advocating a scientific approach to the discipline, an approach now common in many of the systematic branches of the field. Historical and cultural geographers find fewer uses for the methodology. Recent advocates of humanistic and structuralist approaches tend to be particularly critical of the basis of the scientific method and often reject statistical methodology. Even in these fields, however, there are several instances for which statistical methodology can be fruitfully employed in applied research.

One of the first tasks of the statistician is to evaluate the data being used or being proposed in a research inquiry. If suitable data are not already available, or the limitations of existing data sources preclude their confident use, new data must be collected by the researcher. In some instances, an experimental approach is possible. It is mostly in various systematic specialties within physical geography that this is a feasible alternative, and statistical surveys are more commonly used to collect suitable data. Whenever statistical surveys are undertaken, it is necessary to proceed with caution, recognizing the limitations of the data collected in this manner. The greater the control exercised in the design of data collection procedures, the better the data ultimately available to the researcher. It follows that the ability of geographers to make sound judgments in their research often rests on the very first steps taken in their research design. Generating precise, accurate, and valid sets of variables can only assist the development of theory and explanation in geography.

REFERENCES

H. Carey, *Principles of Social Science* (Philadelphia: Lippincott, 1858).
R. Hartshorne, *The Nature of Geography* (Lancaster, PA: Association of American Geographers, 1939).
E. Ravenstein, "The Laws of Migration," *Journal of the Royal Statistical Society* 48 (1885) 167–235.
J.C. Weaver, "Crop Combinations in the Middle West," *Geographical Review* 44 (1954) 175–200.

FURTHER READING

Nearly every introductory statistics textbook for social scientists includes a presentation of much of the material discussed in this chapter. A discussion of the role of statistics and quantification in the discipline of geography can be obtained in Johnston (1983). Currently, survey methods are an extremely popular (some would argue overdone) tool in geographical research. Many of the questionnaires developed by geographers are poorly designed and cannot extract the information they desire. Too much emphasis is often placed on the academic topic and too little on the physical structure of the questionnaire, its language, flow, and ability to be used as a survey instrument. A useful, practical, and especially readable guide to the technique of asking questions is available in Sudman and Bradburn (1983). This book discusses ordering of questions, physical layout, and wording among other considerations.

R. J. Johnston, *Geography and Geographers: Anglo-American Geography since 1945*, 2nd ed. (London: Edward Arnold, 1983).
S. Sudman and N. Bradburn, *Asking Questions: A Practical Guide to Questionnaire Design.* (San Francisco: Jossey Bass, 1983).

PROBLEMS

1. Explain the meaning of the following terms:

   a. Descriptive statistics
   b. Inferential statistics
   c. Statistical population
   d. Population characteristic
   e. Variable
   f. Population census
   g. Sample
   h. Sampling error
   i. Nonsampling error
   j. Representative sample
   k. Random sample
   l. Statistical estimation
   m. Hypothesis test
   o. Inductuve approach
   n. Deductive approach
   p. Internal data
   q. External data
   r. Primary data
   s. Secondary data
   t. Experiments
   u. Surveys
   v. Quantitative and qualitative variables
   w. Scale of measurement
   x. Ordinal, interval, and ratio scales
   y. Measurement accuracy
   z. Measurement validity

2. Make a list of the drawbacks to personal and telephone interviews as a data acquisition method for a large survey. What are the advantages and disadvantages of each?

3. Under what conditions might it be advantageous to undertake a population census rather than a sample?

4. Examine recent issues of academic journals used by geographers and cite three articles that employ statistical techniques. Briefly describe the role of statistical analysis in the study including the nature of the data used and the forms of statistical techniques (descriptive vs. inferential) used.

5. Develop a research project in some systematic field of geography that could be approached using a statistical survey.

6. Explain why human geographers have undertaken so few experimental studies in their research.

7. Examine any national census. Identifying the exact census report you are using, develop a list of 10 separate variables that could be obtained from this source.

8. Locate an article or book by a geographer that employs a structuralist perspective in analyzing an empirical research program. Are statistical techniques used in the study? Could they have been used? Would they have improved the paper, or do you think statistical techniques would be either inappropriate or superfluous to the study? To answer these questions, you must first define the purpose of the study.

9. In the reference room of your college library, locate 10 *different* sources of published data that can be used in statistical analyses. For each source,
   a. Identify the publisher,
   b. Note the frequency of the publication,
   c. Cite two specific forms of information obtainable from the source,
   d. Verify that the information is primary data.

10. Locate a publication that provides statistical data that could be described as secondary data. Where can the true primary data be found?

11. What do you think is the statistical population of interest in the study presented as Example 1-2 in this chapter? What safeguards might be important in developing identifying appropriate house sales to be used in the study?

12. Some research projects are developed using the *case study* approach. Using this approach, a single observation is chosen for detailed study, but there is still the desire to generalize the results of our survey to wider populations. For example, we might want to choose to study one inner city neighborhood, rather than the inner city as a whole, but we are interested in making conclusions relevant across the city or even to other cities. When do you think case studies are potentially more valuable than surveys?

I

DESCRIPTIVE STATISTICS

2

Univariate Descriptive Statistics

In Chapter 1, you were introduced to several types of data geographers encounter in their research. You may not have noticed that we introduced some simple procedures of descriptive statistics in our discussion of data measured at a nominal level. It was found, for example, that a large number of nominally scaled variables can be effectively summarized by recording the frequency, or count, of observations in each nominal class or category. We could even standardize the data by dividing the frequency in each class by the total number of observations to calculate proportions or percentages. The bar chart is a useful summary of data of this type.

In this chapter, we concern ourselves with interval scaled variables. Whether discrete or continuous, interval-scaled variables are not as easily summarized as nominal variables. There is no given set of classes to neatly summarize our counts or frequencies. Often, all the values of a variable are different, or there may be only a small number of repetitions. Consider the following example.

EXAMPLE 2-1. Water Quality One measure of the quality of lake or stream water is the amount of dissolved oxygen (DO). As wastes are discharged into the lake or stream, the amount of DO in the water declines. This oxygen is used in the biochemical reactions by which the nutrients in the waste are broken down. Public water supplies should have a DO content of at least 5 mg/L at any time. Table 2-1 contains a list of DO values taken from a sample of 50 lakes in one region. Notice that the water samples from some lakes have DO levels below the public water supply standard of 5 mg/L, and others have levels above it. A closer examination of these data reveals that there are 25 different DO values in the 50 lakes. It is extremely difficult to scan this list of numbers and get a clear picture of regional water quality. Inasmuch as this is a relatively small data set with only $n = 50$ observations, imagine the difficulty of trying to make sense out of a sample of 500 or even 1000 observations. Our task is to try to reduce this set of numbers to some reasonable collection of descriptive terms in much the same way as we simplified the presentation of a nominal variable.

In Section 2.1 we explore various ways of organizing our data and graphical techniques that can give us an overall picture of the distribution of the variable. When we speak of the *distribution* of a variable, we are referring to the tendency for the variable to take on

TABLE 2-1
Dissolved Oxygen Values for 50 Sampled Lakes, mg/L

Observation No. of lake	DO value	Observation No. of lake	DO value
1	5.1	26	4.9
2	5.6	27	6.6
3	5.3	28	5.7
4	5.7	29	5.4
5	5.8	30	5.9
6	6.4	31	5.6
7	4.3	32	6.7
8	5.9	33	5.4
9	5.4	34	4.8
10	4.7	35	6.4
11	5.6	36	5.8
12	6.8	37	5.3
13	6.9	38	5.7
14	4.8	39	6.3
15	5.6	40	4.5
16	6.4	41	5.6
17	5.9	42	6.2
18	6.0	43	4.2
19	5.5	44	5.2
20	5.4	45	5.8
21	4.4	46	6.1
22	5.1	47	5.1
23	5.6	48	5.9
24	5.8	49	5.5
25	5.7	50	4.7

different values with different frequencies. Although tabular and graphical devices are very useful for obtaining some initial insights into the distribution of a variable, they do not allow us to make precise statements about a variable or to compare several different distributions, except in the simplest of ways. Because the data are expressed as numbers, it is not surprising that mathematical or numerical descriptions of these distributions are the most useful and concise way of analyzing the data. Sections 2.2 to 2.5 discuss various statistical measures that can be used to summarize the important attributes of the distribution of an interval scaled variable. The dissolved oxygen data for the 50 lakes listed in Table 2-1 are used to illustrate all of the descriptive measures discussed in this chapter.

2.1. Tabular and Graphical Techniques

A collection of data can be displayed in several ways. At this time, let us restrict ourselves to methods useful for the description of a single variable such as the DO level of the 50 lakes. In Section III of this book, methods for illustrating and analyzing the *joint* distribution of two variables are described. For example, if the data included the DO level and the number of vacation homes along the shoreline of each lake, we would have two separate variables to analyze. Although we might be interested in each variable in its own right, we would probably also wish to examine the *relationship* between the oxygen level of a lake and the number of homes on its shoreline. Are DO levels lower in lakes with more vacation homes? Such an analysis may lead to setting appropriate standards for the density of homes on lakes. Nevertheless, even when we do have two or more variables to analyze, we normally begin by looking at the distribution of each individual variable.

Frequency Tables

In univariate, or single-variable situations, there are several different ways of displaying our observations. The first step is usually to construct a tally sheet or *frequency table.* Whereas nominal variables have ready-made classes to organize such a table, interval-valued variables do not. An initial decision must be made to define the classes or categories themselves. If one class were to be defined for each 0.1 mg of DO from 4.2 to 6.9, Table 2-2 would be generated. Each different value is associated with a *frequency,* or tally, which is simply the number of times the given value is observed. Because of the gaps in the data, there are 28 categories for the 50 observations! This does not help much. This type of classification scheme might be useful for discrete, or integer-valued variables with a limited range, but not in this case. How can the representation be improved? The answer lies in grouping the observations.

  Grouping is effected by using a different set of categories for the same data. If the oxygen data are grouped into classes using an interval of 0.5 units beginning at 4.0 and ending at 7.0, then six classes are created. Table 2-3 summarizes the frequency table. Frequencies are expressed in two forms. First, they are expressed as raw counts in the fourth column. It is apparent, for example, that 30 of the observations have values between 5 and 6, and smaller numbers of lakes have values below 5 or above 6. In the last column, these frequencies are expressed as *relative frequencies* or proportions of the total number of

TABLE 2-2
An Ungrouped Frequency Table

DO value, mg/L	Frequency	DO value, mg/L	Frequency
4.2	1	5.6	6
4.3	1	5.7	4
4.4	1	5.8	4
4.5	1	5.9	4
4.6	0	6.0	1
4.7	2	6.1	1
4.8	2	6.2	1
4.9	1	6.3	1
5.0	0	6.4	3
5.1	3	6.5	0
5.2	1	6.6	1
5.3	2	6.7	1
5.4	4	6.8	1
5.5	2	6.9	1

TABLE 2-3
A Frequency Table for DO Data

Class	Interval	Midpoint	Frequency count	Relative frequency
1	4.0–4.49	4.25	3	0.06
2	4.5–4.99	4.75	6	0.12
3	5.0–5.49	5.25	10	0.20
4	5.5–5.99	5.75	20	0.40
5	6.0–6.49	6.25	7	0.14
6	6.5–6.99	6.75	4	0.08
			50	1.00

observations. The sum of the column of frequency counts should equal the total number of observations, in this case $n = 50$. The sum of the relative frequency column is always 1.00. Expressed in this form, the oxygen data of Table 2-1 are much easier to interpret.

How are the *interval widths* and the *number of intervals* in a frequency table selected? A few simple rules should be kept in mind, but it should be emphasized that there is *no single correct answer* to this seemingly easy question.

RULE 1 Use intervals with simple bounds.

It is easier to use intervals with a common width of 0.5, 1.0, 10, 100, or 1000 rather than peculiarly selected widths such as 7.28, 97.63, or 729.58. Simplicity is the key. Similarly, the frequency table proves much easier to interpret if the lower bound or class limit of the first interval is a suitably chosen round number. If the interval width is conveniently chosen, this leads to regularly spaced, easy to interpret class limits and midpoints. The values selected in Table 2-3 serve this purpose. A good first step in constructing a frequency table is to order the data from lowest to highest value. Then, select a lower limit on the first class at a round number just below the lowest valued observation and an upper limit just above the highest valued observation. In this case, 4.0 is slightly below 4.2, and 7.0 is slightly above the maximum value of 6.9. Given these limits, we have a range of $7.0 - 4.0 = 3$ units. Potential interval widths of 0.25, 0.5, 0.75, and 1.0 all lead to easily understood class limits and midpoints.

RULE 2 Respect natural breakpoints.

For many variables, certain values within the observable range represent natural values for a class limit. For a series of temperatures expressed in degrees Celsius, it is appropriate to use 0° or 100° as class limits. For some variables expressed as percentages, 50% represents a natural class limit because it is the breakpoint for a majority. For pH values, 7.0 marks the border between acidic and alkaline substances and is a natural breakpoint. In the example of water quality, 5.0 makes a convenient class limit since it is the breakpoint between lakes meeting the public water supply standard of 5.0 mg/L and those lakes not meeting this standard.

RULE 3   The intervals should not overlap and must include all observations.

An observation can appear in one and only one class. Such classes are *mutually exclusive* because placement in one class precludes placement in any other interval. Do not specify overlapping interval widths such as 3–5 and 4–6. An observation of 4.5 would appear in both intervals. Although it is permissible to have categories such as <3 (less than 3) or 5+ (greater than 5, or >5), these should ordinarily be avoided. If they are selected because there is an extreme valued observation well above or below all other observations, then they are surely misused. The difficulty with such open intervals is that it is impossible to tell what the highest or lowest values might be. As we see below, this important characteristic of the data should not be hidden.

To avoid ambiguity and to define mutually exclusive categories, select class limits carefully. Note that the upper bound of the first interval, 4.49, does not equal the lower bound of the second interval, 4.50. An observation of exactly 4.50 must be placed in the second class. If both limits are defined as 4.50, it is not clear exactly where a value of 4.50 is to be placed. Defining limits with one more decimal place than the most precise observation in the data set is a useful general rule.

RULE 4   All intervals should be the same width.

It may seem obvious why this rule is important. It certainly simplifies matters. An even more important reason than simplicity comes to light when a graph of the numbers in a frequency table is constructed. As we see below, very misleading graphical representations of data can be made is this rule is not followed.

RULE 5   Select an appropriate number of classes.

Usually, the most difficult decision to make is the number of classes to be used in the table. A rule stated as "select an *appropriate* number of classes" certainly lacks precision, but it is difficult to develop an exact rule. Some mathematicians have gone so far as to offer rather exacting mathematical rules for determining the optimal number of classes in any given case. One of these rules suggests that the number of classes should be equal to five times the logarithm (to base 10) of the number of observations. For the water quality data, this leads to $5 \times \log_{10} 50 = 8.495$ and either eight or nine classes. Other statisticians have suggested that between 6 and 16 or 25 classes is appropriate depending on the sample size. Clearly, the more observations there are, the more classes that can be constructed. Neither of these rules can be considered inviolate.

Two important things should be kept in mind when deciding on the number of classes. First, when two observations are grouped together, we are saying that they are alike in some way. As a general rule, then, never select a number of groupings or interval widths that forces two unlike observations into the same group. We will break this rule if we place natural breakpoints within an interval and not at a class limit. Setting class limits of –5° to +5° Celsius would be unwise, tantamount to claiming that temperatures above freezing can be grouped with those below freezing.

The second thing to keep in mind is that grouping is a form of data *aggregation*, and there is a tradeoff between the gains from simplicity derived from aggregation and the loss of information that it necessitates. A frequency table with six classes means we have replaced the total number of observations, 50 in the case of our water quality example, with 12 new numbers—the six midpoints and the six frequencies of the classes. (We only require 12 numbers because we can generate the class limits that are always halfway between the midpoints.) There is a considerable gain in simplicity, but some information has been lost. Consider the first interval. If we know only the midpoint of 4.25 and the frequency of 3, then a good estimate of the sum of the observations in the interval is $3 \times 4.25 = 12.75$. Part of the information loss is the exact values of the three observations in the interval. Yet, we are still able to approximate characteristics of the three observations. The three observations in the first interval sum to $4.2 + 4.3 + 4.4 = 12.9$, very close to the estimate made using the frequency table. In this case, the gains in simplicity seem to outweigh the slight loss of information that occurs. Serious problems can occur, however, if we group our observations into too few intervals. This becomes apparent in the analysis of histograms, the graphical counterparts to frequency tables.

Frequency Histograms and Frequency Polygons

Often, it is easier to understand the information given in a frequency table when it is presented in graphical form. There appears to be a grain of truth in the adage "A picture is worth a thousand words." A special form of bar graph known as a *histogram* is used for this purpose. Figure 2-1 depicts the histogram for the data of Table 2-3. The X axis contains the scale of DO levels delimiting the six classes. Notice the small space to the left side of the axis below the lower bound of 4.0 and above the upper bound of 7.0. Bars are then drawn for each class from the X-axis to the height of the frequency of observations in the interval.

The vertical, or Y, axis can be labeled with absolute (or *raw*) frequencies, relative

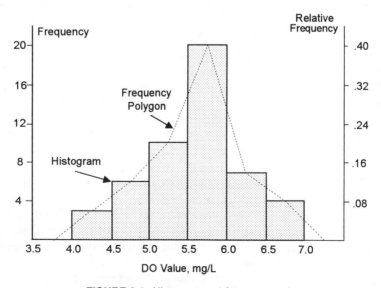

FIGURE 2-1. Histogram and frequency polygon.

frequencies, or both. In Figure 2-1, two Y axes are placed on the chart: absolute frequencies on the left hand side near the lower bound, and relative frequencies on the right hand side just above the upper bound. Most of the commonly used graphical computer packages now allow users to customize these charts in this way as well as to choose colors and fill patterns and similar display options. *Relative* frequency histograms are quite useful for *comparing* the distributions of different variables and can be expressed as either proportions or percentages.

It is important that the intervals be of equal width so the histogram does not give a misleading impression of the data. Suppose, for example, that the last two categories shown in Table 2-3 are grouped together to create a single class with bounds 6.0 to 6.99. The resultant histogram, shown in Figure 2-2 gives the impression that most lakes have sufficient oxygen levels (greater than 5 mg/L), whereas it is clear from Table 2-3 and Figure 2-1 that this is not so. There is another advantage to selecting equal intervals. Assume that the width of classes is standardized to be one unit and the frequencies are expressed as proportions. The total area under the histogram (area = height × width) is

$$.06(1) + .12(1) + .20(1) + .40(1) + .14(1) + .08(1) = 1.0$$

The misleading histogram of Figure 2-2 does not have this property.

There is an alternative way of presenting a frequency distribution. In many instances, bar graphs or histograms give the impression that the variable is not really continuous. For nominal variables, this may not be a problem, and we often graphically summarize such data by using bar graphs with spaces between the bars. For interval-valued variables, we can often give a better representation of the distribution of a continuous variable if we construct a *frequency polygon*. A frequency polygon is obtained by connecting the mid-points at the top of each interval. By convention, we usually extend the endpoints of the

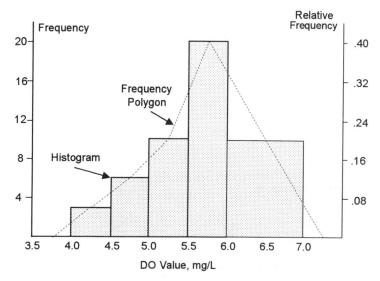

FIGURE 2-2. Histogram with unequal intervals.

frequency polygon beyond the upper and lower bounds to what would normally be the midpoint of the next interval. In Figure 2-2 the frequency polygon begins at 3.75 and ends at 7.25. By using this convention, it is easy to show that the area under the histogram and the area under the frequency polygon must be equal. For every triangular area not in the histogram but under the frequency polygon, there is a comparable triangle above the frequency polygon but beneath the histogram. Thus, both representations have equal areas.

We can think of both these graphical representations as crude approximations to smooth curves. If we are graphing the distribution of a sample, we can think of this sample as an approximation to a more general distribution—the distribution of the variable for the population. As a consequence, it is usual to smooth the histogram and frequency distribution to a simpler representation, as in Figure 2-3. This curve is only a graphical approximation of what we believe the actual population to be. It is an ideal shape. If a large number of observations are in a sample and we construct a histogram with many intervals, we could probably generate a close approximation to this smooth curve. What do histograms or smooth frequency polygons tell us about the variable in question? Characteristic shapes of these histograms can tell us important facts about the variable we are analyzing. A few things can be easily and quickly checked:

1. *Is the variable peaked or rectangular?* A peaked distribution such as Figure 2-1 suggests that many of the observations are centered on some typical value of this variable. A typical level of dissolved oxygen in the 50 sampled lakes seems to be about 5.75. If the histogram has no prominent peaks and the heights of all bars seems more or less equal, then

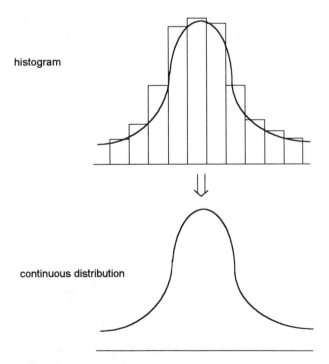

FIGURE 2-3. Histogram as a simplification of a continuous distribution.

the distribution is rectangular. This implies that there is no such typical value, and any value in the range is almost equally likely.

2. *How many different peaks are there?* Obviously, if there is only one peak, there is one typical value. Such a peak is called a *mode,* and a distribution with one mode is said to be *unimodal,* as in Figure 2-3. In some cases there may be a second peak. It may or may not be as noticeable as the first one. Histograms of this type, such as the one in Figure 2-4(a), are *bimodal.* Or, the distribution may be multmodal, as in Figure 2-4(b). Bimodal distributions can occur if two separate factors contribute to the distribution of a variable. For example, the distribution of heights of all 18-year-olds might contain two modes, one for females and for males. The distribution of heights for each sex taken individually would probably be unimodal. Many inferential statistics are invalid when they are applied to variables with bimodal or multimodal distributions.

3. *Is the distribution symmetric about some central point?* A *symmetric* distribution is one in which each half of the distribution is a mirror image of the other half. When this is not the case, the distribution is said to be *skewed.* If the tail of the distribution is on the right, or towards the positive side of the horizontal axis, the distribution is said to be *positively skewed.* If the tail is on the left, it is negatively skewed. These three cases are contrasted in Figure 2-5. For many inferential statistical tests, a symmetric distribution is required, but a few "tricks" can be used to reduce the extent of skewness of many distributions. The histogram of the dissolved oxygen variable illustrated in Figure 2-3 appears to be slightly negatively skewed.

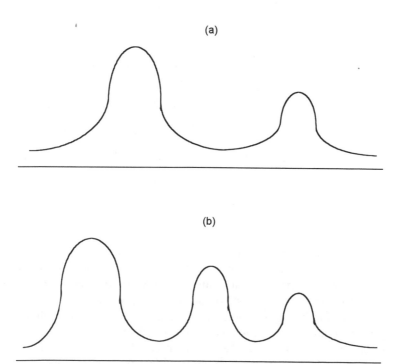

FIGURE 2-4. (a) Bimodal and (b) multimodal distributions.

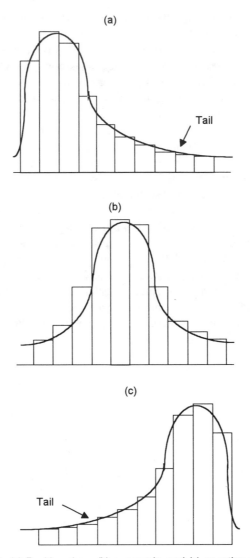

FIGURE 2-5. (a) Positive skew, (b) symmetric, and (c) negative skew distributions.

When the number and width of intervals are chosen, it is important to give a representative impression of the distribution of the variable. The visual impression of a histogram is quite sensitive to the interval width chosen. Poorly selected interval widths can mask important characteristics of the distribution. Modality, skewness, and/or the existence of extreme values should be highlighted by the histogram, not hidden. Such features can be hidden by choosing too few, wide intervals. Consider the two histograms of Figure 2-6. In Figure 2-6(a), the bimodal nature of the variable is clearly illustrated with the use of 12 intervals. In Figure 2-6(b) the same data are grouped into four classes, and the result is a perfectly rectangular distribution. Which is more realistic? Clearly, it is Figure 2-6(a).

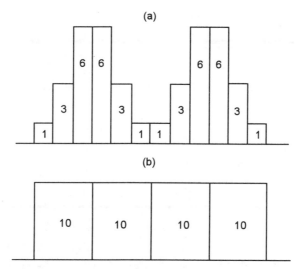

FIGURE 2-6. (a) Bimodal distribution; (b) rectangular distribution.

Similar cases can be constructed to show the reduction of extreme skewness or to hide the existence of an extreme-valued observation.

Cumulative Table and Ogive

For some purposes, it is convenient to summarize and illustrate data in a slightly different way. Suppose, for example, that we are interested in determining the number or proportion of lakes with DO values less than or equal to some given level. How many lakes have DO values less than 5.0? What proportion of the lakes have DO levels less than 5.5? Quick answers to these questions can be derived by portraying the data in cumulative form, as in Table 2-4. The cumulative frequency column is calculated by summing the frequencies of all intervals below and including the interval under consideration. At the end of the first interval, for a value of $X = 4.49$, there are three observations. There are nine observations below a value of 4.99—three in the first interval and six in the second interval. The table is completed in this way. The last entry in the cumulative frequency column is always equal to the total number of observations, or the sample size, n. In Table 2-4 the last entry in the forth column is 50.

Sometimes it is useful to express this cumulative frequency column in relative terms, as in the last column. We calculate the cumulative percentage as the cumulative frequency divided by the total number of observations. For the first row, this is 3/50 = .06 (or 6%); for the second row 9/50 = .18 (or 18%); and so on. Expression of this information as relative frequencies in either proportions or percentages is extremely helpful when we have a large number of observations and the simple cumulative frequencies are not as easily interpretable. Suppose, for example, there are 2200 lakes and the cumulative frequency at less than a DO level of 5.0 is 550. This is much easier to interpret if we say that 25% of observations have DO levels below 5.0 mg/L.

As you may have already guessed, we can also use graphical devices to present the

TABLE 2-4
A Cumulative Frequency Table

Class	Interval	Midpoint	Frequency	Cumulative frequency	Cumulative relative frequency
1	4.0–4.49	4.25	3	3	0.06
2	4.5–4.99	4.75	6	9	0.18
3	5.0–5.49	5.25	10	19	0.38
4	5.5–5.99	5.75	20	39	0.78
5	6.0–6.49	6.25	7	46	0.92
6	6.5–6.99	6.75	4	50	1.00
			50		

3 + 6 = 9
3 + 6 + 10 = 19

information in a cumulative frequency table. Figure 2-7 displays the *cumulative frequency histogram* and the *cumulative frequency polygon,* or *ogive.* The cumulative frequency polygon is drawn in exactly the same manner as a simple histogram except that the height of the bars is proportional to the cumulative frequency at the end of the interval. Normally, the ogive is drawn by connecting the lower bound of the first interval to the cumulative frequency at the end of the first interval, at the end of the second interval, and so on. The last point is always at the end of the last interval with the cumulative frequency equal to the total number of observations. It is often convenient to label two separate vertical axes, one for the *cumulative frequency* and one for the *cumulative relative frequency.* We can use the ogive to derive an estimate of the number or percentage of observations less than some

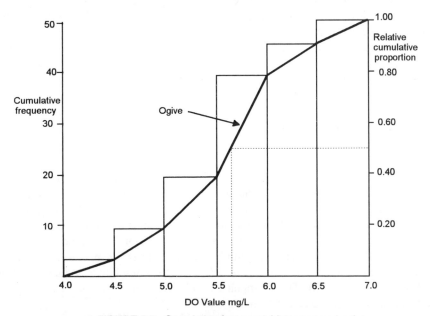

FIGURE 2-7. Cumulative frequency histogram and ogive.

given value. Suppose we are interested in the DO value that is exceeded by 50% of the observations. Extend a line from the relative cumulative proportion axis at .50 until it hits the ogive, and then drop the line vertically from this point to the DO axis. We estimate that 50% of the lakes have DO values less than about 5.6 or 5.7. This procedure is illustrated in Figure 2-7.

Because the ogive and frequency polygon are both derived from the same data, it should not be surprising that the shapes of these curves are closely related. For example, a peaked polygon such as Figure 2-3 is always associated with a more or less S-shaped ogive such as Figure 2-7. Figure 2-8 illustrates generalized or smoothed frequency polygons and the corresponding ogives. Rectangular or near-rectangular frequency polygons have straight-line ogives; peaked distributions have one S-shaped segment for each peak. Figure 2-8(c) illustrates the double S-ogive of a bimodal distribution.

Graphical Devices from Microcomputer Packages

It is now possible to generate simple graphical displays using several different types of software for microcomputers, including statistical packages such as SPSS or SAS, spread-

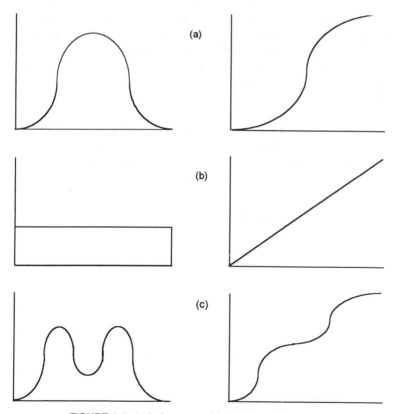

FIGURE 2-8. Left: frequency histograms; right: ogives.

sheet programs such as Lotus 123 or Microsoft Excel, graphics packages such as Harvard Graphics and Freelance Graphics, database programs, and even word processors such as Microsoft Word, WordPerfect, or AmiPro. Vertical and horizontal bar graphs can be generated for interval data, in either two- or three-dimensional formats, are illustrated in Figure 2-9(a) through (d). Nominal variables can also be summarized in two- or three-dimensional pie charts as illustrated in Figure 2-9(e) and (f).

2.2. Measures of Central Tendency

To make more precise statements about the distribution of a variable, quantitative measures or descriptive statistics are used. Although the tabular and graphical devices described in Section 2.1 can tell us a great deal about a variable and even allow us to make simple comparisons about two or more variables, they are never precise. Numerical statistics are an even more compact way of summarizing interval-scale variables, enabling exact comparisons between pairs of variables to be made. One important characteristic of a distribution is the location of the typical value, or center, of the variable. In fact there are several different measures of *central tendency*, each having certain advantages and disadvantages as well as specific mathematical properties. Each measure usually specifies a single number that can be considered the "center" of a set of observations. The fact that we can compute several different measures of central tendency implies we have to fully understand each one in order to know when each is an appropriate summary statistic. Although each measure is in some sense the "average" or center of a set of numbers, none is considered to be *universally* superior to all others. As we shall see, however, there is one that is more commonly used.

Midrange

The *range* of a variable is simply the difference between the highest and lowest valued observations. Let the lowest valued observation be denoted X_{min} and the highest valued be X_{max}. The *midrange* lies exactly halfway between these two observations.

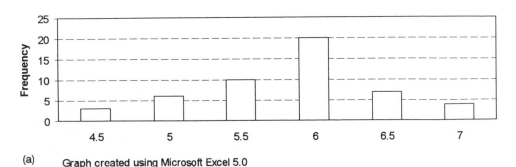

(a)   Graph created using Microsoft Excel 5.0

FIGURE 2-9. Graphical displays from PC software packages.

3-D Bar Chart of Dissolved Oxygen Data

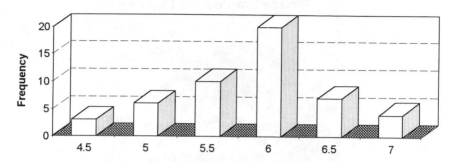

Graph Created Using Microsoft Excel 5.0

(b)

2D Bar Chart of Dissolved Oxygen Data

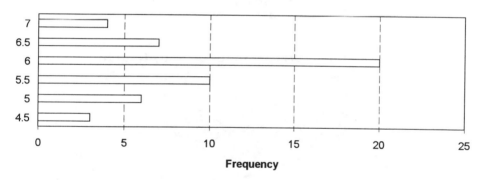

(c)   Graph created using Microsoft Excel 5.0

3-D Bar Chart of Dissolved Oxygen Data

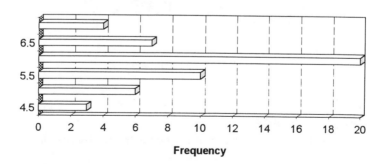

(d)   Graph Created Using Microsoft Excel 5.0

FIGURE 2-9. *continued.*

Service Employment by Region

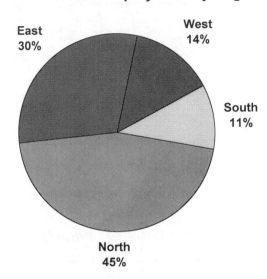

(e)

Manufacturing Employment by Region

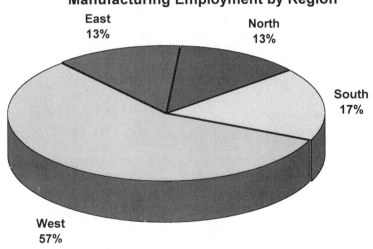

(f)

FIGURE 2-9 *continued.*

DEFINITION: MIDRANGE

The midrange is the arithmetic mean of the two extreme observations:

$$\text{Midrange} = \frac{X_{max} + X_{min}}{2} \qquad (2\text{-}1)$$

Although it is easy to calculate, it has the disadvantage of being inordinately affected by the two extreme observations. Neither of these can be considered typical, so there is no reason to think that any calculation based upon them would be either! For this reason, this measure is seldom used. For the DO data, the maximum is 6.9, and the minimum is 4.2; the range is 2.7, and the midrange is (6.9 + 4.2)/2 = 5.55.

Mode

Strictly speaking, the mode of a distribution is the value of the variable that appears most frequently. For many continuous valued interval scale variables, there may be no exact repeated values in the data. We could say there is no mode, or we might even say that there are *n* different modes. This measure is not very useful in this case. If the data are grouped, the *modal category* or *modal class* is defined as the interval with the highest frequency; the midpoint of the modal class is termed the *crude mode*. Unfortunately, this approach does not define a *unique* mode or interval because the value depends on the number and location of the intervals selected in the grouping procedure. Loosely speaking, the value associated with a peak of a histogram or frequency polygon is termed a *mode*. In this sense, there may be one or more modes of a distribution.

When is the mode useful? Under certain conditions, this measure is useful for nominal variables and discrete-valued interval variables. For nominal variables, the mode is the category with the greatest frequency. Note that the mode is not really a measure of central tendency in this case since a nominal variable is not ordered and hence can have no center. For ordinal or discrete-valued variables, a mode is useful for cases in which there is one value with a dominating frequency. Suppose a set of 200 households yields the following tabulation of household size:

Number of household members	Frequency
1	15
2	30
3	100
4	25
5	30

The mode is three persons per household, and it seems to be a good summary measure for this distribution. To many, it is preferable to saying there are 3.125 members in the average household.

Median

The median of a set of observations is the value of the variable that divides the observations so that one-half are less than or equal to the median and one-half are greater than or equal to it. To compute the median, it is again useful to order the data in ascending order. Table

2-5 contains the ordered list of the 50 DO values. Since the median should divide the set of observations into two equal parts, there should be 25 values in each. The first set contains observations 1 to 25, and the second set contains observations 26 to 50. The median is the $(n + 1)/2$ observations. Since $n = 50$, the median is the value of the 25.5th observation. When n is even, we must have some rule to follow. Normally, the median is defined as the average of the middle two observations. In this case, the median is the average of the 25th and 26th terms, or $(5.6 + 5.6)/2 = 5.6$. When there are an odd number of observations, $(n + 1)/2$ is always a whole number and defines the middle term. Suppose that we have the following five ordered observations: 2, 3, 3, 5, and 8. The median is the middle term, or $(5 + 1)/2 = 3$rd term and is equal to 3.

The median is often called a positional measure because it locates the position of an observation relative to a set of other observations. Just as it is possible to define the median, or middle term, it is also possible to define *quartiles* as the values that divide the set of observations into quarters. The *first quartile* divides the lower half into two equal sets. The *median* is the *second quartile*, and the *third quartile* divides the upper half into two sets. Thus, 25% of the observations are below the first quartile, 50% are below the median, and 75% are below the third quartile. Table 2-5 indicates that the first quartile of the DO data is the value of the $(25 + 1)/2 = 13$th observation and is equal to 5.2. Similarly, the third quartile is found to be 5.9.

We can generalize this procedure to define *deciles*, which divide the observations into tenths, and *percentiles*, which divide the data into hundredths. The median is the second quartile, the fifth decile, and the 50th percentile. The nth percentile is usually specified as P_n, and the three quartiles as Q_1, Q_2, and Q_3. Therefore, $P_{25} = Q_1 = 5.2$, $P_{50} = Q_2 = 5.6$, and $P_{75} = Q_3 = 5.9$. Positional terms are commonly associated with educational testing procedures and are used to locate a student among a large set of student scores for an examination. Percentiles have been infrequently applied in geographical research.

Mean

The *mean* of a set of observations is the most commonly used measure of central tendency. In most instances the term mean is the shortened version of *arithmetic mean*, and it is the sum of the values of all observations, divided by the number of observations. In common use this is simply the *average of a set of observations*. The mean DO value of the 50 lakes is $(4.2 + 4.3 + \ldots + 6.7 + 6.8 + 6.9)/50 = 279/50 = 5.58$. The three dots, or *ellipses*, are shorthand for noting that the summation omits some terms in the set. The middle 45 observations of DO are omitted. To write them out is unnecessary since the meaning of the summation is clear.

To simplify many of the formulas in descriptive and inferential statistics, it is convenient to utilize a system of notation. If the variable is denoted by the uppercase X and a subscript i is used to denote individual observations, then we can express the formula for the mean of a sample as $(X_1 + X_2 + \ldots + X_n)/n$. An even more efficient way of writing this expression makes use of the *summation operator* or what is sometimes called *sigma notation*. The symbol $\Sigma_{i=1}^{n}$ (read "sigma for i equals 1 to n") is a mathematical shorthand that has considerable use in statistics. It is known as the *summation operator* since it means that the items following the symbol are to be summed through all successive values of the

TABLE 2-5
Table 2-1 Expressed in Order of Increasing Value

4.2
4.3
4.4
4.5
4.7
4.7
4.8
4.8
4.9
5.1
5.1
5.1
5.2   ← 1st quartile = 25th percentile = P_{25}
5.3
5.3
5.4
5.4
5.4
5.4
5.5
5.5
5.6
5.6
5.6
5.6   ← Median = 2nd quartile = 5th decile = 50th percentile = P_{50}
5.6
5.6
5.7
5.7
5.7
5.7
5.8
5.8
5.8
5.8
5.9
5.9
5.9   ← 3rd quartile = 75th percentile = P_{75}
5.9
6.0
6.1
6.2
6.3
6.4
6.4
6.4
6.6
6.7
6.8
6.9

index of summation i. Below the sigma, the index of summation specifies the start of the summation, in this case $i = 1$. This is the usual case, but it is possible to start the summation at any point within a series of observations, say $i = 3$, $i = 21$, or a general point such as $i = k$. Above the sigma, the index defines the last observation to be included in the summation; here $i = n$.

Using this notation, we can write

$$\sum_{i=1}^{n} X_i = X_1 + X_2 + \ldots + X_n \tag{2-2}$$

The sigma operator greatly simplifies the specification of many different operations in statistics. Students unfamiliar with summation notation should consult the Appendix to Chapter 2, which contains a brief review of the use of summation notation. Included in this review are several simple rules for summations that are used extensively to simplify many of the formulas encountered in statistical analysis. Even those who are experienced with the use of the summation operator will find this Appendix to be a useful review.

DEFINITION: SAMPLE MEAN $\bar{X}$

The mean of a set of observations X_i, $i = 1, 2, \ldots, n$ is defined as

$$\bar{X} = \frac{\sum_{i=1}^{n} X_i}{n} \tag{2-3}$$

The symbol $\bar{X}$ (read "X bar") denotes the mean of a variable X. At times, we find it necessary to distinguish the mean of a *sample* from the mean of a *population*. For a sample, the observations are numbered consecutively from 1 to n, and the mean is denoted by $\bar{X}$. For a finite population consisting of N observations, the formula and notation is modified.

DEFINITION: POPULATION MEAN μ

The mean of a population with N members is

$$\mu = \frac{\sum_{i=1}^{N} X_i}{N} \tag{2-4}$$

The symbol μ is the lowercase Greek letter mu. These two formulas are *not* interchangeable. There are many different sample means. Depending on which n observations are selected from a population or universe with N members, we get a different value for $\bar{X}$. However, if we compute the mean from all N observations in the population, there is only one possible value for the mean, μ. Generally, a sample size n is only a small portion of the members of the population, so that $n < N$.

Two important conventions have been introduced in these formulas that are followed throughout this text. First, lowercase n always refers to a sample size, and upper case N always refers to the size of a finite population. Second, characteristics of populations are defined by Greek letters such as μ, but samples are not. These distinctions become particularly important when we try to make inferences about a population from a sample. For example, we might wish to use $\bar{X}$ to estimate a value for μ. In particular, we might wish to determine how close we might be to the mean of the population with samples of increasing size. When $n = N$ and all members of the population are included in the sample, $\bar{X} = \mu$. We might expect that, on average, the value of the sample mean will get progressively closer to the population mean as the sample size increases. It turns out that this is exactly the case. We return to this issue of the exact relationship between the sample and population means in Chapter 8.

Choosing an Appropriate Measure of Central Tendency

For the oxygen data of Table 2-1, the following measures of central tendency are obtained:

$$\text{Mean } \bar{X} = 5.58 \qquad \text{Median} = 5.60$$

$$\text{Mode} = 5.60 \qquad \text{Midrange} = 5.55$$

All of these measures are very similar, but this is not necessarily a usual outcome. Let us review the derivation of each measure from a histogram. The mode is associated with a peak of the distribution, and the median is the value that splits the distribution into two equal areas. The midrange divides the length of the distribution into two equal parts. What about the mean? Think of the histogram as a set of weights placed along a board in proportion to the height of the bars. A higher frequency means a higher weight at the point. If we were to place a wedge under the board, then it would balance only if it were placed at the location of the mean. To illustrate this, consider the set of five observations 2, 2, 3, 5, and 8. As shown in Figure 2-10, the mean is the "center of gravity," or balancing point of this distribution.

The mean has one additional advantage over the other measures of central tendency—it is sensitive to a change in any of the observations. Consider again the set 2, 2, 3, 5, and 8. The mean is 4, the mode is 2, and the median is 3. Suppose the last observation is

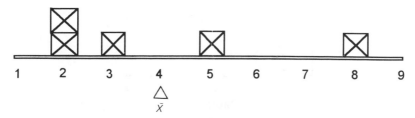

FIGURE 2-10. The mean as the center of gravity of a distribution.

changed from 8 to 18. The median is still 3, the mode is still 2, but the mean changes from 4 to 6. In most instances, any change in the distribution is reflected in the mean, though not necessarily the other measures.

However, there are several instances for which the mean is not a good measure of central tendency:

1. For *bimodal distributions*. Consider Figure 2-11. Note that the mean and median are coincident, but neither picks up the "typical value" of this variable. Neither would the midrange, which is at the same location. Because all these measures provide one value, neither can be useful in this case. The best summary statistic in this case is the *mode*, since two "typical" values are clearly necessary to characterize the distribution.

2. *For skewed distributions.* The mean is not very useful for variables that are either positively or negatively skewed. Figure 2-12 illustrates the relative location of the mean, median, and mode for skewed distributions. For a positively skewed distribution, the mean is greater than the median, which is greater than the mode. For a negatively skewed distribution the reverse is true: the mean is less than the median which is less than the mode. Although the median is pulled toward the tail of the distribution, it is not nearly as sensitive as the mean. Note that for symmetric distributions the three measures are equal. As a distribution deviates from this shape, the three become increasingly different. The median is usually the best measure in this case.

3. *For a distribution with an extreme value (or a small number of extreme values).* We can think of a distribution with an extreme value as highly skewed. The median is also the best measure in this case. Consider again the data 2, 2, 3, 5, and 8. If 8 is changed to 200, then the mode remains at 2, the median stays at 3, but the mean increases to 44. The mean is clearly no longer even close to any of the observations and cannot be described as a typical value. The median is central to four of the five observations.

4. In sum, the mean is the most commonly used measure of central tendency. It is sensitive to all changes in the distribution and makes use of all the information for a set of observations. Unless one of these four conditions applies, it is the best measure of central tendency for a distribution.

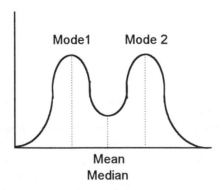

FIGURE 2-11. Measures of central tendency in a bimodal distribution.

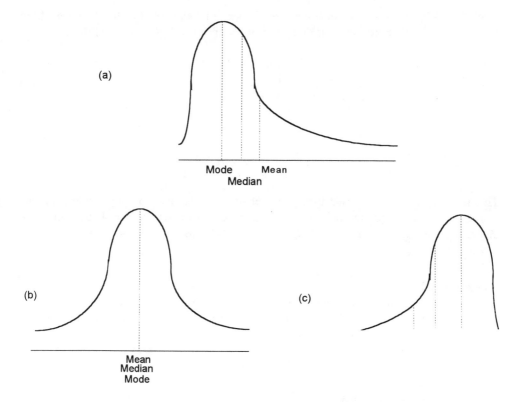

FIGURE 2-12. Measures of central tendency for skewed distributions.

Properties of the Mean and Median

There are three important properties of the mean and median:

PROPERTY 1

The sum of the deviations of each observation from the mean is zero.

$$\sum_{i=1}^{n} (X_i - \bar{X}) = 0 \tag{2-5}$$

This is another sense in which the mean is the middle of a set of observations. The property is easily proved using the simple rules for summation reviewed in Appendix 2. Using Rule 1, we can express Equation (2-5) as

$$\sum_{i=1}^{n} X_i - \sum_{i=1}^{n} \bar{X} = 0 \tag{2-6}$$

And, since $\bar{X}$ is a constant (for any given set of observations), we can write $\Sigma_{i=1}^{n} \bar{X} = n\bar{X}$ from Rule 2. Finally, we note from our definition of the mean in Equation (2-3) that

$$n\bar{X} = n\left(\frac{\sum\limits_{i=1}^{n} X_i}{n}\right) = \sum_{i=1}^{n} X_i \tag{2-7}$$

and Equation (2-6) can be rewritten as

$$\sum_{i=1}^{n} X_i - \sum_{i=1}^{n} X_i = 0$$

This property is easily verified using any set of observations. To keep matters simple, consider once again the five numbers 2, 2, 3, 5, and 8 for which $\bar{X} = 4$. Summing the deviations yields $(2 - 4) + (2 - 4) + (3 - 4) + (5 - 4) + (8 - 4) = -2 + (-2) + (-1) + 1 + 4 = 0$.

PROPERTY 2

The sum of squared deviations of each observation from the mean is a minimum, that is, less than the sum of squared deviations from any other number. Formally, we say

$$\sum_{i=1}^{n} (X_i - M)^2 \tag{2-8}$$

is minimized when $M = \bar{X}$.

This is often called the *least squares property of the mean*. The quantity

$$\sum_{i=1}^{n} (X_i - \bar{X})^2$$

is the *total variation* of variable X. Although it is slightly more difficult to prove, this property is easy to illustrate. Using the same set of five observations, we calculate the sum of squared deviations as follows:

Observation	Mean = 4	Median = 5
2	$(2 - 4)^2 = 4$	$(2 - 3)^2 = 1$
2	$(2 - 4)^2 = 4$	$(2 - 3)^2 = 1$
3	$(3 - 4)^2 = 1$	$(3 - 3)^2 = 0$
5	$(5 - 4)^2 = 1$	$(5 - 3)^2 = 4$
8	$(8 - 4)^2 = 16$	$(8 - 3)^2 = 25$
	Sum = 26	Sum = 31

We see that the sum of squared deviations is 26 for the mean and 31 for the median. Any other value would lead to a sum greater than 26. This property suggests yet another sense in which the mean is the center of a distribution.

A closer examination of Equation (2-9) reveals why the mean is a poor measure of central tendency when the set of observations is highly skewed or contains an extreme value. To minimize Equation (2-9), a value of M must be selected that minimizes the squared deviations. A deviation of $X - M = 1$ is counted as $1^2 = 1$, but a deviation of $X - M = 5$ is counted as $5^2 = 25$ and given 25 times as much weight. Squaring the deviations penalizes larger deviations more than smaller ones. As a result, the mean is "drawn" toward extreme values, or values in the tail of the distribution. In this way, the high penalties associated with large deviations are minimized. Figure 2-12 illustrates this result graphically. The location of the mean is the most extreme of the measures of central tendency in a highly positively or negatively skewed distribution.

PROPERTY 3

The sum of *absolute* deviations is minimized by the median. That is, the expression

$$\sum_{i=1}^{n} |X_i - M| \tag{2-9}$$

is minimized when M is equal to the median of the X values.

This can also be illustrated by using the small sample of $n = 5$ observations 2, 2, 3, 5, and 8. For comparative purposes let us once again compute the sum of absolute deviations for two values: M equal to the mean 4 and M equal to the median 3.

Observation	Median = 3	Mean = 4
2	$|2 - 3| = 1$	$|2 - 4| = 2$
2	$|2 - 3| = 1$	$|2 - 4| = 2$
3	$|3 - 3| = 0$	$|3 - 4| = 1$
5	$|5 - 3| = 2$	$|5 - 4| = 1$
8	$|8 - 3| = 5$	$|8 - 4| = 4$
	Sum = 9	Sum = 10

The sum of absolute deviations from the median is 9 and from the mean is 10. Because it has been drawn toward the extreme value of 8 in order to minimize the squared deviations, the mean is in an inferior position to the median for absolute deviations. Although the median is 5 units away from this extreme observation, it is penalized far less by using this criterion than by using the sum of squared deviations.

Besides explaining why the mean and the median are superior measures of central tendency in particular cases, these properties have interesting interpretations when we consider spatial data in Chapter 4. In particular, they have important implications for locating the center of a two dimensional point pattern.

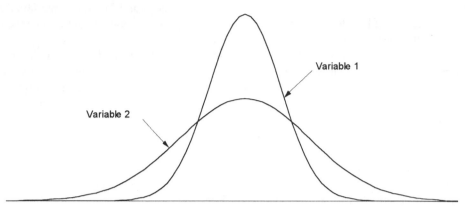

FIGURE 2-13. Variables with the same mean but different dispersions.

2.3. Measures of Dispersion

The measures of central tendency discussed in Section 2.2 are very useful for calculating a summary statistic that represents an average, or typical, value of variable X. However, these statistics measure only one characteristic of a frequency distribution. Compare the two distributions shown in Figure 2-13. Both variables have identical means, but much different *dispersions*, or *spread*, about the mean. Variable 1 is much more compact than variable 2. Just as for the concept of central tendency, several different measures can be used to evaluate the spread or dispersion of a distribution.

Range

In defining the *midrange* of a variable as a measure of central tendency, it was necessary to define the first measure of dispersion of a distribution, the *range*.

DEFINITION: RANGE

The range of a set of observations is the difference between the highest and lowest valued observations:

$$\text{Range} = X_{max} - X_{min}$$

For the dissolved oxygen variable, for example, we find the range to be 6.9 − 4.2 = 2.7. The minimum DO value gives a fairly good estimate of the lower bound on current lake oxygen levels. Over time, the analysis of the lower bounds, upper bounds, and range would provide us with a useful snapshot of variation in lake DO levels. There is one important feature about this statistic that must not be forgotten when it is used as a measure of dispersion: it only takes into account the two most extreme observations in the data set. When one, or both, of these extreme values is unusual and removed from other observations, then the range may be misleading. We should also keep in mind that larger samples almost invariably have greater ranges than smaller samples, since they are more likely to contain the rare or unusual members of a population.

Quartile Deviation

The *quartile deviation* is one-half of the *interquartile range*:

DEFINITION: QUARTILE DEVIATION

$$\text{Quartile deviation} = \frac{P_{75} - P_{25}}{2} \qquad (2\text{-}10)$$

where P_{75} is the 75th percentile or third quartile and P_{25} is the 25th percentile or first quartile.

The difference between the first and third quartiles is often called the *interquartile range*. For the dissolved oxygen data, the quartile deviation is $(5.9 - 5.2)/2 = 0.35$. The percentiles used in this calculation are drawn from Table 2-5. Because this measure ignores the extreme observations and uses the middle half of the observations, it is generally a more stable measure of dispersion. Different samples from the same population may have widely different ranges, but are likely to have fairly similar quartile deviations.

Unfortunately, neither the range nor the quartile deviation takes full advantage of the entire set of observations and their values. Three other measures of dispersion take full advantage of the actual values of all observations. All are based on the deviations of each observation from the mean $\bar{X}$. The mean is used as the center of the distribution because it is the single most satisfactory measure of central tendency.

Mean Deviation

As formalized in Property 1, the sum of the deviations about the mean is zero. This is true because the positive deviations offset the negative deviations. To get around this problem for a measure of dispersion, we take the absolute value of the deviations

---

DEFINITION: MEAN DEVIATION

The mean deviation of a variable X is defined as

$$\text{Mean deviation} = \frac{\sum_{i=1}^{n} |X_i - \bar{X}|}{n} \qquad (2\text{-}11)$$

The calculation of the mean deviation for the dissolved oxygen data is illustrated in Table 2-6. The second column contains the signed deviations and sums to 0.0. The absolute deviations sum to 24.36, leading to a mean (or average) absolute deviation of 24.26/50 =

TABLE 2-6
Work Table for Mean Deviation, Standard Deviation, and Variance

Observation	X_i	$X_i - \bar{X}$	$|X_i - \bar{X}|$	$(X_i - \bar{X})^2$	X_i^2
1	5.1	−0.48	0.48	0.2304	26.01
2	5.6	0.02	0.02	0.0004	31.36
3	5.3	−0.28	0.28	0.0784	28.09
4	5.7	0.12	0.12	0.0144	32.49
5	5.8	0.22	0.22	0.0484	33.64
6	6.4	0.82	0.82	0.6724	40.96
7	4.3	−1.28	1.28	1.6384	18.49
8	5.9	0.32	0.32	0.1024	34.81
9	5.4	−0.18	0.18	0.0324	29.16
10	4.7	−0.88	0.88	0.7744	22.09
11	5.6	0.02	0.02	0.0004	31.36
12	6.8	1.22	1.22	1.4844	46.24
13	6.9	1.32	1.32	1.7424	47.61
14	4.8	−0.78	0.78	0.6084	23.04
15	5.6	0.02	0.02	0.0004	31.36
16	6.4	0.82	0.82	0.6724	40.96
17	5.9	0.32	0.32	0.1024	34.81
18	6.0	0.42	0.42	0.1764	36.00
19	5.5	−0.08	0.08	0.0064	30.25
20	5.4	−0.18	0.18	0.0324	29.16
21	4.4	−1.18	1.18	1.3924	19.36
22	5.1	−0.48	0.48	0.2304	26.01
23	5.6	0.02	0.02	0.0004	31.36
24	5.8	0.22	0.22	0.0484	33.64
25	5.7	0.12	0.12	0.0144	32.49
26	4.9	−0.68	0.68	0.4624	24.01
27	6.6	1.02	1.02	1.0404	43.56
28	5.7	0.12	0.12	0.0144	32.49
29	5.4	−0.18	0.18	0.0324	29.16
30	5.9	0.32	0.32	0.1024	34.81
31	5.6	0.02	0.02	0.0004	31.36
32	6.7	1.12	1.12	1.2544	44.89
33	5.4	−0.18	0.18	0.0324	29.16
34	4.8	−0.78	0.78	0.6084	23.04
35	6.4	0.82	0.82	0.6724	40.96
36	5.8	0.22	0.22	0.0484	33.64
37	5.3	−0.28	0.28	0.0784	28.09
38	5.7	0.12	0.12	0.0144	32.49
39	6.3	0.72	0.72	0.5184	36.69
40	4.5	−1.08	1.08	1.1664	20.25
41	5.6	0.02	0.02	0.0004	31.36
42	6.2	0.62	0.62	0.3844	38.44
43	4.2	−1.38	1.38	1.9044	17.64
44	5.2	−0.38	0.38	0.1444	27.04
45	5.8	0.22	0.22	0.0484	33.64
46	6.1	0.52	0.52	0.2704	37.21
47	5.1	−0.48	0.48	0.2304	26.01
48	5.9	0.32	0.32	0.1024	34.81

TABLE 2-6 (continued)

Observation	X_i	$X_i - \bar{X}$	$|X_i - \bar{X}|$	$(X_i - \bar{X})^2$	X_i^2
49	5.5	−0.08	0.08	0.0064	30.25
50	4.7	−0.88	0.88	0.7744	22.09

$n = 50$
$n - 1 = 49$ $\sum\limits_{i=1}^{n} X_i = 279$ $\sum\limits_{i=1}^{n} (X_i - \bar{X}) = 0$ $\sum\limits_{i=1}^{n} |X_i - \bar{X}| = 24.36$ $\sum\limits_{i=1}^{n} (X_i - \bar{X})^2 = 20.02$ $\sum\limits_{i=1}^{n} X_i^2 = 1576.84$

$$\text{Mean} = \frac{\sum\limits_{i=1}^{n} X_i}{n} = \frac{279}{50} = 5.58$$

$$\text{Mean Deviation} = \frac{\sum\limits_{i=1}^{n} |X_i - \bar{X}|}{n} = \frac{24.36}{50} = .4872$$

$$\text{Variance} = s^2 = \frac{\sum\limits_{i=1}^{n} (X_i - \bar{X})^2}{n - 1} = \frac{20.02}{49} = .4086$$

$$\text{Standard deviation} = s = \sqrt{\frac{\sum\limits_{i=1}^{n} (X_i - \bar{X})^2}{n - 1}} = \sqrt{\frac{20.02}{49}} = .6392$$

Or from computational formulas:

$$s^2 = \frac{1}{n - 1} \left(\sum\limits_{i=1}^{n} X_i^2 - n\bar{X}^2 \right) = \frac{1}{49} [1576.84 - 50(5.58)^2] = .4086$$

$$s = \sqrt{\frac{1}{n - 1} \left(\sum\limits_{i=1}^{n} X_i^2 - n\bar{X}^2 \right)} = \sqrt{\frac{1}{49} [1576.84 - 50(5.58)^2]} = .6392$$

.4872. There is one disadvantage to the mean deviation. Many statistical results depend on easy algebraic manipulation of formulas such as Equation (2-11). Note how we were able to prove that the deviations summed to zero by manipulating Equation (2-5). The algebraic simplification of absolute value terms is quite cumbersome and generally does not lead to simple results. For this and other reasons, an alternative measure of dispersion, the standard deviation, is more commonly used.

Standard Deviation

DEFINITION: STANDARD DEVIATION

The standard deviation of a *sample* of observations is defined as the square root of the mean squared deviations about the mean:

$$s = \sqrt{\frac{\sum\limits_{i=1}^{n} (X_i - \bar{X})^2}{n - 1}} \tag{2-12}$$

DEFINITION: STANDARD DEVIATION

The standard deviation of a *population* is defined as the square root of the mean squared deviations about the population mean μ:

$$\sigma = \sqrt{\frac{\sum_{i=1}^{N} (X_i - \mu)^2}{N}} \qquad (2\text{-}13)$$

The notation is thus $\bar{X}$ and s for samples, and μ and σ for populations.

There are two significant differences between Equations (2-12) and (2-13). First, the squared deviations for the sample standard deviation s are summed over the n different observations, whereas the squared deviations for the population standard deviation are summed over all N members of the population. Second, the denominator for s is $n - 1$, but it is N for σ. When n is very large, (say, 100 or more), the difference between s and σ is quite small; and when $n = N$ the values of s and σ are equal. That is, when the sample contains all the members of the population, there can be no difference between s and σ.

But why do samples have $n - 1$ in the denominator? It turns out that if we divide by n we always underestimate the true value of the population standard deviation σ. By dividing by $n - 1$, the standard deviation becomes larger, and the underestimate is corrected. A more detailed explanation is given in Chapter 8 where we deal with the *statistical estimation* of population parameters. However, this anomaly can also be explained by using two much simpler arguments.

First, let us recall the second property of the mean, which states that $\Sigma_{i=1}^{n} (X_i - M)^2$ is minimized for any set of observations when $M = \bar{X}$. That is, substituting any other number for M other than $\bar{X}$ for M will necessarily result in a larger sum of this expression. Even if we substitute the true population mean μ, at best we can do as well as with $M = \bar{X}$, and generally we will do worse. However, if we know μ, it makes sense to estimate s as

$$\sqrt{\frac{\sum_{i=1}^{n} (X_i - \mu)^2}{n}} \qquad (2\text{-}14)$$

When we do not know μ, we replace it by $\bar{X}$ to obtain

$$\sqrt{\frac{\sum_{i=1}^{n} (X_i - \bar{X})^2}{n}} \qquad (2\text{-}15)$$

But, since the sum of the squared deviations around $\bar{X}$ must be less than that around μ, Equation (2-15) must be smaller than Equation (2-14). It turns out that by replacing n by $n - 1$, we divide by a smaller denominator and get a larger value for s. So, the correction of n to $n - 1$ turns out to be just right.

There is also a second explanation of this correction. One other important property of the mean is that the sum of deviations about the mean is zero: $\Sigma_{i=1}^{n}(X_i - \bar{X}) = 0$. Because of this identity, only $n - 1$ of the n deviations are independent quantities. If we know $n - 1$ of the deviations, we can always use Equation (2-5) to determine the other. Consider once again the five observations 2, 2, 3, 5, and 8. The first four deviations sum to $(2 - 4) + (2 - 4) + (3 - 4) + (5 - 4) = (-2) + (-2) + (-1) + 1 = -4$. For Equation (2-5) to hold, the last deviation must equal 4. This is verified by calculating $8 - 4 = 4$. In general, then, among all the quantities $X_1 - \bar{X}$, $X_2 - \bar{X}, \ldots, X_n - \bar{X}$, there are only $n - 1$ independent quantities. This is another reason why we divide by $n - 1$ and not n. The remaining observation offers no new information. In fact, for estimation purposes, we should also divide the sample mean absolute deviation by $n - 1$.

Equation (2-13) does not represent the most computationally efficient formula for determining the standard deviation. Use of Equation (2-13) requires N different subtractions (one for each observation), N different squares, one division, and one square root operation. By using simple rules for summations, we can show that Equation (2-13) is equivalent to

$$\sigma = \sqrt{\frac{1}{N} \left(\sum_{i=1}^{N} X_i^2 - \bar{X}^2 \right)} \tag{2-16}$$

Equation (2-16) requires $N + 1$ squares (one for each observation and one for the mean), one subtraction, one multiplication, one division, and one square root. This compares favorably with the computational requirements of Equation (2-13). The analogous formula for s is

$$s = \sqrt{\frac{1}{n - 1} \left(\sum_{i=1}^{N} X_i^2 - n\bar{X}^2 \right)} \tag{2-17}$$

The calculation of the standard deviation for the oxygen data is illustrated in Table 2-6. At the bottom of the table, the numerical equivalence of Equations (2-12) and (2-17) is shown for the DO variable.

Variance

A final measure of dispersion is the variance, which is simply the mean or average squared deviation.

DEFINITION: VARIANCE

$$\text{Sample } s^2 = \frac{1}{n - 1} \sum_{i=1}^{n} (X_i - \bar{X})^2 \tag{2-18}$$

$$\text{Population } \sigma^2 = \frac{1}{N} \sum_{i=1}^{N} (X_i - \mu)^2 \tag{2-19}$$

The sample variance has a divisor of $n - 1$, and the population variance has a divisor of N. Variants of these formulas for computational purposes are easily derived as the squares of

Equations (2-16) and (2-17). The variance is seldom used as a descriptive summary statistic because the squaring of the deviations often leads to a value of s^2 well out of the range of the observations. It proves very difficult to interpret. By taking the square root of the variance to obtain the standard deviation, we more or less compensate for the initial squaring of the observations. The standard deviation is usually much easier to interpret. For inferential purposes we will see that the variance is an extremely important measure of a variable.

Interpretation and Use of the Standard Deviation

Both the standard deviation and the variance give an indication of how typical of a whole distribution the mean actually is. The larger these measures, the greater the spread of the observations and the less typical the mean. When all observations are equal to the mean, all deviations are zero and both s and s^2 are zero. Just like the arithmetic mean, the standard deviation is very sensitive to extremes. Because it involves the squares of deviations, the standard deviation gives more relative weight to large deviations. A highly skewed distribution, or one with a few extreme observations, is not effectively summarized by the mean and standard deviation. The median and quartile deviation are more representative summary statistics in this case.

Together, the mean and standard deviation can be used to locate individual observations within the distribution of a variable as a whole. This is an alternative to the percentile method, which relies on ordering our observations. For example, we might speak of an observation or value of X that is between one and two standard deviations above the mean, or between the mean and one standard deviation below the mean. It is very common to identify atypical or extreme observations as those greater than a certain number of standard deviations above or below the mean. Observations located more than two standard deviations from the mean are usually located in the "tails" of the distribution. These observations can thus be considered atypical.

Coefficient of Variation

To compare the dispersion of two frequency distributions having different means, we usually cannot simply compare the two standard deviations. Of course, if the means of the two variables are equal, then the variable with the smaller standard deviation is less dispersed. If, however, the means of the two variables are unequal, then it is misleading to rely on the standard deviation alone. A variable with a mean value in the thousands is likely to have a larger standard deviation than a variable with a mean value less than 100. The *relative variability* of a frequency distribution is measured by the ratio of the standard deviation to the mean.

DEFINITION: COEFFICIENT OF VARIATION
The coefficient of variation (CV) of a distribution with sample mean $\bar{X}$ and standard deviation s is defined as

$$CV = \frac{s}{\bar{X}} \tag{2-20}$$

For populations, an equivalent measure is obtained by substituting μ for $\bar{X}$ and σ for s.

Because both $\bar{X}$ and s are measured in the same units (that is, units of variable X), their quotient must be a dimensionless measure. So, we can use CV to compare variables measured in different units or scales.

To illustrate the advantages of the coefficient of variation, examine the following rainfall data collected from three climatological stations over a 60-yr period.

	Annual rainfall		
Parameter	Station A	Station B	Station C
$\bar{X}$	92.6	97.3	38.8
s	16.6	12.8	9.1
CV	0.179	0.132	0.235

Stations A and B can be directly compared. Their means are almost the same, but Station B has a much lower standard deviation and therefore is less dispersed. The coefficients of variation only verify this. However, even though station C has a lower standard deviation than either stations A or B, an examination of the CV for these three stations indicates that Station C has the most variable rainfall of the three. Relatively less dispersed variables have lower coefficients of variation. The lower limit to this measure is zero.

2.4. Other Descriptive Measures

It is also possible to develop numerical measures for other characteristics of the shape of frequency distributions. Skewness measures the degree of asymmetry in a distribution, and a quite simple measure based on the difference between the values of the mean and median is commonly used (review Figure 2-12):

DEFINITION: PEARSON'S γ

$$\text{Pearson's } \gamma = \frac{3(\bar{X} - \text{median})}{s} \qquad (2\text{-}21)$$

When the distribution is symmetric, the mean and median are equal, and the numerator is zero. When the mean exceeds the median, the distribution is positively skewed, and the measure has a positive value. When the median is larger, it has a negative value. For the dissolved oxygen variable this third measure yields $3(5.58 - 5.60)/.6392 = -.09$. As the histogram suggests, these data are only slightly negatively skewed.

2.5. Summary

In this chapter, we have been concerned with describing a set of observations, using both graphical methods and numerical statistics. These methods are useful for describing data sets whether they represent statistical populations or samples from these populations.

Developing the graphical summaries and generating the descriptive measures outlined in this chapter is a common preliminary step in a research project. In fact, understanding the characteristics of our data is often indispensable to understanding and interpreting the results of the inferential statistical tests undertaken in subsequent steps of the analysis.

Among the graphical techniques, the frequency distribution histogram is the most commonly used method for summarizing a set of data. Other graphical techniques can also be useful as long as they provide both an easily understood and an accurate portrait of the data. A quick glance at these figures often provides a great deal of information about the shape of the distribution. Cumulative displays such as ogives can also be used effectively in certain situations.

The average or typical value of a variable can usually be best summarized by the arithmetic mean, although the mode and the median are preferred in particular situations. For making inferences, the mean is preferred. For purposes of measuring the variability of a data set, the standard deviation is the most commonly used and useful measure, though its square, the variance, is most useful in inferential statistics. Under certain conditions, the quartile deviation proves a better descriptive measure than the standard deviation.

The standard deviation and the mean are the best summary measures of a variable. They can also be used to locate an observation within the frequency distribution. For example, we might speak of an extreme observation as being more than two standard deviations from the mean. The existence of skewness is most easily detected by comparing the values of the mean and median, or by using a simple measure such as Pearson's γ.

Appendix Review of Sigma Notation

Just as we use plus (+) to denote addition or minus (−) to indicate subtraction, we can use the symbol Σ (pronounced sigma) to denote the repeated operation of addition. The symbol Σ is known as the *summation operator*, since it literally means "take the sum of." In conjunction with the summation operator, we normally utilize an *index of summation* to specify the elements to be summed. For example,

$$\sum_{i=1}^{4} X_i$$

is interpreted as the sum of X_1 to X_4, or $X_1 + X_2 + X_3 + X_4$. If $X_1 = 1$, $X_2 = 7$, $X_3 = 9$, and $X_4 = 16$, then $\sum_{i=1}^{4} X_i = 1 + 7 + 9 + 16 = 33$. The index of summation can be placed below the sigma or just to the right of the summation. Also, any lowercase letter can be used as an index of summation. Usually, the variable i is used in a single summation, although j and k are also commonly used.

The sigma operator can also be used when an undetermined number of elements are to be summed. Thus, $X_1 + X_2 + \ldots + X_n$ can be expressed as $\sum_{i=1}^{n} X_i$. The length of the summation depends on the value of n. When there is no possibility of confusion, it is convenient to suppress the index and limits of the summation. We may thus write ΣX when we mean the sum is to be taken over "all" values of X. The specific interpretation of *all* should be evident from the context of the summation. In this text, we find it convenient to suppress these indices when the meaning is $\sum_{i=1}^{n} X_i$.

Several rules for simplifying summations periodically are used in the text. All the following rules can be easily verified using elementary algebra.

RULE 1 If X and Y are two variables, then

$$\sum_{i=1}^{n} (X_i + Y_i) = \sum_{i=1}^{n} X_i + \sum_{i=1}^{n} Y_i$$

To see this, expand the left-hand side:

$$\sum_{i=1}^{n} (X_i + Y_i) = (X_1 + Y_1) + (X_2 + Y_2) + \ldots + (X_n + Y_n)$$
$$= (X_1 + X_2 + \ldots + X_n) + (Y_1 + Y_2 + \ldots + Y_n)$$
$$= \sum_{i=1}^{n} X_i + \sum_{i=1}^{n} Y_i$$

RULE 2 If k is a constant and X is a variable, then

$$\sum_{i=1}^{n} kX_i = k\sum_{i=1}^{n} X_i$$

Again, expand the left-hand side:

$$\sum_{i=1}^{n} kX_i = kX_i + kX_2 + \ldots + X_n$$
$$k(X_1 + X_2 + \ldots + X_n)$$
$$k\sum_{i=1}^{n} X_i$$

RULE 3 For any two variables X and Y,

$$\sum_{i=1}^{n} X_i Y_i = X_1 Y_1 + X_2 Y_2 + \ldots X_n Y_n$$

Note that this does not equal $(\sum_{i=1}^{n} X_i)(\sum_{i=1}^{n} Y_i)$. Consider the following simple example:

i	X	Y
1	2	4
2	6	8
3	10	12

First, we calculate $\sum_{i=1}^{3} X_i Y_i = 2 \cdot 4 + 6 \cdot 8 + 10 \cdot 12 = 176$. Then we calculate $(\sum_{i=1}^{3} X_i)(\sum_{i=1}^{3} Y_i)$ as $(2 + 6 + 10) \cdot (4 + 8 + 12) = 18 \cdot 24 = 432$.

RULE 4 If k is a constant, then

$$\sum_{i=1}^{n} k = nk$$

Note that

$$\sum_{i=1}^{n} k = (k + k + \ldots + k) = nk$$
$$\underline{\qquad n \text{ times} \qquad}$$

RULE 5   If m is an exponent and X is a variable, then

$$\sum_{i=1}^{n} X_i^m = X_1^m + X_2^m + \ldots + X_n^m$$

RULE 6   If m is an exponent and X is a variable, then

$$\left(\sum_{i=1}^{n} X_i\right)^m = (X_1 + X_2 + \ldots + X_n)^m$$

Suppose that $n = 3$, $n = 2$, $X_1 = 2$, $X_2 = 6$, and $X_3 = 10$. Then

$$\sum_{i=1}^{3} X_i^2 = 2^2 + 6^2 + 10^2 = 140$$

$$\left(\sum_{i=1}^{n} X_i\right)^2 = (2 + 6 + 10)^2 = 324$$

In general, we see from Rules 5 and 6 that

$$\sum_{i=1}^{n} X_i^2 \neq \left(\sum_{i=1}^{n} X_i\right)^2$$

We can also work with double summations, i.e., those with two separate subscripts. When we encounter the expression $\sum_{i=1}^{n}\sum_{j=1}^{n} X_{ij}$, we work from the inside out. First we use the j subscript and then the i subscript so that

$$\sum_{i=1}^{n} \sum_{j=1}^{n} X_{ij} = \sum_{i=1}^{n} (X_{i1} + X_{i2} + \ldots + X_{in})$$
$$= (X_{11} + X_{12} + \ldots + X_{1n}) + (X_{21} + X_{22} + \ldots + X_{2n})$$
$$+ \ldots + (X_{n1} + X_{n2} + \ldots + X_{nn})$$

If X is an array or matrix of the form

$i = 1$	X_{11}	X_{12}	$\ldots$	X_{1n}
$i = 2$	X_{21}	X_{22}	$\ldots$	X_{2n}
	$\ldots$	$\ldots$	$\ldots$	$\ldots$
$i = n$	X_{n1}	X_{n2}	$\ldots$	X_{nn}

then the expression $\sum_{i=1}^{n}\sum_{j=1}^{n} X_{ij}$ leads to the sum of all the entries in the matrix. Note that we can also use summation notation to compute the sum of certain rows or columns of this matrix. For example,

$$\sum_{i=1}^{n} X_{2j} = X_{21} + X_{22} + \ldots + X_{2n}$$

specifies the sum of the second row. Similarly, the sum of column 1 is

$$\sum_{i=1}^{n} X_{i1} = X_{11} + X_{21} + \ldots + X_{n1}$$

FURTHER READING

Students often find it useful to consult other textbooks to see a simpler, more detailed, or just a different presentation of some topic covered in this chapter. Three textbooks intended for geography students are listed below. Keep in mind that these textbooks may use a different notation from that used in this text. Equivalent formulas may appear different for this reason only. Examples of the use of descriptive statistics in published research of geographers are included in virtually all journals of geography, particularly those published in Britain and North America. However, in many cases, the use of descriptive statistics may be only a minor part of a discussion of research results that employs other quite sophisticated mathematical or statistical techniques. Some of the problems at the end of this chapter suggest the range of application of these techniques in geography. The range of application is as wide as the discipline itself.

P. J. Taylor, *Quantitative Methods in Geography*, (Prospect Heights, Illinois: Waveland Press, 1983).
D. A. Griffith and C. G. Amrhein, *Statistical Analysis for Geographers*, (Englewood Cliffs, NJ: Prentice-Hall, 1991).
W. A. V. Clark and P. L. Hosking, *Statistical Methods for Geographers*, (New York: John Wiley, 1986).

PROBLEMS

1. Explain the meaning of the following terms:
   a. Frequency table
   b. Distribution of a variable
   c. Median
   d. Percentile
   e. Quartile
   f. Range
   g. Interquartile range
   h. Coefficient of variation
   i. Unimodal, bimodal, and multimodal distributions
   j. Univariate descriptive statistics
   k. Midrange

2. Explain the difference between a frequency polygon and an ogive.

3. When is the mean not a good measure of central tendency? Give an example of a variable that might best be summarized by some other measure of central tendency.

4. Consider the following five observations: −8, 14, −2, 3, 5.
   a. Calculate the mean.
   b. Determine the median.
   c. Show that the Properties 1, 2, and 3 hold for this set of numbers.

5. A large city is divided into 60 police precincts. The number of burglaries in the last 12 months in each precinct is as follows:

200	251	182	191	219	195	224	171	204	205	186	221	193	171	206
225	170	242	200	231	196	188	219	180	224	208	205	184	236	182
207	209	193	225	209	194	219	176	236	234	160	186	203	201	190
201	173	213	258	200	221	180	209	259	172	161	211	241	211	181

   a. Construct a frequency table of these data. Justify your choice of categories.
   b. Draw a frequency polygon and histogram.

   c. Calculate the following descriptive statistics for these data: mean, variance, median, mode, and Pearson's coefficient of skewness.

   d. Write a brief summary of your findings about the distribution of the number of burglaries in these 60 precincts.

6. Corn yields in bushels per acre in 60 counties are as follows:

74	76	67	90	72	74	56	26	96	91	86	96	70	78	65
58	47	35	99	82	81	88	72	82	60	72	67	45	34	98
76	70	85	71	80	83	70	60	42	93	96	75	80	75	92
90	95	92	59	97	81	63	82	79	78	90	88	61	78	79

Repeat Problem 5 for this data.

7. The following data describe the distribution of annual household income in three regions of a country:

Region		s
A	$36,000	$16,000
B	28,000	13,000
C	23,000	11,000

In which region is income the most evenly spread? least evenly spread?

8. In addition to the data given in problem 7, we are now told that the median incomes in the three regions are $34,000, $26,500, and $22,100, respectively. In which region is the income distribution most skewed?

9. Using the data of Problem 5 above, determine
   a. Range
   b. The first and third quartiles
   c. The interquartile range
   d. The 60th percentile
   e. The 10th percentile
   f. The 90th percentile

10. Using the data of problem 6 above, determine
   a. Range
   b. Interquartile range
   c. The first and third (or lower and upper) quartiles
   d. The 70th percentile
   e. The 30th percentile
   f. The 93rd percentile

Spreadsheet exercises:

11. The purpose of this first spreadsheet exercise is to introduce students to spreadsheets and the entry of simple data sets. Using the data of either Problem 5 or 6, enter the data into a spreadsheet

in the format of the accompanying table. A Windows™-based spreadsheet program such as Excel, Quattro Pro, or Lotus for Windows is recommended, but most of the exercise can be completed using a DOS-based spreadsheet program.

a. Enter the title in the home cell and center it over the columns of the worksheet

b. Enter the variable name in cell A2 (e.g., Burglaries), and center it over the column

c. Enter the values for the variable in column A, below the variable name. Format the column so that the variable retains the original format of the variable, e.g., no decimal places for Problem 5.

d. Enter the other text information for the next three columns plus the summary information and percentile sections.

e. Immediately below the last observation in the first column, calculate the sum of the column using the SUM function (note how long a simple formula would be, e.g., A3 + A4 + . . . A63).

f. Calculate the mean by dividing the sum obtained in part (e) by the number of observations. Format this cell to three decimal places.

g. Using the mean of the variable calculated in part (f), enter a formula for the deviation from the mean for the first observation in the second column labeled "Deviations." Format this column to have three decimal places and copy the formula to the remaining observations.

h. At the bottom of the second column, verify that the sum of the deviations about the mean is zero.

i. Enter a formula for the first observation in the third column that squares the deviations calculated in the second column. Format the column to three places. At the bottom of the column, calculate the sum of squared deviations about the mean. Use this value to estimate the population and sample variances, and population and sample standard deviations, using the formulas in the text.

j. In the fourth column, sort the observations from the first column in ascending order to easily isolate the minimum and maximum values of the observations.

k. Use the built-in functions of the spreadsheet program to calculate the descriptive statistics shown in the block of the spreadsheet labeled "Summary Statistics." Verify the values for the mean, variances, and standard deviations developed in part (i).

l. Use the percentile function to calculate the deciles shown in the block labeled "Percentiles." Format the percentiles as percents and the values of the percentiles to two decimal places.

m. Using the format of part (l), enter the two quartiles in the section labeled "quartiles."

n. Print the worksheet.

Advanced spreadsheet functions:

12. Use any print enhancements, including bold, underlining, italics, fonts and font sizes, and color to improve the appearance of the worksheet.

13. Enter the midpoints and frequencies of the intervals used in Problem 5 and use the graphing abilities of the spreadsheet to generate a histogram of the data.

14. If the spreadsheet program has a statistical module, outline the variable in the first column and have it develop a table of summary statistics, and a histogram of the variable. How good is the in-built graphing function at determining appropriate interval widths? If available, examine the documentation for the spreadsheet program to see how the computer program selects the number of intervals and/or interval width for a given set of data.

Chapter 2, Problem 5

Burglaries	Deviations	Sq. Deviations	Ordered	Summary Statistics	
200	(4.033)	16.268	160	204.033	Mean
225	20.967	439.601	161	203.500	Median
207	2.967	8.801	170	160	Min
201	(3.033)	9.201	171	259	Max
251	46.967	2,205.868	171	200	Mode
170	(34.033)	1,158.268	172	552.880	Sample Variance
209	4.967	24.668	173	543.666	Population variance
173	(31.033)	963.068	176	23.513	Sample standard deviation
182	(22.033)	485.468	180	23.317	Population standard deviation
242	37.967	1,441.468	180		
193	(11.033)	121.734	181	Percentiles	
213	8.967	80.401	182	160.00	0%
191	(13.033)	169.868	182	172.90	10%
200	(4.033)	16.268	184	182.00	20%
225	20.967	439.601	186	190.70	30%
258	53.967	2,912.401	186	198.40	40%
219	14.967	224.001	188	203.50	50%
231	26.967	727.201	190	208.40	60%
209	4.967	24.668	191	214.80	70%
200	(4.033)	16.268	193	224.00	80%
195	(9.033)	81.601	193	236.00	90%
196	(8.033)	64.534	194	259.00	100%
194	(10.033)	100.668	195	Quartiles	
221	16.967	287.868	196	186.00	25%
224	19.967	398.668	200	219.50	75%
188	(16.033)	257.068	200		
219	14.967	224.001	200		
180	(24.033)	577.601	201		
171	(33.033)	1,091.201	201		
219	14.967	224.001	203		
176	(28.033)	785.868	204		
209	4.967	24.668	205		
204	(0.033)	0.001	205		
180	(24.033)	577.601	206		
236	31.967	1,021.868	207		
259	54.967	3,021.334	208		
205	0.967	0.934	209		
224	19.967	398.668	209		
234	29.967	898.001	209		
172	(32.033)	1,026.134	211		
186	(18.033)	325.201	211		
208	3.967	15.734	213		
160	(44.033)	1,938.934	219		
161	(43.033)	1,851.868	219		
221	16.967	287.868	219		
205	0.967	0.934	221		
186	(18.033)	325.201	221		
211	6.967	48.534	224		
193	(11.033)	121.734	224		
184	(20.033)	401.334	225		

Burglaries	Deviations	Sq. Deviations	Ordered	Summary Statistics
203	(1.033)	1.068	225	
241	36.967	1,366.534	231	
171	(33.033)	1,091.201	234	
236	31.967	1,021.868	236	
201	(3.033)	9.201	236	
211	6.967	48.534	241	
206	1.967	3.868	242	
182	(22.033)	485.468	251	
190	(14.033)	196.934	258	
181	(23.033)	530.534	259	
12,242	0.000	32,619.933		
204.033		543.666		
		552.880		
		23.317		
		23.513		

3

Descriptive Statistics for Spatial Distributions

The graphical devices and summary measures used in conventional descriptive statistics, as well as the newer techniques developed in exploratory data analysis, can be used to effectively summarize many of the data utilized by geographers. In Chapter 2, for example, a number of techniques of ordinary descriptive statistics are applied to a set of 50 observations of lake-dissolved oxygen values. Although such methods can tell us a lot about the frequency distribution of a variable, in some cases this is only part of the story. There are many instances in geographical research where the *spatial distribution* of a variable is of primary concern. How can we describe *the map* of some variable in an efficient, precise, and simplified way? *Where* are the high values? *Where* are the low values? Is the variable *evenly distributed* across the map or *concentrated* in a single or small number of locations? Is there any *pattern*, or is the map in some sense *random*? To answer these questions, geographers often use several techniques derived from descriptive statistics, but having particular properties or interpretations when applied to spatial data.

Let us begin by identifying some of the different representations of spatial data commonly used by geographers. Four different types are recognized in Figure 3-1; areal data, point data, network data, and directional data. *Areal data* include numerical observations based on areal subdivisions of some region. As a map, areal data can be presented in many ways. The simplest two-dimensional representations are the *choropleth* map, Figure 3-1(a), and the *contour* map, Figure 3-1(b). These two forms can also be given three-dimensional representations. The *stepped statistical surface* illustrated in Figure 3-1(c) is the three-dimensional equivalent of the choropleth map. The value or magnitude of each areal unit is mapped in the third dimension. This often improves the interpretability of the spatial pattern of the map. Where the phenomenon depicted on the map varies continuously, such as rainfall, it is accepted practice that the three-dimensional representation be an *isarithmic map*, such as the continuous smoothed surface in Figure 3-1(d).

A second major class of spatial data is *point data*. The dot map illustrated in Figure 3-1(e) is the most commonly used cartographic display for point data. The analysis of these point distributions, including the search for both pattern and underlying process, has been a topic of continuing interest to geographers. The dots may represent some plant species, farmhouses, towns, villages and cities, or even earthquake epicenters. When it is possible to associate a magnitude or frequency with each dot, a three-dimensional portrayal of the

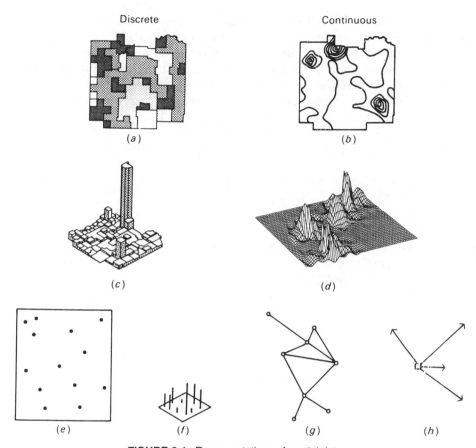

FIGURE 3-1. Representations of spatial data.

information is also possible, as in Figure 3-1(f). Two other types of data sometimes generated in geographical research are *network data*, which arise particularly in the analysis of drainage systems and transport networks, and *directional data*, which occur in studies examining the orientation of geographic phenomena. These two types are illustrated in Figure 3-1(g) and (h), respectively.

The distinction between continuous and discrete areal data and point data is sometimes blurred. Cartographers sometimes argue that a continuous areal representation is appropriate if the phenomenon exists everywhere on the map, both at and between observation points. This argument appears to be valid for many physical phenomena such as rainfall and temperature, but is less compelling for a variable such as population density. To map this variable in a continuous form on an isarithmic map, the assumption must be made that population exists everywhere and not at discrete points. Remember that it is possible to convert data expressed in one way to most other formats, if we are willing to make a few assumptions.

First, let us consider one way in which point data can be converted to areal data. The area around each point is associated to it in such a way that mutually exclusive and collectively exhaustive areal subdivisions are created. A common method is the construction of *Thiessen polygons*. These areas or polygons are drawn in order to satisfy the condition that *any location on the map associated with a point is closer to that point than to any other point on the map.* These polygons were defined by climatologist A. H. Thiessen to create regions around rainfall stations in such a way that station totals could be weighted by their surrounding areas to compute an "average" rainfall for the entire area. His solution is based on the simple logic that observed rainfall at any location is likely to be most similar to the *nearest* location for which a rainfall total is known.

Thiessen polygons can be constructed by using a very simple algorithm:

1. Join each point to all neighboring points.
2. Bisect these lines.
3. Draw the regions.

It is characteristic of the construction that the bisectors created in step 2 will either meet in 3s at a single point or else end at the border of the map. The construction is illustrated for a five-point problem in Figure 3-2. The bisectors meet at three different intersections surrounding the central point, *A*, *B*, and *C*. The other ends of the bisectors terminate at the map boundary. Besides the original application in climatology, Thiessen polygons also find

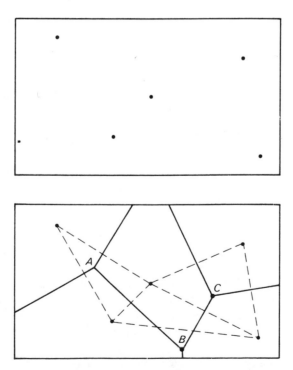

FIGURE 3-2. Conversion of point to areal data by using Thiessen polygons.

application in human geography. Suppose the points represent competing retail outlets in a city. Thiessen polygons define market areas that minimize distance traveled by consumers. Each goes to the closest retail outlet.

Other data conversions are also possible. For instance, areal data can be converted to point data by selecting a representative point in each area. The *center of gravity* is an obvious choice. The implicit assumption is that all the activity or phenomenon is concentrated at this point and not spread throughout the area. Point data can also be converted to a set of isolines by using the process of *interpolation*. The process is illustrated in Figure 3-3. In this figure, the rainfall totals for four points, *A, B, C,* and *P,* are given as well as their locations on a map. The position of the isoline with a value of 50 between point *P* and the three points *A, B,* and *C* can be determined by simple geometry. Since point *A* has a value of 55, the difference between the rainfall at points *A* and *P* is 10. Therefore, on the line, *AP,* the isoline of 50 should lie midway between *A* and *P.* Between *B* and *P,* the split is in the ratio 5:7, and between *C* and *P,* the split along *CP* is in the ratio 5:1. The isoline of 50 is then approximated as a smooth curve running through these points. This is a linear interpolation, since it assumes the value of the phenomenon changes linearly between the observed points. Other *nonlinear* interpolations are also possible.

Although the types of data generated from maps are capable of being translated to other representations, the statistical methods used to analyze such maps developed from different traditions and were used for different purposes. These methods will be presented on the basis of the classification made explicit in Figure 3-3, with the understanding that many of the methods can be applied to data expressed in more than one way. As we shall see, many of these methods can be applied to aspatial data as well.

3.1. Areal Data

Data are frequently published for discrete areal units such as states or provinces, counties, census tracts within cities, and many other administrative areas. This section describes both graphical and statistical summaries of such data. These methods are therefore applicable particularly to choropleth and stepped statistical surface maps.

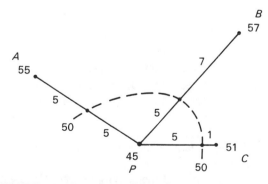

FIGURE 3-3. Linear interpolation of an isoline from point data.

Location Quotients

The location quotient is most frequently used in economic geography and locational analysis, but it has much wider applicability. The location quotient (LQ) is an index for comparing an area's share of a particular activity with the area's share of some basic or aggregate phenomenon. Suppose, for example, that the following data are taken from a larger data set which details the employment structure of a region divided into four areas:

Area	Manufacturing		Service		Total	
	Employment	Percentage	Employment	Percentage	Employment	Percentage
A	5	5	40	10	150	15
B	70	70	220	55	500	50
C	15	15	60	15	200	20
D	10	10	80	20	150	15
	100	100	400	100	1000	100

Is employment in manufacturing or services concentrated in some area(s), or is it evenly distributed across the map? Location quotients compare the distribution of an activity to some *base*, or *standard*, in this case to total employment. The question can be reexpressed as follows: Is manufacturing or service employment more or less concentrated than total employment?

DEFINITION: LOCATION QUOTIENT

The location quotient for a given activity for area i is the ratio of the percentage of the total regional activity in area i to the percentage of the total base in area i. If A_i is equal to the level of the activity in area i and B_i is the level of the base, then

$$LQ_i = \frac{A_i/\Sigma A_i}{B_i/\Sigma B_i} \tag{3-1}$$

The numerator of Equation (3-1) is the percentage of the activity in area i, and the denominator is the percentage of the base. A location quotient is thus the ratio of two percentages and is therefore dimensionless.

The location quotient for manufacturing in area A is $5/100 \div 150/1000 = 5/15 = 0.333$. That is, region A has 5 percent of the manufacturing employment of the region but 15 percent of the total employment in all sectors. On the basis of its share of total employment, we would expect region A to have 15 percent of the manufacturing employment. It has only 5 percent, or 0.333, as much as would be expected. Location quotients can be interpreted by using the following conventions:

1. If LQ > 1, this indicates a relative concentration of the activity in area i, compared to the region as a whole.
2. If LQ = 1, the area has a share of the activity in accordance with its share of the base.

3. If LQ < 1, the area has less of a share of the activity than is more generally, or regionally, found.

The complete table of location quotients for these two employment sectors indicates which areas of the map fall into these categories:

	Location quotient	
Area	Manufacturing	Services
A	0.333	0.667
B	1.400	1.100
C	0.750	0.750
D	0.677	1.333

For manufacturing, the location quotients reveal a concentration in area B, and less than expected shares in each of the three other areas. For services, however, the distribution is much less concentrated, with relative concentrations in areas B and D and less than expected shares in A and C. Spatial patterns can be revealed by mapping these location quotients.

The selection of the base or standard distribution used in the denominator of location quotients is subject to choice. Usually, if the activities are part of some aggregate, then the aggregate is used as the base. In this example, it makes sense to compare the concentration of various industrial sectors to the total employment in all sectors. However, it is also possible to use area populations as the standard of comparison. In this instance, we would be comparing the areal distribution of sectoral employment to the areal distribution of population. Or, the base activity could also be defined as the actual land area in each of the areas of the region.

Coefficient of Localization

One of the drawbacks to the use of the location quotient is that one value is calculated for each area in the region being analyzed. For a city with perhaps 300 or more census tracts, this would be a very inefficient form of summary, even if the location quotients themselves were mapped. As an alternative, the coefficient of localization (CL) can be calculated. It describes the relative concentrations of an activity by a single number.

DEFINITION: COEFFICIENT OF LOCALIZATION

The coefficient of localization (CL) is a measure of the relative concentration of an activity in relation to some base. To calculate CL:

1. Calculate the percentage share of the regional activity in each area, $A_i / \Sigma_{i=1}^{n} A$.
2. Calculate the percentage share of the regional base in each area, $B_i / \Sigma_{i=1}^{n} B$.
3. Subtract the value in step 2 from that in step 1, and add *either* all the positive differences or all the negative differences.
4. Divide by 100.

Using the sectoral employment in manufacturing or services, we can calculate the coefficient of localization as follows:

Area	Percentage of manufacturing employment	Percentage of total employment	Difference +	Difference −
A	5	15		10
B	70	50	20	
C	15	20		5
D	10	15		5
	100	100	20	20

Therefore, the CL for manufacturing is equal to 20/100 = .20. Similarly, the coefficient of localization for services is .10. Note that the CL calculation requires that the areal percentages of both the activity and the base to add to 100.

The CL ranges from 0 to 1. This differs from the location quotient which has a lower limit of zero but an upper limit approaching positive infinity. If CL = 0, the percentage distribution of the activity is evenly spread over the region in exact accordance with the base. The only numerical way this can occur is if the areal percentages for the activity are exactly equal to the areal percentages for the base. As CL approaches 1, the activity becomes increasingly concentrated in one region. The results indicate that both services and manufacturing are relatively evenly spread over the areas, but, as we might expect, service employment is more evenly spread.

Lorenz Curve

The Lorenz curve is another way to index the distribution of a variable among spatial units, although it is most commonly used to measure the extent of inequality of a variable distributed over aspatial categories. For example, if all residents of a country have the same income, then income should be distributed perfectly equally. To the extent that there exist some rich and some poor individuals, there is some inequality in income distribution. There is an obvious analogy to areal data. An activity may be concentrated in one or a few areas, or it may be spread throughout the region. The Lorenz curve is a graphical display of the degree of inequality.

As with location quotients and the coefficient of localization, the Lorenz curve compares the areal distribution of some activity to some base distribution. The Lorenz curve is constructed by using the following rules:

1. Calculate the location quotients for the various areas in a region. Reorder the areas in decreasing order of their location quotients.
2. Cumulate the percentage distributions of both the activity and the base in the order determined in step 1.
3. Graph the cumulated percentages for the activity and the base, and join the points to produce a Lorenz curve.

The necessary calculations for the manufacturing employment data are summarized in Table 3-1, and the Lorenz curves for both manufacturing and service employment are illustrated in Figure 3-4.

There are several important properties of a Lorenz curve. If the activity is evenly distributed across the areas of the region, that is, in proportion to the base, then the Lorenz curve is a straight line following the 45° diagonal shown in Figure 3-4. The more concentrated the activity, the further the Lorenz curve is from the diagonal. In the limiting case, the curve follows the X axis to the point (100, 0) and then proceeds vertically to the point (100, 100). Except for the case where the activity is perfectly evenly distributed, the slope of the Lorenz curve is always increasing. This follows directly from the ordering of the areas by LQ. The Lorenz curves in Figure 3-4 suggest that manufacturing employment is more concentrated than service employment. The Lorenz curve is favored by many who prefer graphical summaries to statistical measures.

The most common summary measure of inequality used in conjunction with the Lorenz curve is the *Gini coefficient*, or, as it is sometimes called, the *index of dissimilarity*.

DEFINITION: GINI COEFFICIENT, OR INDEX OR DISSIMILARITY
The Gini coefficient, or index of dissimilarity, is defined graphically as the maximum vertical deviation between the Lorenz curve and the diagonal.

The range of the Gini coefficient is from 0 to 100 percent. There are two other ways of calculating the value of this index. First, it can be determined by identifying the largest difference between the cumulated percentages of the activity and the base. From Table 3-1, the maximum difference between the two cumulative columns for manufacturing and total employment is $70 - 50 = 20$. This is exactly equal to the vertical deviation between the

TABLE 3-1
Work Table for Lorenz Curve Calculations for Manufacturing Data

Step 1.   The location quotients for the four areas are: A, 0.333; B, 1.400; C, 0.750; D, 0.667. So the order of the areas in the cumulative table should be B, C, D, and A.

Step 2.

Area	LQ	Percentage of manufacturing employment	Percentage of total employment	Cumulative percentage	
				Manufacturing	Total
B	1.400	70	50	70	50
C	0.750	15	20	85	70
D	0.667	10	15	95	85
A	0.333	5	15	100	100
		100	100		

Step 3.   See Figure 3-4.

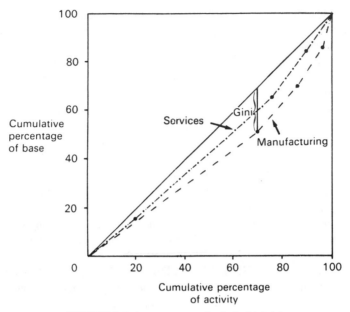

FIGURE 3-4. Lorenz curves for industrial data.

Lorenz curve and the diagonal at this point. The point (70, 50) differs from the diagonal (70, 70) by 20 percent.

Second, we might note that the CL for manufacturing was previously calculated as .20. The Gini coefficient can be easily obtained by multiplying the coefficient of localization by 100. For manufacturing we see that (.20)(100) = 20, and for service employment (.10)(100) = 10. Another equivalent method for calculating the Gini coefficient is to take one-half of the sum of the *absolute value* of the differences between the uncumulated percentage distributions of the activity and the base. Using the data for the manufacturing sector given in Table 3-1, we calculate the Gini coefficient as $\frac{1}{2}(|70 - 50| + |15 - 20| + |10 - 15| + |5 - 15|) = \frac{1}{2}(20 + 5 + 5 + 10) = 20$. Smaller values of both the Gini coefficient and the CL indicate similarity between the areal distribution of the base and the activity. As they increase, the similarity between the base and the activity decreases. This is why the Gini coefficient is sometimes called the index of dissimilarity.

The Lorenz curve and Gini coefficient can also be used to measure the degree of similarity of the percentage distributions of any two activities, neither of which is necessarily the base. For example, it is possible to compute the Gini coefficient between manufacturing and service employment distributions. Using the original data introduced for calculating location quotients, we see the Gini coefficient is $\frac{1}{2}(|5 - 10| + |70 - 55| + |15 - 15| + |10 - 20|) = \frac{1}{2}(30) = 15$. This application of the Gini coefficient is utilized in urban social geography to compare the areal distributions of ethnic groups in cities. The similarity of the areal distribution of ethnic groups is a useful indicator of their degree of integration or assimilation into the host society. Over time, as assimilation occurs, the residential segregation of many ethnic groups becomes less pronounced. Gini coefficients can be used to test this hypothesis.

All the procedures described for areal data have one common problem which limits their utility in assessing the similarity of two maps: They are inextricably tied to the exact areal subdivisions used in their calculations and can be interpreted only in this light. As we see in Section 3.4, this leads to several different specific problems. The values of the coefficients are very sensitive to the size and areal definitions used as the basis for their calculations. These issues reappear in the topics presented in Chapters 11 and 12.

3.2. Point Data

All the techniques used to describe areal distributions presented in Section 3.1 are applicable to spatial as well as aspatial data. Unfortunately, each of the measures fails to explicitly incorporate the spatial dimension through a variable related to one of the fundamental spatial concepts—distance, direction, or relative location. This is not true of the statistical methods designed for the analysis of point data. Distance is either explicitly or implicitly included within these measures. This branch of statistics is termed, appropriately enough, *geostatistics.*

The first step in analyzing a set of point data is to overlay the map with a Cartesian grid and determine the coordinates of each point on the map. Figure 3-5 illustrates this procedure for a point distribution with nine observations. Each of the nine points is given an X and Y coordinate on the basis of a 100×100 grid placed over the map. The origin of the grid is usually placed at the southwest corner of the map, so that all the coordinates have positive values. Observation 1, for example, has coordinates (20, 40). The set of nine observations can thus be represented by the following point coordinate data:

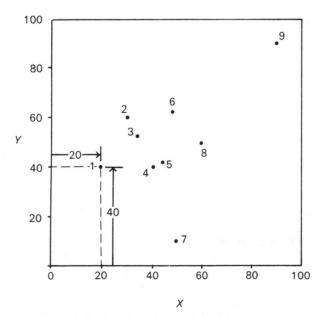

FIGURE 3-5. Obtaining grid coordinates from a dot map.

Observation No.	X Coordinate	Y Coordinate
1	20	40
2	30	60
3	34	52
4	40	40
5	44	42
6	48	62
7	50	10
8	60	50
9	90	90

Depending on the type of point distribution involved, we might add a third variable to these data. For example, if the dots represent towns, villages, and cities, we might associate a "weight," such as the population, to each of the points. In any case, at least two, and possibly three, variables characterize the dot map. The first step in geostatistics is to summarize this map in an efficient way. The standard statistical concepts of central tendency and variability are commonly used for this purpose. Each concept has a distinctly spatial interpretation.

Measures of Central Tendency

The first question to answer is, What is the center, or middle, or average, of this point distribution? Translated into geostatistics, the question is more appropriately phrased, Where on the map is the center of this point distribution? Five different measures can be used to identify the center of the map:

1. Mean center
2. Weighted mean center
3. Manhattan median
4. Euclidean median
5. Weighted medians

Since the properties of these five measures are so often incorrectly stated and insufficiently identified in the geographical literature, they are discussed in detail in this section.

Mean Center

The mean center can be thought of as the "center of gravity" of a point distribution and is a simple generalization of the familiar arithmetic mean. It is remarkably easy to calculate.

DEFINITION: MEAN CENTER

Let (X_i, Y_i), $i = 1, 2, \ldots, n$, be the coordinates of a given set of n points on a map. The mean center of this point distribution is defined as $(\bar{X}, \bar{Y})$ and is given by

$$\bar{X} = \sum_{i=1}^{n} \frac{X_i}{n} \quad \text{and} \quad \bar{Y} = \sum_{i=1}^{n} \frac{Y_i}{n} \tag{3-2}$$

Note that the mean center defines a point or location on the map with coordinates $(\bar{X}, \bar{Y})$. Using the coordinate data for the set of nine points of Figure 3-5, we find the mean center has coordinates

$$\bar{X} = \frac{20 + 30 + 34 + 40 + 44 + 48 + 50 + 60 + 90}{9} = 46.22$$

$$\bar{Y} = \frac{40 + 60 + 52 + 40 + 42 + 62 + 10 + 50 + 90}{9} = 49.56$$

This location is identified by the circular symbol in Figure 3-6.

One simple application of the mean center is to trace the center of gravity of a population distribution over time. To do this, the mean center of the population dot map is calculated for a series of maps of the same region at regular intervals of time. The movement of the mean center over the study period summarizes the change in the distribution of population in the region. For the hypothetical region illustrated in Figure 3-7, for example, the movement of the mean center indicates a systematic northerly trend.

WEIGHTED MEAN CENTER

The mean center can also be generalized to include the case for which each of the points on the map has an associated frequency, magnitude, or "weight." For example, suppose

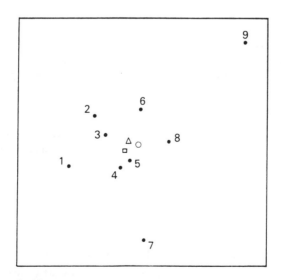

○ Mean center
△ Manhattan median
□ Euclidean median

FIGURE 3-6. Measures of central tendency for point pattern of Figure 3-5.

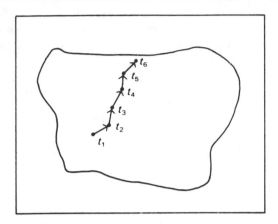

FIGURE 3-7. Tracing the movement of the mean center over six periods.

each of the points in Figure 3-5 represents a city with a given population. The location of the mean center should be drawn toward those points with the largest populations and away from those points with the smallest populations. If the populations of the cities represented by the points in Figure 3-5 are 10, 20, 10, 20, 10, 80, 10, 90, and 100, respectively, we expect the mean center to be drawn toward the locations of observations 6, 8, and 9. Since these three observations all lie in the northwest corner of the map, the weighted mean center will be drawn in this direction. The values of both the X and Y coordinates of the center should be greater than the coordinates of the mean center, (46.22, 49.56).

DEFINITION: WEIGHTED MEAN CENTER

Let (X_i, Y_i), $i = 1, 2, \ldots, n$, be the coordinates of a set of n points, and let w_i be the weight attached to the ith point. The weighted mean center has coordinates $(\bar{X}_w, \bar{Y}_w)$ given by

$$\bar{X}_w = \frac{\sum_{i=1}^{n} w_i X_i}{\sum_{i=1}^{n} w_i} \quad \text{and} \quad \bar{Y}_w = \frac{\sum_{i=1}^{n} w_i Y_i}{\sum_{i=1}^{n} w_i} \tag{3-3}$$

The weighted mean center for the nine points of Figure 3-5 is determined in the following way:

$$\bar{X}_w = \frac{10 \cdot 20 + 20 \cdot 30 + 10 \cdot 34 + 20 \cdot 40 + 10 \cdot 44 + 80 \cdot 48 + 10 \cdot 50 + 90 \cdot 60 + 100 \cdot 90}{10 + 20 + 10 + 20 + 10 + 80 + 10 + 90 + 100}$$

$$= \frac{21{,}120}{350} = 60.34$$

and

$$\bar{Y}_w = \frac{10 \cdot 40 + 20 \cdot 60 + 10 \cdot 52 + 20 \cdot 40 + 10 \cdot 42 + 80 \cdot 62 + 10 \cdot 10 + 90 \cdot 50 + 100 \cdot 90}{10 + 20 + 10 + 20 + 10 + 80 + 10 + 90 + 100}$$

$$= \frac{21,900}{350} = 62.57$$

As expected, the weighted mean center with coordinates (60.34, 62.57) is located to the northwest of the mean center (46.22, 49.56).

The formulas for calculating the mean center, Equation (3-2), is equivalent to the formulas used to calculate the arithmetic mean, Equation (2-3). There is also one important property of the arithmetic mean which is shared with the mean center and has interesting implications for a spatial distribution. Recall that the arithmetic mean of a set of numbers minimizes $\Sigma_{i=1}^{n}(X_i - \bar{X})^2$. This property, the so-called least squares property of the mean, has been explained in Chapter 2. The equivalent property for the mean center is that $(\bar{X}, \bar{Y})$ minimizes

$$\sum_{i=1}^{n} (X_i - \bar{X})^2 + (Y_i - \bar{Y})^2 \tag{3-4}$$

Note that, by the Pythagorean theorem, the distance between points (X_i, Y_i) and $(\bar{X}, \bar{Y})$ is

$$d = \sqrt{(X_i - \bar{X})^2 + (Y_i - \bar{Y})^2}$$

Therefore, the mean center has the property that it minimizes the sum of *squared distances* from itself to the n other points. Also, like the mean, the mean center is very sensitive to the existence of extreme observations. Extreme observations in a point distribution are distant points set apart from the others. To see this sensitivity, let us augment the nine points of Figure 3-5 with a tenth point having coordinates (1000, 1000). The new mean center is (141.6, 144.6). This point lies between the existing observations and the new extreme point.

MANHATTAN MEDIAN

The concept of the median can also be applied to point distributions. For a set of n observations, the median is defined as the "middle," or $[(n + 1)/2]$th observation in an ordered array of values of X. How can we find the middle, or median, of a point distribution? The spatial median in this sense is the point of intersection of two perpendicular lines, one which divides the distribution of points in a north–south direction into two equal parts and the other which divides it into two equal parts in an east–west direction. Consider again the nine points of Figure 3-5. In the north–south direction, the line must pass through observation 8 since it has four points below it (observations 1, 4, 5, and 7) and four observations above it (observations, 2, 3, 6, and 9). A horizontal line through observation 8 can thus be drawn. In an east–west direction, observation 5 is the middle point. A vertical line is drawn through this observation. The intersection of these two vertical lines is shown as the triangular symbol in Figure 3-6. It is close to, but not coincident with, the mean center.

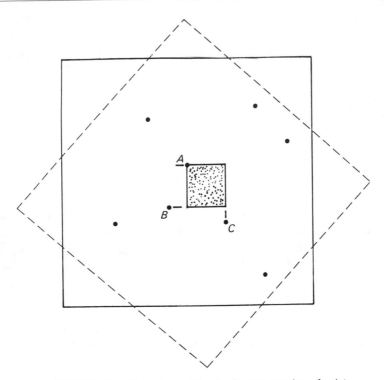

FIGURE 3-8. Manhattan median for an even number of points.

There are two distinct problems with the Manhattan median defined above. First, let us see what happens when we try to determine the median for a set with only eight observations, such as in Figure 3-8. Using this definition, we would have to find a line in the north–south direction having four observations on each side. As in Figure 3-8, *any vertical line* between points A and C will have four points on either side. Similarly, *any horizontal line* between points A and B divides the set into two equal parts. The shaded area in Figure 3-8 encloses all points having the property of the Manhattan median. It is obviously not a unique point. What went wrong? It turns out that the Manhattan median is *never* unique when there are an even number of points and is *always* unique when there are an odd number of points.

The second problem with this measure of central tendency can also be illustrated with Figure 3-8. Suppose the coordinate system of the map is shifted to the position shown by the dashed lines. What happens to the shaded area defining the median? It must be rotated in the same way. This will define another rectangular area—for the same points! The location of the Manhattan median is therefore not unique under axis rotation. This is somewhat undesirable from a statistical point of view.

The Manhattan median shares one important property with the median of a frequency distribution. From (2-11), we know that the median minimizes the sum of the absolute deviations between itself and the other n points in the frequency distribution. The equivalent property for the Manhattan median is formalized as follows:

DEFINITION: MANHATTAN MEDIAN

Let (X_i, Y_i), $i = 1, 2, \ldots, n$, be the coordinates of a set of n points. The Manhattan median (X_m, Y_m) minimizes

$$\sum_{i=1}^{n} |X_i - X_m| + |Y_i - Y_m| \qquad (3\text{-}5)$$

That is, (X_m, Y_m) minimizes the sum of the absolute Manhattan deviations from itself to the other n points of the distribution.

Why is it called the Manhattan median? Imagine we are trying to locate some facility in a city where travel is limited to the north–south and east–west directions. It is impossible to travel "as the crow flies." As shown in Figure 3-9, the Manhattan "distance" between the two points (X_i, Y_i) and (X_m, Y_m) is $|X_i - X_m| + |Y_i - Y_m|$. The absolute value signs are used to ensure that the calculated distance is nonnegative. This measure, or *metric*, of spatial separation bears a close resemblance to the movement possibilities in a dense rectangular grid street network of a large city, in particular Manhattan. Hence the name *Manhattan metric* and the term *Manhattan median*. Travel is possible only along the north–south and east–west directions. Note that the Manhattan distance between two points is always greater than the distance calculated by using the Pythagorean theorem, except when the two points lie on a north–south or east–west line. We term the distance metric calculated by using the Pythagorean theorem the *Euclidean metric.*

EUCLIDEAN MEDIAN

Knowing that the Manhattan median minimizes the sum of Manhattan distances from itself to the n points of a spatial distribution, we might ask, Which point minimizes the Euclidean distances from itself to these same n points?

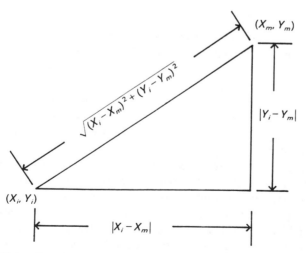

FIGURE 3-9. Distances between points in the Manhattan and Euclidean metrics.

DEFINITION: EUCLIDEAN MEDIAN

Let (X_i, Y_i), $i = 1, 2, \ldots, n$, be the coordinates of a set of n points. The Euclidean median has coordinates (X_e, Y_e) and is defined as the location that minimizes

$$\sum_{e=1}^{n} \sqrt{(X_i - X_e)^2 + (Y_i - Y_e)^2} \qquad (3\text{-}6)$$

Unfortunately, there is no direct method of determining the coordinates (X_e, Y_e). Appendix 3A describes one iterative numerical algorithm developed by Kuhn and Kuenne (1962) which can be used to solve this problem. Only small problems can be solved without the aid of a minicomputer or programmable calculator. For the nine points of Figure 3-5, the location of the Euclidean median is found to be (43.98, 42.05). As illustrated in Figure 3-6, the location of the Euclidean median (the square symbol) is very close to, but distinct from, the mean center and Manhattan median.

WEIGHTED MEDIANS

Just as the mean center can be generalized to the weighted mean center, so, too, can the Euclidean and Manhattan medians. The weighted Manhattan median is defined to have equal weights above and below, to the left and to the right. Unlike its unweighted counterpart, problems of nonuniqueness are not nearly so common. The weighted Euclidean median has drawn much more attention from geographers.

DEFINITION: WEIGHTED EUCLIDEAN MEDIAN

Let (X_i, Y_i), $i = 1, 2, \ldots, n$, be the coordinates of a set of n points distributed in the plane. To each of these points there is an attached weight w_i. The weighted Euclidean median (X_{we}, Y_{we}) minimizes

$$\sum_{i=1}^{n} w_i \sqrt{(X_i - X_{we})^2 + (Y_i - Y_{we})^2} \qquad (3\text{-}7)$$

In other words, the distances between the median and each point are weighted by the value w_i. These weights are defined in the context of the problem at hand. This problem can also be solved by using the iterative algorithm proposed by Kuhn and Kuenne. The locations of the weighted mean center, Manhattan median, and Euclidean median for the points in Figure 3-5 are illustrated in Figure 3-10. The weights used in each case are the populations set for the weighted mean center problem. Again, all three locations are reasonably close, although we would expect divergences when there are extreme points (or weights) in the spatial distribution.

Interpreted in one way, the weighted Euclidean median can be shown to be the solution to the classical location problem of Alfred Weber. *Weber's problem* is to find the best, or optimal, location for a factory—one which minimizes the sum of transport costs between the factory and two sources of raw materials and between the factory and the market. The situation is illustrated in Figure 3-11. If we make the additional assumptions that (1) transportation costs are a linear function of distance traveled, (2) the weights attached to the points represent the weights of raw material required to produce 1 ton of the finished product, and

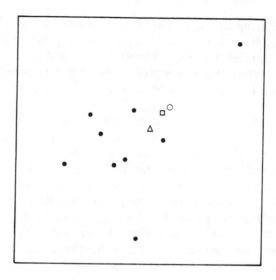

○ Weighted mean center
△ Weighted Manhattan median
□ Weighted Euclidean median

FIGURE 3-10. Weighted medians for point distribution of Figure 3-5.

(3) the weight at the market is 1, then the optimal location for the factory coincides with the weighted Euclidean median. The location of the factory represents the resolution of the three forces "pulling" the factory toward each of the three locations.

The problem can also be viewed as a public facility location problem. Given the locations of groups of facility users, for example, school children, where is the best location for a public facility such as a school? The weights might be defined as the number of

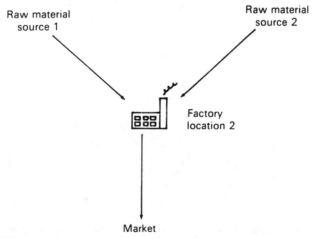

FIGURE 3-11. Weber's problem as the weighted Euclidean median.

school-age children living on a block. The objective of minimizing (3-7) can be interpreted as minimizing the total distance traveled by school children. Note that this also minimizes the average distance traveled by the children. The weighted mean center minimizes the square of the distances traveled by the school children. Since we are interested in the length of their walk to school, not the *square* of the length of their trip, the Euclidean median is the best solution.

The problem of locating a spatial median is a cornerstone of many private and public facility location problems. One generalization that has drawn a great deal of attention from geographers is the complex problem of locating an entire system of facilities, say five schools, within a spatial distribution of potential users. This is one variant of what are now termed *location-allocation problems*. Once a set of facilities is *located*, consumers or patrons are *allocated* to the appropriate, usually the closest, facility. The optimal locations for a system of five facilities are illustrated in Figure 3-12. This simple extension of the Weber problem is known as the *multiple-source Weber problem*. There is now a rich body of literature on both the theoretical aspects of location-allocation problems and their applications to school, hospital, and other facility systems.

Measures of Dispersion

The second important characteristic of a spatial distribution is its dispersion. Measures of dispersion in descriptive statistics are usually based on the notion of a deviation, the differences in value of an observation from a central value such as the mean or median. For spatial distributions the notion of a deviation is the actual distance between an observation and the central point.

STANDARD DISTANCE

The *standard distance* (SD) is the spatial equivalent to the standard deviation. Distances between each observation and the mean center are squared and summed, and this sum is divided by the number of observations. The easiest way to calculate the standard distance of a point distribution is to use the coordinates of each point directly in the following formula:

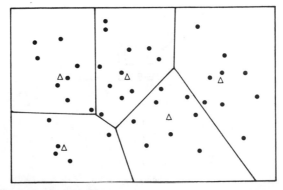

FIGURE 3-12. Optimal locations for a system of five facilities.

$$SD = \sqrt{\frac{\Sigma_{i=1}^{n} (X_i - \bar{X})^2}{n} + \frac{\Sigma_{i=1}^{n} (Y_i - \bar{Y})^2}{n}} \tag{3-8}$$

This formula can be rewritten as

$$SD = \sqrt{\sigma_x^2 + \sigma_y^2}$$

More dispersed point patterns will have large standard distances. For example, consider the two point distributions in Figure 3-13. The distribution in Figure 3-13(b) is clearly more dispersed than the point distribution in (a). A useful graphical device is to draw circles with a radius equal to the standard distance around the mean center of the distribution. It is also possible to use weighted observations by computing a standard distance around the weighted mean center. Together with the mean center, the standard distance can be used to compare and contrast point distributions. Both measures are very sensitive to extreme observations.

QUARTILIDES

Dispersion about the Manhattan median center can be graphically displayed by using the spatial equivalent of the interquartile range. A *quartilide* divides a point distribution into

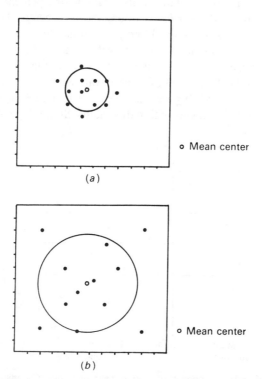

(a)

(b)

FIGURE 3-13. Dispersion of point patterns using standard distance.

quarters. Surrounding the median are eastern, western, northern, and southern quartilides. Together these define a rectangle, whose size clearly depends on the dispersion of the set of points. An example is shown in Figure 3-14 for a distribution with $n = 16$ points. In this case, problems of nonuniqueness arise because the median of a set of points with an even number of observations is an area. The median should have eight points to the east and eight to the west. Many lines satisfy this criterion; the one selected here lies *midway* between the two central observations. Similarly, the line locating the median in a north–south direction lies midway between the central points in this direction. The same convention is used to define the quartilides. The eastern quartilide lies midway between the fourth and fifth easternmost points, the western quartilide lies midway between the fourth and fifth westernmost points, and so on. The larger the rectangle defined by these quartilides, the more dispersed the set of points. The construction seems to have been used seldom in the geographical literature.

AN EMPIRICAL APPROACH

A useful empirical approach to describe the dispersion of a point pattern is to graph the cumulative frequency of points around some central location in bands of distance. For example, the point data of Figure 3-15(a) are summarized in this manner in Figure 3-15(b). It is easy to see that 80 percent of the points lie within 6 mi of the mean center, 20 percent within 2 mi, and so on.

One useful benchmark is the uniform distribution—the percentages that would be expected if the points were *evenly distributed* around the mean center. The number of points within a given distance from a point should be proportional to the area within that given distance. Since the area of a circle is proportional to the radius squared, the uniform distribution is not represented as a straight line on the graph in Figure 3-15(b). What would a uniform distribution having 80 percent of its observations within 6 mi be like? Since the area within a distance of 4 mi is $\pi(4^2) = 16\pi$ and the area with 6 mi is $\pi(6^2) = 36\pi$, we would expect $16\pi/(36\pi) = 4/9$ as many points within this radius. Similarly, a uniform distribution

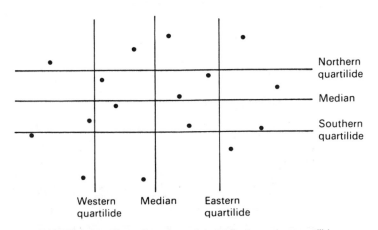

FIGURE 3-14. Dispersion of a point distribution using quartilides.

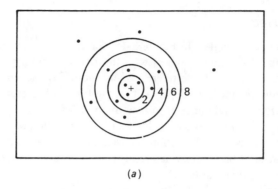

(a)

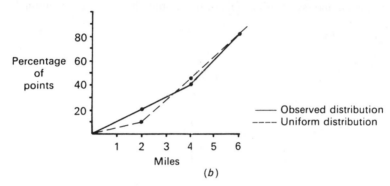

(b)

FIGURE 3-15. Summarizing the dispersion of a point distribution. (a) Concentric circles with distances; (b) graphical comparison to uniform distribution.

would have one-ninth as many points within 2 mi as it does within 6 mi. By comparison, then, the observed point distribution is more concentrated than a uniform distribution within the first 3 mi and less concentrated within the next 3 mi.

Although the measures of central tendency and dispersion can help us to make simple comparisons between different point distributions, they are incapable of assessing point *patterns.* Some point patterns are uniform, consisting of a more or less even distribution of points on the map. Other patterns are very clustered, with many points located in one area of the map and large areas without any points. A random pattern contains elements of both clustered and uniform patterns. More sophisticated methods can be used to assess the pattern of a point distribution along this continuum.

3.3. Directional Statistics

Direction is another fundamental measurement that can be extracted from maps. Just as there exist some unusual characteristics in statistical measures based on distance, so are there peculiar properties of directional measurements.

Measurement of Direction

Directions are measured as angles. In Figure 3-16, for example, the direction A from the origin is based on the 30° angle it makes with the north bearing. Or, equivalently, the direction of A could just as easily be referenced by the 60° angle with the east bearing. Direction is thus a *relative* measure dependent on the frame of reference used to take the bearing. The usual frame of reference is to take true north as 0° and record all directions as bearings from true north in a clockwise direction. The direction of A is therefore 30°, B is 110°, C is 235°, and D is 320°. Direction is based on a circular frame of reference with limits [0, 360]. Because of the circular nature of this measurement system, two directions which are remarkably similar can have widely different bearings. Note that the directions represented by 31° and 32° are similar, but so are the two directions 0° and 359°. There is a 1° difference in both pairs of directions. It is thus impossible to translate these bearings into the simple number systems capable of being summarized by conventional descriptive statistics.

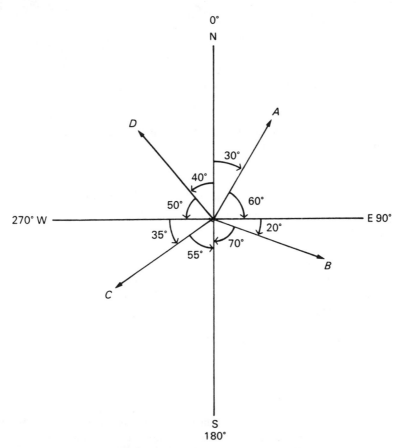

FIGURE 3-16. Measurement of direction.

Graphical Summaries of Directional Data

Directional data can be presented graphically in five different ways. For raw or ungrouped data, the directions may be illustrated as points on a unit circle, or as vectors drawn from a common origin. Suppose the data consist of the following 10 bearings taken from a single location:

Observation No.	Bearing, 0° = N
1	60°
2	45°
3	230°
4	50°
5	220°
6	55°
7	225°
8	30°
9	40°
10	320°

These observations are summarized graphically in Figure 3-17. In (*a*) each direction is illustrated as a vector of unit length. Clearly the bearings tend to be predominantly northeasterly or southwesterly. This representation is more striking than the alternative representation in (*b*). Whenever there are a large number of observations, these summaries tend to be difficult both to draw and to interpret. In such instances it is common to group the observations and present the data in a *conventional histogram*, a *circular histogram*, or a *rose diagram*.

The difficulty with the usual linear representation of the histogram is that the advantage of a circular display for what is inherently circular data is lost. This feature is illustrated with the wind direction data in Chapter 1. Two circular formats tend to be superior devices for portraying directional data. To illustrate their construction and interpretation, consider the following set of grouped data:

Observation No.	Compass direction	Number of migrants	Percentage
1	N	0	0
2	NNE	50	5.0
3	NE	100	10.0
4	ENE	200	20.0
5	E	125	12.5
6	ESE	75	7.5
7	SE	50	5.0
8	SSE	25	2.5
9	S	0	0
10	SSW	15	1.5
11	SW	35	3.5
12	WSW	275	27.5
13	W	25	2.5
14	WNW	25	2.5
15	NW	0	0
16	NNW	0	0
		1000	100.0

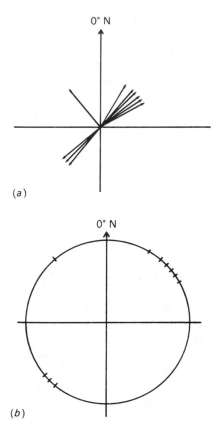

(a)

(b)

FIGURE 3-17. Summarizing ungrouped directional data.

The directions are aggregated to the usual 16 compass directions (N, NNE, NE, etc.), $22\frac{1}{2}°$ subdivisions of the 360° in the compass. The frequencies represent the number of migrants from places within the compass heading to some particular city. Is there a directional bias to this migration stream?

To construct a circular histogram, first draw a reference unit circle, as in Figure 3-18(a). Intervals are placed on the perimeter of this reference circle. Just as in the usual linear histogram, there is some choice of interval width. In this case, the 16 compass directions in the tabulated data are used without further aggregation. Blocks are extended from the perimeter of the circle for each interval. The height of each block is scaled to the observed frequency (or percentage) in the category. The width of each block is equal to the length of the chord connecting the two points on the reference circle which delimit the interval. For purposes of interpretation, a series of concentric circles are drawn outside the reference circle and labeled with the appropriate frequency or percentage. In this example, concentric circles are drawn for 10, 20, and 30 percent of all migrants. The raw frequencies could also have been used in this histogram.

For the rose diagram of Figure 3-18(b), the class limits are radii extended from the origin. The *length* of each radius is proportional to the class frequency or percentage. The limit of the sector for each class is an arc centered on the origin. Or, if it is desired to have

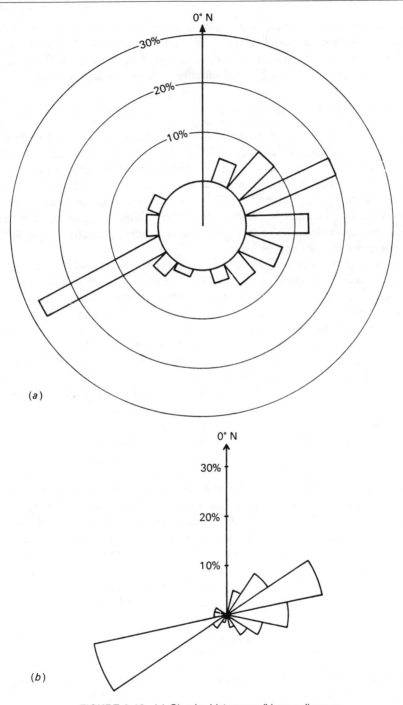

FIGURE 3-18. (a) Circular histogram; (b) rose diagram.

the *area* of the sector proportional to the class frequency or percentage, the square root of the frequency or percentage should be used to define radii lengths. A quick examination of Figure 3-18(a) or (b) reveals the strong directional trends of these migration flows. Migrants to this city come from places along a northeasterly to southwesterly corridor.

Directional Mean and Circular Variance

We have already noted that the usual linear number sequence cannot be used in making many calculations for directional data. Consider again the two pairs of angles (31°, 32°) and (0°, 359°). If we treat these as real numbers, we can compute the means of these pairs as 31.5° and 179.5°, respectively. This average seems intuitively reasonable for the first pair, but not for the second pair. The mean of 179.5° is almost due south, but the two directions are almost due north! Fortunately, we can solve this problem by using the concept of a vector and some elementary trigonometric relations. Appendix 3B briefly reviews sufficient elementary trigonometry for this problem.

   A *vector* is completely defined by a magnitude (or length) and a direction. Let us assume that all our directional measurements have the identical length of 1 unit, that is, they are *unit vectors*. The *mean* of a set of vectors is defined as the *resultant* of vector addition. To locate the resultant, we simply place the vectors end to end, preserving their directions. The vector connecting the origin to the end of the last vector in the sequence is the resultant vector. The angle that this resultant makes with 0°N bearing is the average, or mean, direction. The graphical solution to determine the directional mean is illustrated in Figure 3-19 for the four directional measurements of Figure 3-16. From the end of vector **A**, a vector of unit length is placed, with the bearing of vector **B**. Since the direction of vector **B** is 110°, **B** is drawn 20° below the horizontal (or east bearing of 90°). Similarly, **C** is added to **B**, and finally **D** is added to **C**. The resultant **R** is found by connecting the origin to the end of this vector. It is shown as the dashed line in Figure 3-19. The directional mean, if read from a protractor on an accurate diagram, is almost true north, in fact 358.2°.

   The difficulty with this method is obvious. It is extremely inefficient where there are a large number of directions to be added, and problems of precision usually arise. A direct method of calculating the directional mean uses some simple arguments from trigonometry.

   Consider Figure 3-20. Two vectors **A** and **B** and a resultant **R** are drawn in the conventional *XY* plane. Vector **A** can be defined by the coordinates (X_A, Y_A) and vector **B** by the coordinates (X_B, Y_B). Clearly the coordinates of the resultant are $(X_A + X_B, Y_A + Y_B)$. Since both **A** and **B** are unit vectors, **OA** = **OB** = 1. Simple trigonometric relations can be used to determine θ_R, the direction of the resultant. Note that tangent of the resultant is given by the ratio of the opposite to the adjacent sides of the right-angle triangle based on θ_R:

$$\tan \theta_R = \frac{Y_A + Y_B}{X_A + X_B} \tag{3-9}$$

Since **OA** = **OB** = 1, we know $X_A = \cos \theta_A$, $X_B = \cos \theta_B$, $Y_A = \sin \theta_A$, and $Y_B = \sin \theta_B$. Thus, (3-9) can be rewritten as

$$\tan \theta_R = \frac{\sin \theta_A + \sin \theta_B}{\cos \theta_A + \cos \theta_B} \tag{3-10}$$

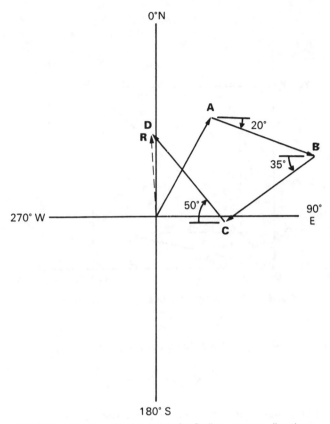

FIGURE 3-19. Graphical solution for finding average direction.

and it follows that

$$\theta_R = \arctan \frac{\sin \theta_A + \sin \theta_B}{\cos \theta_A + \cos \theta_B} \tag{3-11}$$

Note that this arctangent is the ratio of the distance of both vectors in a *northerly direction* to the distance in an *eastward direction*.

For a set of n directions, the definition of directional mean can be formalized by a simple generalization of (3-11):

DEFINITION: DIRECTIONAL MEAN

Let $\theta_1, \theta_2, \ldots, \theta_n$ be a set of bearings or directions taken from a single origin. The directional mean, or bearing, of the resultant is

$$\theta_R = \arctan \frac{\sum_{i=1}^n \sin \theta_i}{\sum_{i=1}^n \cos \theta_i} \tag{3-12}$$

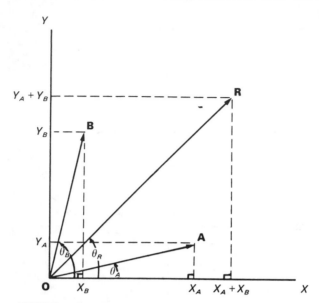

FIGURE 3-20. Trigonometric relations of a vector resultant.

Interestingly, it can be shown that the directional mean has the property

$$\sum_{i=1}^{n} \sin(\theta_i - \theta_R) = 0 \qquad (3\text{-}13)$$

That is, the sum of the sines of the angular deviations from each observation to the resultant is zero. This property is equivalent to a well-known property of the arithmetic mean given in Chapter 2: The sum of the deviations about the mean is zero.

In determining the arctangent given by (3-12), it is also necessary to fix the quadrant of the resultant bearing by observing the signs of $\sum_{i=1}^{n} \sin \theta_i$ and $\sum_{i=1}^{n} \cos \theta_i$. The easiest method of determining the resultant bearing is to compute θ_R by using the absolute value of the ratio and then to use the following corrections to specify the quadrant in which it lies:

1. If $\sum_{i=1}^{n} \sin \theta_i > 0$ (+) and $\sum_{i=1}^{n} \cos \theta_i > 0$ (+), then the ratio is +/+ and the value of θ_R can be used directly for the directional mean since the resultant must be in the first quadrant (i.e., between 0° and 90° or to the northeast).
2. If the ratio is +/−, then the bearing of the resultant is 180° − θ_R, or to the southeast, in the second quadrant.
3. If the ratio is −/−, then the bearing of the resultant is 180° − θ_R, or to the southwest, in the third quadrant.
4. If the ratio is −/+, then the bearing is 360° − θ_R or the the northwest, in the fourth quadrant.

The calculation required to determine the directional mean of the four vectors in Figure 3-16 are given in Table 3-2. Note that the result confirms the answer obtained by using the graphical vector addition technique. Computing the directional mean for grouped data requires only a slight modification of (3-12):

$$\theta_R = \arctan \frac{\sum_{i=1}^{n} f_i \sin \theta_i}{\sum_{i=1}^{n} f_i \cos \theta_i} \tag{3-14}$$

In this formula f_i is the frequency of the ith interval.

The directional mean shares many properties with the arithmetic mean. It is a poor measure of central tendency for a bimodal distribution. For a set of bimodal directional data, the directional mean usually falls between the two peaks. This can be a serious problem for directional data having one orientation, that is, two peaks differing by 180°. The migration data illustrated in Figure 3-18 have this characteristic, with strong peaks in the ENE and WSw directions. The directional mean lies between these two peaks of the distribution.

The variability of a sample of directional measurements is indicated by the length of the resultant vector **R**. In Figure 3-20, for example, the tip of the resultant vector lies far from the origin. This reflects the similarity of the directions **A** and **B**. If **A** and **B** were identical, the length of **R** would be 2, the sum of the lengths of the two unit vectors **A** and **B**. If **A** and **B** were in opposite directions, the resultant would be at the origin and the length of **R** would be 0. The length of the resultant vector **OR** can be determined in Figure 3-20 by using the Pythagorean theorem. It is the hypotenuse of the largest right-angle triangle and is equal to

$$\sqrt{(X_A + X_B)^2 + (Y_A + Y_B)^2} = \sqrt{\sin A^2 + \sin B^2 + \cos A^2 + \cos B^2}$$

TABLE 3-2
Work Table for Calculation of Mean Direction

	Bearing $0° = N$	Cosine	Sine
A	30°	.86603	.50000
B	110°	−.34202	.93969
C	235°	−.57358	−.81915
D	320°	.76604	−.64279
		.71647	−.02225

$$\theta = \arctan\left(\frac{-.02225}{.71647}\right)$$

$$= \arctan .03106 = 1.8°$$

Mean direction $= 360° - 1.8° = 358.2°$

In general, for n vectors

$$OR = \sqrt{\left(\sum_{i=1}^{n} \sin \theta_i\right)^2 + \left(\sum_{i=1}^{n} \cos \theta_i\right)^2} \tag{3-15}$$

There is one problem with using this measure in this unstandardized form. Larger sample sizes can have longer resultant lengths than smaller samples without having less variability. To develop a standardized measure of variability it is necessary to account for differing sample sizes.

DEFINITION: CIRCULAR VARIANCE

Let θ_i, $i = 1, 2, \ldots, n$, be a set of directional measurements. The circular variance is defined as

$$S_0 = 1 - \frac{OR}{n} \tag{3-16}$$

where **OR** is the resultant length and n is the sample size.

What are the bounds on S_0? When all the directional measurements are coincident, there is zero variability, $OR = n$ and therefore $S_0 = 0$. When the resultant length $OR = 0$, there is maximum variability and $S_0 = 1$. The bounds on S_0 are $[0, 1]$. Using the information in Table 3-2 for the directions in Figure 3-16, we see that $OR = \sqrt{(.71647)^2 + (-.02223)^2}$ = .7168 and $S_0 = 1 - .7168/4 = .8208$. This indicates a high degree of variability and confirms the visual impression of Figure 3-16. The most common use of S_0 is to compare the directional variability of different samples.

In some geographical studies, orientation, and not direction, is important. We say that a road has a north–south orientation, but traffic on the road has either a north or a south direction. If all angles less than 180° are doubled, then the directional data can be converted to orientation data. Then the directional mean and circular variance can be applied to these modified data. This is common practice in applications of directional statistics in physical geography such as drumlin or pebble orientations.

3.4. Descriptive Statistics and Spatial Data: Four Problems

There are four recurring problems which arise when virtually any descriptive statistic, including those specifically designed for the purpose, is applied to spatial data. So it is necessary to be extremely careful in interpreting the results of a statistical analysis based on locational observations. In fact, many of the recent advances in spatial statistics have been developed in response to the need to overcome the limitations imposed by these four problems. These four problems are often referred to as the boundary problem, the scale problem, the problem of modifiable units, and the problem of pattern.

Boundary Problem

The location of the boundary of the study area, as well as the placement of the internal boundaries in an areal design, is often a crucial question in geographical research. Poorly chosen designs can lead to several problems. Let us illustrate the potential difficulties with two examples. First consider the two point patterns of Figure 3-21. Both are identical and would yield identical values for any measure of central tendency or dispersion. However, these patterns are clearly different. The distribution in (a) could be described as a reasonably regular, dispersed pattern. In (b), the pattern could be described as clustered. This means that the standard distance or any other measure of dispersion cannot be interpreted independent of the study area.

As a second example, consider the problem of locating areal boundaries within some study area. Suppose the location of some phenomenon, for example a particular ethnic group in a city, is depicted as the shaded area in the two maps of Figure 3-22. Despite the fact that the "map is the same in each case, the values of location quotients, coefficients of localization, and Gini coefficients would be markedly different for these two cases. All measures for Figure 3-22(a) would indicate significant areal concentration of the ethnic group, since the boundary of one zone completely encloses the neighborhood occupied by the ethnic group. For the boundaries in (b), these same measures would tend to indicate a rather even distribution of the ethnic group in this study area. In short, the map in (a) suggests ethnic *segregation*, and the map in (b) suggests *integration*. Although the results may not always be as dramatic as in this case, boundary locations can mask certain map patterns. The moral is clear—summary statistics can be interpreted only for the particular areal divisions upon which their calculation is based. If this framework is poorly chosen,

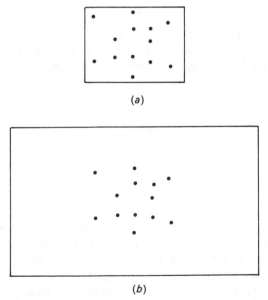

(*a*)

(*b*)

FIGURE 3-21. Boundary problem for point patterns.

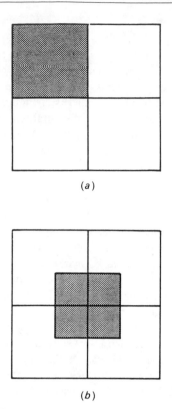

FIGURE 3-22. Boundary problem for areal patterns.

the results may be at the minimum misleading and possibly even false. Where there is some control over the placement of areal boundaries, these problems can sometimes be minimized by the use of a fine areal breakdown within an appropriately delimited study area.

Scale Problem

A second problem is termed the *scale problem*, or the *areal aggregation problem*. Quite simply, the values for many descriptive statistics can vary systematically when increasingly aggregated areal data are used. An example of the effects of areal aggregation is in Figure 3-23. In (a), the areal distribution of some variable X is shown in an area with 16 cells. The mean value of the map is $\mu = 7.5$, and the variance is $\sigma^2 = 9.75$. Aggregate these data by joining neighboring cells to create eight zones, as in Figure 3-23(b). In each two-cell aggregation, the value of X for each zone is the average value of the two smaller cells from which it has been created. For example, the zone in the northwest corner of (b) has a value of $(4 + 8)/2 = 6$. What does change with this aggregation? The mean remains constant at 7.5, but the variance declines significantly, to $\sigma^2 = 2$. Much of the variation in X is now lost. Since geographers are often interested in spatial variation, this is particularly unfortunate. All observations on the eight-zone map are within a range $6 < X < 9$, but the values on the original map have a wider range of $2 < X < 12$. At an even higher level of aggregation, note

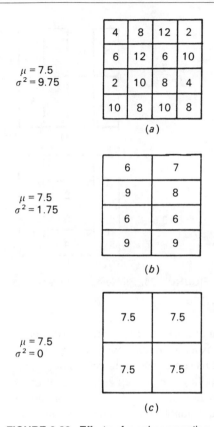

FIGURE 3-23. Effects of areal aggregation.

that the four-district representation in Figure 3-23(c) has a variance of zero! This is an obvious distortion of the real situation.

In many instances, then, spatial aggregation tends to reduce the variation depicted on a map. Comparisons of maps of a variable at different levels of aggregation must take into account the ramifications of this variance reduction. The problem becomes even more acute when we try to examine the *relationship* between two maps of different variables. The difficulties of aggregation in this problem of *correlation* are discussed in Chapter 12.

Problem of Modifiable Units

Even at the same scale of analysis, different areal definitions can also have a substantial impact on the values of most descriptive statistics. To see this, suppose the aggregation of Figure 3-23(a) into eight zones from 16 cells is accomplished by joining contiguous north–south rather than east–west neighbors, as in Figure 3-24(a), or by using a mixed pattern of north–south and east–west aggregations, as in Figure 3-24(b). Again, both systems lead to the same map mean of 7.5. However, although each of these aggregations has the same variance, it is higher than that of the eight-zone map of Figure 3-23(b). The effects of using modified areal units are not nearly so predictable as the effects of aggrega-

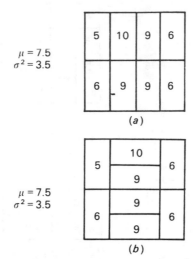

FIGURE 3-24. Problem of modifiable units.

tion. Although all aggregations decrease the variance (it is possible to maintain the variance if the zones joined are exactly alike), some systems will result in significantly lower variances, others not. In this sense, the aggregations in Figure 3-24 are superior to the one in Figure 3-23(b). They both are more representative of the more detailed map in Figure 3-23(a). But what if the map in Figure 3-23(a) were not known? Can we have faith that the eight-zone map used in our analysis is truly representative of the actual variation of the variable across the map? We cannot be sure.

   One conclusion we can draw from this analysis is that it is always better to join similar zones when we must aggregate the data. This will preserve the variation in the original map as much as possible. Since one of the "laws" of geography is that closer places are more alike than distant places, contiguous areal aggregations are likely to be less disruptive than aggregations of areas which are not close together.

Problem of Pattern

All the methods described in this chapter share one shortcoming: They are generally incapable of assessing the type of pattern which exists on a map. Consider the two contrasting patterns of Figure 3-25(a) and (b). Again, suppose the shaded areas represent residential concentrations of some particular ethnic group in this city. The coefficient of localization, Gini coefficient, or Lorenz curve would indicate a significant level of areal concentration of this ethnic group in both maps. There are eight zones completely populated by the group and eight zones where they are completely absent. Moreover, and even more important, because these measures are insensitive to pattern, both maps would have identically valued coefficients of areal concentration. But are these maps similar? Do both maps have about the same degree of concentration of this ethnic group? One could argue that there is much more segregation in the city mapped in Figure 3-25(b). Not only does the host population exclude them from their neighborhoods, but also the host population excludes

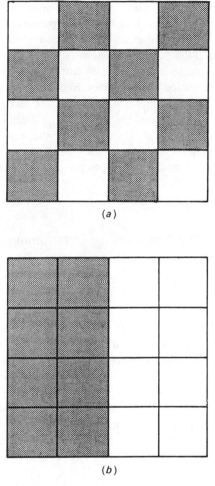

(a)

(b)

FIGURE 3-25. Illustrating the problem of map pattern.

them from nearby neighborhoods as well. Techniques capable of distinguishing between areal patterns are discussed in Chapter 12. Pattern in point distributions is discussed in Chapter 6.

3.5. Summary

Geographers share a common interest in the analysis of spatial distributions. Whether the data they collect are for areas, points, directions, or networks, specialized techniques can be used to summarize the maps of spatially varying phenomena. In fact, some of the techniques geographers often use to analyze spatial patterns are used to analyze aspatial variables. For areal data, the relative concentration of a variable across a map can be measured by the location quotient, the Gini coefficient, the coefficient of localization, or the Lorenz curve.

For point data, measures of central tendency and dispersion such as the mean center and standard distance can be used to summarize these distributions. These measures have similar properties to conventional statistical measures such as the mean and standard deviation. Directional data, because of their circular framework, require special handling.

Virtually all these methods have several important shortcomings. First, they are extremely sensitive to the choice of boundaries. Second, they are not independent of the scale or degree of areal aggregation. Third, the values of many statistics can vary significantly depending on the choice of areal units depicted on the map. In fact, one could almost take some maps and determine the areal subdivisions that result in any desired value for some statistics! Finally, most methods are not capable of assessing the nature of the pattern exhibited on the map. More specialized methods presented in later chapters address several of these issues.

Appendix 3A. An Iterative Algorithm for Determining the Weighted or Unweighted Euclidean Median

One of the most efficient algorithms for solving the weighted or unweighted Euclidean median problem was developed by Kuhn and Kuenne (1962). The following algorithm is a simple variant of their method.

Let (X_i, Y_i), $i = 1, 2, \ldots, n$, be the set of n given points with weights w_i, $i = 1, 2, \ldots, n$. The location of the current estimate of the median in the tth iteration is (X^t, Y^t). As the algorithm proceeds, (X^t, Y^t) gradually converges to (X_e, Y_e). The algorithm stops whenever the estimates of the coordinates in successive iterations are less than some predetermined tolerance level TOL. For example, if we wish the coordinates of (X_e, Y_e) found by the algorithm to be within .01, we set TOL = .01. Since it is an approximating algorithm, the method is more efficient if a "good" starting point is selected. A good point is one close to the answer. The bivariate mean $(\bar{X}, \bar{Y})$ and weighted mean $(\bar{X}_w, \bar{Y}_w)$ are good and obvious choices for a starting point. So $(X^1, Y^1) = (\bar{X}, \bar{Y})$. The algorithm can be described as the following sequence of steps:

1. Calculate the distance from each point (X_i, Y_i) to the current estimate of the median location

$$d_i^t = \sqrt{(X_i - X^t)^2 + (Y_i - Y^t)^2}$$

   where d_i^t is the distance from point i to the median during the tth iteration.

2. Determine the values K_i^t from $K_i^t = w_i/d_i^t$. Note that in the unweighted case all values of w_i are set equal to 1.

3. Calculate a new estimate of the median from

$$X^{t+1} = \frac{\sum_{i=1}^n K_i^t X_i}{\sum_{i=1}^n K_i^t} \qquad Y^{t+1} = \frac{\sum_{i=1}^n K_i^t Y_i}{\sum_{i=1}^n K_i^t}$$

4. Check to see whether the location has changed between iterations. If $|X^{t+1} - X^t|$ and $|Y^{t+1} - Y^t| \le$ TOL, stop. Otherwise, set $X^t = X^{t+1}$ and $Y^t = Y^{t+1}$ and go to step 1.

The algorithm usually converges in a small number of iterations. For the nine points of Figure 3-5, for the unweighted case, the following steps summarize the results of the algorithm initialized with TOL = .10.

Iteration	X^t	Y^t
1	44.867	45.517
2	43.865	44.060
3	43.672	43.204
4	43.710	42.742

The location of the unweighted median is thus estimated as (43.71, 42.74).

Appendix 3B. Trigonometric Review for Directional Statistics

Consider the angle θ defined in relation to the coordinate system of Figure 3-26. Here OP is a segment of length r, identifying a point P with coordinates (x, y) and making an angle of θ with the x axis. Since OAP is a right-angle triangle with $OA = x$ and $AP = y$, by the Pythagorean theorem $r = \sqrt{x^2 + y^2}$. Trigonometric functions of an angle θ are defined in relation to the three quantities x, y, and r. These are known as the adjacent, opposite and hypotenuse, respectively, to the angle θ. The three principal trigonometric functions are defined as the following ratios:

$$\sin \theta = \frac{y}{r} \quad \text{(sine of } \theta\text{)}$$

$$\cos \theta = \frac{x}{r} \quad \text{(cosine of } \theta\text{)}$$

$$\tan \theta = \frac{y}{x} \quad \text{(tangent of } \theta\text{)}$$

As the angle of θ increases and P moves around the origin from 0° to 360°, or equivalently from 0 to 2π radians, these three functions take on a range of values shown in Figures 3-27,

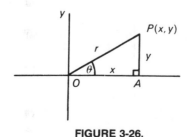

FIGURE 3-26.

3-28, and 3-29, respectively. Notice that when $\theta = 0°$, $y = 0$, and therefore $\sin \theta = 0$. At $\theta = 90°$, y equals r and therefore $\sin \theta = 1$.

The reciprocals of these three functions define three other important trigonometric functions:

$$\csc \theta = \frac{1}{\sin \theta}$$

$$\sec \theta = \frac{1}{\cos \theta}$$

$$\cot \theta = \frac{1}{\tan \theta}$$

where $\csc \theta$ is the cosecant of θ, $\sec \theta$ is the secant of θ, and $\cot \theta$ is the cotangent of θ.

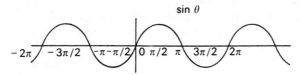

FIGURE 3-27.

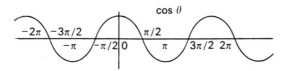

FIGURE 3-28.

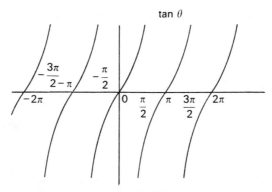

FIGURE 3-29.

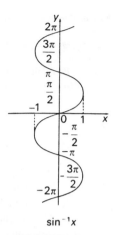

$\sin^{-1} x$

FIGURE 3-30.

Inverse trigonometric functions are specified as $\sin^{-1} x$ (or arcsin x), $\cos^{-1} x$ (or arccos x), and so on. For $\theta = \sin^{-1} x$, for example, we are directed to find the angle θ with a sine of x. An example of an inverse trigonometric function is shown in Figure 3-30. Since the trigonometric functions are cyclical over a period of 360°, inverse trigonometric functions are many-valued functions and often quite difficult to work with.

REFERENCES

C. D. Harris, "The Market as a Factor in the Localization of Industry in the United States," *Annals, Association of American Geographers* 34 (1954) 315–348.

H. W. Kuhn and R. E. Kuenne, "An Efficient Algorithm for the Numerical Solution of the Generalized Weber Problem in Spatial Economics," *Journal of Regional Science* 4 (1962) 21–33.

FURTHER READING

Virtually all elementary statistical textbooks for geographers include a discussion of at least part of the material presented in this chapter. A useful first reference for geostatistics is a monograph by Neft (1966). Many of the early applications of geostatistics and debates on the interpretation of specific measures are reviewed in considerable detail. An introduction to location-allocation problems can be found in Abler, Adams, and Gould (1971) and in a recent review by Hodgart (1978). Advances in the statistical analysis of directional data are documented in the influential textbook of Mardia (1972), but this is an extremely advanced treatment for the beginning student. Geographical applications of graph theory and network analysis are reviewed in Tinkler (1979) and Haggett and Chorley (1969).

R. Abler, J. S. Adams, and P. R. Gould, *Spatial Organization: The Geographer's View of the World* (Englewood Cliffs, NJ: Prentice-Hall, 1971).

P. Haggett and R. J. Chorley, *Network Analysis in Geography* (London: Edward Arnold, 1969).

R. L. Hodgart, "Optimizing Access to Public Services: A Review of Problems, Models, and Methods of Locating Central Facilities," *Progress in Human Geography* 2 (1978) 17–48.

D. V. Mardia, *Statistics of Directional Data* (London: Academic, 1972).

D. Neft, *Statistical Analysis for Areal Distributions* (Philadelphia: Regional Science Research Institute, 1966).

K. J. Tinkler, "Graph Theory," *Progress in Human Geography* 3 (1979) 85–116.

PROBLEMS

1. Test your knowledge of this chapter by explaining the meaning of the following terms:

   a. Spatial distribution of a variable   g. Mean center
   b. Thiessen polygon   h. Standard distance
   c. Location quotient   i. Circular histogram
   d. Coefficient of localization   j. Rose diagram
   e. Lorenz curve   k. Directional mean
   f. Gini coefficient   l. Circular variance

2. Differentiate between the concept of a Manhattan median and Euclidean median.

3. Consider the following 12 coordinate pairs and weights:

X Coordinate	*Y* Coordinate	Weight
50	80	7
30	70	4
50	70	5
60	70	6
40	60	4
70	70	6
80	60	2
40	50	1
60	50	1
0	50	1
50	40	1

   a. Locate each point, using the given coordinates on a regular piece of graph paper.
   b. Calculate (i) the mean center (assume all weights equal to 1), (ii) the weighted mean center, and (iii) the Manhattan median. Locate each on the graph paper.
   c. Calculate the standard distance. Draw a circle with radius equal to the standard distance centered on the mean center.
   d. (*Optional*) Use the Kuhn and Kuenne algorithm (Appendix 3A) to determine the Euclidean median.

4. Consider a geographic area divided into four regions, north, south, east, and west. The following table contains the distribution of employment in these four regions for three sectors of the economy:

| | Economic sector | | | |
	Primary	Manufacturing	Services	Total
North	2000	100	500	2600
South	100	300	200	600
East	300	800	600	1700
West	100	400	300	800
Total	2500	1600	1600	5700

a. Calculate location quotients for each region and each sector of the economy.
b. Calculate the coefficient of localization for each sector.
c. Draw Lorenz curves for each of the three sectors, superimposing them on a single graph.
d. Write a brief paragraph describing the spatial distribution of these activities, using the results of parts (a), (b), and (c).

5. Solve parts (a) to (d) of Problem 3, using the following coordinates and weights:

X Coordinate	Y Coordinate	Weight
20	100	1
70	90	1
100	100	1
40	70	2
80	70	1
60	60	2
40	40	4
50	50	3
70	40	6
20	20	5
50	20	2
100	20	1

6. The bearings ($0° = N$) from a house to the retail outlets chosen by members of that household on their last 10 shopping trips are given in the following table:

Observation No.	Bearing, deg
1	150
2	320
3	180
4	300
5	120
6	340
7	170
8	360
9	90
10	30

   a. Illustrate these bearings as vectors and as points on a unit circle in two separate graphs.
   b. Calculate the directional mean.
   c. Find the circular variance.

7. From the census of the United States or of Canada (or an equivalent source), generate the value of some variable for a region containing at least 30 subareas.
   a. Calculate the mean and variance for the variable for these base units.
   b. Group contiguous subareas in pairs so that there are 15 or more zones; calculate the mean and variance in the regions, using these larger zones.
   c. Group contiguous zones identified in part (b) to get eight or more areas; find the mean and variance for this distribution.
   d. Continue this procedure until there are only two subregions. At each step calculate the mean and variance.
   e. The mean of the variable at each step should be equal. Why?
   f. Construct a graph of the variance of the variable (Y axis) versus the number of subareas (X axis). What does it reveal?

8. For the same data used in Problem 7(b), generate five different groupings in which the subareas are joined in pairs. Calculate the variance for each grouping. Do your results confirm the existence of the problem of modifiable units?

9. What is the smallest possible value for a location quotient? When can it occur? What is the largest possible value for a location quotient?

10. Generate an example illustrating the sensitivity of both location quotients and coefficients of localization to the areal units on which their calculations are based. (*Hint*: Try a grouping experiment such as in Problem 8.)

11. What happens to a coefficient of localization if two areas with identical location quotients are jointed into a single larger zone? What happens if two zones with different location quotients are grouped?

4

Concepts of Time Series Analysis

Geographers and other social scientists often deal with processes that vary over time. Observations on a process of this type when arranged in a time sequence are called a *time series*. Examples of time series are temperatures at 2:00 P.M. at a particular weather station for the past 20 years and annual coal production in a country since 1850. The techniques used to analyze data in this form are known collectively as *time series analysis*.

Time series are analyzed to describe, understand or explain, predict, and sometimes even control the underlying process generating the observations in the time series. The analysis of a time series normally involves the study of various *components* of the time series, for example, the long-term trend, the cyclical component, or the seasonal component. The particular component in which we are interested often depends on the nature of the problem being studied. For example, a climatologist interested in long-term climatic change may focus on the trends in a series and ignore seasonal and daily cycles in the data. On the other hand, an analyst interested in the demands for electricity may be interested in long-term trends, those related, say, to population growth and technological change, cyclical components related to expansionary and recessionary periods in the economy, and also seasonal components related especially to differential summer/winter requirements.

In this chapter, we consider the basic aspects of time series analysis and discuss some descriptive methods that are widely used. After formally defining a time series, we discuss three examples of time series drawn from various specializations within geography. We then move on to discuss the basic goals of time series analysis so that we can place the elementary material presented in this chapter within a broader context. As we shall see, the description of time series is only the first step in time series analysis. Following this, we present some elementary techniques used to *smooth* time series. Smoothing what appear at first sight to be highly irregular patterns in time series often helps to uncover underlying patterns within the series. Finally, we show how it is possible to decompose a time series into *components*. It is often best to attempt to analyze these complex series by isolating the long-term trend from the rest of the series, or the seasonal trends from the longer cycles. In Chapter 15, we extend our study of time series to include more sophisticated modeling procedures.

4.1. What is a Time Series?

Let us begin by introducing a formal definition for a time series:

DEFINITION: TIME SERIES

A time series is a sequence of observations $Y_1, Y_2, \ldots, Y_t, \ldots, Y_n$ that are collected or recorded over successive increments of time.

Notice that the definition of a time series specifies measurements or readings taken at equally or almost equally spaced time intervals. Very little theoretical work has been done on time series observed at unequal intervals, though it is not rare to see a time series analysis with a few missing observations or with an extended gap somewhere in the series. This might occur, for example, in a situation in which data collection procedures were interrupted as a result of the breakdown of some data-gathering device. It is fairly common to *interpolate* these few missing values so that the data can be treated as a conventional time series. When the data series is quite long and contains hundreds or even thousands of observations, this should have an insignificant impact on the results of the analysis and any interpretations we make. However, a caution is in order. Extreme observations can often have a substantial impact on time series analysis, just as they can lead to significant changes in the values of any simple descriptive statistics. While we may have some justification in eliminating extreme observations that we suspect to be invalid, it is important to remember that *every* observation in a data set tells us something about the variable being analyzed. Replacing unusual observations just because they look strange is unwise. Appropriate methods of interpolating values in a time series are discussed in Section 4.4. These methods are based on the concept of smoothing. In short, we replace an observation by the average value of similar observations. In a time series this normally means we generate a missing value from our neighboring observations. In some instances this procedure itself may be biased. For example, we would not wish to replace an average monthly rainfall figure for July with, say, the average of June and August; it may be better to use average values of the previous and subsequent July. A visual examination of the time series is usually an important guide in deciding what kind of substitution to perform.

Time series data normally arise in one of two ways:

1. First, we can generate such a series by taking measurements of some specific continuous variable at equally spaced intervals of time. Such series are often defined as *instantaneously recorded.* For example, if we were to take the ambient air temperature at 3:00 P.M. every day for a year, we would generate a time series of 365 observations of a continuous time series.
2. The second way we can generate a discrete time series is through *accumulation* or *aggregation.* Specific examples of accumulated series include monthly rainfall at Santiago, Chile; quarterly industrial production in the United States; the number of tourists visiting China; the number of armed robberies in Chicago; and the reported cases of rubella by week at a health clinic in New York City.

Although we do not define a times series in continuous terms, continuous data are often transformed into discrete data so they can be analyzed using the methods presented in this chapter.

It is also useful to distinguish between a time series *process* and a time series *realization*. The observed data in our time series represents an actual realization of an underlying time series process. By a realization, we mean a sequence of observed data points, not merely a single observation. Quite often in time series analysis, we use some specific model that has properties analogous to the time series process and see how well it describes an actual time series realization. In Chapter 15 of this text, we will explore a few time series *models* and address the underlying theoretical processes they emulate.

4.2. Examples of Time Series

Time series occur in a variety of fields within geography and social sciences in general, but also in physics and engineering, the biological sciences, business, and medicine and public health. The following three examples illustrate typical instances where time series analysis has been successfully applied.

EXAMPLE 4-1. Epidemiological Statistics   Within the fields of medicine and public health, a number of familiar examples of time series can be found. Electrocardiograms and electroencephalograms are well-known medical procedures that generate time series of elements of bodily activity. Time series data that trace the paths of various epidemics are also extremely common. Analyzing the incidence of HIV infection or deaths due to AIDS may yield insights into the requirements for providing public health facilities and allocating resources to limit the spread of this disease.

Figure 4-1 illustrates a time series generated from epidemiological research, tracing

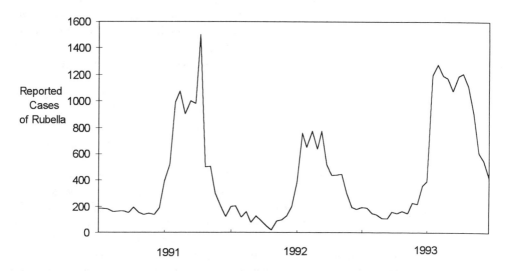

FIGURE 4-1.   Reported cases of rubella from 1990 to 1992, biweekly.

the number of reported cases of rubella over a 3-yr period. This continuous series has been developed by counting all of the cases reported to all public health clinics in a given jurisdiction in 26 two-week periods for each of the years from 1990 to 1992.

What can we see from this graph?

1. First, we note that there are reported rubella cases in each two-week period. This suggests that rubella never completely disappears from the population.
2. The average number of reported cases is not a realistic indicator of the time series. There are no long stretches of time in which the average number of reported cases in a two-week period is constant at the mean.
3. There are periods during which the number of reported cases of rubella in each two-week period is low, less than 100. Moreover, this seems to characterize the second half of the year.
4. There are three identifiable outbreaks of rubella within the 3-yr period, one in each year. Upon closer examination, we can see that each outbreak occurs in the first half of the year.
5. The duration of each outbreak is less than six months, with peak numbers of cases reported remaining high for perhaps three months.

Even this cursory examination of the rubella data suggests that a more sophisticated analysis may reveal other patterns not apparent in the graph of the raw data. We might be interested in applying time series analysis to help us identify the typical pattern of an outbreak, to examine the impact or viability of various public intervention programs such as public education or mass immunization, or to aid the authorities in estimating required storage levels of vaccines.

EXAMPLE 4-2. Economics and Business Series   We are all aware of the periodic ups and downs in the economy. For long periods of time, the economies of many regions have been undergoing growth, though it may be at a very low rate. Superimposed on this long-term trend are cycles of very rapid growth and sometimes even downturns in the economy. Within these longer term movements, there may be important seasonal influences, often due to weather patterns. For example, people use more heating oil in the winter or more electricity for air conditioning in the summer months.

In addition to these patterns, the data may also contain a number of nuisance effects. First, there are a different number of days in the months of the year. One way around this is to convert calendar month production to a standard month of 30 days. For example, we would multiply February values by 30/28 and December values by 31/30. The simple fact that a month may contain four or five weekends will also sometimes influence the observed time series. Finally, holidays that are moved may also add confusion, if the day, week, or month in which they occur is not fixed. The analyst may be confronted with serious outliers in the series. In many instances, we may be able to identify the cause of outliers, if, for example, they are related to a holiday or labor dispute. Because we sometimes smooth our data in the course of our analysis, many of these effects may not be serious. Unfortunately, no hard and fast rules exist on when and how to "clean" our data or whether we should use the raw data values. Before proceeding, however, we may wish to examine our data for potential outliers.

Figure 4-2 illustrates a time series of industrial production measured in millions of dollars over the 1986–1992 period. The dotted line represents the mean monthly production of the entire period. The series as a whole shows evidence of a trend; although production over the whole period averages about $300 million, we would certainly predict a higher series production level in 1993 than in 1992. Even a cursory glance at this graph suggests a significant, possibly linear trend in the series. In addition, the sequence of monthly production levels exhibits a distinct seasonal pattern. In each year, production is at a peak in November and is lower in the March–July period of each year. Finally, we note that some parts of the time series can't be explained as part of the overall trend or the seasonal pattern. These variations may be due to any number of factors, but can be grouped together and treated as a more or less random component.

Were we interested in forecasting this time series, we would try to exploit the patterns apparent in the trend and seasonal components to provide a good estimate of future production. To do this, we make one critical assumption: the future behavior of the system will be governed by the same forces that underlay the movements in the available time series data. Sometimes this may not be obvious. We may fail to identify long-term cycles in the series if the length of our existing data series is insufficient. Or, the data series at our disposal may not be typical. In either case, the quality of our forecasts would suffer from the missing information.

EXAMPLE 4-3. Environmental Time Series Time series frequently arise in environmental and physical sciences. Monitoring air and water pollution variables is necessary to establish baselines before the introduction of policy initiatives and to estimate the impact of these changes on key environmental indicators. Because it is an indicator of the greenhouse effect, CO_2 levels are monitored at several observation points around the world. If these points are at locations remote from urban areas and large-scale local producers of CO_2, it may be feasible to measure the changes in the greenhouse effect over longer periods.

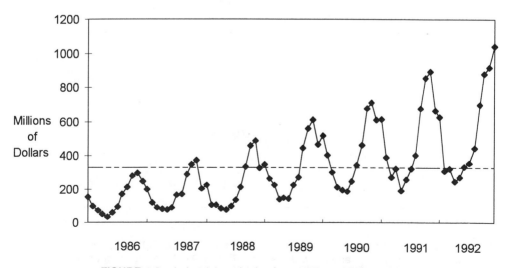

FIGURE 4-2. Industrial production from 1986 to 1992, monthly.

Figure 4-3 illustrates the levels of CO_2 in parts per million (ppm) observed at the Mauna Loa observatory in Hawaii. Because it is located at a very high elevation, observed concentrations will measure the changes in the gaseous content of the upper atmosphere. The series illustrates monthly changes in the observed CO_2 levels over the 35-yr period from 1958 to 1993.

There are three important characteristics of this time series:

1. First, variations occur over the year in the range of 3–4 ppm.
2. There has been a dramatic increase in CO_2 levels over the observation period suggesting that greenhouse gases are increasing in concentration at this location. If it is representative of the composition of the upper atmosphere, this may be one indicator of the change in the greenhouse effect over this period.
3. Finally, there appears to be a possibility that the rate of increase in the composition of CO_2 has leveled off in the last few years of the series.

Time series analysis may be useful in isolating the systematic component of this series and evaluating the significance of both the long-term change in CO_2 and the recent drop-off in the rate of increase.

4.3. Objectives of Time Series Analysis

We normally study time series data for one of four reasons:

1. To obtain a concise *description* of the features of a particular time series. This is normally the first step in time series analysis and is analogous to the importance given to descriptive statistics generally in statistical analysis (review Figure 1-2);

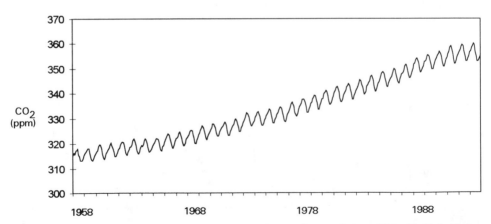

FIGURE 4-3. Monthly atmospheric CO_2, Mauna Loa, Hawaii. Source: Data published in T. A. Boden, D. P. Kaiser, R. J. Sepanski, and F. W. Stoss, eds., *Trends '93: A Compendium of Data on Global Change.* (Washington, DC: U.S. Dept. of Energy, 1994).

2. To construct a model to *explain* the time series behavior in terms of other variables and to relate the observations to some structural rules;

3. To use the knowledge gained from the results of (1) and (2) to *forecast* the future behavior of the time series based on our knowledge of the past;

4. To *control* the process generating the time series by examining what the outcome on the observed data series would be if certain model parameters were changed. For example, we might wish to predict the impacts on stream flows below a dam if certain water release regimes were put in place.

In this chapter we primarily restrict ourselves to the first of these four objectives, focusing on the simple methods developed for describing patterns within time series.

4.4. Smoothing of Time Series

Smoothing of a time series is a statistical procedure used to dampen out the fluctuations in a time series to obtain a more regular series. The smooth component then reflects the *systematic* movement in the series. The difference between the smoothed series and the actual series is often taken to be the random component. More often than not, it is the systematic component in which we are interested, not the apparently random fluctuations around it. Let us consider the topic of smoothing within the context of a simple example.

EXAMPLE 4-4. A Time Series of Traffic Counts   Traffic counts were recorded at an intersection in a residential neighborhood where complaints of excessive traffic noise had been registered over the previous several months. City officials decided to take vehicle counts over the 5:00 to 5:30 P.M. period for an initial baseline period of four weeks. At this time, a traffic amelioration scheme was introduced, and the survey continued for four additional weeks. Data on the daily vehicle counts over the entire 8-wk survey period are shown in Figure 4-4. A transportation geographer was asked to analyze the data with a view

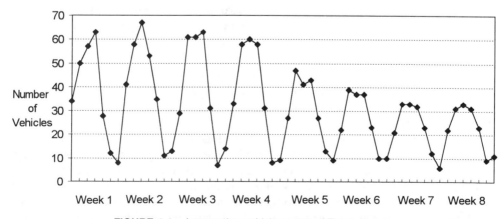

FIGURE 4-4.   Intersection vehicle counts of Example 4-4.

to evaluating the effectiveness of the traffic diversion. Two characteristics of the vehicle count series are apparent in Figure 4-4:

1. There are marked fluctuations within each seven-day period. There is significantly less traffic on weekends. To some extent, Mondays and Fridays seem to have lower traffic volumes than midweek. Upon investigation, the analyst found that this was due to the flex-time scheduling at a nearby major manufacturing enterprise where employees typically program their off day on either side of the weekend to maximize the benefits of their flex time.
2. A downward shift in the underlying direction of the series *seems* to coincide with the introduction of the traffic reduction program.

To present the evidence effectively, the analyst wished to average out the observations in the series in order to obtain the clearest view of the overall pattern of traffic during the study period.

Moving Average

One of the simplest ways of smoothing the fluctuations in a time series is to calculate the average of successive, overlapping groups of observations.

DEFINITION: MOVING AVERAGE

A moving average of a time series is developed by replacing each successive sequence of L observations by the mean of the sequence. The first sequence contains the observations $Y_1, Y_2, \ldots, Y_k$; the second sequence contains the observations $Y_2, Y_3, \ldots, Y_{k+1}$, and so on. The value of L is the *term* of the moving average.

The original series is thus replaced by a new series of means that run or move as we go down the series from the first to the last observation.

For the traffic count data, the calculations leading to a seven-term moving average are shown in Table 4-1. The actual vehicle counts over the study period are given in the third column. We begin the calculation of the moving average at $t = 4$, since this is the first value of t over which seven surrounding values can be found. In the fourth column, the sum of the seven values is shown. For $t = 4$, the value 252 is equal to $34 + 50 + 57 + 63 + 28 + 12 + 8$. The moving average is calculated by dividing this sum by 7. Thus, for $t = 4$, the seven-term moving average is calculated as $252/7 = 36.0$. Note that the seven-term moving average series ends at $t = 53$, because there are no longer seven surrounding observations for $t = 54$, 55, and 56.

What does the smooth sequence look like? Figure 4-5 superimposes the smoothed sequence on the original data series. As we might expect, the smoothed series has markedly less variation than the original time series, and it is much easier to see the trend of traffic over the eight-week survey period. First, it is clear that traffic flows decrease exactly when the traffic amelioration scheme was introduced at the end of week 4. Second, there appears to be another slight decline in the second week, as more and more drivers find the reroute

TABLE 4-1
Calculation of Seven-term Moving Average for Vehicle Count Data

Week and day	t	Number of vehicles	Seven-term total	Seven-term moving average	Fluctuating component
1M	1	34			
T	2	50			
W	3	57			
T	4	63	252	36.00	27.00
F	5	28	259	37.00	−9.00
S	6	12	267	38.14	−26.14
S	7	8	277	39.57	−31.57
2M	8	41	267	38.14	2.86
T	9	58	274	39.14	18.86
W	10	67	273	39.00	28.00
T	11	53	278	39.71	13.29
F	12	35	266	38.00	−3.00
S	13	11	269	38.43	−27.43
S	14	13	263	37.57	−24.57
3M	15	29	273	39.00	−10.00
T	16	61	269	38.43	22.57
W	17	61	265	37.86	23.14
T	18	63	266	38.00	25.00
F	19	31	270	38.57	−7.57
S	20	7	267	38.14	−31.14
S	21	14	266	38.00	−24.00
4M	22	33	261	37.29	−4.29
T	23	58	261	37.29	20.71
W	24	60	262	37.43	22.57
T	25	58	257	36.71	21.29
F	26	31	251	35.86	−4.86
S	27	8	240	34.29	−26.29
S	28	9	221	31.57	−22.57
5M	29	27	206	29.43	−2.43
T	30	47	202	28.86	18.14
W	31	41	207	29.57	11.43
T	32	43	207	29.57	13.43
F	33	27	202	28.86	−1.86
S	34	13	194	27.71	−14.71
S	35	9	190	27.14	−18.14
6M	36	22	184	26.29	(4.29)
T	37	39	180	25.71	13.29
W	38	37	177	25.29	11.71
T	39	37	178	25.43	11.57
F	40	23	177	25.29	(2.29)
S	41	10	171	24.43	(14.43)
S	42	10	167	23.86	(13.86)
7M	43	21	162	23.14	(2.14)
T	44	33	162	23.14	9.86
W	45	33	164	23.43	9.57
T	46	32	160	22.86	9.14
F	47	23	161	23.00	0.00
S	48	12	159	22.71	(10.71)
S	49	6	159	22.71	(16.71)

(*continues*)

TABLE 4-1 (continued)

Week and day	t	Number of vehicles	Seven-term total	Seven-term moving average	Fluctuating component
8M	50	22	158	22.57	(0.57)
T	51	31	158	22.57	8.43
W	52	33	155	22.14	10.86
T	53	31	160	22.86	8.14
F	54	23			
S	55	9			
S	56	11			

to be slow and seek alternative paths. Finally, there is some evidence that traffic patterns have stabilized. The path of the seven-term moving average becomes horizontal for weeks 3 and 4. This suggests that there may be no further declines in traffic volumes and that further measures will have to be introduced if this traffic volume is considered to be unacceptable.

Once we have determined the smoothed part of the series using the moving average, we can define the fluctuating component of the series as the *difference* between these values and the actual values given in column 3. This represents the day-to-day fluctuation in traffic levels and illustrates the significance of both midweek peaks, weekend troughs, and the significance of lower volumes on Mondays and Fridays. This fluctuating component is graphed in Figure 4-6. The series is horizontal over the entire eight-week study period, though the overall level of the fluctuation is reduced in the "after" period.

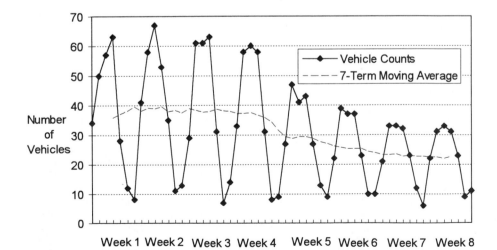

FIGURE 4-5. Seven-term moving average of vehicle counts of Example 4-4.

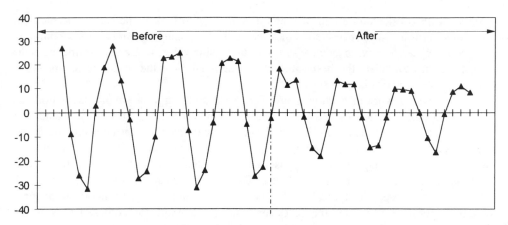

FIGURE 4-6. The fluctuating component of the traffic count series.

Effect of Term on Moving Average

The path of the moving average is particularly sensitive to the length of the term chosen for the calculation. It was natural to use a seven-term moving average for the vehicle count data; three- or five-term moving averages would not be as effective in portraying the systematic component of the series. Figure 4-7 illustrates the three-, five-, and seven-term moving averages for the vehicle count data. Note that the three- and five-term series do not smooth the series as well as the seven-term moving average series. In general, the best choice for the term of the moving average series is the length of the fluctuating component. The normal weekly pattern of traffic in a North American city leads us to the selection of the $L = 7$-term moving average series. Unfortunately, many time series in the social sciences

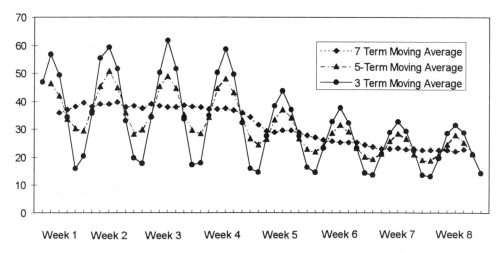

FIGURE 4-7. Three-term, five-term, and seven-term moving average of traffic example.

oscillate with considerable frequency, and it is difficult to look at the series and pick out the *single* term length that best smoothes the series. Typically, one must use a computer program to generate moving average series for increasing values of L and examine the paths of these series to isolate the best value of L. The smallest value of L that dampens the fluctuations in a series is then selected.

However, a word of caution is in order. An usually large or small observation can introduce oscillations into a series where none exists in the original series. Why? Remember that any times series observation Y_t enters L successive terms of the moving average series. As an example, let us change the value of a single original observation and compare the resulting moving average series. Suppose the traffic count for Wednesday in the second week is changed from 67 to eight vehicles. This value is clearly unusual for a midweek observation and resembles typical weekend traffic levels in the neighborhood for this time of day. Figure 4-8 illustrates the seven-term moving average for both series. Note the dramatic change in the series during the second week. Because we are using a seven-term moving average, seven consecutive data points are drawn dramatically downwards. The movement of the series is now markedly different, and the overall impact of the traffic reduction scheme is not nearly so obvious. If there were a second unusual observation in the third week, there would be two U-shaped segments in the series, and it would be even more difficult to interpret the impact of the traffic amelioration plan. The point is clear: where unusual observations exist in a time series, the paths of many smoothed moving average series may have erratic sequences.

Centered Moving Average

When an even number of terms is required in a moving average, e.g., a four-term or a 12-term calculation, a problem arises. Consider the following case in which a four-term moving average is taken:

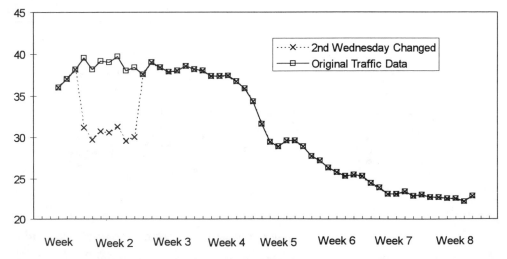

FIGURE 4-8. The effect of an extreme observation on a moving average series.

t	1	2	3	4	5	6	7
Y_t	2	7	6	8	4	6	2
Moving total $L = 4$			23	25	24	20	
Moving average $L = 4$			5.75	6.25	6.00	5.00	

The four-term totals and averages, placed at the centers of their sequences, are located at $t = 2.5, 3.5$, etc. The difficulty with these locations is that we can't easily and directly compare these moving averages to the original observations. We remedy this situation by centering the moving averages as follows:

t	1	2	3	4	5	6	7
Y_t	2	7	6	8	4	6	2
Moving total $L = 4$			23	25	24	20	
Moving average $L = 4$			5.75	6.25	6.00	5.00	
Centered moving average			6.0	6.125	5.5		

For example, the first centered moving average of 6.0 at $t = 3$ is the average of the two moving averages of $L = 4$; that is, $6.0 = (5.75 + 6.25)/2$. For $t = 4$, we find the second centered average to be $(6.25 + 6.0)/2 = 6.125$. Note that we shorten the number of values in the series when we use centered moving averages. In general, we lose moving average values for the first and last $L/2$ time periods. In this series $L = 4$, and we lose the first two and the last two elements in the time series.

It is possible to view a centered moving average as a *weighted moving average* when the observations in the series are given unequal weights. The centered moving average of 6.0 for $t = 3$ can be calculated by weighting the first five observations by the sequence 1, 2, 2, 2, 1:

$$[1(2) + 2(7) + 2(6) + 2(8) + 1(4)] / 8 = 48/8 = 6.0$$

Similarly, two-term centered moving averages can be calculated as weighted three-term moving averages, and a six-term centered moving average can be determined as a weighted seven-term moving averages.

4.5. Time Series Components

We have already seen from our vehicle data that a time series can be decomposed into a *smooth* component following the systematic movement of the series, and a *fluctuating* component representing irregular, erratic, or random movements in the data sequence. Unfortunately, smoothing methods make no attempt to identify *individual* systematic components within this basic underlying pattern. In many instances, the time series can be broken down or *decomposed* into several separate, often additive, components. This decomposition often aids us in better understanding the overall behavior of the series and also usually facilitates improved forecasting. For example, the underlying pattern may consist of a long term trend of growth (or decay), an added seasonal component, plus another less predictable random component. Theoretically, of course, a series might consist of any number of different simple components with different periodicities, the composite effect of which is an extremely complex relationship. The question is: How do we isolate these individual patterns from each other when all we have is the aggregate movement of the variable expressed as a time series?

For many business or economic series, it is common to break down the series into four components:

---

1. THE TREND COMPONENT
The trend component describes the long-term movement of the series. It may be increasing, decreasing, or remain unchanged over time.

For many phenomena, this trend or long term pattern can be assumed to follow a simple linear growth pattern or else some simple curve. Examples of several different trend curves are illustrated in Figure 4-9. The trend can be linearly increasing as in (a), linearly decreasing as in (b), increasing exponentially as in (c), or following a form of S-shaped growth as in (d). In theory, just about any functional form could be the governing trend for

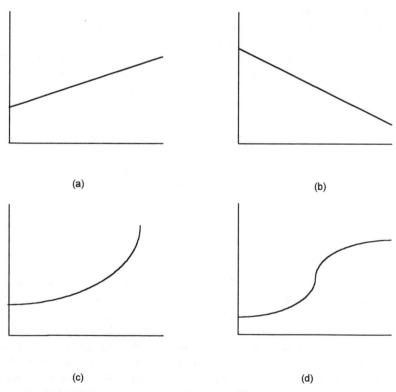

(a)

(b)

(c)

(d)

FIGURE 4-9. Possible forms of the trend component of a time series.

some empirically derived time series. It is not just the fit of the trend curve to the data that tells us which trend curve we should apply to any given series, particularly if we wish to use the curve for forecasting purposes. For example, Figure 4-10 illustrates two potential trend curves for a given time series. While both do a reasonable job within the available data, the forecast of the trend beyond the data series leads to markedly different trajectories. If we choose trend A, we would seriously overpredict the actual course of the series if it fails to sustain the rate of growth and levels off as in trend B. On the other hand, using trend B may put us in a difficult position if the series happens to maintain the growth rate of trend A. The choice is often one of judgment. If other indicators point to a leveling off of the growth trend, we might choose trend B; if not, trend A.

2. THE CYCLICAL COMPONENT
The cyclical component of a time series refers to the recurring movements of the data above and below the trend of the time series.

Many time series contain cycles of periodic growth and decay that follow the general form of Figure 4-11. These fluctuations are usually longer than one year. The *duration* of a cycle is measured from successive peaks in the series. The duration of the cycles in a series need not be constant, though many physical series have regular periodic cyclical patterns governed by specific physical processes.

We also know that shorter term movements in some economic series are affected by the overall ups and downs in the economy. When interest rates rise, many variables decrease; when interest rates fall, they increase. The time series of these variables thus contains a component that rises and falls regularly with the expansion and contraction of the economy. The difficulty with interpreting and modeling these components of a time series is that the cycles often have different amplitudes and duration over the course of the

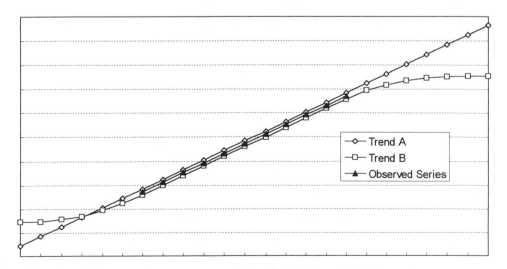

FIGURE 4-10. The difficulty of forecasting a time series trend component.

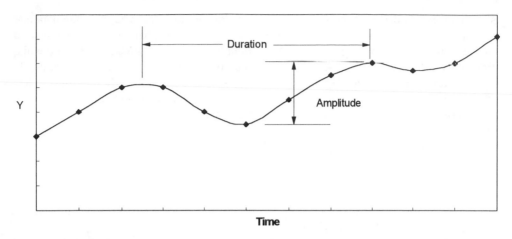

FIGURE 4-11.   General form of a series with a cyclical component.

time series. Furthermore, different factors may be responsible for the movements of the cycles at different times. The position within a cycle is often the most important factor in making short-term forecasts of the time series. For example, a forecast for two to five years ahead for many economic variables depends especially on whether we are likely to be entering a recession, are in the midst of one, or are entering a period of growth. Unfortunately, because the duration and amplitude of these cycles varies over time, it is not easy to predict when the turning points in these cycles will occur. One of the ways economic forecasters handle this problem is to seek out variables that tend to identify turning points in the cycle. Such variables are often called *economic indicators.*

---

3. THE SEASONAL COMPONENT
The seasonal component consists of patterns of change with a period of less than one year and is associated with the passage of time according to a clock or calendar.

Seasonal effects are normally distinguished from the cyclical component of a time series on two grounds. First, many choose to limit seasonal components to cycles that complete themselves within a calendar year. This is obvious for many physical processes such as climatic time series, but also is a standard component of many economic time series, especially those that follow quarterly series (Jan-Mar; Apr-Jun; Jul-Sep; and Oct-Dec). Demand for many goods (and also transportation of them) often exhibits regular seasonal patterns. Demand may be high in the winter, low in the summer, or vice versa. Second, these seasonal patterns are much more regular than cyclical components, particularly their regular *duration.* For example, an economic series might contain two growth cycles: one lasting 10 years and relatively weak, the other strong but lasting only four years; and two recessions, a short, deep recession of three years and a longer, mild recession of seven years. This pattern is much more irregular than is characteristic of seasonal patterns—though of course we can still have extremely mild winters or cool summers!

   That seasonal effects tend to be very regular is illustrated by the methods developed to seasonally "adjust" many time series. The seasonally adjusted unemployment rate is one example in which the regularities in the movement of unemployment over a year are used to adjust the rate so that a judgment can be made as to whether a particular value is abnormal for a certain month.

4. THE IRREGULAR OR RANDOM COMPONENT

This is the residual movement in a series remaining after the other components have been taken into account. They are not recurrent and cannot be attributed to either seasonal, trend, or cyclical factors.

Rarely are we able to account for all the variations in a time series by seasonal, cyclical, and trend factors. We are left with an *irregular* or *random* component. Usually, this is assumed to be the difference between the combined effect of the first three components and the actual or observed data series. Hopefully, we will be able to explain much of the movement of the series by the first three components so that the magnitude of the irregular component is relatively small. This is extremely important if we are interested in forecasting the future path of the series.

Decomposition

The methods known collectively as *decomposition* assume that the path of the observed data series Y_t is made up of two parts:

$$\text{Data} = \text{pattern} + \text{trend}$$

In turn, the pattern itself is composed of several subpatterns. Several alternative approaches may be taken to isolate these components.

In mathematical terms, the decomposition approach can be represented by the following general equation:

$$Y_t = f(T_t, S_t, C_t, E_t)$$

where T_t is the trend component at time t; S_t is the seasonal component at time t; C_t is the cyclical component at time t; and E_t is the error component at time t.

Many functional forms are possible within this general framework. We will examine one particular model of the process in Chapter 15 of this text. We could combine any of the trend formats illustrated in Figure 4-9 with various seasonal patterns and complicated cyclical movements to obtain any number of functional forms. The paths of the times series illustrated in the three examples described in Section 2 of this chapter are indicative of the apparent complexity of actual time series. Yet, surprisingly, some of these series can be decomposed into quite simple functions, which together describe much of the movement in the path of the variable being analyzed.

The easiest way to view the principles of decomposition is to perform the reverse process: constructing a series from simple assumptions about the nature of each of the individual components. To begin, we will make the simplest of all assumptions about the nature of the long-term trend: let us assume it follows a straight line. To make matters even more simple, let us suppose that the nature of this trend is such that the value of Y will increase by 0.4 units in each time period, beginning with a value of 0 at time $t = 1$. The path of T_t is shown in the second column of Table 4-2 and is illustrated in Figure 4-12. For ease of interpretation, let us assume that the time periods represent seasons (or quarters) of a year, and label them W [Winter], Sp [Spring], Su [Summer], and F [Fall].

Now, in addition to this long term trend, we also have a seasonal component that operates in the following way. We will assume that seasonality provides the following additions to the observed value in the four seasons:

Season (or quarter)	Increment
W	0
Sp	1
Su	3
F	2

The sequence of increments for the seasonal component is shown in the third column of Table 4-2 and shown in Figure 4-13.

Again, for purposes of simplicity, let us assume that there is no additional cyclical component. The addition of such a component would only unnecessarily complicate our analysis. We can introduce sufficient richness into the series by adding a random element to our two initial components. We select a random number between 0 and 5 to add to the trend and seasonal series. Details of the concept of randomness and generating random numbers are left to Chapter 7. For now, consider a random number in a simple nontechnical way: when we select a number in the range of 0 to 5, each number has an equal chance of being selected. These numbers are shown in the fourth column of Table 4-2, and the graph of this component of the time series is shown in Figure 4-14.

While each of these individual components is easily understood and leads to a relatively simple graph, they generate a more complex-looking series when we add them together and consider the time series

$$Y_t = T_t + S_t + E_t$$

Figure 4-15 illustrates the path of Y_t. Knowing how the series has been constructed, we can probably identify the trend and seasonal components of this series by a visual examination of the graph. In other instances, this may not be so apparent. First, we have yet to consider the impact of a complicated cyclical component on the series. Second, if the error component E_t were to be a larger component of the series, it is sometimes more difficult to visually pick out the trend or seasonal components of the time series.

4.6. Summary

A time series is a collection of observations generated sequentially through time. The special features of this series are that the observations are *ordered* with respect to time, and

TABLE 4-2
Construction of a Time Series through Composition

Time period		Trend	Season	Random	Y
1	W	0.0	0.0	2.4	2.4
2	Sp	0.4	1.0	0.3	1.7
3	Su	0.8	3.0	2.5	6.3
4	F	1.2	2.0	0.3	3.5
5	W	1.6	0.0	5.0	6.6
6	Sp	2.0	1.0	5.0	8.0
7	Su	2.4	3.0	4.5	9.9
8	F	2.8	2.0	4.4	9.2
9	W	3.2	0.0	2.9	6.1
10	Sp	3.6	1.0	0.8	5.4
11	Su	4.0	3.0	1.1	8.1
12	F	4.4	2.0	0.8	7.2
13	W	4.8	0.0	2.1	6.9
14	Sp	5.2	1.0	3.5	9.7
15	Su	5.6	3.0	5.0	13.6
16	F	6.0	2.0	2.2	10.2
17	W	6.4	0.0	3.1	9.5
18	Sp	6.8	1.0	1.3	9.1
19	Su	7.2	3.0	4.2	14.4
20	F	7.6	2.0	0.6	10.2
21	W	8.0	0.0	2.3	10.3
22	Sp	8.4	1.0	2.4	11.8
23	Su	8.8	3.0	2.9	14.7
24	F	9.2	2.0	1.9	13.1
25	W	9.6	0.0	3.5	13.1
26	Sp	10.0	1.0	4.6	15.6
27	Su	10.4	3.0	4.8	18.2
28	F	10.8	2.0	2.3	15.1
29	W	11.2	0.0	1.2	12.4
30	Sp	11.6	1.0	3.6	16.2
31	Su	12.0	3.0	0.6	15.6
32	F	12.4	2.0	3.0	17.4
33	W	12.8	0.0	0.9	13.7
34	Sp	13.2	1.0	0.2	14.4
35	Su	13.6	3.0	4.2	20.8
36	F	14.0	2.0	4.7	20.7
37	W	14.4	0.0	0.2	14.6
38	Sp	14.8	1.0	0.8	16.6
39	Su	15.2	3.0	4.6	22.8
40	F	15.6	2.0	1.5	19.1
41	W	16.0	0.0	1.7	17.7
42	Sp	16.4	1.0	4.8	22.2
43	Su	16.8	3.0	2.7	22.5
44	F	17.2	2.0	2.5	21.7
45	W	17.6	0.0	4.2	21.8
46	Sp	18.0	1.0	1.3	20.3
47	Su	18.4	3.0	1.5	22.9
48	F	18.8	2.0	3.5	24.3

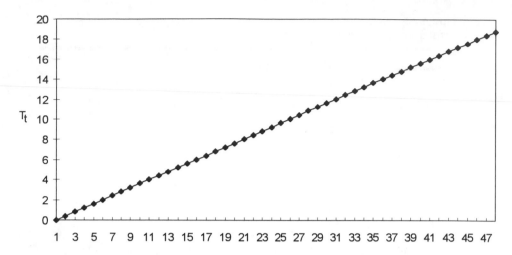

FIGURE 4-12. The trend component of the constructed time series.

that successive observations are usually expected to be dependent. In addition, the observations are at equally (or almost equally) spaced time intervals. *Discrete* time series occur when the time series is generated by instantaneously recording the value of the series at specific predetermined times. Many scientific instruments are designed to accommodate the generation of such time series, for example, instruments used to sample and record air quality measurements. Many other time series are generated through aggregation of continuous data series. For example, a business measures the total value of its receipts each day, or a climatologist records the total amount of rainfall in an hour, a day, or a month.

Some time series can contain nuisance effects that can introduce small distortions into our analysis: calendar months are not equal, holidays are not on identical dates in successive

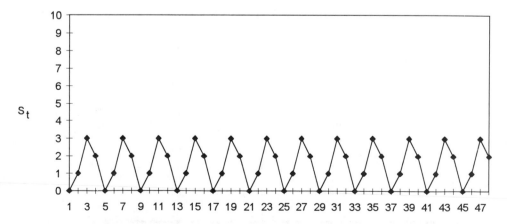

FIGURE 4-13. The seasonal component of the constructed time series.

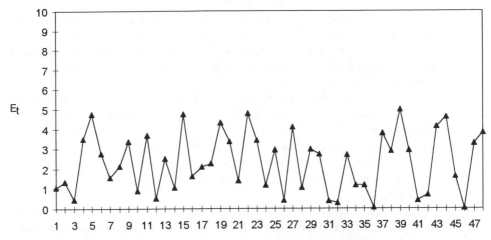

FIGURE 4-14. The random component of the constructed time series.

years, months have different numbers of weekends. Also, some time series may be interrupted by instrument failure. Or, the definition of the variable being measured is changed,
and a discontinuity is introduced into the data series. The measurement of a consumer price
index may be changed because we alter the mix of the products sought by consumers to
reflect changes in tastes. Finally, there may be any number of peculiar events that are not
repeated but cause dramatic changes in a certain portion of the time series.

The first step in any time series analysis is to plot the available observations against
time. This helps us to identify various types of data patterns that appear to exist in the
data:

1. A trend exists when there is a long-term systematic movement in the series. It may
   take on any form whatsoever but is usually more apparent when it takes a simple

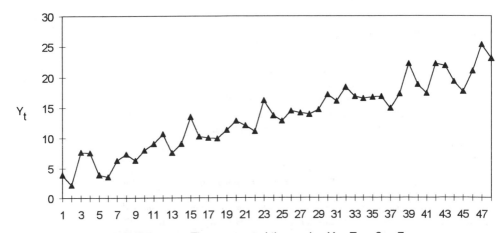

FIGURE 4-15. The constructed time series $Y_t = T_t + S_t + E_t$

form. A linear trend, either increasing or decreasing, is the simplest form of this component, but more complex relations can be estimated and removed.

2. A seasonal pattern is apparent when a data series is influenced by seasonal or calendar-related factors. Figure 4-13 illustrates the typical form of a seasonal component in a time series.

3. A cyclical pattern exists when the data are influenced by longer term fluctuations, especially those associated with the business cycle. Cyclical components can be distinguished from seasonal patterns on the basis of their length (they are usually longer than one year) and their irregularity (the first cycle may be 4 yr, the second 12 yr, and third 7 yr. Also, what appears to be part of a trend component in a short time series may turn out to be part of a long cycle when a longer time series is available.

4. A random or error component is also apparent in most time series, and represents the residual component of observed time series once the first three components have been removed.

Because of the complexity of many time series, it is often convenient to smooth the data series before analysis. This step often helps us to uncover patterns in the data and aids us in the next step of time series analysis, constructing a model of the series. We normally remove these data patterns from our times series before the modeling exercise. In Chapter 15, we show how to construct models of a time series and use them to forecast the future path of the time series.

One-dimensional time series of the sort described in this chapter are obviously less complex than spatial series. As geographers, we know that spatial series exist in at least two dimensions and the translation of many of the patterns described in this chapter becomes increasingly complex in two or more dimensions. Second, causality in time series is unidirectional: past events can affect future ones, but not vice versa. In spatial series we might expect events at one location to affect events at other places located in any direction!

FURTHER READING

Very few textbooks on time series analysis include extensive introductory sections on the nature and use of time series and elementary techniques such as smoothing. A useful alternative is to consult the appropriate chapters introductory applied statistics textbooks such as Pfaffenberger and Patterson (1981) or Neter, Wasserman, and Whitmore (1978). Two of the more useful references on time series at an intermediate level are Makridakis and Wheelwright (1978) and Bowerman and O'Connell (1979).

B. L. Bowerman and R. T. O'Connell, *Time Series and Forecasting: An Applied Approach* (Boston: Duxbury Press, 1979).

J. Neter, W. Wasserman, and G. Whitmore, *Applied Statistics* (Boston: Allyn and Bacon, 1978).

R. Pfaffenberger and J. H. Patterson, *Statistical Methods* (Homewood, Ill.: Richard D. Irwin, 1981).

PROBLEMS

1. Explain the meaning of the following terms:
   a Time series
   b. Smoothing
   c. Moving average
   d. Centered moving average
   e. *Term* of a moving average
   f. Trend component of a time series
   g. Cyclical component of a time series
   h. Seasonal component of a time series
   i. Random component of a time series
   j. Duration of a cycle
   k. Amplitude of a cycle

2. Explain the two different ways in which time series data is obtained.

3. What can we do if our time series of data contains a small number of interruptions so that the observed series is not continuous?

4. For what reason might we decompose a time series into a single smooth component and a fluctuating component?

5. How does a centered moving average differ from a simple moving average?

6. Differentiate between the four basic components of a time series that are identified in a time series decomposition.

7. Why is the selection of the form of the trend component of a time series so important if we are using the model to predict into the future?

8. Consider the following set of data, which can be considered a time series developed over 45 consecutive time periods (read the data by rows):

569	416	422	565	484	520	573	518	501	505	468	382	310	334	359
372	439	446	349	395	461	511	583	590	620	578	534	631	600	438
516	534	467	457	392	467	500	493	410	412	416	403	433	459	467

   a. Smooth the data using a three-term moving average.
   b. Smooth the data using a four-term moving average.
   c. Graph the original data and the two smoothed sequences from (a) and (b).
   d. Graph the fluctuating component from each of the two sequences.

9. The production of coal in millions of metric tonnes over a 30-yr period is given below (again, read the data by rows):

59	51	51	57	57	46	54	70	60	50
59	48	50	50	60	60	60	70	70	56
69	70	95	53	66	70	56	54	70	60

   a. Smooth the data using a three-term moving average.
   b. Smooth the data using a four-term moving average.
   c. Suppose that the production in year 10 was actually 95 million metric tonnes. How does this change the two series?
   d. Graph the two sequences with and without the data correction.

10. Using a spreadsheet program and the data of Problem 8, develop a table that generates smoothed values of the sequence for moving averages of term lengths $L = 3$ to 9. Superimpose a few graphs of these smoothed sequence over the original series. Is any term length superior to the others?

11. The following 30 observations represent the total annual miles (expressed in hundreds of thousands) traveled by a fleet of trucks from the ABC shipping company:

115.18	67.62	104.48	126.06	154.17	174.86	193.99	186.99	223.89	251.29
261.78	232.87	266.13	308.05	283.71	324.67	422.23	387.27	448.08	517.24
558.97	588.05	581.69	685.47	744.77	792.75	826.26	900.89	1,026.24	1,093.86

You are asked to present this information to management within the context of helping management to better forecast the path of this variable over time. In particular, management would like to set in place a procedure whereby they could plan the necessary size of the fleet for the next five years, knowing that some vehicles in the fleet will need replacement.

a. Graph the path of total miles traveled.
b. Smooth the sequence in a way that effectively portrays the movement of the variable. (Remember that a longer sequence may show the pattern more effectively even if it is not as smooth as a shorter sequence with a longer term!)
c. What pattern of growth is apparent? Which of the four patterns in Figure 4.9 best describes this pattern?

II

INFERENTIAL STATISTICS

5

Elementary Probability Theory

Everyone has some idea of what the word probability means. Everyday conversation contains numerous references to it: The *chances* of rain this holiday weekend are 50%. The *odds* are small that a Super Bowl champion will repeat. Or, an unmarried male driver between the ages of 18 and 25 will *probably* have more accidents than an unmarried female in the same age bracket. All these statements invoke a notion of probability, although a very vague one. In this chapter, we discuss some basic notions of probability theory in a more precise way. We show how to rigorously define the term *probability* as it is used in conventional statistical inference. Then we explain in detail the methods used to calculate the probability of an event occurring. Finally, we show how to add and multiply probabilities when two or more events are involved.

Probability theory is of interest in its own right, since it is the basis for a great number of decision-making activities. When several courses of action are open to a decision maker, but the outcome of making a specific decision is not known with complete certainty, we can use probability theory to define rational choices or courses of action. This is known as *statistical decision theory*. In addition, and more central to our concerns, the theory of probability provides the logical foundation for *statistical inference*. Because these inferences are made from only a sample (or small portion of the total members of some population), all statistical inferences are necessarily probabilistic. Therefore, a study of probability is essential to our understanding of the process of statistical inferences about statistical populations.

5.1. Statistical Experiments, Sample Spaces, and Events

The basic concepts of probability theory are based on the notion of a statistical experiment, or random trial.

DEFINITION: STATISTICAL EXPERIMENT
A statistical experiment, or random trial, is a process or activity in which one outcome from a set of possible outcomes occurs. Which outcome occurs is not known with complete certainty beforehand.

Consider, for example, the process of sampling from a population. Each time we select a single member of a population and record the value of some variable, we are performing a statistical experiment. The values of the variable for different members of the population represent the possible outcomes from the experiment. Using some type of randomization device, we select one member of the population and record the value of the variable. Which member is selected, and therefore what value is recorded, is not known until the drawing actually occurs. Repeated trials of this experiment would yield a sample from the population.

DEFINITIONS: ELEMENTARY OUTCOMES AND SAMPLE SPACE

Each different outcome of an experiment is known as an elementary outcome, and the set of all elementary outcomes constitutes the sample space.

Consider the statistical experiment of selecting a card from a shuffled deck of cards. The 52 elementary outcomes are the 52 individual cards in the deck. Together, the 52 outcomes represent the sample space of this experiment. Sometimes, it is not so easy to enumerate all possible outcomes of a statistical experiment. In such cases, a convenient way of generating the sample space is to use an *outcome tree.* Suppose we define an experiment as the toss of two coins. The elementary outcomes are shown in Figure 5-1. The first set of branches in the tree consists of the possible outcomes for the first coin, heads (H) or tails (T). The second set of branches contains the possible outcomes for the second coin. In total, there are four possible outcomes to this experiment. The outcome tree or tree diagram is a useful graphical device for portraying the sample spaces of small experiments. For larger experiments, such as drawing two cards from a shuffled deck, this technique is neither practical nor helpful. In these instances, some of the counting rules discussed in Section 5.3 are especially useful.

We have some choice in how we define the elementary outcomes of an experiment. In the coin toss experiment, we could just as easily define the outcomes as (1) no heads, (2) one head, and (3) two heads. As we shall see in Section 5.2, it is easier to work with elementary outcomes that have an equal chance of occurring. Defined as {HH}, {HT}, {TH}, and {TT}, the four elementary outcomes are equally likely. When expressed as the number of heads, the probabilities of all three outcomes are not equal. There is a one-in-four, or 25%, chance of getting either zero heads or two heads, but a two-in-four, or 50%, chance of getting only one head.

After the elementary outcomes of an experiment have been defined, it is possible to examine collections of elementary outcomes, or *events.*

DEFINITION: EVENT

An event is a subset of the sample space of an experiment, a collection of elementary outcomes.

The particular subset used to define an experiment can include one or more elementary outcomes. Returning to the card-drawing experiment, we might define event *A* to be "the card drawn is a spade," event *B* to be "the card drawn is an ace," and event *C* to be "the

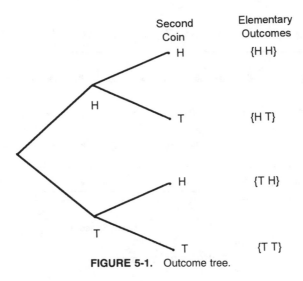

FIGURE 5-1. Outcome tree.

card drawn is the ace of spaces." Event A contains 13 elementary outcomes—the 13 spades in the deck; event B contains the four aces; and event C consists of only one elementary outcome, the ace of spades. Because event C contains only one elementary outcome, the ace of spades is both an elementary outcome and an event. A null event, denoted $\varnothing$ is an event that contains no elementary outcomes. It cannot occur.

Set Notation and Terminology; Venn Diagrams

Another useful graphical device is the Venn diagram. Together with the terminology and notation of set theory, a Venn diagram can be used to portray the elementary outcomes of an experiment, define events, and illustrate the relationship between events. Let us denote the sample space of an experiment as S. The set S contains all the elementary outcomes of the experiment and can be represented as $S = \{E_1, E_2, \ldots, E_n\}$, where $E_1, E_2, \ldots, E_n$ are the n elementary outcomes of the experiment. In a Venn diagram, the sample space encloses all the elementary outcomes. As an example, consider the experiment in which a single draw is made from a deck of shuffled cards. The Venn diagram of Figure 5-2 illustrates the sample space for this problem. It contains all 52 elementary outcomes.

DEFINITION: EVENT SPACE

The event space of an experiment contains all elementary outcomes that constitute the event.

The three event spaces for events A, B, and C of the card selection example are illustrated in Figure 5-2.

Sometimes we are interested in the elementary outcomes that are *not* part of some event A. In the card selection example, we might be interested in the set of elementary outcomes that are not aces.

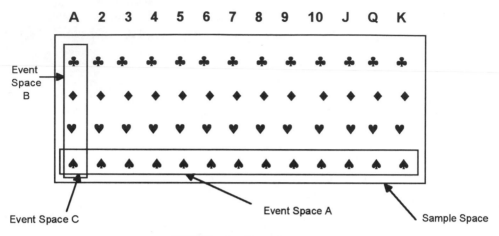

FIGURE 5-2. Venn diagram.

DEFINITION: COMPLEMENTARY EVENT
The set of elementary outcomes not in an event space A constitute the complementary event to A. This complementary event is denoted $\bar{A}$.

As we shall see, the problem of determining the probability of some event A can occasionally be solved more simply by first determining the probability that event A does not occur. To do this, we employ the complementary event $\bar{A}$.

Relationship between Events

Some questions of probability require us to examine the probability of two or more events, at least one of several events, or neither of two or more events occurring. To answer such questions, it is necessary to understand the possible ways in which two or more events can be related. Events can be compounded in two ways.

DEFINITION: EVENT INTERSECTION
The set of elementary outcomes that belong to both events A and B of a sample space is called the intersection of A and B and is denoted by $A \cap B$.

Figure 5-3 illustrates a Venn diagram in which two events G and H are defined over a sample space consisting of the elementary outcomes $\{E_1, E_2, \ldots, E_8\}$. Event G contains the elementary outcomes $\{E_1, E_2, E_3\}$, and event H consists of the elementary outcomes $\{E_3, E_4, E_5\}$. The intersection $G \cap H$ consists of the single elementary outcome E_3. The shaded area enclosing E_3 in Figure 5-3 is a graphical representation of $G \cap H$. As a second example, consider the two events A and B defined in the Venn diagram of Figure 5-2. The intersection $A \cap B$ contains all the elementary outcomes that are both aces and spades. The only

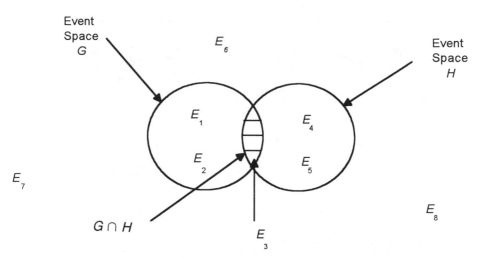

FIGURE 5-3. Venn diagram for the intersection of two events.

elementary outcome meeting both criteria is the ace of spades. We can thus say that $C = A \cap B$, since event C is defined as the selection of the ace of spades.

If the intersection of two events contains no elementary outcomes, that is, if it defines a null event, then the events are said to be *mutually exclusive.*

DEFINITION: MUTUALLY EXCLUSIVE EVENTS

If events A and B are defined over a sample space and have no elementary outcomes in common, A and B are said to be mutually exclusive. In this case, $A \cap B = \varnothing$, where $\varnothing$ is the null event.

Venn diagrams for mutually exclusive and non-mutually exclusive events are illustrated in Figure 5-4.

Events A and B of the card experiment defined in Figure 5-2 are not mutually exclusive events because they share one outcome, the ace of spades. We have already noted that $A \cap B \ne \varnothing$. Let us now define event D as "the card drawn is a diamond." It is easily seen that $A \cap D = \varnothing$, so that these two events are mutually exclusive. Note also that $B \cap D \ne \varnothing$ and these two events are not mutually exclusive. The ace of diamonds is common to both events B and D.

What is the intersection of an event and its complement? Because an event and its complement are by definition mutually exclusive, it follows that $A \cap \bar{A} = \varnothing$.

There is a second way in which two events can be related.

DEFINITION: UNION

Let A and B be two events defined over a sample space S. The union of A and B, denoted $A \cup B$, is the set of all elementary outcomes that belong to *at least one* of the events A and B.

Mutually exclusive events

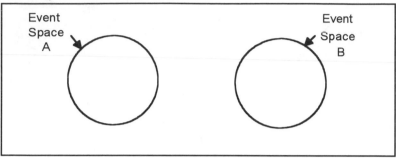

Non-mutually-exclusive events

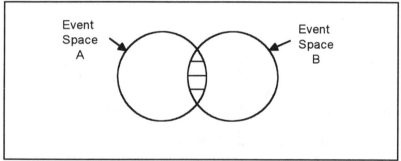

FIGURE 5-4.   Venn diagrams for exclusivity.

Venn diagrams for the union of both mutually exclusive and non-mutually exclusive events are shown in Figure 5-5. Consider once again the card selection example. The union $A \cup B$ contains 16 elementary outcomes—the 13 spades and the aces of hearts, clubs, and diamonds. (The ace of spades is only counted once!) The union $A \cup D$ contains 26 elementary outcomes, 13 spades and 13 diamonds.

5.2. Computing Probabilities in Statistical Experiments

The outcome of a statistical experiment will be one, and only one, of the elementary outcomes of the sample space. To determine the probability that an event occurs in a single trial of an experiment, it is necessary to attach a value to each elementary outcome E_i of the experiment. This value is a *probability*—a measure of how likely it is that outcome E_i is the realized outcome of the experiment. Leaving aside for the moment the thorny question of how these values are obtained, let us define three postulates that formalize certain properties of probabilities.

Non-mutually-exclusive events

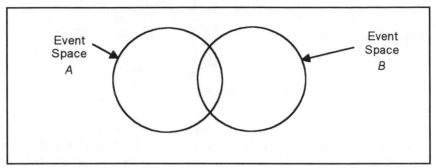

Mutually exclusive events

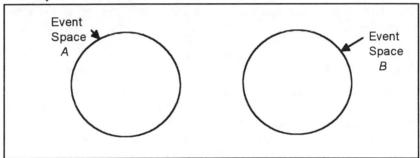

FIGURE 5-5. Venn diagram for the union of two events.

Probability Postulates

Denote the n elementary outcomes of the experiment as the set $S = \{E_1, E_2, \ldots, E_n\}$ and their assigned probability values as $P(E_1), P(E_2), \ldots, P(E_n)$.

POSTULATE 1

$$0 \leq P(E_i) \leq 1 \qquad \text{for } i = 1, 2, \ldots, n \qquad (5\text{-}1)$$

For every elementary outcome in the sample space, the assigned probability must be a nonnegative number between zero and one inclusive. There are several important definitions included in this seemingly simple postulate. First, probabilities are nonnegative. Second, if we say some outcome, for example outcome j, is *impossible*, we are saying $P(E_j) = 0$. If we say some outcome, for example, outcome k, is *certain* to occur, then we are saying that $P(E_k) = 1$.

POSTULATE 2

$$P(A) = \sum_{i \in A} P(E_i) \qquad (5\text{-}2)$$

The second postulate states that the probability of any event A occurring is the sum of the probabilities assigned to the individual elementary outcomes that constitute event A.

POSTULATE 3

$$P(S) = 1$$

$$P(\varnothing) = 0 \tag{5-3}$$

The third postulate says that the probability value associated with the sample space must equal one, and the probability value of a null event must be zero. Although these two statements are almost intuitive, it is possible to deduce a number of important results from them.

First, it is easily shown that the sum of the probabilities of all the elementary outcomes must equal one:

$$\sum_{i=1}^{n} P(E_i) = 1 \tag{5-4}$$

This follows directly from Postulates 2 and 3. Since the sample space contains all the elementary outcomes, that is, $S = \{E_1, E_2, \ldots, E_n\}$, and from Postulate 2 we know that $P(S) = \sum_{i=1}^{n} P(E_i)$, it follows from Postulate 3 that $\sum_{i=1}^{n} P(E_i) = 1$.

Second, since any event A must contain a subset of the elementary outcomes of S, and since from Postulate 1 the probability of any elementary outcome occurring is between zero and one, it follows that

$$0 \leq P(A) \leq 1 \tag{5-5}$$

for any event A. In other words, every event has a probability value between zero and one. Also, if events A and B are mutually exclusive, then from Postulate 3

$$P(A \cap B) = 0 \tag{5-6}$$

This follows from the fact that the intersection of two mutually exclusive events must be the null set: $A \cap B = \varnothing$.

Definitions of Probability

The three probability postulates tell us that probability is a measure that, for any elementary outcome or event, is a number between zero and one. But how do we interpret this probability value? How do we determine these values in practice? Let us examine two different interpretations of probability. First, probability can be interpreted or defined using a *subjective* method. In using a subjective method, the term probability refers to an individual's "degree of belief" that some event might occur. In this way we can assign probability values to events that occurred in the past, events that may occur in the future, and even unique events. For example, a doctor might suggest that a certain patient has a .5 probability of surviving another year. The problem with subjective probabilities such as this is that the value assigned to the event varies with the person making the judgment. Another doctor might suggest that the probability of the patient surviving another year is .9. Subjective probabilities depend on the information available to the person and the manner in which he or she evaluates that information. Because each person can arrive at his or her own answer for the probability value for an event, many parts of probability theory cannot utilize subjective probabilities. In some instances, however, we may be forced to rely on probability values estimated in this way.

In *objective* interpretations of probability, the probability value of an event is determined by examining the relative frequency of occurrence of the event in a repeated statistical experiment. Suppose, for example, we wish to determine the probability of a head occurring in the toss of a "fair" coin. An experiment is designed in which a coin is tossed a large number of times and the result recorded. If the coin is fair, we would expect the proportion of heads to be about $\frac{1}{2}$ or .5. The larger the number of trials, the closer this observed proportion should be to .5. This objective interpretation of the term *probability* can be formalized in the following definition of relative frequency.

DEFINITION: RELATIVE FREQUENCY INTERPRETATION OF PROBABILITY

Let event A be defined in some experiment. If the experiment is repeated N times and event A occurs in n of these trials, then $P(A) = n/N$. The relative frequency definition $P(A) = n/N$ defines the true probability of event A occurring in the limit, that is, when the number of trials approaches infinity.

Since experiments cannot be conducted indefinitely, the observed relative frequency in an extended set of trials can be used as an estimate of the probability of an event occurring.

Another objective method of determining probabilities uses deductive logic to assess the probabilities, usually based on arguments of symmetry or geometry. For example, the probability of getting a head in the toss of a fair coin might be deduced to be $\frac{1}{2}$ on the basis of the supposed symmetry of the coin. Similarly, the probability of drawing a single card in one draw from a shuffled deck of cards must logically be 1/52. Whether based on the relative frequency or on deductive logic, the probabilities arrived at are objective. Because each person using the approach would come to the same answer, these objective methods are fundamentally different from subjective interpretations of probability.

Computing Probabilities from Sample Spaces

Once the probability values are determined for the elementary outcomes of an experiment by an objective method, it is possible to compute the probability of occurrence of any event defined over the sample space. Very simply, the probability of any event occurring is the sum of the probabilities of the elementary outcomes that constitute the event. For example, once we have assigned the probability of 1/52 to each elementary outcome of the card selection experiment, we can compute the probability of getting an ace as the sum of the probabilities of the four elementary outcomes: P(ace of diamonds) + P(ace of spades) + P(ace of clubs) + P(ace of hearts). This is equal to 1/52 + 1/52 + 1/52 + 1/52 = 4/52 = 1/13.

Whenever all the elementary outcomes of an experiment are equally likely, that is, they have the same probability value, the probability of an event occurring can be more easily determined. Suppose there are n equally likely elementary outcomes in the experiment, and event A occurs if any one of these elementary outcomes occurs. Then

$$P(A) = m/n \qquad (5\text{-}7)$$

For example, the sample space S for the card selection experiment consists of the 52 elementary outcomes illustrated in Figure 5-2. The event space for event A, the selection of a spade, contains $m = 13$ elementary outcomes. Since $m = 13$ and $n = 52$, the probability of drawing a spade is equal to $13/52 = 1/4$. Similarly, the probability of drawing an ace is $4/52 = 1/13$ since $m = 4$ and $n = 52$.

    If Equation (5-7) cannot be used to calculate event probabilities, then other, more time-consuming, methods must be used. First, it is necessary to identify and count all of the elementary outcomes in the sample space. Then, the subset of the sample space that comprises the event space must be specified. Finally, the addition of the probabilities of the elementary outcomes in the event space will yield the event probability. The simplification of this operation by using Equation (5-7) is obvious, particularly when there may be thousands or even millions of elementary outcomes in the sample space and/or event spaces. In fact, it may even be difficult to generate a list of the elementary outcomes in a sample space. In such cases, we often use *counting rules* to compute both m and n.

5.3. Counting Rules for Computing Probabilities

To use Equation (5-7) to compute event probabilities, it sometimes proves useful to employ one or more counting rules. These counting rules are used to quickly enumerate the number of elementary outcomes in a sample or event space, without specifically listing each one. Let us consider four of the more commonly used counting rules.

Product Rule

RULE 1: THE PRODUCT RULE

Suppose there are r groups of objects. In group 1 there are n_1 objects, in group 2 there are n_2 objects, and so on. If we define the experiment to be the selection of one object from each of the r groups, then there are $n_1 \cdot n_2 \cdot \ldots \cdot n_r$ elementary objects in the experiment.

To illustrate the use of this counting rule, consider the following simple example.

EXAMPLE 5.1.   A planner is interested in surveying retail patronage at three types of shoe stores: factory outlets, discount stores, and specialty stores. He decides to sample one store of each type from the outlets in the town. If there are 10 factory outlets, seven discount stores, and three specialty stores in the town, how many different possible combinations of stores are there?

There are $r = 3$ groups of objects, with $n_1 = 10$, $n_2 = 7$, and $n_3 = 3$. If one store must be selected from each category, then there are $(10)(7)(3) = 210$ possible combinations of stores from which the planner may choose.

Combinations Rule

The second and third counting rules are concerned with a slightly different problem. The experiment consists of selecting a subset of r objects from a set of n objects, where $r \le n$.

EXAMPLE 5-2. Suppose a city is considering adding two new libraries to its existing system. To minimize expenses, the libraries will be constructed on two of the five vacant properties currently owned by the city. How many different locational plans for the system of libraries are there?

Label the five available sites as A, B, C, D, and E. The possible outcomes of the experiment are

$$
\begin{array}{cccc}
AB & BC & CD & DE \\
AC & BD & CE \\
AD & BE \\
AE
\end{array}
$$

Note that each of the possible plans contains two different sites; these are *combinations*.

DEFINITION: COMBINATION
A combination is a set of r distinguishable objects, irrespective of order.

The counting rule for combinations is concerned with the number of different combinations of size r that can be drawn from a master set of n objects

---

RULE 2: COMBINATIONS RULE
The number of combinations of r objects taken from n different objects is

$$C_r^n = \frac{n!}{r!(n-r)!} \tag{5-8}$$

Sometimes the number of combinations C_r^n is written $C(n,r)$ or $\binom{n}{r}$. The symbol $n!$ is read "n factorial" and is defined as $n! = n \cdot (n - 1) \cdot (n - 2) \cdot \ldots \cdot 2 \cdot 1$. For example, $5! = 5 \cdot 4 \cdot 3 \cdot 2 \cdot 1 = 120$ and $2! = 2 \cdot 1 = 2$. Be definition, the term $0!$ is defined to be 1.

For the library example, $n = 5$ and $r = 2$; therefore

$$C_2^5 = \frac{5!}{2!(5-2)!} = \frac{5 \cdot 4 \cdot 3!}{2 \cdot 1 \cdot 3!} = \frac{20}{2} = 10$$

These are the 10 combinations listed above. As you can imagine, the combinations rule can save a great deal of time in enumerating the elementary outcomes of an experiment.

Permutations Rule

Let us now consider a slightly different problem. Suppose the city has five sites for two new libraries, one that will house the main library, and one that will house video, cassette, and other special collections. The sites are labeled A to E as before. Now consider the outcome AB. This is interpreted as meaning site A is to be used for the main library (the first letter of the pair) and site B (the second letter) is to be used for special collections. Notice that outcome AB is not the same as outcome BA, because the two libraries are placed on different sites. In BA, the main library is located at site B while in AB the main library is on site A. In this instance, the *order* of the elements in the set is important.

DEFINITION: PERMUTATION

A permutation is a distinct ordering of r objects.

The counting rule for permutations defines the number of different ordered arrangements of size r that can be drawn from a set of n objects.

RULE 3: PERMUTATIONS RULE

The number of permutations of r objects taken from a set of n different objects is

$$P^n_r = \frac{n!}{(n-r)!} \tag{5-9}$$

Sometimes P^n_r is written as $P(n,r)$. In the library example, $n = 5$ and $r = 2$, so that

$$P^5_2 = \frac{5!}{(5-2)!} = 20$$

That is, there are twice as many permutations as combinations in a problem of this size. For each combination AB there is the reverse arrangement BA. This is not a general result. The formal relation between the number of combinations and the number of permutations is given by

$$C^n_r = \frac{1}{r!} \cdot P^n_r$$

Whenever $r = 2$, there are twice as many permutations as combinations.

Hypergeometric Rule

The final rule combines the combinations rule with the product rule.

RULE 4: HYPERGEOMETRIC RULE

A set of n objects consists of n_1, of type 1, n_2 of type 2, where $n_1 + n_2 = n$. From this group of n objects, define an experiment in which r_1 of the first type and r_2 of the second type are to be chosen. The number of different groups that can be drawn is

$$C^{n_1}_{r_1} \cdot C^{n_2}_{r_2} \tag{5-11}$$

As an example, consider the following counting problem. A survey is to be undertaken of neighborhoods in some city. There are 10 suburban neighborhoods and eight inner-city neighborhoods. A decision is made to survey three suburban and two inner-city neighborhoods. How many different combinations of neighborhoods are possible for the survey? Clearly, the hypergeometric rule applies. There are two groups with $n_1 = 10$ and $n_2 = 8$. From these two groups $r_1 = 3$ and $r_2 = 2$, objects are to be drawn. From Equation (5-11), the number of combinations of neighborhoods that could be drawn is

$$C_3^{10} \cdot C_2^8 = \frac{10!}{3!(10-3)!} \cdot \frac{8!}{2!(8-2)!} = 120(28) = 3360$$

5.4. Basic Probability Theorems

Several useful theorems follow directly from the probability postulates outlined in Section 5.2 and the event relationships described in Section 5.1. Together with the counting rules presented in Section 5.3, these theorems can be used to determine probabilities for quite complicated problems.

Addition Theorem

The addition theorem is used to obtain the probability that at least one of two events occurs.

ADDITION THEOREM

Let A and B be any two events defined over the sample space S. Then

$$P(A \cup B) = P(A) + P(B) - P(A \cap B) \qquad (5\text{-}12)$$

where $A \cup B$ represents the union and $A \cap B$ represents the intersection of events A and B.

Let us consider the problem of determining the probability of drawing a spade or an ace in a single draw from a deck of shuffled cards. As before, let event A be the selection of a spade and event B be the selection of an ace. For these two events we know

$$P(A) = \frac{13}{52} \qquad P(B) = \frac{4}{52} \qquad P(A \cap B) = \frac{1}{52}$$

Event A contains 13 outcomes (the 13 spades); event B contains four outcomes; and one outcome, the ace of spades, is common to both events A and B. Substituting these known values into Equation (5-12) yields $P(A \cup B) = 13/52 + 4/52 - 1/52 = 16/52 = .308$. It should be clear why $P(A \cap B)$ must be subtracted from $P(A) + P(B)$ to calculate $P(A \cup B)$. Since the elementary outcome the ace of spades is common to both events A and B, that outcome is counted twice when $P(A)$ and $P(B)$ are added. To avoid double counting, $P(A \cap B)$ is subtracted. This can easily be appreciated by closely examining Figure 5-2.

The addition theorem can be simplified when the two events A and B are mutually exclusive. We know that $P(A \cap B) = 0$ for mutually exclusive events [from Equation (5-6)], so that the addition theorem reduces to

$$P(A \cup B) = P(A) + P(B) \qquad (5\text{-}13)$$

For events A and D of the card selection problem, there are no elementary outcomes in common, so the two events must be mutually exclusive. A spade and a diamond cannot be simultaneously drawn. Therefore $P(A \cup D) = P(A) + P(D) = 13/52 + 13/52 = 26/52 = .50$. There is a 50% chance of drawing either a spade or a diamond.

The addition theorem for mutually exclusive events can be generalized for any number of events. For three mutually exclusive events A, B and C, the addition rule is written as

$$P(A \cup B \cup C) = P(A) + P(B) + P(C) \qquad (5\text{-}14)$$

Generalizing the addition rule for more than two non-mutually exclusive events is much more complicated. Equation (5-12) would have to be modified to include the intersections between all pairs of events included in the addition.

Complementation Theorem

It is sometimes more difficult to find the probability of event A occurring than to find the probability of the complementary event $\bar{A}$ occurring. By using the complementation theorem, it is possible to derive $P(A)$ from $P(\bar{A})$.

COMPLEMENTATION THEOREM

For any event A and its complement A of a sample space S

$$P(A) = 1 - P(\bar{A}) \text{ or } P(\bar{A}) = 1 - P(A) \qquad (5\text{-}15)$$

The complementation theorem follows directly from a few simple arguments. First, by definition, we know that $A \cup \bar{A} = S$. That is, the two events A and $\bar{A}$ exhaust the sample space. Therefore, $P(S) = P(A \cup \bar{A})$, and in turn $P(A) + P(\bar{A}) = 1$, by the addition rule for mutually exclusive events. The complementation theorem is a simple rearrangement of this last equation.

Let event E be defined as the selection of any card but a club in a single draw from a deck of shuffled cards. Event $\bar{E}$ is thus the selection of a club in a single draw. From Equation 5-15,

$$P(E) = 1 - P(\bar{E}) = 1 - \frac{13}{52} = \frac{39}{52} = .75$$

Conditional Probabilities and Statistical Independence

To find the joint probability of two events occurring (that probability that both occur), it is necessary to introduce the notion of a *conditional probability*. A conditional probability is the probability of an event A occurring given that a second event B has occurred.

DEFINITION: CONDITIONAL PROBABILITY
If events A and B are defined over a sample space S, then the conditional probability
of event A, given B, denoted $P(A|B)$, is defined as

$$P(A|B) = \frac{P(A \cap B)}{P(B)} \tag{5-16}$$

The vertical line separating the A and B in $P(A|B)$ denotes a conditional probability. The
term $P(A|B)$ is read "the probability of A, given B."

To illustrate the concept of a conditional probability, let us consider a slightly
different experiment. Suppose the experiment is to roll a perfectly balanced single die and
record the number of dots appearing on the upward face. The sample space S for this
experiment consists of six elementary outcomes. Define event A as "the upward face
contains an odd number of dots" and event B as "the upward face contains less than or equal
to three dots." These two events are illustrated in the Venn diagram of Figure 5-6. What
is $P(A|B)$? It is the probability of the upward face containing an odd number of dots given
that it has three or fewer dots. From Figure 5-6 we can see

$$P(A) = \frac{3}{6} = \frac{1}{2}$$

$$P(B) = \frac{3}{6} = \frac{1}{2}$$

$$P(A \cap B) = \frac{2}{6} = \frac{1}{3}$$

and from Equation (5-16), $P(A|B) = 1/3 \div 1/2 = 2/3$. The probability that the upward face
of a die contains an odd number of dots is 2/3, given that it must be less than three. This
results can also be directly inferred from the Venn diagram of Figure 5-6. If event A occurs,
then one of the elementary outcomes in the left hand column must be the result. Of these
three elementary outcomes, two have fewer than or equal to three dots. Therefore, $P(A|B)$
= 2/3. Note that in this case, $P(A) \neq P(A|B)$.

Two events are statistically independent if the probability that one event occurs is
unaffected by whether or not the other event has occurred. This is formalized in the
following definition.

DEFINITION: STATISTICALLY INDEPENDENT EVENTS
Two events A and B defined over a sample space S are said to be independent if and
only if

$$P(A|B) = P(A) \quad \text{or equivalently} \quad P(B|A) = P(B) \tag{5-17}$$

If $P(A|B) \neq P(A)$, then events A and B are said to be dependent. For the two events A and
B defined for the single roll of a die, we have already shown that $P(A|B) = 2/3$ and $P(A)$

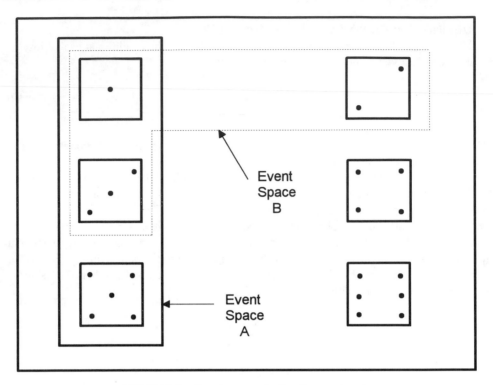

FIGURE 5-6. Event spaces for the die experiment.

= 1/2, so that these two events are statistically dependent. That is, knowing that the die roll is less than or equal to three affects the probability that it is odd. For events A and B defined in the card selection experiment, $P(A) = 13/52$, $P(B) = 4/52$ and $P(A \cap B) = 1/52$. From (5-16) it can be shown that $P(A|B) = 13/52 = 1/4 = P(A)$, so that these two events are independent. This implies that knowing that the outcome is an ace does not affect the probability of the card drawn being a spade.

Multiplication Theorem

The multiplication theorem is used to find the joint probability that two events occur. The multiplication theorem can be derived by rearranging Equation (5-16), the definition of a conditional probability.

MULTIPLICATION THEOREM

For any two events A and B defined over a sample space S,

$$P(A \cap B) = P(A|B) \cdot P(B) \quad \text{or equivalently} \quad P(A \cap B) = P(B|A) \cdot P(A) \quad (5\text{-}18)$$

Here $P(A \cap B)$ is the probability that both events A and B occur. For events A and B of the card selection experiment,

$$P(A \cap B) = P(A|B) \cdot P(B) = \frac{1}{4} \cdot \frac{1}{13} = \frac{1}{52}$$

The probability of getting both an ace and a spade is $1/52$. The single elementary outcome satisfying both event definitions is the ace of spades. This can also be verified by examining the Venn diagram for this experiment, shown as Figure 5-2. For the two events of the experiment calling for the roll of a single die, $P(A \cap B) = P(A|B) \cdot P(B) = 2/3(1/2) = 1/3$. Two of the six elementary outcomes in this experiment are both odd and have fewer than three dots.

For independent events, the multiplication theorem can be simplified. Since $P(A|B) = P(A)$ for statistically independent events, Equation (5-18) can be reduced to

$$P(A \cap B) = P(A) \cdot P(B) \qquad (5\text{-}19)$$

The multiplication theorem for independent events expressed as Equation (5-19) can be generalized to include n independent events. For three events, the joint probability is given by

$$P(A \cap B \cap C) = P(A) \cdot P(B) \cdot P(C) \qquad (5\text{-}20)$$

5.5. Illustrative Probability Problems

To solve probability problems, we frequently must resort to the event space–sample space rule (5-7), the counting rules presented in Section 5.3, and the compounding rules presented in Section 5.4. Sometimes two different methods can be used to solve the same problem. The following problems illustrate the use of these techniques in answering probability questions.

EXAMPLE 5-3. A local instrument supply firm received a shipment of 40 net pyra-diometers and sold two to the local university for use in its climatology laboratories. A month after the instruments arrived at the university, the supply firm received a letter from the manufacturer stating that eight of the 40 instruments were incorrectly calibrated. What is the probability that the university purchased two of the incorrectly calibrated instruments?

Solution. Let event A be fined as "both pyradiometers purchased by the university are incorrectly calibrated." It is possible to rewrite event A as the compound event $A_1 \cap A_2$ where event A_1 is "the first pyradiometer is incorrectly calibrated" and event A_2 is "the second pyradiometer is incorrectly calibrated." $P(A)$ can be determined in two ways. First, $P(A)$ can be computed using Equation (5-7) and by determining the number of elementary outcomes in the sample space n, and the number in the event space, m, from counting rules. The number of elementary outcomes in the sample space is simply the number of ways of selecting two pyradiometers from the shipment of 40:

$$n = C_2^{40} = \frac{40!}{2!(40-2)} = \frac{40 \cdot 39}{2} = 780$$

The set of pyradiometers from which the two purchased by the university were chosen can be divided into two types—correctly calibrated and incorrectly calibrated instruments. We can therefore use the hypergeometric rule to determine the number of ways of getting two incorrectly calibrated instruments in the university order.

$$m = C_2^8 \cdot C_0^{32} = \frac{8!}{2!(8-2)!} \cdot \frac{32!}{0!(32-0)!}$$

$$= \frac{8(7)}{2} \cdot (1) = 28$$

Here C_2^8 is the number of ways of drawing two defective instruments from the eight instruments known to be incorrectly calibrated. This is multiplied by C_0^{32}, the number of ways of drawing zero defective pyradiometers from the remaining 32. Therefore, $P(A) = m/n = 2/270 = .036$.

The second way of finding $P(A)$ is to use the multiplication rule, Equation (5-18), on the compound event $A_1 \cap A_2 : P(A_1 \cap A_2) = P(A_1) \cdot P(A_1|A_2)$. Notice that these two events are not independent, and so the conditional probability must be used. If the first delivered instrument is not correctly calibrated, then this fact affects the probability that the second instrument is not correctly calibrated (since there will be one less instrument to choose from, and one less defective instrument. The required probability is therefore

$$P(A_1 \cap A_2) = P(A_1) \cdot P(A_1|A_2) = \frac{8}{40} \cdot \frac{7}{39} = .036.$$

Note that the probability of selecting the second incorrectly calibrated instrument is different from the first. Both this method and the method based on the event space–sample space ratio gives the same result. There is often more than one way to attack many probability problems.

EXAMPLE 5-4.   In a survey of landscape preferences, respondents were asked to express a preference for one of two landscape photographs. The gender of the respondent was recorded at the time of the interview. The results of the survey are summarized as follows:

	Prefer landscape 1	Prefer landscape 2	Total
Male	15	5	20
Female	25	5	30
Total	40	10	50

For example, it was found that 20 respondents were male, 40 preferred landscape photograph 1, and 25 were female *and* preferred landscape photograph 1. This table provides a convenient format for counting the elementary outcomes of an experiment in a sample space. The experiment consists of selecting one of the respondents from the survey at random and recording the gender and landscape preference.

Let event A be that the gender of the chosen respondent is male, and let event $\bar{A}$ be that the gender is female. Similarly, let event B be that the chosen subject expresses a preference for landscape 1, and event $\bar{B}$ is a preference for landscape 2. Determine $P(A)$, $P(B)$, $P(A \cap B)$, and $P(B|A)$.

Solution. There are 50 elementary outcomes in the experiment—the 50 individuals whose preferences and gender are cross-classified in the table. To find the event probabilities defined using Equation (5-7), it is necessary to calculate the number of elementary outcomes in the appropriate event space, m. First, $P(A) = 20/50 = .4$ since 20 of the respondents are male. By the complementation theorem, $P(\bar{A}) = 1 - .4 = .6$. Similarly, $P(B) = 40/50 = .8$ and $P(\bar{B}) = 1 - .8 = .2$.

The probability of the compound event $P(A \cap B)$ refers to the probability that a randomly selected respondent will be both male *and* prefer landscape photograph 1. The number of elementary outcomes in this event space is $m = 15$, so $P(A \cap B) = 15/50 = .3$. The conditional probability $P(B|A)$ refers to the probability that the selected respondent prefers landscape 1, given that the respondent is a male. All the information necessary to compute this probability is contained in the first row of the table. There are 20 males, 15 of whom prefer landscape photograph 1. Therefore $P(B|A) = 15/20 = .75$. This conditional probability could also have been computed using the definitional formula

$$P(A|B) = \frac{P(A \cap B)}{P(A)} = \frac{.3}{.4} = .75$$

Again, two approaches to solving a probability problem lead to an identical result. Inferential questions about the relationships between two variables cross-tabulated in this way are discussed extensively in Chapter 11. For example, we might be interested in knowing whether the differences in preferences between the genders was significant, or whether one type of landscape is generally preferred to another type.

EXAMPLE 5-5. The proportion of people living in various regions of a country and owning their homes is given in the following table:

Region	Proportion of population	Proportion Owning homes
East	0.4	0.2
West	0.2	0.4
North	0.3	0.8
South	0.1	0.6

What is the probability that a person randomly selected from the population will own his or her home?

Solution: Let $P(A)$, $P(W)$, $P(N)$, and $P(S)$ be the probability that the person selected at random is a resident of the east, west, north, and south, respectively. These values can be read directly from the table as .44, .19, .28, and .09. The probability that the randomly selected person lives in the east and owns her home is $P(E) \cdot P(O|E)$, where $P(O|E)$ is the probability the person owns a home given that he or she lives in the east. Thus $P(E) \cdot P(O|E)$ = .44(.2) = .088. Similarly, in the west, the required probability is $P(W) \cdot P(O|W)$ = .19(.4) = .076. In the north, the required probability is .28(.8) = .224, and in the south it is .09(.6) = .054. The probability that the randomly selected person owns a home is thus .088 + .076 + .224 + .054 = .442. These four probabilities can be added, because residence in each of the four regions must be considered to be mutually exclusive.

5.6. Summary

When it is used in statistics, the term *probability* is a measure of the likelihood of an event occurring. It is a measure having a value between zero and one. The basic concepts of probability are founded on the notions of a *statistical experiment*. To compute event probabilities, it is usual to determine the elementary outcomes of the experiment. These elementary outcomes should constitute the set of mutually exclusive and collectively exhaustive outcomes of the experiment. Together, these outcomes represent the *sample space* of the experiment. If all elementary outcomes are equally likely, the probability that an event occurs in the experiment can be determined by taking the ratio of the number of outcomes in the *event space* to that of the sample space. The enumeration of the elementary outcomes in either the event or the sample space is often made easier by the application of one or more counting rules.

The identification of the probabilities associated with complex events can be determined once the relationship between each of the individual events that constitute it are known. Whether events are *mutually exclusive* or are *independent* are the two most important considerations in deciding which probability laws should be applied to solve a particular problem. Frequently, there are several different ways to approach any given probability question. In the next chapter, the basic concepts of probability are related to the procedures of statistical inference.

PROBLEMS

1. Explain the meaning of the following terms:
   a. Statistical experiment or random trial
   b. Sample space
   c. Elementary outcome of an experiment
   d. Event
   e. Complementary event
   f. Event space
   g. Event intersection and event union
   h. Relative frequency definition of probability
   i. Conditional probability

2. Differentiate between an *objective* and a *subjective* interpretation of the concept of probability. Give an example of each. Why are objective interpretations preferred in statistical analysis?

3. Draw a Venn diagram and illustrate the event space for the following problems:
   a. The experiment is the selection of a single card from a deck of cards; event A is the selection of a jack, event B is the selection of a jack *or* a queen, and event C is the selection of the king or queen of hearts.

b. The experiment is defined as the roll of a single die with the outcome being the number of dots showing on the upward face; event A is the roll yields a one or a two, event B is the roll yields an even number of dots, and event C is the number of dots is less than four.

4. Differentiate between events that are (a) mutually exclusive and those that are not and (b) statistically independent events and those that are not.

5. Residents of a city are asked to rank the desirability of five different neighborhoods A, B, C, D, and E. How many different ways can they be ranked, assuming there can be no ties?

6. A traveling salesperson beginning a trip at city A must visit (in order) cities X, Y, and Z before returning to city A. Several highway routes connect each pair of cities. There are four different ways of traveling between cities A and X, three routes between X and Y, five between Y and Z, and two between Z, and A. By how many different routes can the salesperson complete the entire trip?

7. Construct an outcome tree outlining the elementary outcomes for three tosses of a coin. What is the probability that three consecutive tails occur, assuming the coin is fair?

8. The probability that a student selected at random from an introductory course in geography at a certain university is male is .55. The probability that the student is unmarried is .80, and the probability that the student is both married and male is .05. If one student in the class is selected at random, calculate the probability that the student is (a) female, (b) married, (c) male or unmarried, and (d) female and unmarried. *Hint*: Set up an appropriate contingency table.

9. The license plates in a certain jurisdiction have six alphanumeric digits. The first digit is A, B, or C. The second digit is N, S, E, or W. The last four digits are restricted to the integers 0, 1, 2, $\ldots$, 9. How many different license plates are possible?

10. How many license plates are possible in a six-digit license plate system if (a) all six digits are letters, (b) all six digits are numbers, and (c) there are three letters and three numbers?

11. Given $P(A) = 1/3$, $P(B) = 2/3$ and $P(A \cup B) = 1/3$, (a) find $P(A \cap B)$, (b) find $P(B|A)$, and (c) draw a Venn diagram of the two events. (d) Are events A and B independent?

12. Draw Venn diagrams for the following sets of events and develop the addition rule for the problem.
    a. Of the three events A, B, and C, only A and C are not mutually exclusive.
    b. Of the three events A, B, and C, only B and C are not mutually exclusive.
    c. All events A, B, and C are not mutually exclusive when they are taken together as a group of three or in any pair.

13. The probability that it rains on a given day in July is .10.
    a. Assuming independent trials, what is the probability that it does not rain for three consecutive days?
    b. Again assuming independent trials, what is the probability that one day of rain is followed by two nonrainy days?
    c. Is the assumption of independent trials in this experiment reasonable?

14. A retail geographer surveys 200 shoppers after each has visited one of three shopping centers A_1,

A_2, and A_3. He records whether each made a purchase, with B_1 = yes and B_2 = no. The survey results are as follows:

Center	B_1	B_2	Total
A_1	25	25	50
A_2	10	50	60
A_3	65	25	90
Total	100	100	200

a. Find $P(A_3)$.
b. Find $P(A_1 \cup B_2)$.
c. Find $P(A_2 \cup B_1)$.
d. Find $P(A_1 \cap' A_3) \mid B_2)$.
e. Explain in your own words the meanings of parts (a) to (d).

15. Given events A and B where $P(A)$ = .30 and $P(B)$ = .50 and $P(B|A)$ = .40, find (a) $P(A|B)$ and (b) $P(A \cap B)$.

6

Random Variables and
Probability Distributions

In Chapter 5, the basic concepts of probability were developed in considerable detail. Although most people are familiar with notions of probability in games of chance, calling heads or tails in a coin toss and winning or losing a bet, the fundamental concepts of probability are frequently quite perplexing and difficult to comprehend on first exposure. Definitions of statistical experiments, statistical independence, and mutual exclusivity seem far removed from the context of these games of chance. Yet we know that probability theory lies at the core of statistical inference because all statistical judgments are necessarily probabilistic. The purpose of this chapter is to help clarify the link between statistical judgments, on the one hand, and probability theory, on the other.

Central to these arguments is the notion of a *random variable*. In Section 6.1 a random variable is formally defined and related by example to the process of population sampling. We also differentiate between the two major classes of random variables—those that are discrete and those that are continuous. Sections 6.2 and 6.3 are devoted to the specification of several probability distribution models for these two classes of random variables. In Section 6.4, we extend our knowledge of random variables and probability distributions to include *bivariate* random variables and bivariate probability distributions. Bivariate probability distributions describe the *joint* distribution of *two* variables. The various models are summarized and compared in Section 6.5.

6.1. Concept of a Random Variable

Suppose we consider any measurable characteristic of the units that constitute some typical population. Because this characteristic can assume one or more values in the population, we normally refer to it as a variable. To appreciate the concept of a *random variable*, let us consider a simple example. Assume that a population consists of eight different households and that the variable under consideration is the total number of persons in the household. If we select one household at random and note its size, then we could consider the variable called size of household to be a random variable. Thus, in Table 6-1, we treat the eight persons as elementary outcomes from a random sampling experiment. The values of the random variable called size of household are limited to two, three, four, and five. The

TABLE 6-1
The Concept of a Random Variable

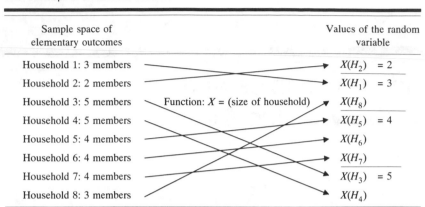

Sample space of elementary outcomes	Values of the random variable
Household 1: 3 members	$X(H_2)$ = 2
Household 2: 2 members	$X(H_1)$ = 3
Household 3: 5 members Function: X = (size of household)	$X(H_8)$
Household 4: 5 members	$X(H_5)$ = 4
Household 5: 4 members	$X(H_6)$
Household 6: 4 members	$X(H_7)$
Household 7: 4 members	$X(H_3)$ = 5
Household 8: 3 members	$X(H_4)$

sampling experiment defines one and only one numerical value for each household. If the first household is selected, then the value of the random variable is three; if the second household is chosen, it is two; and so on. The variable is random not because a household makes a random decision to include a certain number of persons, but *because our sampling experiment selects a household randomly.*

Let us define the random variable as X and denote any particular value that the random variable make take as x.

DEFINITION: RANDOM VARIABLE
A random variable is any numerically valued function that is defined over a sample space.

A random variable always has numerical values and probabilities associated with these values. The term *probability distribution* or *probability function* refers to a listing or graph of all values of a random variable and their corresponding probabilities.

DEFINITION: PROBABILITY DISTRIBUTION OR FUNCTION
A table, graph, or mathematical function that describes the potential values of a random variable X, and their corresponding probabilities is a probability function.

As noted in this definition, there are several ways of presenting the probability function: (1) in tabular form as in Table 6-2; (2) in graphical form such as Figure 6-1; and (3) as a mathematical formula such as Equation (6-7). The probabilities of Table 6-2 are relative frequency probabilities of the random variable size of household for the population described in Table 6-1. The graph in Figure 6-1 is constructed from the same data. Henceforth, we will use the notation $P(X)$ to denote the probability distribution of the random variable X and $P(X = x)$ or $P(x)$ to denote the probability that the random variable X takes on any particular value x. For example, $P(2)$ is the probability that $X = 2$, in this case the probability

TABLE 6-2
Relative Frequency Probabilities for the
Random Variable Called Household Size
of Table 6-1

x_i	$P(x_i)$ or $P(X = x_i)$
2	1/8 = .125
3	2/8 = .250
4	3/8 = .375
5	2/8 = .250
	1.0

that the household size is two persons. Because the population values are limited to two, three, four, and five, the probability distribution is also limited to these values. Of course we know that other household sizes are possible, but these sizes are not represented in this set of households.

Classes of Random Variables

There are two major classes of random variables, discrete and continuous.

DEFINITION: DISCRETE RANDOM VARIABLE
A random variable is said to be discrete if the set of values it can assume is finite or countably infinite.

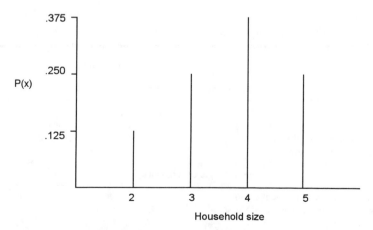

FIGURE 6-1. A probability mass function.

DEFINITION: CONTINUOUS RANDOM VARIABLE
A random variable is said to be continuous if the set of values it can assume constitute the real number line, or an interval on the real line.

One important class of discrete random variables are *counting variables*, a variable restricted to the value of zero and the positive integers, 1, 2, . . . , ∞. Counting variables typically arise when we are observing or recording the number of occurrences of some event in a particular time or space interval. The number of shoppers patronizing some store, the number of seedlings germinating in a given experimental plot, and the number of accidents at a specific road intersection are all examples of counting variables.

Many variables analyzed in physical geography are continuous. Temperature, weight, speed and distance theoretically are all continuous variables, even if the instruments used to record them can only display a finite number of alternatives. For example, a digital thermometer may only be able to display integer-valued temperatures.

Discrete Random Variables

Table 6-3 lists several different examples of discrete random variables. Size of household, gender, and many rating scales can all be considered discrete random variables if the particular value taken by the random variable is governed by a random sampling experiment. The outcome of a coin toss experiment, the roll of a six-sided die, and the selection of a playing card from a shuffled deck are also discrete random variables, since they result from statistical experiments.

In many instances, the discrete outcomes of a random variable are qualitative. To be random variables, each must be assigned numerical values because a random variable is by definition a numerically valued function. The actual numerical assignments to the qualitative classes is purely arbitrary, though very simple assignments are the most often utilized. For example, Table 6-3 indicates that we could assign the numerical values male

TABLE 6-3
Examples of Discrete Random Variables

Random variable X	Values of random variable
From random sampling	
Size of household	1, 2, 3 . . .
Gender	0, 1; or 1, 2 or any two integers
Rating scale (strongly disagree, disagree, neutral, agree, strongly agree	1, 2, 3, 4, 5 or −2, −1, 0, 1, 2
From simple experiments	
Number of heads in N tosses of a coin	0, 1, 2, . . . , N
Number of dots on the upward face of a die on a single row	1, 2, 3, 4, 5, 6

= 0 and female = 1 or male = 1 and female = 2 to the variable called gender. The 0-1 distinction is more common. For rating scales, the scale $-2, -1, 0, 1, 2$ is useful, since negative attitudes are associated with negative values, zero with neutral attitudes, and positive attitudes with positive values.

The probability distribution for a discrete random variable is specified by a *probability mass function*. This function assigns probabilities to the values taken by the discrete random variable. Formally, we say that a probability mass function assigns probabilities to the k discrete values of a random variable X with the following two provisions:

$$0 \le P(x_i) \le 1 \qquad i = 1, 2, \ldots, k$$

$$\sum_{i=1}^{k} P(x_i) = 1$$

where x_i, $i = 1, 2, \ldots, k$ are the k different values of the discrete random variable. The first condition simply requires that all probabilities be nonnegative and less than one. The second condition requires the sum of all probabilities to be equal to one. The probability mass function for the variable called household size illustrated in Figure 6-1 and summarized in Table 6-2 satisfies both conditions. Why is the function called a mass function? It is called a *mass function* because the probabilities are *massed* at discrete values of the random variable, not spread across an interval.

We can also portray probability mass functions in cumulative form in much the same way that frequency tables and histograms can be cumulated. For a discrete random variable with values ordered from lowest to highest at $x_1 < x_2 < x_3$, the cumulative mass function $F(x_i) = P(X \le x_i)$ is determined by summing the probabilities that $X = x_1, X = x_2, \ldots, X = x_i$. That is,

$$F(x_i) = \sum_{x=x_1}^{x=x_i} P(x) = P(x_1) + P(x_2) + \ldots + P(x_i) \qquad (6\text{-}1)$$

Consider again the discrete random variable called household size of Table 6-2. To cumulate these values, note $F(2) = P(X = 2) = .125$; $F(3) = P(X = 2) + P(X = 3) = .125 + .250 = .375$; and so on. The complete cumulative mass function is listed in Table 6-4 and illustrated in Figure 6-2. Their construction and display mirrors that of the cumulative frequency table and cumulative frequency histogram.

Just as an empirically derived variable has a mean value and variance, so too does

TABLE 6-4
Cumulative Mass Function

x	$P(x_i)$	-	$F(x_i)$
2	0.125		0.125
3	0.250		0.375
4	0.375		0.750
5	0.250		1.000

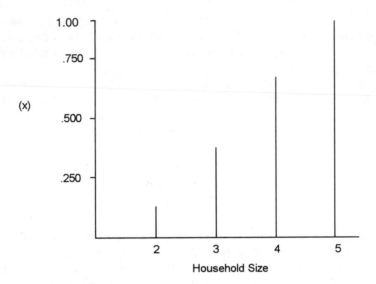

FIGURE 6-2. A cumulative probability mass function.

a random variable. The average of the values of a random variable is known as its *expected value.*

DEFINITION: EXPECTED VALUE OF A DISCRETE RANDOM VARIABLE

For a discrete random variable with values $x_1, x_2, \ldots, x_k$, the expected value of X, denoted $E(X)$ is defined as

$$E(X) = \sum_{i=1}^{k} P(x_i) \cdot x_i \qquad (6\text{-}2)$$

For the household size data we find the following:

x_i	$P(x_i)$	$x_i\, P(x_i)$
2	0.125	0.250
3	0.250	0.750
4	0.375	1.500
5	0.250	1.250
		3.750

Thus, the mean or expected value of household size is 3.75. Note that this need not be one of the original discrete values of the variable X. Equation (6-2) is a "weighted" mean where the weights are defined as the relative frequencies or probabilities of each discrete value.

The variability of a discrete random variable is measured by the *variance $V(X)$*. The standard deviation is defined as the square root of the variance. For a discrete random variable, then

DEFINITION: VARIANCE $V(X)$ OF A DISCRETE RANDOM VARIABLE

For a discrete random variable, the variance of X, denoted $V(X)$ is defined by the following equation:

$$V(X) = \sum_{i=1}^{k} [x_i - E(X)]^2 P(x_i) = \sum_{i=1}^{k} x_i^2 P(x_i) - [E(x)]^2$$

The second form for the variance given in $V(X)$ is a more efficient computational formula. For the variable called household size,

x_i	$P(x_i)$	x_i^2	$x_i^2(P(x_i))$
2	0.125	4	0.500
3	0.250	9	2.250
4	0.375	16	6.000
5	0.250	25	6.250
			15.000

and $V(X) = 15.000 - (3.75)^2 = 15.000 - 14.0625 = .9375$. The standard deviation is calculated as the square root of the variance: $\sqrt{.9375} = .9682$.

Continuous Random Variables

A random variable is said to be continuous if it can assume all the real number values in some interval of the real number line. The number of different values that a continuous random variable can assume is therefore infinite. Table 6-5 lists several different variables that can be considered to be continuous random variables if they are selected by some random-sampling procedure. For example, if annual rainfall totals are available from 1000 meteorological stations in North America, then annual rainfall can be considered a random variable if we randomly select one of these stations. This is not to say that the amount of rainfall at any station is determined randomly! It is the random sampling procedure that

TABLE 6-5
Example of Continuous Random Variables

Random variable X	Values of random variable
Annual rainfall	$X \geq 0$
Distance traveled to some facility	$X \geq 0$
Stream discharge	$X \geq 0$
pH value of a water sample	$0 \leq X \leq 14$
Pseudo-continuous variables	
Population of city, region, or census tract	$X \geq 0$
Interaction between two places	$X \geq 0$

introduces the notion of randomness, not anything related to the process that causes variation in the observed values of the variable.

In addition, many discrete variables are often modeled as continuous variables. For example, the population of a city or region or census tract can be treated as if it were a continuous variable even though, in the strictest sense, it is a discrete variable. This is a common simplification for many discrete variables with large ranges.

The probability distribution of a random continuous variable is represented by a *probability density function.* Let X be a continuous random variable defined over some interval of the real number line, say from a to b. A probability density function of X denoted $f(X)$, satisfies two conditions:

1. $f(x) \geq 0$ for $a \leq x \leq b$.
2. The area under $f(x)$ from $x = a$ to $x = b$ must be equal to 1.

These two conditions are formally identical to the two conditions defined for discrete random variables. In the discrete case, the probabilities are said to be *massed* at the discrete values of the random variable. In the continuous case, we say the probability is *spread densely* over the range of the random variable.

The probability density function for a random continuous variable is illustrated as Figure 6-3. Just as heights of the vertical bars in a probability mass function of a discrete variable sum to one, *the area under a probability density function must also equal one.* It is possible to define random continuous variables over an infinite-length real number line, that is from $-\infty$ to $+\infty$. If $f(x)$ is asymptotic to the x-axis as it approaches $-\infty$ and $+\infty$, then $f(x)$ may still define a random continuous variable if the area under $f(x)$ is 1. The most famous of all continuous distributions, the normal distribution, is defined for a continuous variable with potential values in the range $[-\infty, +\infty]$.

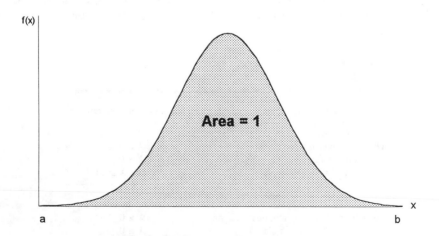

FIGURE 6-3. A probability density function.

How can we compute probabilities by using these probability distributions? Suppose we are interested in computing $P(c \leq x \leq d)$ for some random variable X. If X is discrete, then this is simply a matter of summing the probabilities of X taking on any discrete value between (and including) c and d:

$$P(c \leq x \leq d) = \sum_{x=c}^{x=d} P(x) \qquad (6\text{-}4)$$

For a continuous random variable, we use the convention that the probability that X takes on a value between c and d is equal to the area under the probability density function $f(x)$ between c and d. This is illustrated in Figure 6-4. Now, depending upon the shape of the probability density function, this may not be an easy matter. However, it is not necessary to know advanced calculus to comprehend the concept. For computational purposes, tables of probabilities are provided at the end of this text; they can be used to determine these values for all the continuous-valued probability distributions encountered in this book.

A simple example illustrates the procedure for computing probabilities of a continuous variable. Suppose the probability density function is given by

$$f(x) = \begin{cases} 0.5 & 0 \leq x \leq 2 \\ 0 & x < 0 \; or \; x > 2 \end{cases} \qquad (6\text{-}5)$$

as illustrated in Figure 6-5. The total area under this function can be calculated as the area of a rectangle with base = 2 and height = 0.5. The area under this function is $0.5 \times 2 = 1$.ote also that $f(x) \geq 0$ for all x. To calculate $P(0.5 \leq x \leq 1.5)$, simply find the area under the density function between these two limits. The base is $1.50 - 0.50 = 1.00$, and the height is 0.5, so the total area, and thus the probability, is $0.50 \times 1.0 = .50$. As illustrated in Figure 6-5, this is one-half the area under the probability density function. Unfortunately, it is not

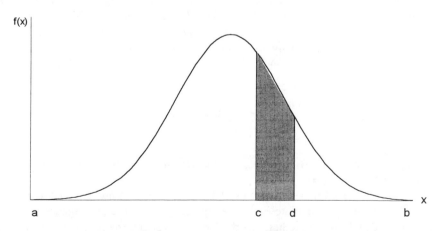

FIGURE 6-4. The probability that $c \leq x \leq d$.

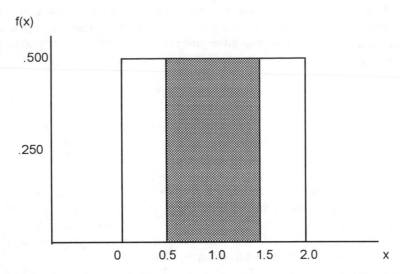

FIGURE 6-5. Computing a probability from a simple random continuous probability distribution.

always possible to use such simple arguments from elementary geometry for more complex probability density functions.

Although there are many similarities between continuous and discrete variables, there is one key difference. Consider the expression $P(X = r)$. For a discrete variable, this is simply the probability that the variable takes on the discrete value r. This probability can be determined directly from the probability mass function as the height of the bar graph for the value $X = r$. This operation is not possible for a continuous random variable. For a continuous random variable, $P(X = r) = 0$. Why? Assume that the variable X is defined on the real number line from $X = a$ to $X = b$. If we were to let $P(X = r) = f(r)$, the height of the density function at $X = r$, then this would have to apply to all other values in the interval $[a,b]$. Since there are an infinite number of such values, say $r_1, r_2, \ldots, r_n$, the sum of the probabilities $f(r) + f(r_1) + f(r_2) + \ldots + f(r_n)$ would undoubtedly exceed 1.0. It is therefore not possible to interpret $f(r)$ directly as a probability. In fact, there is nothing in the definition of a probability density function that prohibits any single value of $f(r)$ from exceeding one. The height of $f(r)$ is only constrained to be nonnegative. Consider the probability density function

$$f(x) = \begin{cases} 2 & 0 \leq x \leq 0.5 \\ 0 & x < 0 \ or \ x > 0.5 \end{cases} \tag{6-6}$$

Clearly, $f(0) = f(0.25) = f(0.5) = 2$. However, the area under the probability density function, $P(0 \leq x \leq 0.5) = 0.5 \times 2 = 1$.

There is another explanation of why $P(X = r)$ for any continuous variable must equal zero. Let us suppose we calculate $P(c \leq x \leq d)$ and derive some nonnegative probability. As we move c and d closer together, this probability must get smaller as the area under $f(x)$

declines. In the limit, $c = d$, and there is no area under the curve. The probability that a continuous random variable assumes a particular value must always be zero, regardless of the value of the underlying probability distribution $f(x)$.

Computing the expected value $E(X)$ and variance $V(X)$ of a continuous random variable is not nearly so straightforward as the computation for discrete variables. The derivation of these quantities requires a background in calculus well beyond the scope of this text. The expected value and variance of the probability distributions described in Section 6.3 are presented without derivations. Students with a background in calculus may wish to consult the Appendix at the end of the chapter for a more complete specification of the derivation of $E(X)$ and $V(X)$ for continuous variables.

The properties of continuous and discrete random variables are summarized in Table 6-6. Both types of variables must satisfy an identical set of basic properties including (1) nonnegativity and (2) the sum of probabilities equal to one. The major difference arises from the number of different values that the random variable can assume. For a discrete random variable, this is always a finite number. By definition, a continuous random variable can assume an infinite number of different values. This distinction leads to several important differences in the specification of the properties of these two classes of random variables.

6.2. Discrete Probability Distribution Models

Discrete random variables can be represented by a number of different probability mass functions. If a complete listing of the *population* of interest is available, then the relative frequency distribution is the appropriate probability mass function. Figure 6-1 represents the probability mass distribution for the variable called household size as distributed in the

TABLE 6-6
A Comparison of Continuous and Discrete Random Variables

	Discrete random variable	Continuous random variable
Number of values that can be assumed	Finite (or countably infinite) say k	Infinite, from a to b or $-\infty$ to $+\infty$
Graphical summary	Probability mass function as linear bar chart	Probability density function $f(x)$, area under curve
$P(X = a)$	$P(a)$	0
$P(c \leq x \leq d)$	$\sum_{x=c}^{d} P(x)$	$\int_{c}^{d} f(x)dx$
Conditions on $P(x)$ (1)	$0 \leq P(x_i) \leq 1$ $i = 1, 2, \ldots, k$	$f(x) \geq 0, a \leq x \leq b$ or $f(x) \geq 0, -\infty \leq x \leq +\infty$
(2)	$\sum_{i=1}^{k} P(x_i) = 1$	Area under $f(x) = 1$, that is $\int_{a}^{b} f(x)dx = 1$ or $\int_{-\infty}^{+\infty} f(x)dx = 1$

population listed in Table 6-1. For cases in which the available data are a sample from some larger population, this procedure is not feasible. Instead, it is usual to see how well the available data fit certain probability distribution models derived from specific statistical experiments. For discrete variables, three different probability distributions are often used in statistical analyses:

1. Discrete uniform distribution,
2. Binomial distribution,
3. Poisson distribution.

It is not unusual to find close parallels between the assumptions underlying the statistical experiments governing these distributions and the nature of some empirical statistical problem. In Chapter 11, methods are presented for testing the fit of an empirically derived frequency distribution to any probability distribution model.

Discrete Uniform Distribution

One of the simplest probability distributions is the discrete uniform distribution. It occurs frequently in games of chance. Examine the probability of getting a head or a tail in the single toss of a fair coin. Each outcome has an equal probability—0.5. In a single draw from a shuffled deck of playing cards, the probability of drawing a card of any suit is .25. Each of these experiments defines a discrete uniform random variable. In such a distribution, the probability of occurrence for each outcome or value of the random variable is equal. For two outcomes, each has a probability of .5; for three outcomes, each has a probability of .33; and so on.

The uniform distribution also has applications in statistical decision making in situations of complete uncertainty. Suppose, for example, someone is lost and arrives at an unmarked fork in the road. There is no reason to suspect that either road leads to the desired destination. Which road should be taken? In the absence of any other information, each road could be assigned an equal probability. A coin toss could be used to make the decision. Many other decision making situations are similar to this example. All involve the notion of complete uncertainty.

To formalize this probability mass function, assign a single integer from $1, 2, \ldots, k$ to each of the k different outcomes that can be assumed by the random variable X.

DEFINITION: DISCRETE UNIFORM DISTRIBUTION
If X is a discrete uniform random variable, then

$$P(x) = \frac{1}{k} \quad x = 1, 2, \ldots, k \tag{6-7}$$

That is, each state or value of the random variable has an equal (hence the name *uniform*) probability of occurrence. Figure 6-6(a) illustrates the general form of the probability mass function of a discrete uniform variable. Figure 6-6(b) and (c) illustrate the probability mass functions for the coin toss and card selection examples.

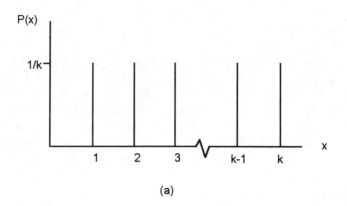

(a)

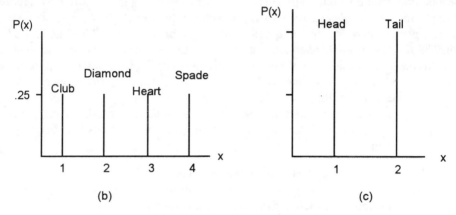

(b)   (c)

FIGURE 6-6.   Discrete uniform random variables.

What are the mean and variance of a discrete uniform variable? Using (6-2), we can calculate the expected value as

$$E(X) = \sum_{x=1}^{k} xP(x) = \sum_{x=1}^{k} x\left(\frac{1}{k}\right)$$ (6-8)

Now, since $1/k$ is a constant, it can be brought outside the summation (recall Rule 2 of the Appendix to Chapter 2). The expression $\sum_{x=1}^{k} x$ is the sum of the first k integers and can be simplified to $k(k+1)/2$. Therefore,

$$E(X) = \frac{1}{k}\left[\frac{k(k + 1)}{2}\right] = \frac{k + 1}{2}$$ (6-9)

Although it is somewhat more difficult to derive, the variance of a discrete uniform variable can be shown to be

$$V(X) = \frac{k^2 - 1}{12} \tag{6-10}$$

As an example, consider again the outcome of a single coin toss. If we let $x = 1$ be the result of getting a head and $x = 2$ the result of getting a tail, then $P(1) = P(2) = .5$ for the $k = 2$ potential values of the random variable. The mean or expected value is $(2 + 1)/2 = 3/2 = 1.5$. Note once again that this average is not one of the two values that can be assumed by X. It is midway between each. From a decision-making point of view, this tells us that neither a head nor a tail is a preferable choice in a coin toss. Using (6-3), we find that $V(X) = (3^2 - 1)/12 = .25$. This has no immediate interpretation.

Binomial Distribution

A second important distribution of a discrete random variable is the *binomial distribution*. It is applicable to variables having only two possible outcomes. A person is either male or female, a person answers a certain question yes or no, or the coin yields either a head or a tail. The simplest example of a binomial variable is the operation of a coin toss. Suppose we toss a coin n times and record the number of heads occurring in the n tosses. If the tosses are all independent events and the random variable X is defined to the number of heads occurring in the n tosses, then X is a binomial random variable.

A binomial random variable is produced by a statistical experiment known as a *Bernoulli* process, or a set of *Bernoulli* trials. A set of Bernoulli trials is defined by the following four conditions:

1. There are n independent trials of the experiment.
2. The same pair of outcomes is possible on all trials.
3. The probability of each outcome is the same on all trials.
4. The random variable is defined to be the number of "successes" in the n trials.

Let us relate this definition to the example of a coin toss experiment. Our interest lies in the number of heads, or "successes," that occur in an experiment in which a coin is tossed

TABLE 6-7
Possible Outcomes of the Coin Toss Experiment

$n = 1$				T		H			
$n = 2$			TT		TH HT		HH			
$n = 3$		TTT		TTH THT HTT		THH HTH HHT		HHH		
$n = 4$	TTTT		TTTH TTHT THTT HTTT		TTHH THTH HTTH THHT HTHT HHTT		THHH HTHH HHTH HHHT		HHHH

n times. The probability of a head, or a success, is labeled π, and the probability of a tail, or failure, is $1 - \pi$. Each trial is represented by one toss of the coin. Table 6-7 lists the possible outcomes from this experiment in which a coin is tossed $n = 1, 2, 3$, or 4 times.

Consider the first row of the table. There are only two possible outcomes. Either we get a head (H) or tail (T). We may get either zero heads or one head, depending on the outcome. The probability of no heads is $1 - \pi$, and the probability of one head is π. Since $\pi + (1 - \pi) = 1$, this is a legitimate probability distribution for a random variable. What if the coin is tossed twice? Now, there are three different possible outcomes: zero, one, or two heads. Although there is only one way of getting zero heads or two heads, there are two ways of obtaining one head. To get two heads requires a head on each of the $n = 2$ tosses. This is listed as outcome {HH}. Similarly, only one sequence leads to two tails, {TT}. But, to obtain one head, two different sequences are possible: {HT} and {TH}.

For $n = 3$ tosses there are now eight possible outcomes. For example, there are now three ways of obtaining one head. It can appear on the third, second, or first toss of the coin. These three outcomes are listed as {TTH}, {THT}, and {HTT}. Note that these three alternatives can be derived from the sequences for $n = 2$, which are listed immediately above. The sequence {TTH} results from the addition of a head to the sequence {TT}, and {THT} and {HTT} derive from the sequences {TH} and {HT} with the addition of a tail. Further, the three sequences with two heads—{THH}, {HTH}, and {HHT}—can be generated by the addition of the appropriate result to the sequences listed above as {TH}, {HT}, and {HH}. The entire list of outcomes in Table 6-7 is constructed in such as way that the sequences in any row can be generated from the immediately preceding row with the addition of the appropriate outcomes. The size of this triangle expands quite rapidly as the number of tosses increases.

One way of summarizing these results is to calculate the different number of sequences that lead to any outcome for a given number of tosses. For example, there are three ways of getting one head in $n = 3$ tosses. These numbers are termed the binomial coefficients and are displayed in Table 6-8 for $n = 1$ to $n = 6$ Bernoulli trials. This representation is known as *Pascal's triangle* in recognition of the role of the French mathematician Blaise Pascal in the development of these results. Note that any number in this table can be generated by adding the number of sequences above the entry in the immediately preceding row. The six in the fourth row is $3 + 3$, the sum of the two numbers directly above it. How are these binomial coefficients to be interpreted? Each entry represents *the number of different ways of obtaining exactly x heads in n tosses of a coin.* There are exactly six ways of obtaining three heads in four tosses of a coin. This is denoted as

TABLE 6-8
Pascal's Triangle of Binomial Coefficients

$n = 1$					1		1			
$n = 2$				1		2		1		
$n = 3$			1		3		3		1	
$n = 4$		1		4		6		4		1	
$n = 5$	1		5		10		10		5		1
$n = 6$	1		6		15		20		15		6		1

C_2^4, since there are this many combinations of heads and tails leading to the desired result. From Equation (6-8), C_2^4 is calculated as $4!/[2!(4 - 2)!] = 24/4 = 6$ combinations. These six combinations are listed in the third column of the $n = 4$ row of Table 6-7. Remember that the five numbers in the $n = 4$ row refer to the appearance of 0, 1, 2, 3, or 4 heads in the $n = 4$ tosses of the coin.

Now, consider the probability that any of these individual sequences occurs. We know that the probability of a head is π, the probability of a tail is $1 - \pi$, and the trials are independent. For example, what is the probability of each of the six sequences {TTHH}, {THTH}, {HTHT}, {THHT}, {HTHT}, and {HHTT}? Each contains $x = 2$ heads in $n = 4$ tosses. It is rather easy to show that the probabilities of each of these sequences are equal and are given by $\pi^2(1 - \pi)^2$. For the first sequence, the independence of trials allows us to rewrite this as

$$P(\text{TTHH}) = P(\text{T}) \cdot P(\text{T}) \cdot P(\text{H}) \cdot P(\text{H})$$

Since $P(\text{T}) = 1 - \pi$ and $P(\text{H}) = \pi$, this simplifies to $(1 - \pi) \cdot (1 - \pi) \cdot (\pi) \cdot (\pi)$ or, with rearrangement, $\pi^2 \cdot (1 - \pi)^2$. All sequences can be rearranged in this form.

Our ultimate concern is determining the probability of obtaining exactly x heads in n tosses of a coin. To continue the same example, what is the probability of obtaining exactly two heads in four tosses of a coin? Since there are six different sequences with exactly two heads, each with a probability of $\pi^2(1 - \pi)^2$, the required probability is $6\pi^2 (1 - \pi)^2$. The probability of obtaining x heads in n tosses of a coin for up to $n = 4$ trials is listed in Table 6-9. For example, for $n = 4$ and $x = 2$, the five coefficients are 1, 4, 6, 4, and 1. These are the entries of Pascal's triangle, Table 6-8, for the row $n = 4$. For this reason the entries in Pascal's triangle are known as the binomial coefficients. Also, the exponents of

TABLE 6-9
Binomial Distributions

Number of trials n	Number of heads	Probability of number of heads
1	0	$1 - \pi$
	1	π
2	0	$(1 - \pi)^2$
	1	$2\pi(1 - \pi)$
	2	π^2
3	0	$(1 - \pi)^3$
	1	$3\pi(1 - \pi)^2$
	2	$3\pi^2(1 - \pi)$
	3	π^3
4	0	$(1 - \pi)^4$
	1	$4\pi(1 - \pi)^3$
	2	$6\pi^2(1 - \pi)^2$
	3	$4\pi^3(1 - \pi)$
	4	π^4

π is the number of heads and the exponent of $1 - \pi$ is the number of tails. In general, the probability of obtaining x heads in n tosses of a coin is equal to the product of the number of sequences of length n with x heads and the probability that the sequence contains x heads with $n - x$ tails. Let us now formally define the probability mass function for a binomial random variable.

DEFINITION: BINOMIAL RANDOM VARIABLE

If X is a binomial random variable, then the probability mass function is given by

$$P(x) = C(n, x)\pi^x(1 - \pi)^{n-x} \quad x = 0, 1, 2, \ldots, n \tag{6-11}$$

where n is the number of trials, x is the number of successes and π is the probability of a success in each trial.

The specific form of the binomial distribution depends on the values of the two parameters π and n. Given values of π and n, we can easily compute the probability for any number of successes using Equation (6-11). If $\pi = .2$ and $n = 4$ and we wish to compute the probability of three successes, then by Equation (6-11), $P(3) = C(4,3)(.2)^3(.8)^1 = 4(.008)(.8)$ = .0256. Figure 6-7 illustrates the probability mass functions for four members of the *family* of binomial probability distributions. If the coin we are tossing is a fair coin, then $\pi = .5$ and the distribution of successes is always symmetric. If $\pi < .5$, then the distribution is negatively skewed; if $\pi > .5$ the distribution is positively skewed. When n is very small, the probability mass function for a binomial random variable (Equation 6-11) can be used directly to compute any probabilities required for a specific problem. But when n is large, it is more convenient to consult statistical tables that compile binomial probabilities for many different values of π and n. Statistical tables for a few members of the binomial family are included at the end of this text.

Using Equations (6-3) and (6-3), it is possible to determine the mean and variance of a binomial probability distribution

---

DEFINITION: MEAN AND VARIANCE OF A BINOMIAL RANDOM VARIABLE

If X is a binomial random variable, then the mean and variance of X are given by

$$E(X) = \sum_{x=0}^{n} xC(n, x)\pi^x(1 - \pi)^{n-x} = n\pi \tag{6-12}$$

$$V(X) = \sum_{x=0}^{n} [x - E(X)]^2 C(n, x)\pi^x(1 - \pi)^{n-x} = n\pi(1 - \pi) \tag{6-13}$$

EXAMPLE 6-1: Suppose a student taking a course in statistics is required to take an examination consisting of 10 multiple choice questions. The questions are designed in such a way that the probability of correctly guessing the answer to any question is .2, as they consist of five equally plausible choices. If the student can *only* guess at the answers to each question (never attended lectures, labs, nor read the textbook), what is the expected number of correct answers? What is the standard deviation of the number of correct answers? Assuming a grade of 50 percent is required to pass the test, what is the probability that the student passes?

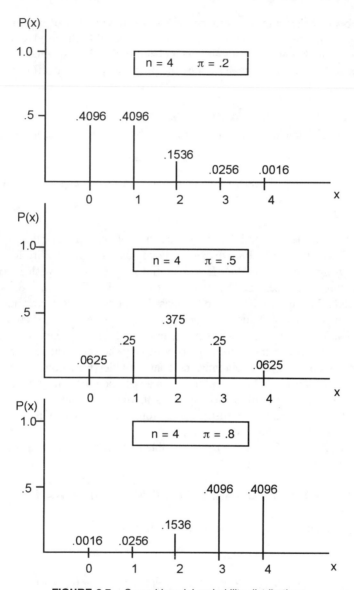

FIGURE 6-7. Some binomial probability distributions.

Solution. Let the binomial random variable X be the number of correct answers on the test. There are $n = 10$ trials, and it is known that $\pi = .2$. The appropriate probability mass function is

$$P(x) = C(10, x)(.2)^x(.8)^{10-x} \qquad x = 0, 1, \ldots, 10 \qquad (6\text{-}14)$$

where x is the number of correct answers. The expected number of correct answers is obtained from (6-12):

$$E(X) = n\pi = 10(.2) = 2 \tag{6-15}$$

and $V(X)$ is calculated from (6-13):

$$V(X) = n\pi(1 - \pi) = 10(.2)(.8) = 1.6 \tag{6-16}$$

Therefore, the standard deviation is $\sqrt{1.6} = 1.26$ correct answers. To determine the probability that the student passes the test, it is necessary to calculate $P(5) + P(6) + P(7) + P(8) + P(9) + P(10)$, since all these outcomes lead to passing grades:

$$P(5) = C(10, 5)(.2)^5(.8)^5 = .026$$
$$P(6) = C(10, 6)(.2)^6(.8)^4 = .006$$
$$P(7) = C(10, 7)(.2)^7(.8)^3 = .001$$
$$P(8) = C(10, 8)(.2)^8(.8)^2 = .000$$
$$P(9) = C(10, 9)(.2)^9(.8)^1 = .000 \quad \left.\right\} \text{(to three places)}$$
$$P(10) = C(10, 10)(.2)^{10}(.8)^0 = .000$$

The sum of these probabilities is .033, and thus there is a 3.3% chance a student could pass this test just by guessing.

In our discussion of the binomial distribution, we assumed that the values of the two parameters π and n are known. With these values we can specify the appropriate probability mass function, calculate the expected value and variance, and solve many probability problems. What if they are not known? Then it is necessary to estimate these parameters from some sample data. We see how this problem in estimation is solved in Chapter 8. In Chapter 11 we see how to solve the related problem of testing the fit of the binomial distribution to sample data. Such methods are extremely useful for determining which particular probability mass function best describes empirical data that we wish to analyze.

Poisson Probability Distribution

The final discrete distribution discussed in this chapter is the *Poisson probability distribution*. This distribution proves to be very important in the analysis of geographic point patterns. Point pattern analysis involves a variety of techniques that describe the spatial distribution of phenomena represented by points on the traditional dot map. Figure 6-8 illustrates a typical dot map. Each dot represents the location of one observation of some item of interest. These dots might represent settlements, individual dwelling units, households of some particular ethnic group, or specific geographic features such as drumlins, erratics, or members of specific plant species. To derive the Poisson probability distribution and to develop the notion of a Poisson random variable, consider the analysis of such a dot map. Let us suppose that the dots represent drumlins and that these drumlins are randomly distributed over the area depicted on the dot map. Suppose it is also known that there is, on average, one drumlin per square mile.

One way of characterizing this dot map is to record the number of dots appearing in *quadrats* of the map. Quadrats are simply areal subdivisions of a map. Usually they are regular in size and square. The region depicted in Figure 6-8 is divided into 100 quadrats. Initially, let us analyze this map as a Bernoulli process in which the examinations of

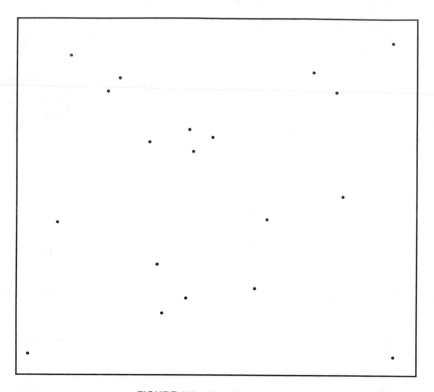

FIGURE 6-8.   A typical dot map.

quadrats are treated as "trials" of a statistical experiment. When examining a quadrat, we define a success as the existence of a drumlin and a failure as the absence of a drumlin. Because there is, on average, one drumlin per square mile, the expected number of drumlins in any quadrat is one.

Now suppose each 1-mi² quadrat is divided into four equal-sized 0.25-mi squares as in Figure 6-9(a). Also, let us make the additional assumption that no more than one drumlin can occur in any 0.25-mi quadrat. Let us examine these four 0.25-mi squares as a set of Bernoulli trials. Each quadrat either contains a drumlin or does not. Because the expected number of drumlins per square mile is one, the probability that there is a drumlin in any of the four quadrats is 1/4. Examining each of these quadrats as four trials of a Bernoulli process leads to five possible outcomes—there can be zero drumlins, one drumlin, or two, three, or four drumlins in these four quadrats. Clearly this is a binomial random variable. The probability that the four quadrats contains exactly two drumlins is

$$P(2) = \frac{4!}{2!(4-2)!} \left(\frac{1}{4}\right)^2 \left(\frac{3}{4}\right)^2 = .2109375 \tag{6-17}$$

This expression is derived by the application of Equation (6-11) with $\pi = 1/4$, $1 - \pi = 3/4$, $n = 4$, and $x = 2$.

To use Equation (6-11), we must make the assumption that the probability we might find two drumlins (or even more) in any 0.25-mi quadrat is zero. This violates the assumption that the distribution of points on the map is random. If the distribution is truly random, then there must be some probability that any quadrat could contain two, three, or even all dots of the map. To get a closer approximation, let us divide these four 0.25 mi quadrats into 16 quadrats of 1/8 mi square, as in Figure 6-9(b). This time we assume that there can be at most one drumlin in each of these 16 quadrats. The possible outcomes of the set of Bernoulli trials based on these 16 quadrats are 0, 1, 2, . . . , 16 drumlins. The probability of exactly two drumlins is

$$P(2) = \frac{16!}{2!(16-2)!} \left(\frac{1}{16}\right)^2 \left(\frac{15}{16}\right)^{14} = .1899 \tag{6-18}$$

since for this set of trials $\pi = 1/16$, $1 - \pi = 15/16$, $n = 16$, and $x = 2$.

The estimate of $P(2)$ given by Equation (6-18) is more accurate than Equation (6-17) since the probability of more than one drumlin in any quadrat is smaller. This is because the quadrats themselves are smaller. To get an even better estimate, let us divide our 16 quadrats into 64 quadrats, as in Figure 6-9(c). The probability that any $1/8$-mi^2 quadrat contains exactly two drumlins is given by

$$P(2) = \frac{64!}{2!(64-2)!} \left(\frac{1}{64}\right)^2 \left(\frac{63}{64}\right)^{62} = .1854 \tag{6-19}$$

Because 64 quadrats are a better basis for the approximation than 16 quadrats, we might ask what would happen if we examined this process as the number of quadrats approached infinity, and the probability that any quadrat contained a drumlin approaches zero. Examining the sequence of our estimates $P(2)$ with increasing trials, or equivalently with increasingly smaller quadrats, we note the following:

n	4	16	64
$P(2)$	0.2109	0.1899	0.1854

To determine the expression for $P(2)$ for an infinite number of trials requires an evaluation of the limit as n approaches infinity of the following expression:

$$P(2) = \frac{n!}{2!(n-2)!} \left(\frac{1}{n}\right)^2 \left(\frac{n-1}{n}\right)^{n-2} \tag{6-20}$$

Although some calculus is required to derive the result, it can be shown that the limit converges to

$$P(2) = \frac{e^{-1}(1)^2}{2!} = .1839$$

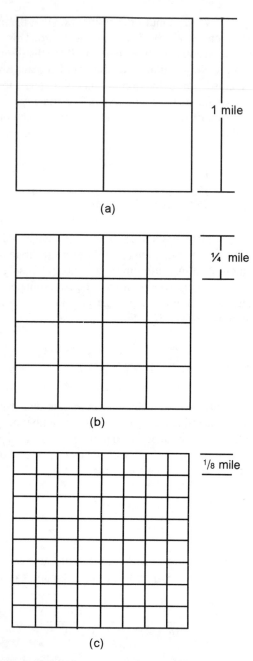

FIGURE 6-9. Subdividing quadrats of a map.

where e is the base of the natural logarithms, or 2.71828. The limit of $P(2)$ thus converges to .1839, and the increasing accuracy of our estimate with the increasing number of quadrats is clearly illustrated.

This result can now be extended to determine the limits for $P(0)$, $P(1)$, $P(2)$, $P(3)$ and the probabilities for the rest of the positive integers. Let λ be the average number of drumlins per unit area. In the example above, $\lambda = 1$ drumlin per square mile. The random variable X is the number of drumlins found in a randomly selected quadrat, and X may take on the values $x = 0$, $x = 1$, $x = 2$, or any other positive integer. The probability that a quadrat contains exactly x drumlins is a Poisson random variable.

DEFINITION: POISSON RANDOM VARIABLE

If X is a Poisson random variable, then the probability mass function is given by

$$P(x) = \frac{e^{-\lambda}\lambda^x}{x!} \qquad x = 0, 1, 2, \ldots \qquad (6\text{-}22)$$

There is only one parameter for this distribution, λ.

To generalize, the experiment generating a Poisson random variable is described in the following way. Let the random variable X be the number of occurrences $x = 0, 1, 2, \ldots$ of some specific event in a given continuous interval. The interval can be a time interval, a spatial interval (such as a quadrat), or even the length of some physical object such as a rope or a chain. The random variable X is a Poisson random variable if the experiment generating the values of X satisfies the following conditions:

1. The number of occurrences of the event in two mutually exclusive intervals is independent.
2. The probability of an occurrence of the event in a small interval is small and proportional to the length of the interval, i.e., the event is rare.
3. The probability of two or more occurrences of the event in a small interval is near zero.

For the point pattern problem, the event is the existence of a dot in the quadrat. Because the pattern is random, the number of dots in any two quadrats is independent. Finally the probability of two or more dots in one quadrat is extremely small.

There is a family of Poisson probability mass functions, each member of which is specified by selecting a particular value of λ. Figure 6-10 illustrates the probability mass functions for three members of this family. Note that all distributions are truncated and depict probabilities only for small values of x. But, as is indicated in Equation (6-22), the probability mass function of a Poisson random variable assigns probabilities to all positive integers. The probabilities for larger values of x are extremely small and have been omitted for clarity. A comparison of these three probability mass functions indicates that the probabilities of a Poisson random variable become increasingly spread over the values of X as λ increases. Compact tables for a few values of λ are provided at the end of the text. An exercise to compute Poisson probabilities using a spreadsheet program is given at the end of the chapter.

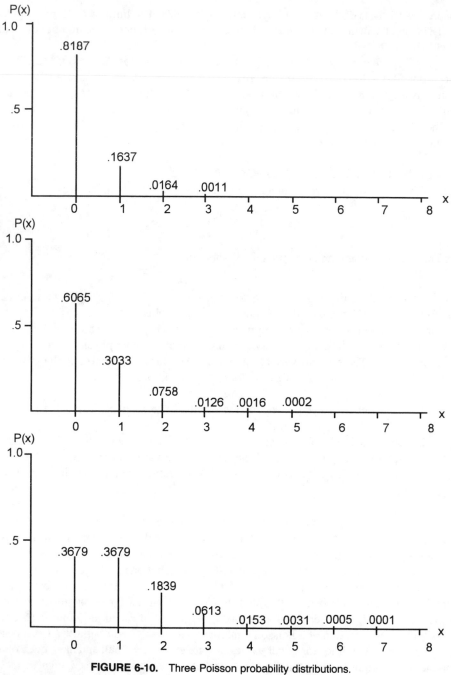

FIGURE 6-10. Three Poisson probability distributions.

The mean and variance of a Poisson random variable can be shown to be

$$E(X) = \sum_{x=0}^{\infty} xP(x) = \sum_{x=0}^{\infty} x\frac{e^{-\lambda}\lambda^x}{x!} = \lambda \tag{6-23}$$

$$V(X) = \sum_{x=0}^{\infty} [x - E(X)]^2 \frac{e^{-\lambda}\lambda^x}{x!} = \lambda \tag{6-24}$$

Thus, a characteristic feature of a Poisson random variable is the equality of the mean and variance. This is a useful check for examining the applicability of the Poisson random variable to some sample data.

EXAMPLE 6-2. The number of arrivals to a wilderness park is found to be Poisson distributed with a mean of 2.5 camping groups per day. On a particular day during the summer period, what is the probability that no groups arrive at the park? What is the probability that between one and three groups arrive? What is the most likely number of arriving camping groups?

Solution. The appropriate probability mass function is

$$P(x) = \frac{e^{-2.5}(2.5)^x}{x!} \qquad x = 0, 1, 2 \ldots \tag{6-25}$$

where x is the number of camping groups arriving at the park per summer season day. To find the probability that no groups arrive, we simply compute $P(0)$ from (6-25):

$$P(0) = \frac{e^{-2.5}(2.5)^0}{0!} = .0821 \tag{6-26}$$

We see that there is an 8% chance that no groups arrive on a particular day. This computation can be checked by consulting the Poisson statistical table at the end of the text for the value $x = 0$ and $\lambda = 2.5$. To determine the probability that between one and three groups arrive, $P(1 \leq x \leq 3)$, we can either consult the table or calculate $P(1) + P(2) + P(3)$ from (6-25). Using the table , we get $P(1 \leq x \leq 3) = P(1) + P(2) + P(3) = .2052 + .2565 + .2138 = .6755$.

Let us examine the Poisson table for the value of $\lambda = 2.5$. We note that

$$P(0) = .0821 \qquad P(2) = .2565$$
$$P(1) = .2052 \qquad P(3) = .2138$$

The mode is $x = 2$ because it occurs with the greatest probability. The most likely number of arriving camping groups is therefore two.

The Poisson distribution is applied extensively in problems related to the modeling of the distribution of the number of persons joining a queue, or line. Examples of this situation include arrivals at service stations or traffic facilities such as toll booths or ferries. These types of models are then used to make decisions about how many servicing units (for example toll booths or ferries with a given capacity) to provide to keep the length of the queue manageable.

The experiment that leads to the value of a random variable should always be compared to the experimental conditions generating a Poisson random variable. In the example of the wilderness park, the random variable X is the number of camping groups arriving in a one-day period. The possible values that X can assume are $x = 0, 1, 2, \ldots, \infty$. If we consider the one-day interval to be composed of many small subintervals, each 10 min long, then the conditions for X to be considered a Poisson random variable appear to be satisfied. In a 10-min interval:

1. The number of arriving camping groups is independent of the number arriving in any other 10-min interval of the day.
2. The probability that a camping group arrives in a 10-min interval is small and proportional to the length of the interval.
3. The probability of more than one camping group arriving in a 10-min interval is extremely small.

The Poisson distribution appears to be a reasonable model for the random variable in this case. A specific member of the family of Poisson distributions is chosen by the estimation of λ from some sample data.

6.3. Continuous Probability Distribution Models

Unlike a discrete random variable, a continuous random variable can assume an infinite number of values. Although the distribution of a random continuous random variable can also take on many forms, two distributions are presented in this text. First, the uniform continuous distribution is defined. It is the continuous version of the uniform discrete probability model described in Section 6.2. The second distribution, a bell-shaped curve discovered by Gauss and now known as the normal distribution, is the most important distribution in conventional statistical analysis.

Uniform Continuous Distribution

Just as we can define a uniform discrete distribution in which each value of X is equally probable, we can define a uniform continuous distribution in which all values of X are equally likely. Let the random variable X be defined over the range $a \leq x \leq b$. The probability density function for a uniform random variable is

$$f(x) = \begin{cases} \dfrac{1}{b - a} & a \leq x \leq b \\ 0 & \text{otherwise} \end{cases} \tag{6-27}$$

An example of a uniformly distributed random variable is shown in Figure 6-11. In this example, both a and b are positive, but there is no restriction on the sign of a or b. The mean $E(X)$ and variance $V(X)$ of a uniform random variable are given by

$$E(X) = \frac{b + a}{2} \qquad (6\text{-}28)$$

$$V(X) = \frac{(b - a)^2}{12} \qquad (6\text{-}29)$$

It is not possible to determine the probability that a continuous random variable assumes a particular value of X, but it is possible to determine the probability that X assumes any value between c and d by finding the area under the probability density function from c to d. Suppose a uniform random variable is defined over the interval from 0 to 10. What is the probability that X assumes a value between three and five? The density function is illustrated in Figure 6-12, where the area under the density function between three and five is shaded. To determine the required probability, it is necessary to calculate the ratio of the shaded area to the total area under the probability density function. By definition, the total area under the distribution is one. The required probability is $(5 - 3)/(10 - 0) = 2/10 = .2$. In general, the probability that a uniform random variable takes on a value between c and d, where $a \le c \le d \le b$, is $(d - c)/(b - a)$. The simplicity of this formula means that it is not necessary to have extensive tables for the uniform random distribution. They can be calculated from this formula whenever they are required. For example, the probability that x assumes a value between 6 and 15 in a uniform continuous distribution defined over the range 2 to 32 is $(15 - 6)/(32 - 2) = 9/30 = 0.3$.

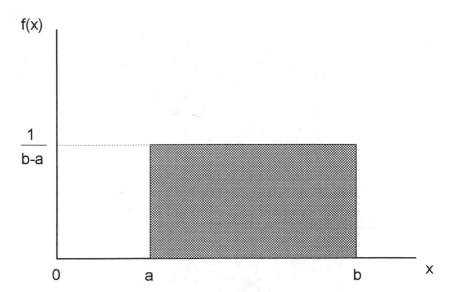

FIGURE 6-11. Uniform continuous probability distribution.

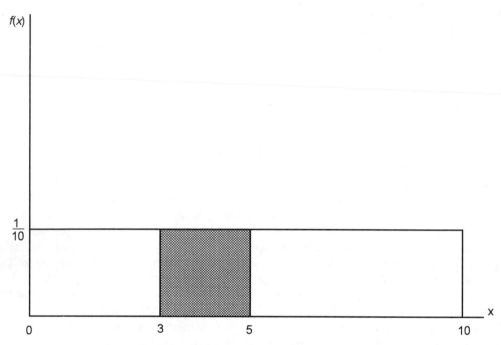

FIGURE 6-12. A uniform continuous probability distribution.

Normal Probability Distribution

For many continuous random variables, the governing probability distribution is a specific bell-shaped curve known as the *normal distribution.* Originally developed by the German mathematician Gauss, this distribution is sometimes termed the *Gaussian distribution.* The basic shape of the normal distribution is a simple bell. Three different normal distributions are illustrated in Figure 6-13. Why is this distribution so common? One reason is that this basic bell shape can take on any of the forms illustrated in Figure 6-13. The center of the distribution can be located at any point along the real number line. The distribution can be very flat, or it can be very peaked. Notice that distribution *A* in Figure 6-13, is centered on the lowest value of *X* and distribution *C* on the highest value of *X*. However, distribution *C* is the flattest and *B* the most peaked of the three.

What do all three distributions have in common? All randomly distributed normal variables follow a particular mathematical expression that defines the characteristic bell shape. Specifically, we say that a continuous random variable *X* is normally distributed if its probability density function is given by

$$f(x) = \frac{1}{\sigma\sqrt{2\pi}}\exp\left[-\frac{1}{2}\left(\frac{x-\mu}{\sigma}\right)^{2}\right] \qquad (6\text{-}30)$$

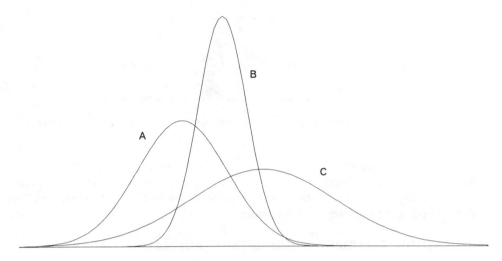

FIGURE 6-13. Three normal distributions.

where $-\infty < x$, $\mu < +\infty$, and $\sigma > 0$. The unknowns π and e are the mathematical constants defined as 3.14159 and 2.71828, respectively. There are two parameters of the normal distribution, μ and σ. Although the derivation requires a knowledge of calculus beyond the scope of this text, it can be shown that the expected value and variance of a random normal variable are given by

$$E(X) = \mu \qquad (6\text{-}31)$$

$$V(X) = \sigma^2 \qquad (6\text{-}32)$$

The parameter μ is the mean of the distribution and locates the number along the real number line. The variance σ^2 controls the dispersion of the values of X around this center. It should now be clear why the members of the normal family come in such a variety of shapes. Two *independent* parameters can be used to combine any degree of dispersion with a central value located anywhere along the real number line. Although there is a great diversity on the appearance of many random normal distributions, all are symmetric about μ. Thus, all normal variables are characterized by the equality of the mode, the median, and the mean. Not only is the highest point on the probability density function located at μ, but it also divides the distribution into two equal parts. Exactly 0.5000 of the area under the probability density function lies on either side of μ.

To compute the probability that a random normal variable assumes a value between two values of X, say c and d, we must calculate the area under the probability density function given by Equation (6-30) between $X = c$ and $X = d$. This area is illustrated in Figure 6-14. For the uniform continuous distribution, these areas could be determined from simple geometry. Unfortunately, this is not possible for the normal distribution. What is the alternative? Of course, we could calculate and tabulate these areas for any given values of

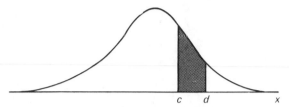

FIGURE 6-14. Probability that a random normal variable takes on a value in the interval $[c,d]$.

μ and σ, but there are an infinite number of normal distributions, each with a unique combination of μ and σ. Fortunately, we can convert *any* normally distributed random variable to a single probability distribution known as the *standard normal distribution*. This is done by using the following transformation

---

DEFINITION: STANDARD NORMAL TRANSFORMATION

$$z = \frac{x - \mu}{\sigma} \tag{6-33}$$

This transformation converts a variable measured in units of x to a standard normal variable measured in units of z. When measured in units of x, the variable as a mean of μ and a standard deviation of σ. When it is converted to a standard normal variable, it has a mean of zero and a standard deviation (and also a variance) of one.

DEFINITION: THE STANDARD NORMAL DISTRIBUTION
The probability density function for the standard normal distribution is given by

$$f(z) = \frac{1}{\sqrt{2\pi}} e^{-1/2z^2} \tag{6-34}$$

For example, consider the two random normal variables X_1 and X_2, the first with a mean of $\mu_1 = 100$ and $\sigma_1 = 10$ and the second with a mean of $\mu_2 = 10$ and a standard deviation of $\sigma_2 = 2$. Both can be converted to standard normal variables by taking the original measurements and coding them according to Equation (6-33). Consider the first variable X_1 illustrated in Figure 6-15(a). The value $X = 110$ is converted to a standard z value by calculating $z_{110} = (110 - 100)/10 = 1.0$. In what units is z measured? It is always measured in *standard deviation units* of the variable being considered. The value $z = 1.0$ implies that a value of $x = 110$ is one standard deviation above the mean $\mu_1 = 100$. Similarly, $x = 90$ is one standard deviation ($\sigma_1 = 10$) *below* the mean of $\mu_1 = 100$, so that $z_{90} = (90 - 100)/10 = -1.0$.

Variable X_2 of Figure 6-15(b) can be transformed in the same way. Consider first a measurement of $x = 14$. Expressed as a *standard score* or *z-score*, $z_{14} = (14 - 10)/2 = 2.0$. That is, the value of $x = 14$ lies two standard deviations above the mean of $\mu_2 = 10$. As another example, we know that $x = 4$ lies three standard deviations *below* the mean of 10. To verify this, we calculate the z-score as $z_4 = (4 - 10)/2 = -3.0$. Any value of x can be converted to an equivalent value of z and vice-versa. For variable X_2, what value of x corresponds to $z = -2.0$? From Equation (6-33) we know that $(x - 10)/2 = -2.0$. Solving

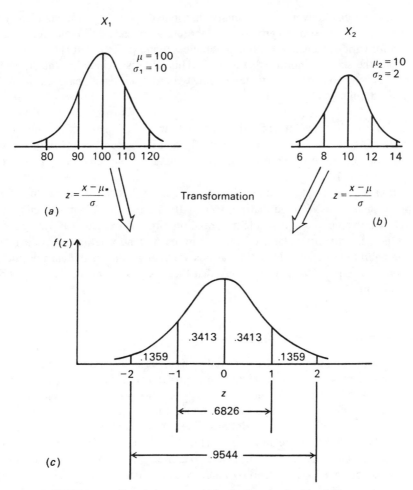

FIGURE 6-15. Converting normal distributions to the standard normal.

for x leads to $x = 6$; that is, the value of six lies two standard deviations below the mean of 10. Thus, we can now see that both variables X_1 and X_2 can be converted to the form of the standard normal distribution illustrated in Figure 6-15(c).

When this transformation is made, it is easier to see the characteristics shared by all random normal variables. As illustrated in Figure 6-15(c), exactly 68.26 percent or .6826 of the values of a random normal variable lie within one standard deviation of the mean. For the standard normal, this is between the values of $z = 1.0$ and $z = -1.0$. For variable X_1, 68.26% on the area under the probability density function lies between one standard deviation below the mean, 90, and one standard deviation above the mean, 110. For variable X_2, the corresponding limits are $x = 8$ and $x = 12$. Also, 95.44% of the area under the probability density function of the standard normal lies within two standard deviations of the mean. For X_1 these limits are 80 and 120, and for X_2 the limits are 6 and 14. For all random normal variables the limits to the distribution are $+\infty$ and $-\infty$, but virtually 100% of the area under the curve (in fact 99.72%) lies within three standard deviations of the mean.

Because we can convert any normally distributed variable to the standard normal distribution, it is not necessary to provide an elaborate set of statistical tables for the normal distribution for various values of the two parameters μ and σ. Rather, it is usual to provide a single table, the standard normal distribution. This table is included in the Appendix at the end of this text. To access this table, we must convert our data measured in units of x to standardized units of z.

EXAMPLE 6-4.   The travel time for the journey to work in a metropolitan area averages 40 min with a standard deviation of exactly 12.5 min. Assuming these distances are normally distributed, what is the probability that a randomly selected person's journey to work is longer than 1 hr?

Solution. To use the standard normal table, we must first convert the value of $x = 60$ min (1 hr) to units of z. We note that $z = (60 - 40)/12.5 = 1.6$. We must determine the probability that a value of z in a standard normal distribution exceeds 1.6. Using the third column of the standard normal table (Table A-3 in the Appendix at the end of the text), we find this probability to be .055. Note the various ways in which the standard normal table can be expressed. Equivalently, we can say that there is a .945 chance that the work trip *is less than* 60 min.

6.4. Bivariate Random Variables

When two or more random variables are jointly involved in a statistical experiment, their outcomes are generated by a *multivariate probability function. Bivariate probability functions* are a class of multivariate functions in which two random variables are jointly involved in the outcome of a statistical experiment. In turn, bivariate probability functions can be classed as either discrete or continuous. The concepts of bivariate random variables are most easily explained in relation to discrete distributions without resorting to the mathematical complications inherent in continuous models. The relevant features of continuous models can then be explained by analogy to the discrete case.

Bivariate Probability Functions

The *joint* or *bivariate probability mass function* for two discrete variables is a function that assigns probabilities to joint values of X and Y such that two conditions are satisfied:

1. The probability that random variable X takes on the value $X = x$ *and* that random variable Y takes on the value $Y = y$, denoted $P(x, y)$, is nonnegative and less than or equal to 1. That is, $0 \le P(x, y) \le 1$ for all (x, y) pairs of the discrete random variables X and Y.
2. The sum of the probabilities $P(x, y)$ taken over all discrete values of X and Y is 1. That is $\Sigma_x \Sigma_y P(x, y) = 1$.

To understand these two conditions, consider the relationship between the size of a household and the number of cars owned by the household. Suppose 100 households are surveyed, and the number of members of the household and the number of cars owned by that household are determined. The responses of the households are classified in Table 6-10.

TABLE 6-10
Car Ownership and Household Size

Household size	Cars owned 0	1	2	3	Total
2	10	8	3	2	23
3	7	10	6	3	26
4	4	5	12	6	27
5	1	2	6	15	24
Total	22	25	27	26	100

Of the 100 survey households, 10 have two members and no cars available, eight have two members and one car available, and so on. When households are categorized by size, 23 have two members, 26 have three members, 27 have four members, and 24 have five members. Similarly, the totals by number of cars owned are given in the last row of the table.

These data can be converted to a bivariate probability distribution by dividing each entry in Table 6-10 by the number of households in the survey, $n = 100$. The bivariate probability function $P(X, Y)$ for these two variables is given in Table 6-11. In the last column of the table is the *marginal probability distribution* of random variable X, the probability distribution of random variable X taken alone. The marginal probability distribution of X can be calculated from the relation $P(x) = \Sigma_y P(x, y)$. Note that, for example, the probability of a randomly selected household having exactly three members is $P(X = 3) = .07 + .10 + .06 + .03 = .26$. Similarly, the marginal probability distribution of Y can be determined from $P(y) = \Sigma_x P(x, y)$. The probability that a household has no cars is obtained by summing the first column: $P(Y = 0) = .10 + .07 + .04 + .01 = .22$. It is thus possible to construct the marginal probability distributions $P(X)$ and $P(Y)$ from a joint or bivariate probability function $P(X, Y)$.

The joint probability distribution $P(X, Y)$ can also be used to determine the *conditional probability functions* for variables X and Y. What is the conditional probability distribution of Y given that X takes on some specific value $X = x$? For example, what is the conditional probability distribution of car ownership, given that a household contains three

TABLE 6-11
Probabilities of Car Ownership and Household Size

Size	0	1	2	3	$P(x)$
2	0.10	0.08	0.03	0.02	0.23
3	0.70	0.10	0.06	0.03	0.26
4	0.04	0.05	0.12	0.06	0.27
5	0.01	0.02	0.06	0.15	0.24
$P(y)$	0.22	0.25	0.27	0.26	1.00

members? The conditional probability distribution given $X = x$ is calculated from the relation

$$P(y|x) = \frac{P(x, y)}{P(y)} \tag{6-35}$$

and the conditional distribution of X given $Y = y$ is

$$P(x|y) = \frac{P(x, y)}{P(y)} \tag{6-36}$$

The conditional probability distribution of car ownership for three-person households can be computed from Equation (6-35) with the data of Table 6-12. The conditional probability function $P(y|x = 3)$ is

$$P(y = 0|x = 3) = \frac{.07}{.26} = .269$$

$$P(y = 1|x = 3) = \frac{.10}{.26} = .385$$

$$P(y = 2|x = 3) = \frac{.06}{.26} = .231$$

$$P(y = 3|x = 3) = \frac{.03}{.26} = .115$$

Note that this is a valid probability distribution because it sums to one. To obtain this distribution, we simply divide each entry in the row for $x = 3$ by the row sum $P(3)$ given in the last column. It is not possible to use the entries in the row as they stand. Why? Because they do not sum to one. The division by .26 standardizes the distribution so that it sums to one. Similarly, the conditional probability distribution of household size given that the household has three cars is

$$P(x = 2|y = 3) = \frac{.02}{.26} = .077$$

$$P(x = 3|y = 3) = \frac{.03}{.26} = .115$$

$$P(x = 4|y = 3) = \frac{.06}{.26} = .231$$

$$P(x = 5|y = 3) = \frac{.15}{.26} = .577$$

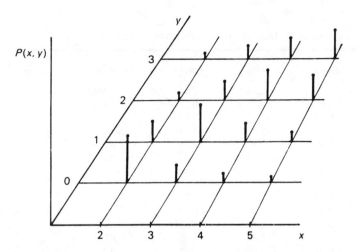

FIGURE 6-16. Graph of bivariate probability distribution of Table 6-11.

It is also possible to illustrate a bivariate probability function by using a simple three-dimensional graph. The vertical axis is used for $P(x, y)$, and the two random variables are depicted in the XY plane. Where there are only a few discrete values for the variables, such illustrations give a visual impression of the joint distribution of the two random variables involved. Figure 6-16 illustrates the bivariate probability distribution for the data in Table 6-11. It is apparent, for example, that larger households tend to own more cars than smaller households. In fact, there appears to be a direct relation between the size of the household and the number of cars owned. How can the strength of this relation be measured? The answer lies in a measure known as covariance.

Covariance of Two Random Variables

For a single random variable, it is possible to calculate the mean or expected value. For two random variables, a useful statistical measure is the *covariance*. The covariance of two random variables is defined as

$$C(X, Y) = E\{[X - E(X)]\} = \sum_x \sum_y P(x, y)[X - E(X)][y - E(Y)] \qquad (6\text{-}37)$$

or

$$C(X, Y) = E(X, Y) - E(X)E(Y) = \sum_x \sum_y xyP(x, y) - E(X)E(Y) \qquad (6\text{-}38)$$

Here $C(X, Y)$ is the covariance, and E is the expectation operator. Recall from Equation (6-2) that $E(X) = \sum_{i=1}^{k} x_i P(x_i)$. The covariance is a direct statistical measure of the degree to which two random variables X and Y tend to vary together. Whenever large values of X tend to be associated with large values of Y, and small values of X with small values of Y,

TABLE 6-12
Computation of Covariance for Data of Table 6-11 by Using Equation (6-37)

(x, y)	$P(x, y)$	$x-E(x)$	$y-E(Y)$	$[x-E(x)][y-E(y)]$	$P(x, y)[x-E(X)][y-E(Y)]$
2, 0	0.10	−1.52	−1.57	2.3864	0.2386
2, 1	0.08	1.52	−0.57	0.8664	0.0693
2, 3	0.03	−1.52	0.43	−0.6536	−0.0196
2, 3	0.02	−1.52	1.43	−2.1736	−0.0435
3, 0	0.07	−0.52	−1.57	0.8162	0.0571
3, 1	0.10	0.52	−0.57	0.2964	0.0296
3, 2	0.06	−0.52	0.43	−0.2236	−0.0134
3, 3	0.03	0.52	1.43	−0.7436	−0.0223
4, 0	0.04	0.48	−1.57	−0.7536	−0.0301
4, 1	0.05	0.48	−0.57	−0.2736	−0.0137
4, 2	0.12	0.48	0.43	0.2064	0.0248
4, 3	0.06	0.48	1.43	0.6864	0.0412
5, 0	0.01	1.48	−1.57	2.3236	−0.0232
5, 1	0.02	1.48	−0.57	−0.8436	−0.0169
5, 2	0.06	1.48	0.43	0.6364	0.0382
5, 3	0.15	1.48	1.43	2.1164	0.3175

$$C(X, Y) = .6336$$

then $C(X, Y)$ has a large positive value. When large values of X are associated with small values of Y and small values of X with large values of Y, then $C(X, Y)$ is a large negative number. Whenever there is no pattern, $C(X, Y) = 0$ or is close to zero.

Consider the data of Table 6-11. Although Equation (6-38) is the most efficient way to calculate the covariance, Equation (6-37) gives the identical result. This equivalence is illustrated by the results of the computations using each formula summarized in Tables 6-12 and 6-13. The calculated value of $C(X, Y)$ indicates that the two variables tend to covary, or "run together." Because $C(X, Y)$ is positive, household size and car ownership are positively correlated. Unfortunately, the covariance is not an easily interpretable measure of correlation. The numerical value of $C(X, Y)$ is completely dependent on the magnitudes of the two random variables X and Y. If either of these two variables is measured in different units, then $C(X, Y)$ must change. As we shall see in Chapter 12, Pearson's product–moment correlation coefficient standardizes the covariance and overcomes this problem.

First, however, we must introduce the notion of independence of two random variables X and Y. Two variables are said to be independent if

$$P(x, y) = P(x)P(y) \tag{6-39}$$

for all possible combinations X and Y. To show that two variables are dependent, it is sufficient to show that Equation (6-39) does not hold for any one combination of X and Y. Clearly, household size and car ownership are not independent but covary. To see this, examine the relation given by (6-39) for the values $x = 3$ and $y - 1$. From Table 6-11

TABLE 6-13
Computation of Covariance for Data of Table 6-11 by Using Equation (6-38)

x	$P(x)$	$xP(x)$	y	$P(y)$	$yP(y)$	x, y	$P(x, y)$	$xyP(x, y)$
2	0.23	0.46	0	0.22	0.00	2, 0	0.10	0.00
3	0.26	0.78	1	0.25	0.25	2, 1	0.08	0.16
4	0.27	1.08	2	0.27	0.54	2, 2	0.03	0.12
5	0.24	1.20	3	0.26	0.78	2, 3	0.02	0.12
		$E(X) = 3.52$			$E(Y) = 1.57$	3, 0	0.07	0.00
						3, 1	0.10	0.30
						3, 2	0.06	0.36
						3, 3	0.03	0.27
						4, 0	0.04	0.00
						4, 1	0.05	0.20
						4, 2	0.12	0.96
						4, 3	0.06	0.72
						5, 0	0.01	0.00
						5, 1	0.02	0.10
						5, 2	0.06	0.60
						5, 3	0.15	2.25
								6.16

$$C(X, Y) = 6.16 - (3.52)(1.57) = .6336$$

$$P(x = 3) = .26 \qquad P(y = 1) = .25$$

$$P(x = 3, y = 1) = .10$$

so that $P(x = 3)P(y = 1) = (.26)(.25) = .065 \neq P(x = 3, y = 1) = .10$.

Independence and covariance are closely related. If two random variables X and Y are independent, then their covariance must equal zero. However, it is not necessarily true that the two variables are independent if the covariance is zero. It is possible for two variables to have a covariance $C(X, Y) = 0$, yet still not satisfy the demanding requirements of Equation (6-39).

Bivariate Normal Random Variables

For continuous bivariate random variables, the joint probability distribution is specified by using a density function $f(X, Y)$. The most important distribution is the *bivariate normal distribution*. Figure 6-17 contains a graphical representation of a bivariate normal distribution. For every pair of (x, y) values, there is a density $f(x, y)$ represented by the height of the surface at that point. The surface is continuous and probability corresponds to the volume under the surface. If two random variables are jointly normally distributed, then

1. The marginal distributions of both X and Y are univariate normal.
2. Any conditional distribution of X or Y is also univariate normal.

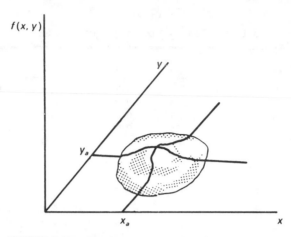

FIGURE 6-17. Bivariate normal probability density function.

The implication of item 2 is that the cross section of a "slice" through the bivariate normal for a given value of X, say $X = x_a$, has the characteristic shape of the normal curve. A slice at $Y = y_b$. has the same characteristic. Both cases are illustrated in Figure 6-17.

There are five parameters to the bivariate normal density function: μ_x and μ_y, and σ_x and σ_y, and ρ_{xy} (pronounced "row sub XY"). The first four parameters are the means and standard deviations of the marginal distributions for X and Y. The parameter ρ_{xy} is the correlation coefficient between random variables X and Y and is defined as

$$\rho_{xy} = \frac{C(X, Y)}{\sigma_x \sigma_y} \tag{6-40}$$

The correlation coefficient is a pure number, and it can take on any value between −1 and +1 inclusive. If variables X and Y are independent, then $C(X, Y) = 0$ and therefore $\rho_{xy} = 0$. If the two variables are perfectly positively correlated, then $\rho_{xy} = 1$. If $\sigma_{xy} = -1$, then the perfect linear relation is a negative or inverse one.

It is common to portray a bivariate normal distribution by using a contour diagram. A contour diagram is created by taking horizontal slices through the bivariate normal distribution, as in Figure 6-18. A contour is composed of all (x, y) pairs having a constant density $f(x, y)$. The contour curves of all bivariate normal distributions are elliptical, except where $\rho_{xy} = 0$ and $\sigma_x = \sigma_y$. In this single situation the contours will appear as concentric circles. Several examples of bivariate normal distributions are shown in Figure 6-19. In (a) note that $\rho_{xy} = 0$ and the two standard deviations are equal, so all contours are circular. In (b), X and Y are independent since $\rho_{xy} = 0$, but the contours are elliptical since $\sigma_x > \sigma_y$. In both (c) and (d), variables X and Y are positively related, and $\rho_{xy} > 0$. The principal axis of each ellipse has a positive slope, indicating that the surface tends to run along a line with positive slope. Similarly, when X and Y are negatively related and $\rho_{xy} < 0$, the principal axis has a negative slope. This case is illustrated in (e) and (f). In all cases in Figure 6-19, the

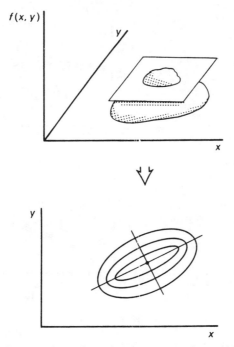

FIGURE 6-18. Three- and two-dimensional representations of bivariate normal distribution.

centers of the ellipses are at coordinates (μ_x, μ_y). The roles of the parameters are clearly illustrated by these eight cases. The two means μ_x and μ_y control the location of the center of the surface, and the two standard deviations and the correlation coefficient control the shape.

6.5. Summary

In this chapter the notion of a random variable has been formally introduced. We recognize that there are two different types of random variables, continuous and discrete. Five important probability distribution models for random variables were introduced in Sections 6.2 and 6.3. The parameters, mean, and variance of these five distributions are summarized in Table 6-14. In addition, Section 6.4 developed the ideas underlying bivariate probability distributions and introduced the bivariate normal distribution. The material presented in Section 6.4 will prove useful when we begin to study the relationships between variables in Chapters 12, 13, and 14.

Appendix

There is a direct correspondence between the area under a probability density function between any two limits c and d and the probability that a continuous random variable X

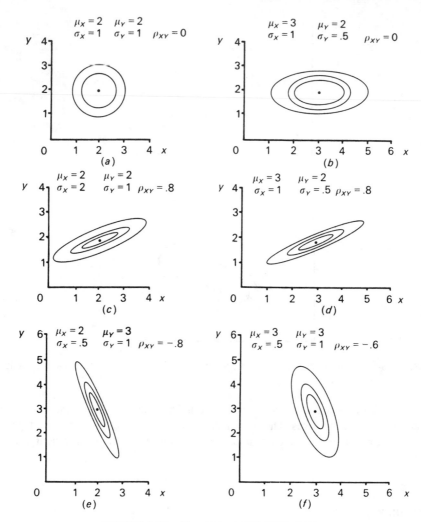

FIGURE 6-19. Bivariate normal distributions.

TABLE 6-14
Summary of Five Probability Distributions

Distribution	Continuous (C) or discrete (D)	Parameters	Mean	Variance
Discrete uniform	D	k	$(k + 1)/2$	$(k^2 - 1)/12$
Binomial	D	n, π	$n\pi$	$n\pi(1 - \pi)$
Poisson	D	λ	λ	λ
Continuous uniform	C	a, b	$(a + b)/2$	$(a - b)^2/12$
Normal	C	μ, σ	μ	σ

assumes a value between these two limits. Students familiar with calculus will recognize that the area under $f(x)$ between c and d is given by

$$\int_c^d f(x)dx$$

where the symbol $\int$ represents the integration operator. It is directly analogous to the summation operator Σ used for discrete variables. The requirement that the sum of the probabilities assumable by a random variable equal to one can be stated as

$$\int_a^b f(x)dx = 1$$

where a and b are the limits of the values assumable by the random variable. For continuous random variables defined over the infinite length of the real number line, the appropriate equation is

$$\int_{-\infty}^{+\infty} f(x)dx = 1$$

A second important property is that the probability of a continuous variable assuming a particular value, say $X = c$, is always zero. That is,

$$\int_c^c f(x)dx = 0$$

because any definite integral over the limits c to c must be zero. The expected value of a continuous random variable is

$$E(X) = \int_a^b xf(x)dx$$

and the variance $V(X)$ is

$$V(X) = \int_a^b [x - E(X)]^2 f(x)dx = \int_a^b x^2 f(x)dx - [E(X)]^2$$

As an example, let us consider the uniform continuous random variable defined by

$$f(x) = \begin{cases} \dfrac{1}{10} & 0 \le x \le 10 \\ 0 & \text{otherwise} \end{cases}$$

First, to show that the area under $f(x)$ equals one, we note

$$\int_0^{10} \frac{1}{10} dx = \frac{1}{10}(10{-}0) = 1$$

To calculate the expected value, we write

$$E(X) = \int_0^{10} x\left(\frac{1}{10}\right) dx = \frac{1}{10}\int_0^{10} x\,dx$$

$$= \frac{1}{10}\frac{x^2}{2}\Big|_0^{10}$$

$$= \frac{1}{10}\left(\frac{10^2}{2} - \frac{0^2}{2}\right) = 5$$

By solving the general case

$$E(X) = \int_a^b x\left(\frac{1}{b-a}\right) dx$$

it is possible to derive the general formula $E(X) = (b + a)/12$, given as (6-28). The variance is

$$V(X) = \int_0^{10} x^2\left(\frac{1}{10}\right) dx - [E(X)]^2$$

$$= \frac{1}{10}\frac{x^3}{3}\Big|_0^{10} - 5^2$$

$$= \frac{1}{10}\left(\frac{1000}{3}\right) - 25 = 8.33$$

By solving the general case

$$V(X) = \int_a^b x^2\left(\frac{1}{b-a}\right) dx - [E(x)]^2$$

the variance of a uniform continuous random variable can be shown to be $V(X) = (b - a)/12$. This is given in the text as Equation (6-29).

FURTHER READING

Introductory statistics textbooks for geographers seldom discuss the concept of random variables in detail, nor do they give an extended treatment of probability distribution models. Introductory statistics textbooks for business and economics students and some texts for applied statistics courses usually give a more complete presentation of these topics. Three representative examples are Neter, Wasserman, and Whitmore (1982); Pfaffenberger and Patterson (1981); and Winkler and Hays (1975). The development of the Poisson probability distribution using the analogy to a dot map only hints at the widespread use of probability models in this situation. Additional material on quadrat analysis and a related technique known as nearest neighbor analysis is provided in the problems below. Extended treatment of these topics can be found in Getis and Boots (1978), Taylor (1977), and Unwin (1981).

A. Getis and B. Boots, *Models of Spatial Processes* (Cambridge, England: Cambridge University Press, 1978).
J. Neter, W. Wasserman, and G. Whitmore, *Applied Statistics* (Boston: Allyn and Bacon, 1982).
R. Pfaffenberger and J. Patterson, *Statistical Methods for Business and Economics* (Homewood, Ill.: Richard Irwin, 1981).
P. Taylor, *Quantitative Methods in Geography* (Boston: Houghton Mifflin, 1977).
D. Unwin, *Introductory Spatial Analysis* (London: Methuen, 1981).
R. Winkler and W. L. Hays, *Statistics: Probability Inference and Decision Making*, 2d ed. (New York: Holt, 1975).

PROBLEMS

1. Explain the meaning of the following terms:
   a. Random variable
   b. Probability distribution
   c. Expected value of a random variable
   d. Variance of a random variable
   e. Bernoulli trial
   f. Binomial coefficients
   g. Pascal's triangle
   h. Point pattern analysis
   i. Standard normal distribution
   j. Standard score
   k. Correction for continuity
   l. Bivariate probability function
   m. Covariance
   n. Correlation coefficient

2. Differentiate between the following:
   a. A discrete and a continuous random variable
   b. A probability mass function and a probability density function
   c. A normal distribution and the standard normal distribution
   d. A marginal probability distribution and a conditional probability distribution
   e. A probability mass function and a cumulative mass function

3. Explain what is meant when we say two random variables X and Y are (a) dependent and (b) independent.

4. Let X be a random variable with the following probability distribution:

x	$P(x)$
0	0.40
1	0.30
2	0.15
3	0.15

   a. Verify that this is a valid probability distribution model.
   b. Determine $E(X)$
   c. Determine $V(X)$
   d. What is the mode of X?

5. A census of all households in a town is undertaken, and the number of trips made by members of each household on a given day is recorded. The trip frequencies follow the probability distribution specified in this table:

Trips per day x	Probability of making x trips
0	0.13
1	0.14
2	0.21
3	0.18
4	0.10
5	0.08
6	0.07
7	0.05
8	0.03
9	0.01

   a. Portray this distribution as a graph.
   b. Graph the cumulative mass function.
   c. Find $E(X)$ and $V(X)$.
   d. What proportion of the values of x are within two standard deviations of the mean?

6. Graph the probability mass functions for the following discrete probability distribution models:
   a. Uniform: $k = 3$, 5, and 10
   b. Binomial: $n = 5$, $\pi = .20$; $n = 5$, $\pi = .50$; $n = 5$, $\pi = .70$
   c. Poisson: $\lambda = .20$, .40, and .80

7. The number of persons X in a camping party arriving at a wilderness park is a uniform discrete random variable. The maximum number in a party is nine.
   a. Determine $P(3)$ and $P(7)$.
   b. Find $P(X < 6)$.
   c. Determine $P(X > 6)$.
   d. What is the expected number of persons in a camping party?
   e. What is $V(X)$?

8. Suppose the number of persons in a camping party is a Poisson random variable with a mean of $\lambda = 3$. Solve (a) to (e) of Problem 7.

9. The maximum temperature reached on any day can be classified as above freezing (a success!) or below freezing (a failure?). In a certain city of eastern North America, January weather statistics indicate the probability a January day will be above freezing is .30. Use the binomial distribution to determine these probabilities:
   a. Exactly two of the next seven January days will be above freezing.
   b. More than five of the next seven days will be above freezing.
   c. There will be at least one day above freezing in the next seven days.
   d. All seven days in the next week will be above freezing.
   e. Is this a reasonable application of the binomial distribution? Why or why not?

10. Using a spreadsheet package, generate a series of tables for the Poisson and binomial probability distributions.

a. For the Poisson, use a column of the table for values of λ between 0 and 1, with an increment of 0.1. Use a row for each value of x from 0 to 7 and format the probabilities to four decimal places.

b. For the binomial, generate a table for the values $n = 10$ (and therefore values of x from 0 to 10) as the rows and values of π from 0.05 to 0.50 as the columns. Format the probabilities to four decimal places.

The remaining problems involve *quadrat analysis*. In addition, material is introduced, and problems are defined for a closely related technique known as nearest-neighbor analysis

11. A grid of squares has been placed over a map, and the number of points (say, houses) failing in each square is counted. The number in each quadrat of the 25-cell map illustrated below represents this frequency.

0	1	1	1	0
0	2	2	1	2
1	0	0	2	0
1	1	1	2	3
0	3	0	1	0

a. Construct a frequency distribution of points per quadrat.
b. Determine the mean number of points per quadrat.
c. What would the frequency distribution of points be if the point pattern were random (i.e., given by the Poisson distribution) with mean given by the value calculated in part (b)?
d. Compare the frequency distribution generated in part (a) with the frequency distribution predicted by the Poisson distribution for these 25 quadrats. Does the Poisson distribution appear to give a reasonable fit?

12. Complete parts (a) to (d) of Problem 11 for the following map.

1	0	0	0	0	0	0	2	0	1
0	0	0	0	0	0	0	1	0	0
1	0	0	1	1	0	0	0	2	0
0	0	0	0	0	0	0	0	0	1
0	1	0	0	0	0	1	1	0	1
0	0	0	0	0	0	0	1	2	0
0	2	0	1	0	1	0	1	0	0
1	0	0	0	1	1	0	0	0	0
0	0	0	0	2	0	0	0	0	0
0	1	0	0	0	0	0	1	0	1

13. One problem in quadrat analysis is that the results are highly dependent on the size of the quadrats chosen. Conclusions drawn from a study based on a certain quadrat size may be contradicted by conclusions of a second study of the same data based on a different quadrat size. This phenomenon might be called the *scale problem*, by analogy to the problem identified in descriptive geographical statistics of Chapter 4. To examine the impact of quadrat size, complete parts (a) to (d) of Problem 12, using the same base but with the quadrats grouped into fours, forming 25 larger square quadrats.

14. A second problem in quadrat analysis is that markedly different point patterns can give rise to identical frequency distributions of points by quadrats. This is because the frequency distributions cannot indicate whether quadrats with zero points are located close to or away from other quadrats with zero points. In Chapter 3 we called this the pattern problem. To see the significance of this problem, perform the following experiment. Construct two 5 × 5 grids of 25 quadrats each. Into each grid place a distribution of points with the following frequency distribution:

Number of points	Number of quadrats
0	12
1	9
2	3
3	1

In the first map, place the points in the grid so that the distribution appears random. In the second map, place the points so that the map appears clustered.

15. Quadrat analysis also suffers from the effects of the boundary problem and the modifiable areal units problem identified in Chapter 3. Generate dot maps that expose the significance of each of these problems.

16. In this chapter, we have shown that an independent random spatial process can be modeled by the Poisson probability distribution. If a map has a random distribution of points, the Poisson distribution should provide a good fit to the frequency distribution of points per quadrat. However, there are many potential point distributions other than a random one. In fact, a random distribution can be considered to be an intermediate pattern, a mix of two other types: dispersed and clustered. Consider these five point patterns:

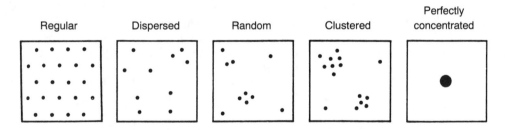

First, there are *dispersed* patterns. These are distributions of points more or less evenly spread over the map. The term describing the limiting case is *regular* and describes a pattern in which each point is equidistant from its neighbors. This is known as a *triangular lattice* of points. *Clustered* patterns have one or more groups of points in clusters and large areas of the map without any points. In the limiting case, all points share one location, and we have a perfectly clustered pattern. A random pattern has elements of both clustered and dispersed patterns. One way of classifying point patterns as clustered, dispersed, or random is to examine the variance/mean ratio of the observed frequency distribution of points per quadrat. A random pattern will have a variance/mean ratio of exactly one, since it can be described by the Poisson distribution for which $V(X) = E(X) = \gamma$. Hence, $V(X) = E(X) = \lambda = 1$. Dispersed patterns will have small variance/mean ratios, and clustered patterns will have variance/mean ratios larger than one.

a. Briefly explain why the variance/mean ratio of dispersed patterns will be low, and conversely why clustered patterns will have high variance/mean ratios.

b. Classify the following five frequency distributions of quadrat counts as clustered, random, or dispersed based on their variance/mean ratios:

Number of points in a quadrat, x	Number of quadrats				
	Map A	Map B	Map C	Map D	Map E
0	18	1	15	61	10
1	6	5	25	30	20
2	1	12	25	7	40
3	0	5	30	1	20
4	0	1	4	1	5
5	0	1	1	1	5
	25	25	100	100	100

Hint: Use the following equations to estimate the mean and variance for grouped data

$$\text{Mean } \bar{X} = \frac{\sum\limits_{k=1}^{K} f_k X_k}{n} \qquad \text{Variance } s^2 = \frac{1}{n-1}\left(\sum\limits_{k=1}^{K} f_k X_k^2 - \frac{\left(\sum\limits_{k=1}^{K} f_k X_k\right)^2}{n}\right)$$

where f_k is the frequency of the kth class.

17. One alternative to quadrat analysis that avoids some of the difficulties described in Problems 12 to 14 is nearest-neighbor analysis. By using this technique, it is also possible to classify point patterns. The method is quite straightforward. The distance d_{ij} between each pair of points i and j in a given point pattern is calculated by using Pythagoras's theorem. For each point $i = 1$, $2, \ldots, n$, the closest point is determined, that is, $\min_j d_{ij}$. The mean, or average, of all these nearest-neighbor distances is denoted $\bar{d}_a$. Unfortunately, this measure cannot be used to compare maps because it depends especially on the size of the area depicted on the map. What is needed is some standard against which this measure can be compared. The obvious standard is the expected average distance between nearest neighbors in a random point pattern. It turns out that the expected mean nearest-neighbor distance for a random point pattern is given by

$$\bar{d}_e = \frac{1}{2\sqrt{n/A}}$$

where n is the number of points in the pattern and A is the area of the study region. Therefore, the ratio $R = \bar{d}_a/\bar{d}_e$ is always equal to one if the point pattern being analyzed is random. Where there is a clustered point pattern, distances to nearest neighbors will be small, $\bar{d}_a$ is less than $\bar{d}_e$, and thus $R < 1$. For a perfectly regular point pattern, R attains the maximum possible value of 2.149. Thus, for dispersed point patterns $1 < R < 2.149$. The minimum value of R is zero, and it occurs when the point pattern is perfectly concentrated and all points share the same location. In this case $\bar{d}_a = 0$ and $R = 0$. So for clustered patterns $0 < R < 1$. Coordinates for a set of $n = 20$ points of five maps A, B, C, D, are given below. The coordinates are based on a 10×10 km grid laid over an area $A = 100$ km^2.

	Map coordinates				
Point	A	B	C	D	E
1	(2, 1)	(0.5, 0.5)	(2, 1)	(1, 3)	(2, 4)
2	(2, 3)	(4, 0.25)	(2, 3)	(1, 5)	(2, 4.5)
3	(2, 5)	(7, 0.75)	(1, 2)	(1, 9)	(2, 5)
4	(2, 7)	(9.5, 0.5)	(3, 2)	(2, 1)	(2, 5.5)
5	(2, 9)	(5, 2)	(5, 4)	(2, 4)	(2, 6)
6	(4, 1)	(2, 2.25)	(1, 5.5)	(3, 2)	(1.5, 5)
7	(4, 3)	(4, 4)	(4, 5.5)	(3, 6)	(1.5, 1.5)
8	(4, 5)	(5, 5)	(3, 6)	(3, 10)	(3,5)
9	(4,7)	(8,4)	(3, 7)	(4,8)	(3,5.5)
10	(4, 9)	(8.5, 3.5)	(3.5, 6)	(5, 1)	(1.5, 6)
11	(6, 1)	(8.5, 4.5)	(6, 5)	(5, 4)	(7, 4)
12	(6, 3)	(2, 6.5)	(6, 5.5)	(6, 6)	(7, 4.5)
13	(6, 5)	(0.5, 9.5)	(6, 6)	(6, 9)	(7, 5)
14	(6, 7)	(3, 9)	(6, 6.5)	(7, 2)	(7, 6)
15	(6, 9)	(5, 9.5)	(9, 2)	(7, 4)	(6.5, 4)
16	(8, 1)	(9.5, 9.5)	(8, 9)	(8, 7)	(6.5, 6)
17	(8, 3)	(4, 7.5)	(8, 8)	(8, 9)	(8, 4)
18	(8, 5)	(4.5, 7)	(8, 7)	(9, 3)	(8, 5)
19	(8, 7)	(4.5, 8)	(9, 8.5)	(9, 8)	(8, 6)
20	(8, 9)	(5, 7.5)	(7, 8.5)	(10, 5)	(7, 5.5)

a. Draw maps of the five-point patterns using graph paper.
b. Calculate distance between each point and its nearest neighbor. A spreadsheet program is a useful vehicle for this analysis. Use a 20 by 20 range to calculate the distances and the MIN function to isolate the nearest neighbor to each point.
c. Determine $\bar{d}_a$ for each map, the actual mean distance to the nearest neighbor. If you are using a spreadsheet program, use the AVG or AVERAGE function to determine this value.
d. Find $\bar{d}_e$ for a map with $n = 20$ points and $A = 100$ km^2. Calculate R for each map. Classify each map as clustered, dispersed, or nearly random.
e. Suppose the actual area on the map from which the coordinates are taken is 200 km^2. Does this alter any of your conclusions?

7

Sampling

In Chapter 1, statistical methodology was conveniently divided into *descriptive statistics* and *inferential statistics*. In inferential statistics, a descriptive characteristic of a sample is linked with probability theory so that a researcher can generalize the results of a study of a few individuals to some larger group. This idea was made more explicit in Chapter 6, where the notions of a *random variable* and its *probability distribution* were defined. At the core of inferential statistics is the distinction between a *population* and a *sample*. From a statistical universe or population, a small subset of individuals is selected for detailed study. This sample is used to estimate the value of some population characteristic or to answer a question about a particular characteristic of the population. However, to make such inferences, the sample must be collected in a specific way. It is not possible to make statistically reliable inferences from any sample. Whereas street corner interviews, for example, tend to make interesting news items, they may not reflect the views of the population they are supposed to represent.

Ideally, it would be best to have a sample that is a good representation of the population from which it has been drawn. High-quality inferences are made by using high-quality samples. Unfortunately, unless we know everything about the population, say from a census, we have no way of knowing if we do have a representative sample. The very act of sampling thus introduces some uncertainty into our inferences, simply because the sample may *not* be representative of the population. This is known as *sampling error.* Suppose, for example, we wished to sample the students at a university and determine the number of hours the average student spends studying in any given week. By mere chance, we may just select a sample that includes more industrious students than average students and thus overestimate the amount of time spent studying. We might be led to believe that the average student spent more time studying than is, in fact, the case. Our only safeguard against such sampling error is to select a larger sample. The larger the sample, the more likely it includes a true cross section of the population; that is, the more likely it is *representative* of the population. Notice that sampling error is not a "mistake" such as choosing the "wrong" sample or some other methodological failing. All samples deviate from the population in some way; thus, sampling error is always present. The associated uncertainty is the price one pays for using a subset of the population rather than the entire population. The appeal of statistics is not that it removes uncertainty, but rather that it permits inference in the presence of uncertainty.

DEFINITION: SAMPLING ERROR

Sampling error is uncertainty that arises from working with a sample rather than with the entire population.

Besides sampling error, there is another reason why our sample may not be representative of the population. This could occur if the way in which the sample is collected is itself biased. This is known as *sampling bias*. In the example of university student study habits, a sample would surely be biased if it were selected by interviews of students leaving the university library late in the evening!

DEFINITION: SAMPLING BIAS

Sampling bias occurs when the procedures used to select the sample tend to favor the inclusion of individuals of the population with certain population characteristics.

Sampling bias can usually be avoided, or at least minimized, by selecting an appropriate sampling plan. Errors in recording, editing, and processing sample data can likewise be limited by various checks. When data are collected through mail questionnaires, a form of bias due to *nonresponse* often occurs. The respondents to the questionnaire may not be representative of the overall population. Many studies have found these respondents to be more highly educated, wealthier, and more interested in the subject of the questionnaire than members of the population at large. Since the quality of the inferences from a sample depends so much on the sample itself, clearly any researcher must carefully select a sampling plan capable of minimizing, or at least controlling to acceptable limits, both sampling error and sampling bias.

Sampling techniques with this characteristic are the focus of this chapter. First, the advantages of sampling are enumerated in Section 7.1. Why do we favor a sample over a complete census of a population? In Section 7.2 an extremely useful four-step procedure for sampling is outlined. The tasks defined in these steps are encountered in every sampling problem. Various types of samples are identified in Section 7.3. Only specific types of samples can be used to generalize to a population—with a *known* degree of risk. The most commonly used type, the *simple random sample*, is explained in Section 7.4. It is then compared to a few other sample designs. In Section 7.5 the concept of a sampling distribution is introduced. The sampling distribution of sample statistics such as the mean $\bar{X}$ and proportion P is central to both the estimation and the hypothesis testing procedures of statistical inference. Finally, issues arising in geographic sampling are presented in Section 7.6.

7.1. Why Do We Sample?

Seldom must we collect information from all members of a population in order make reliable statements about the characteristics or attributes of that population. Often a sample constituting only a small percentage of the total population is sufficient for such inferences. There are several reasons for choosing a *sample* rather than a census of an entire population.

1. Usually it is not necessary to take a complete census. Valid, reliable generalizations about the characteristics of a population can be made with a sample of modest size—if the sample is properly taken. The uncertainty inherent in generalizing from the few to the many not only is within acceptable limits, but also sometimes is even less than the uncertainties that arise when we try to precisely control the enormous amount of data generated from a complete enumeration of an extremely large population. It is simply far easier to check the data of a small sample than those of a large population.

2. The time, cost, and effort of collecting data from a sample are usually substantially less than are required to collect the same information from a larger population. The workforce or available financial resources usually constrain a researcher from taking a full census.

3. The population of interest may be infinite, and therefore sampling is the only alternative. We could, for example, consider the population to be the water temperature at a certain depth at a given reach of a particular stream. There are an infinite number of times when we could record the water temperature—even in a small time interval. Because space itself can be treated as a continuous variable, there are an infinite number of places in any area where a set of sample measurements could be taken. This issue is explored more fully in Section 7.6.

4. The act of sampling may be destructive. To estimate the mean lifetime of light bulbs, for example, any light bulb in a sample must be tested until it is no longer of use. A census of the light bulbs produced by a manufacturer would destroy the entire production!

5. The population may be only hypothetical. In the case of the light bulb manufacturer, the real population of interest is the set of bulbs that *will* be produced by the manufacturer in the future. At the time any sample is taken, this population is not observable.

6. The population may be empirically definable, but not practically available to a researcher. Not all slopes in a study region may be accessible to a geomorphologist interested in studying the dynamics of freeze-thaw weathering. Even an experienced climber may find only a few suitable sites for study.

7. Information from a population census can be quickly outdated. Given the volatility of political polls, it would certainly be unwise to determine the party supported by each member of a population. *Repeated* censuses of this type would be impossible, and sufficient accuracy can be obtained by using only a small proportion of voters. Repeated polls of this type are a usual occurrence now.

8. When the study requires an in-depth study of individuals in the population, only a small sample may be possible. By restricting attention to only a few individuals, extremely comprehensive information can be collected. A study of the residential mobility of residents of a large city might require detailed questions concerning the history of moves, characteristics of the current and past residences, motivations for each move, the search process used to locate new residences, and characteristics of the household itself. Probing for sufficient detail in all these areas precludes the possibility of a complete census of the population. Such a task would certainly be beyond the resources of most institutions. For

a variety of reasons, then, many research questions must be answered through the use of a small sample from a population. Providing the sample is collected properly, valid conclusions about key characteristics of the population can be drawn, with only a surprisingly small degree of uncertainty.

7.2. Steps in the Sampling Process

Having decided that no suitable data exist to answer some research question, and concluding that a sample is the only feasible method of collecting the necessary data, the researcher must specify a sampling plan. Rushing out to collect the data as quickly as possible is often the *worst* thing that could be done. It is far better to follow the simple five-step sampling procedure illustrated in Figure 7-1. Many potential problems not anticipated by the researcher can be addressed and successfully solved *before* a considerable effort has been put into the actual task of data collection. In literally hundreds, if not thousands, of studies, insufficient time and care in devising a sampling plan have led to the collection of large data sets with only limited possibilities for statistical inference. Let us consider each of these five steps in turn.

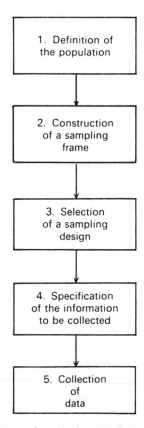

FIGURE 7-1. Steps in the sampling process.

Definition of the Population

The first step is to define the population. What at first glance might appear to be a trivial task often proves to be an extremely difficult chore. It is easy to conceive of a statistical population as a collection of individual elements that may be individual people, objects, or even locations. However, to actually identify which individuals should be included in the population and which should be excluded is not so simple. To see some of the potential issues and difficulties, consider the problem of defining the population for a study of the elderly in some particular city. A number practical questions immediately surface:

1. How will we distinguish the elderly from the nonelderly? By age? If so, what age? Age 60? Age 65?
2. Or, will the elderly be defined by occupational categories? Should we restrict ourselves to retired persons? Or should we restrict the population to persons over 65 years of age *and* retired?
3. Will we include all elderly or those living independently, that is, not in some long-term care home?
4. Is the study concerned with elderly persons or households of the elderly? What about mixed households with both elderly and nonelderly members?

As you can see, even if we can conceptualize the population of interest, arriving at a strict, operationally useful definition may require considerable thought and difficult choices. In the study of the elderly, it is still necessary to define a time frame and a geographical limit to the study region.

Construction of a Sampling Frame

Once we have chosen the specific definition to be used in identifying the individuals of a population, it is necessary to construct a *sampling frame.*

DEFINITION: SAMPLING FRAME

A sampling frame (also called a *population frame*) is an ordered list of the individuals of a population.

There are two key properties of the sampling frame. First, it must include all individuals in the population; that is, it must be exhaustive. Second, each individual element of the population must appear once and only once on the list. Obtaining a sampling frame for a particular population may itself be a time-consuming task. It is usually easy to compile a list of all current students at a university in a given academic year from existing academic records or transcripts. But what if the population of interest is not regularly monitored in any way? Where, for example, could a list of all the elderly residents of a city be obtained? It may be possible to extract a fairly complete list of the elderly by examining the list of recipients of social security or old-age assistance from a government agency. But would the list include all the elderly? What about noncitizens or residents otherwise ineligible for this type of aid?

As a second example, consider the use of telephone surveys for evaluation of voter preferences for political parties. Although the population of interest is all eligible voters, the population actually sampled is composed of those residents with telephones—or, more accurately, the set of persons who answer these phones. There is clearly a great deal of overlap between these two groups, but they are not exactly the same. Restricting ourselves to those with listed telephone numbers may exclude some relatively wealthy residents with unlisted numbers as well as some poorer households without telephones. So it is useful to distinguish the *target* population from the *sampled* population.

DEFINITION: TARGET AND SAMPLED POPULATIONS

The target population is the set of all individuals relevant to a particular study. The sampled population consists of all the individuals listed in the sampling frame.

Obviously, it is desirable to have the sampled and target populations as nearly identical as possible. When they do differ, it is extremely important to know the particular way(s) in which they differ, since this is a form of sampling bias. It is sometimes necessary to qualify the inferences made by using a sampled population that differs in significant ways from the target population. This is equivalent to recognizing the limitations imposed on the study by the sampling frame.

Selection of a Sample Design

Next we must decide how we are going to select individuals from the sampling frame to include in the sample.

DEFINITION: SAMPLE DESIGN

A sample design is a procedure used to select individuals from the sampling frame for the sample.

There are several ways that this could be done. We could select a sample simply by taking the first *n* individuals listed in the sample frame. Or we could select the last *n* individuals or every kth individual on the list until we get *n* members for the sample. There are many types of samples and sample designs. Because of the importance of this step, it is described in depth in Sections 7.3 and 7.4. At this point it is sufficient to note that a *random* sample is an extremely useful design in statistical analysis. Individuals to be included in the sample are chosen by using some procedure incorporating chance. The mechanical devices used in many lotteries are one example. An urn is filled with identical balls, one for each member of the sampled population. The sample is chosen by selecting balls from a well-mixed urn, one at a time.

The important characteristic of this type of sample is that we know the probability that each individual in the population is included in the sample. In this case, each individual has an equal chance of being included. A number of variations of this design are explained in Section 7.4.

Specify the Information to Be Collected

This step can usually be accomplished at *any* point before the commencement of data collection. The particular format used to collect data must be rigorously defined, and pretested by using a pilot sample.

DEFINITION: PILOT SAMPLE OR PRETEST

A pilot sample, or pretest, is an extended test of data collection procedures to be used in a study in advance of the main data collection effort.

In a field study, the pretest can be used to check instruments, data loggers, and all other logistics. For surveys—mail, telephone, or personal interview—the pretest can sometimes reveal deficiencies for any of the following reasons: difficulty in locating population members; dealing with an abnormally high percentage of refusals or incomplete questionnaires; problems in questionnaire wording, question sequence, or format; unanticipated responses; or inadequately trained interviewers.

Collection of Data

Once all the problems indicated in the pretest have been successfully solved, the ultimate task of data collection can begin. At this stage, careful tabulation and editing are particularly important if we wish to minimize nonsampling error.

7.3. Types of Samples

In this section, we expand on the ideas discussed in the third step of the sampling process, the selection of a sampling design. Sampling designs can be conveniently divided into two classes: probability samples and nonprobability samples. Simple random sampling is one type of probability sample.

DEFINITION: PROBABILITY SAMPLE

A probability sample is one in which the probability of any individual member of the population being picked for the sample can be determined.

Because we know only the probability that an individual is included in the sample, an element of chance, or uncertainty, is introduced into any inferences made from the sample. Expressed simply, it could happen that a particular random sample is quite unrepresentative of the population it is supposed to reflect. The advantage of a probability sample is that we can determine the *probable* accuracy of our results. The four principal types of probability samples are shown in Figure 7-2. Because of their importance in statistical inference, these samples are discussed in depth in the following section.

Nonprobability samples may also be excellent or poor representations of the population. The difficulty is that whether it is a good or bad sample can never be determined. Four

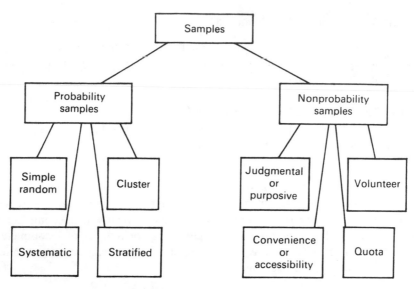

FIGURE 7-2. Types of samples.

types of nonprobability samples are sometimes used to collect sample data: *judgmental*, *convenience*, *quota*, and *volunteer*.

DEFINITION: JUDGMENTAL, OR PURPOSIVE, SAMPLE

A judgmental, or purposive, sample is one in which personal judgment is used to decide which individuals of a population are to be included in the sample. These are individuals that the investigator feels can best serve the purpose of the sample.

Obviously, a very skillful investigator with considerable knowledge of a population can sometimes generate extremely useful data from such a sample. If the deliberate choices turn out to be poor, however, poor inferences are made. The risk is not known. Purposive samples are sometimes selected in pretest, or pilot, samples. A range of respondents, including both "typical" and "unusual" individuals, is chosen. This sample is used to get an idea of what the full range of questions in the survey should be. In addition, it is not unusual to use these interviews to preview the types of answers likely to be given by the respondents in the actual survey. Many times, this information can be used to significantly improve the survey instrument. It would be completely erroneous to utilize such a sample to draw conclusions about the whole population.

DEFINITION: CONVENIENCE, OR ACCESSIBILITY, SAMPLE

A convenience, or accessibility, sample is one in which only convenient, or accessible, members of the population are selected.

Because they include only easily identifiable members of a population, convenience and accessibility samples are subject to sampling bias. These individuals are almost always special or different from other population members in some way. Street corner interviews rarely reflect overall opinion; only certain persons will allow themselves to be interviewed by the media.

There are two other types of nonprobability samples. Before probability samples were widely accepted in empirical surveys, a pseudoscientific approach known as quota sampling was common.

DEFINITION: QUOTA SAMPLING

A quota sampling is an attempt to obtain a representative sample by instructing interviewers to obtain data from given subgroups of the population.

The basic idea behind quota sampling is sound. Suppose, for example, that we were attempting a voters' poll and knew voter preference varied by whether the respondent was male or female, resided in an urban or rural area, and was young or old. If we knew the breakdowns of the population among these categories, we would instruct one of our interviewers to obtain a specific number of young male voters in some given rural area. Similar instructions could be given to other interviewers for other categories. Some judgment is involved because the interviewer must select the actual respondents in that category. A more sophisticated form of quota sampling known as *stratified* random sampling is discussed in the following section.

Another type of sample that fails to take adequate safeguards against sampling bias is the *volunteer sample.*

DEFINITION: VOLUNTEER SAMPLE

Individuals who volunteer to take part in a survey are self-selected from the population.

Such samples are rarely representative. Usually, these persons are more motivated and have a higher interest in the topic than the general population. Because the respondents to mail questionnaires sometimes have these characteristics, this type of survey is often subject to nonresponse bias.

The sampled population units in a probability sample are always chosen according to some probability plan. This minimizes the sampling bias. Even though the resulting sample need not be representative of the population, we can still determine the probable accuracy of our results. For these reasons, a probability sample is the preferred way in which a sample should be drawn from a population.

7.4. Simple Random Sampling and Related Probability Designs

In this section, we develop the notion of one kind of probability sample, and we illustrate its operation by using the random number table. Three other probability sampling techniques—stratified, systematic, and cluster sampling—are also discussed and compared to simple random sampling.

In most social science applications of simple random sampling, the population from which we are sampling is finite. Nevertheless, it is useful to distinguish between two definitions for simple random sampling, one that applies when the population is finite in size and the other that applies when the population is infinite.

DEFINITION: SIMPLE RANDOM SAMPLE AND FINITE POPULATION

A simple random sample from a finite population is one in which each possible sample of a given size n has an equal probability of being selected.

To illustrate this definition, consider a population consisting of the elements A, B, C, and D. We wish to draw a sample of size $n = 2$ from this population according to the definition given. First, how many samples of size $n = 2$ are there? As illustrated in Figure 7-3, there are $C_2^4 = 6$ samples of size $n = 2$, and each must have an equal chance of being selected. Notice that the samples AA, BB, CC, and DD are not included in this list. When each unit of the population is selected for the sample, it is not replaced, so it cannot be drawn a second time. We restrict ourselves to such sampling, known as *sampling without replacement.* In fact, by not putting the unit first drawn back into the population, we must be aware that sampling without replacement does not yield a *statistically independent* sample. For such a sample, we would require that the act of choosing any observation from our sample not affect the probability of choosing any other element of the population. To develop a statistically independent sample, we would have to put each unit chosen back into the population, so that it could be drawn again. In most instances, the difference between these two samples is insignificant. The procedure based on sampling without replacement is thus used in this case.

It is relatively easy to select a simple random sample. For most populations and samples of modest size, it is not feasible to develop a list of all possible samples as in Figure 7-3. For a population of size $N = 10,000$ and a sample of $n = 100$, there are an impossibly large number of samples to list. Instead, we select one observation for the sample at a time from the sampling frame. To illustrate this procedure, consider the problem of selecting a sample of size $n = 10$ from a population with $N = 87$ members. The procedure described for this case can be easily generalized to larger samples from larger populations.

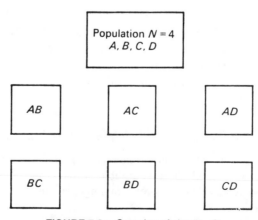

FIGURE 7-3. Samples of size $n = 2$.

To perform this procedure, we number each individual in the sampling frame consecutively from 1 to 87. Using some type of randomization device, we select the first person from the population. There are several ways that this can be done. One way is to take a set of 87 identical balls, numbered from 1 to 87, and select one from an urn in which the balls have been thoroughly mixed. This is often done mechanically in lottery draws. We then note the individual on the list corresponding to this ball, and we include that element in the sample. Nine more balls are drawn to complete the sample of $n = 10$ elements. The selected balls are not replaced before the next draw.

This somewhat cumbersome procedure can be replaced by another randomization device—a *mathematical* random number generator. Mathematical random number generators are actually commonly encountered in everyday life. They are available on hand calculators, embedded in many video games, and can be found as user functions in most statistical programs and computer spreadsheets, to name just a few examples.

DEFINITION: MATHEMATICAL RANDOM NUMBER GENERATOR TABLE

A mathematical random number generator is an equation or mathematical procedure that produces sequences of random digits.

Before describing how a random number generator works, we need to explain what the term "random" means in this context. In other words, what properties do we require "random" numbers to possess? Though simple to phrase, this is a surprisingly difficult question to answer; we will cover just two issues, *uniformity* and *independence* (see Knuth, pp. 127–155 for a complete discussion). Let us suppose that we have a population of size 2000, so that we require four-digit random numbers. A successful random number generator will provide numbers that are:

1. *Uniform.* We want all individuals to have an equal chance of being included in the sample; thus, no value produced by a random number generator should be any more likely than any other. This means that the probability distribution of the numbers produced must be uniform. Equivalently, all 10 digits must have an equal probability of occurring in each of the four positions.

2. *Independent.* As numbers are generated, we want successive numbers to be statistically independent of one another. In other words, there should be no possibility of predicting the next number before it appears. Note that this is not the same as being uniform. A generator that produces the integers 1 to 1000 in order is giving uniform numbers, but not independent numbers. Knowing that the last number was 534, we can predict the next number with certainty. Imagine the consequences if winning lottery numbers were chosen in this fashion! Thinking more about independence, it seems clear that once again independence in the digits of the number is required. If the hundreds digit cannot be predicted by means of knowing any of the other digits, then the sequence of four-digit numbers will be independent.

The above suggests that if we can generate a long sequence of random single digits, then we can form random numbers of any length. What works for four-digit numbers will work for 10-digit numbers, and so forth. We will just pull out four-digit or 10-digit clumps as needed. Can we think of any examples of long random numbers? The digits of π are one such example. There is no pattern (yet discovered) to those digits. One can start at any point in the sequence of digits and find a random number. Knowing what came before provides no clue about what follows, and no digit appears more often than any other. If we wanted to, we could use π as a source of random numbers. Notice that, though random, the method is deterministic in the sense that it always produces exactly the same sequence from the same starting point. Because of this, such generators are more properly called *pseudo-random* number generators.

The most popular mathematical generator works somewhat like our imaginary π-based method. It is called the *linear-congruential method*, and works by dividing one number by another and using the *remainder* as the random number. If it doesn't sound like this could produce a random number, recall that π is the ratio of a circle's circumference to its diameter. The digits in the remainder, the fractional part of π, are random. The general procedure is:

1. Choose a starting, or initial number X_0.
2. Form a new number $Y = aX_0 + b$, where a and b are constants.
3. Divide Y by another constant c, and let the remainder be the first random number X_1.
4. Return to step (2) above, this time using X_1 to form Y.

Of course, the constants a, b, c must be carefully selected—dividing four by two does not give a random remainder—but otherwise the method is just this simple. Notice that the sequence is determined completely by the starting value, so the method gives pseudo-random numbers. Table 7-1 shows a sequence of numbers generated by this technique (similar tables are often printed as appendices in statistics texts).

The next step in the random sampling procedure is to associate one or more random numbers with each member of the population listed in the sampling frame. For any population with up to $N = 100$ members, it is possible to associate one two-digit random number with each of these individuals. The random number pair 01 refers to the first individual listed in the sampling frame, the pair 02 refers to the second individual, and so on. The random number pair 00 is assigned to the last, or $N = 100$th, individual of the

TABLE 7-1
200 Random Designs

61	56	24	90	10	28	59	01	45	47
28	61	34	93	93	02	83	73	99	29
42	40	63	36	82	63	15	01	28	36
68	02	63	83	41	24	30	10	18	50
41	05	14	37	40	64	10	37	57	18
04	59	34	82	76	56	80	00	53	36
69	16	43	43	46	04	35	64	32	05
74	90	09	63	92	36	59	16	05	88
25	29	81	66	99	22	21	86	79	64
43	12	55	80	46	31	98	69	22	22

population. Any random number we draw thus leads us to a particular individual in the population. For a population with up to $N = 1000$ members, a three-digit random number can be associated with each individual in the sampling frame.

To select the sample, choose a place in the random number table to begin—for example, by looking away from the page and placing a finger somewhere in table. Suppose this technique leads to the entry in the fourth row and fourth column, 83. The 83rd individual in the sampling frame is the first element chosen for the sample. Proceeding by columns (rows or even diagonals are also possible), the following individuals will be selected for the first seven members of the sample:

Sample No.	Sampling frame identification
1	83
2	37
3	82
4	43
5	63
6	66
7	80

Thus, moving to the next column, we see the eighth member of the sample is identified as individual 10 on the sampling frame. The next random number of the list is 93. Since this is larger than 87, it is discarded, and sampling is continued. The last two individuals are then selected by using the random numbers 82 and 41.

This simple procedure can be followed in all cases where random sampling from a finite population is required. There is no fundamental difference whether the numbers come from a table, as above, or are generated directly in a computer program. In both cases we obtain an identifying number, uniquely tied to an individual in the population.

Whenever the population from which we are sampling is infinite, it is necessary to modify the previous definition of simple random sampling. There are an infinite number of potential samples for any given sample size n.

DEFINITION: SIMPLE RANDOM SAMPLE IN AN INFINITE POPULATION
From an infinite population, a simple random sample is selected in such a way that all observations chosen are statistically independent.

As an example, consider the problem of selecting a sample of temperatures in a lake or sand grains in a beach dune. Conceptually, we think of temperature as applying to a point in space; hence, there are an infinite number of "individuals" (points) that might appear in the sample. A simple random sample of temperature will be one in which the first observation in the sample is independent of the second, etc., etc. To appreciate what this means, imagine that we draw repeated samples from the same population. Obviously, the individual appearing in the first position will vary from sample to sample, as will the second, and so forth. We see that there is a probability distribution associated with position one, two, and so on. In other words, X_1, X_2, etc. are random variables. Independence in sampling means that the value of X_1 provides no information about the value of X_2 or any other observation. Now, if the observations are independent and come from the same population, it must be that the probability distributions $f(X_1), f(X_2)$, etc., are identical to each other.

This is a great advantage, because it means that when we operate on observations in a random sample, we will be operating on independent and identically distributed variables. As we have seen in Chapter 5, independence greatly simplifies probability theory. Many important theorems in statistics will therefore call for a random sample, not because they are so easy to obtain, but because the theory is so much more tractable.

Returning to the sand grain example, some students will notice that the population cannot truly be infinite, because the dune is obviously finite in spatial extent. Nevertheless, given the huge number of grains in the dune, we are able to treat the population as if it were infinite for the purposes of sampling. Accounting for the differences between sampling with replacement and without replacement will not change probabilities significantly, but will greatly complicate the analysis. There are many other instances in both physical and human geography where we will make this approximation; indeed, we will do so whenever the population is vastly larger than the sample size.

Systematic Sampling

The *systematic random sample* is a variation on simple random sampling. This method is generally easier to operationalize than simple random sampling in terms of both time and cost.

DEFINITION: SYSTEMATIC RANDOM SAMPLE

To draw a systematic random sample, every kth element of the sampling frame is chosen, beginning with a randomly chosen point.

Consider a population with $N = 200$ elements from which we wish to draw a random sample of size $n = 10$. This sample can be drawn by selecting every $k = N/n = 200/10 = 20$th individual listed on the sampling frame. The next thing to be decided is *where* on the sampling frame selection is to begin. The easiest procedure is to draw a single random number between 01 and 20. The possible samples include the following:

01	21	41	$\cdots$	181
02	22	42	$\cdots$	182
03	23	43	$\cdots$	183
$\cdots$	$\cdots$	$\cdots$	$\cdots$	$\cdots$
19	39	59	$\cdots$	199
20	40	60	$\cdots$	200

Why is a systematic random sample *not* a simple random sample? Notice there are not C_{10}^{200} different samples that could be selected—only 20. For example, it is impossible to collect a sample that includes individuals 2, 3, and 27. Only a small fraction of the potential samples of size n may be drawn.

The systematic sample is likely to be as good as a random sample provided that the arrangement of the individuals in the sampling frame is random. Where there are regularities or periodicities in the sampling frame, these will be picked up in the systematic sample. To see the potential problems this may create, consider a sampling frame that includes all the apartments in some high-rise block. Suppose there are 10 floors with 20 apartment units per floor. As is usual in such cases, apartment units above and below any given apartment are identical in floor plan, size, and other features. A systematic sample of size $n = 10$ would probably include 10 identical units. Furthermore, we might expect them to be occupied by roughly similar-sized households. Our systematic sample might include 10 households, each living in a two-bedroom unit. This would hardly be a representative sample if the building contained, one-, two-, and three-bedroom units. As we shall see in Section 7.6, this potential for bias also exists in spatial sampling. Before a systematic sample is drawn, it is always necessary to check for randomness in the sampling frame itself. If the list has been collected in an orderly manner, then one must be sure that such an ordering will not lead to a biased sample through systematic sampling. Many obvious periodicities exist in data ordered in time sequence or by regular spatial intervals.

Stratified Sampling

One of the drawbacks of simple random sampling is that there is always the possibility that the individuals selected for the sample may be unrepresentative of the population. Unlike with simple random sampling, the composition of a stratified random sample is not left entirely to chance.

DEFINITION: STRATIFIED RANDOM SAMPLE

A stratified random sample is obtained by forming classes, or strata, in the population and then selecting a simple random sample from each.

The basic idea behind stratified sampling is illustrated in Figure 7-4. The population is divided into a set of strata before sampling takes place. The divisions must be both

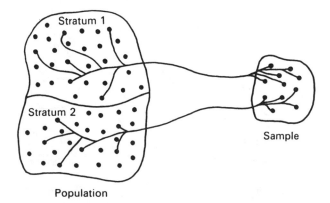

FIGURE 7-4. Stratified random sampling from a population with two strata.

exhaustive and mutually exclusive. That is, each member of the population must appear in one and only one class. Simple random samples are drawn independent of class. Population units may be stratified on the basis of one or more characteristics. Social surveys often stratify the population by age, income, and location of residence.

Why do we stratify? If it is properly organized, stratified sampling can use the additional control over the sampling process to reduce sampling error. Put simply, stratified sampling can decrease the likelihood of obtaining an unrepresentative sample. To reduce sampling error, we define homogeneous strata, that is, strata consisting of individuals that are very much alike in terms of the principal characteristic of the study. Suppose we are interested in the trip-making behavior of the households in a metropolitan area. Knowing that households with two or more cars make more total trips, make different types of trips, and are less likely to use public transport means that they are less variable than the population as a whole. By defining subpopulations or substrata with less *internal* variability, we effectively reduce sampling error. A stratified sample is preferred whenever two conditions are met: First, we must know something about the relationship between certain characteristics of the population and the problem being analyzed. Second, we must actually be able to achieve stratification, that is, to identify population members in each class. Sometimes we may not be able to develop a sampling frame for this purpose. Where, for example, could we obtain a list of households according to the number of cars owned? Various post hoc procedures of classifying households after they are sampled are usually less than satisfactory. Because of the advantages of stratification, it should be undertaken whenever possible. Virtually all political polls employ stratified sampling techniques, particularly by geographic area.

Cluster Sampling

The notion of cluster sampling is illustrated in Figure 7-5. First, the population is divided into mutually exclusive and exhaustive classes, as in the initial step in stratified sampling.

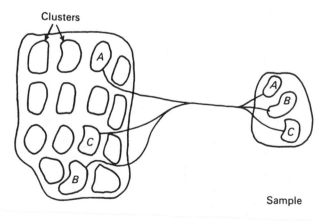

FIGURE 7-5. Cluster sampling.

These classes are usually defined on the basis of convenience, rather than according to some variable thought to be important to the problem under study. Next, certain clusters are selected for detailed study, usually by some random procedure. However, a random sample is *not* drawn from *within* each cluster. Cluster sampling is completed either by taking a complete census of those clusters selected for sampling or by selecting a random sample from these clusters. These units are then combined and constitute the complete cluster sample. Ideally, each cluster should be internally heterogeneous. This is in direct contrast to the situation in stratified sampling where it is desirable to define the strata to be as internally homogeneous as possible.

Cluster sampling can often give very poor results. The sampling error from a cluster is usually higher than the sampling error from a simple random sample or a stratified sample. A cluster sample is effective only when the classes or clusters defined for the problem are each representative of the population as a whole. If they are not representative, then the clusters selected for the sample are likely to be biased and create considerable sampling error. Randomly or arbitrarily defined clusters are unlikely to be representative. Obviously, great care should be taken in specifying the clusters into which any population is to be divided.

Why are cluster samples taken? Usually, the reason is the efficiencies in time and cost. These efficiencies are particularly important whenever the population to be sampled is geographically dispersed over a large area. Suppose, for example, that the problem under study dictates that a personal interview be taken from a sample of such a population. If we were to select a random or systematic sample, this would require a considerable amount of traveling and hence would be a very costly collection procedure. But by restricting our sampling to a small number of clusters, each covering only a small area, these costs are significantly reduced.

Choice of Sampling Design

In terms of efficiency alone, stratified samples are best and cluster samples are worst. Random samples usually lie somewhere between these two extremes. That is, for a given level of precision, stratified samples require the fewest respondents and cluster samples require the most respondents. Unfortunately, sampling efficiency alone does not dictate the type of design chosen in any study. The selection of a stratified design presumes that the relevant information necessary to define the strata and construct the sampling frame is available. Also, the time and cost of collecting the data must be considered. If the sample requires a survey of some human population, then another concern is whether the information is to be collected through a personal interview, telephone interview, or a mail questionnaire.

Finally, many sampling designs used in practice are hybrids that combine elements of several different designs. A sample of housing in a metropolitan area might use stratification to ensure adequate representation of black, Asian, and other ethnic groups. Within these strata, cluster sampling might be used to select particular neighborhoods and/or blocks within these neighborhoods. A simple random or systematic random sample might be used to select the individual households for the sample.

7.5 Sampling Distributions

In Chapter 2, we introduced several numerical measures of data sets, such as the mean, median, standard deviation, and variance. At that time, we made a distinction between numerical measures calculated for a data set that could be considered a *population* and measures for a data set that is a *sample*. Let us now clarify the distinction between such numerical measures. If the measure is computed from a population data set, it is called the *value of the population parameter*; if it is calculated from a sample data set, it is called the *value of the sample statistic*. For a finite population, the *population mean parameter* and the *population standard deviation parameter* are defined in the following way.

DEFINITION: POPULATION MEAN AND STANDARD DEVIATION PARAMETERS (FINITE POPULATION)

Let $x_1, x_2, \ldots, x_N$ be the values of some population characteristic X in a finite population of N elements. The value of the population mean parameter, denoted μ, is

$$\mu = \frac{\sum_{i=1}^{N} x_i}{N} \tag{7-1}$$

and the value of the population standard deviation parameter, denoted σ, is

$$\sigma = \sqrt{\frac{\sum_{i=1}^{N} (x_i - \mu)^2}{N}} \tag{7-2}$$

Only *one* population mean parameter and one population standard deviation parameter exist in any population. All population *parameters* are denoted by Greek letters.

The equivalent numerical measures in the sample, the values of the *sample statistics* for the mean and standard deviation, are defined as follows.

DEFINITION: SAMPLE AND STANDARD DEVIATION STATISTICS

Let $x_1, x_2, \ldots, x_n$ be a set of values of the n elements taken from a population. The values of the sample mean statistic $\bar{x}$ is defined as

$$\bar{x} = \frac{\sum_{i=1}^{n} x_i}{n} \tag{7-3}$$

and the value of the sample standard deviation statistic is defined by

$$s = \sqrt{\frac{\sum_{i=1}^{n} (x_i - \mu)^2}{n - 1}} \tag{7-4}$$

There are two principal differences between these two formulas. First, sample statistics are calculated by using the n values of a sample, and population parameters are calculated by using the N values of the population, $n < N$. Second, the divisor within the radical of σ is N, but for s it is $n - 1$. However, there is another more important difference. Whereas there is only one value for a population parameter such as μ, there are many different possible values for $\bar{x}$. Why? Each different sample of size n drawn from a finite population of N elements can have a different value for $\bar{x}$. In fact, *if we choose a random sample from the population, then the value of the sample mean must itself be a random variable.*

In Chapter 4, we distinguished a random variable by an uppercase letter X and a specific value of this random variable by the corresponding lowercase symbol x. This distinction is also useful in the sampling context. When we refer to a random sample of size n drawn from a population *before* a value has been drawn, the sample is designated X_1, $X_2, \ldots, X_n$. *After* a single sample has been drawn, the *values* of these random variables are known. These n numbers constitute the sample $x_1, x_2, \ldots, x_n$, which is a set of random values. They are random because the value of any of these variables depends on which member of the population is drawn while sampling. As noted earlier, randomness is usually introduced into sampling through the use of a random number generator or table. We must emphasize that saying X_1, X_2, etc. are random is not to say that that all values are equally likely. That would imply X is uniformly distributed, something that will be true only in special circumstances.

Sample Statistics

The sample mean is also a random variable since its value is not known before the sample is actually drawn. Before to taking the sample, the sample mean is a random variable

$$\bar{X} = \frac{\sum\limits_{i=1}^{n} X_i}{n} = \frac{X_1 + X_2 + \ldots + X_n}{n} \tag{7-5}$$

Formally, we say that $\bar{X}$ is a *sample statistic*.

DEFINITION: SAMPLE STATISTIC
A sample statistic is a random variable based on the sample random variables X_1, $X_2, \ldots, X_n$. The *value* of this random variable can be determined once the values of the observations in a specific random sample have been drawn from the population.

Just as there are many numerical measures of a data set, so there are many sample statistics. In fact, each measure has an associated sample statistic. Thus, for example, the sample mean statistic $\bar{X}$ given by Equation (7-5) is the random variable, and its value is given by (7-3). This notation, which differentiates $\bar{X}$ from $\bar{x}$, is necessary so we appreciate that the statistic $\bar{X}$ will vary from sample to sample for a fixed sample size n.

Given that $\bar{X}$ is a random variable, the obvious question is, What is the distribution of this random variable?

Sampling Distributions

DEFINITION: SAMPLING DISTRIBUTION OF A STATISTIC

A sampling distribution is the probability distribution of a sample statistic. The sampling distribution of a statistic can be developed by taking all possible samples of size n from a population, calculating the value of the statistic for each sample, and drawing the distribution of these values.

To illustrate this concept, let us generate the sampling distribution of the sample means $\bar{X}$ for the following population consisting of five elements:

Element	Values of x
A	$x_1 = 6$
B	$x_2 = 6$
C	$x_3 = 5$
D	$x_4 = 4$
E	$x_5 = 4$

Now first let us calculate the population mean and standard deviation, using formulas (7-1) and (7-2):

$$\mu_X = \frac{\sum_{i=1}^{5} x_i}{5} = \frac{6 + 6 + 5 + 4 + 4}{5} = 5 \tag{7-6}$$

$$\sigma_X = \sqrt{\frac{\sum_{i=1}^{5} (x_i - \mu)^2}{5}} \tag{7-7}$$

$$= \sqrt{\frac{(6 - 5)^2 + (6 - 5)^2 + (5 - 5)^2 + (4 - 5)^2 + (4 - 5)^2}{5}} = \sqrt{\frac{4}{5}} = 0.89$$

Notice the divisor in each case is equal to 5, the number of elements in the population.

Suppose a simple random sample of size $n = 3$ is to be taken from this population. How many samples are possible? There are $C_3^5 = 10$ different samples of size 3. These possibilities are listed in Table 7-2, along with the value of the statistic $\bar{X}$. The sampling distribution of the statistic $\bar{X}$, the distribution of sample means, is illustrated in Figure 7-6.

Computing the mean and standard deviation of this distribution yields

$$\mu_{\bar{X}} = \frac{\sum_{i=1}^{10} \bar{x}_i}{10} = \frac{5.67 + 5.33 + \ldots + 4.33}{10} = 5 \tag{7-8}$$

TABLE 7-2
Possible Samples of Size $n = 3$ from Population $N = 5$

Elements in sample	Values of X	Mean $\bar{x}$
ABC	6, 6, 5	5.7
ABD	6, 6, 4	5.3
ABE	6, 6, 4	5.3
ACD	6, 5, 4	5.0
ACE	6, 5, 4	5.0
ADE	6, 4, 4	4.7
BCD	6, 5, 4	5.0
ACE	6, 5. 4	5.0
BDE	6, 4, 4	4.7
CDE	5, 4, 4	4.3

$$\sigma_{\bar{x}} = \sqrt{\frac{\sum_{i=1}^{10} (x_i - \mu)^2}{10}} \tag{7-9}$$

$$= \sqrt{\frac{(5.67 - 5)^2 + (5.33 - 5)^2 + \ldots + (4.33 - 5)^2}{5}} = 3.65$$

The notation used to denote the mean of $\bar{X}$ is $\mu_{\bar{x}}$. Note that this is *not* the same as μ_x even though $\mu_{\bar{x}} = \mu_x$. Notice that the divisor for μ_x in (7-5) is 5, the number of elements in the population, but it is 10 for $\mu_{\bar{x}}$ in (7-8), the number of samples. When we think about how $\bar{X}$ will vary from sample to sample, we would expect it to be distributed about the population mean, μ. Indeed, one of the important properties of the sample mean statistic is that its mean is equal to the population mean.

To compute the sample standard deviation, the mean of each sample of size 3 (given in the last column of Table 7-2) is subtracted from the mean of $\bar{x}$ and squared. The sum is then divided by 10, the number of samples of size 3 that can be drawn from the population with $N = 5$ elements. The fact that $\sigma_{\bar{x}} < \sigma_x$ should not be surprising. We should expect less variability in the distribution of the means $\bar{X}$ than in the population itself. Why? Because

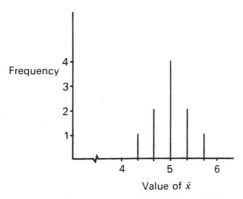

FIGURE 7-6. Sampling distribution of $\bar{X}$.

when we take a sample of items from the population, we tend to select samples in which large values of the variable are averaged with small values, and the result lies close to the population mean, μ. To distinguish σ_X from $\sigma_{\bar{x}}$, it is common to call σ_X the population standard deviation and $\sigma_{\bar{x}}$ the *standard error of the mean*. By convention, the standard deviation of a sampling distribution is known as its standard error.

Inferential statistics would be a time-consuming process if it were necessary to manually generate sampling distributions in this way. Fortunately, this is not the case. A set of theorems can be used to specify the nature of the sampling distributions of many important sample statistics under quite general conditions. We illustrate the principal theorem involved, using as an example the sampling distribution of the sample mean $\bar{X}$.

Central Limit Theorem

DEFINITION: CENTRAL LIMIT THEOREM

Let $X_1, X_2, \ldots, X_n$ be a random sample of size n drawn from a population with mean μ and standard deviation σ. Then for large n, the sampling distribution of $\bar{X}$ is approximately normally distributed with mean μ and standard deviation $\sigma/\sqrt{n}$. In the special case where X is normal, the distribution of $\bar{X}$ is exactly normal regardless of sample size.

This is an extremely powerful theorem. It says that no matter what the distribution of X might be, we can know that the distribution of $\bar{X}$ is approximately normal. Moreover, as we will see later, the approximation is very good for even moderately sized samples—$\bar{X}$ will be nearly normal for n as small as 10 or 20. The implications of the central limit theorem are illustrated in Figure 7-7 for a normal random variable.

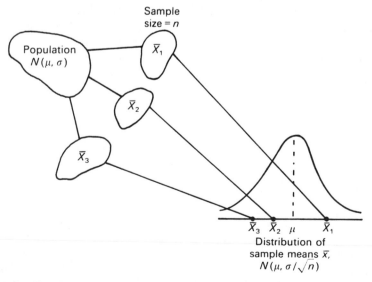

FIGURE 7-7. Central limit theorem and the distribution of sample means.

We need to make several important observations from this theorem. The most important observation is the *inverse* relation between sample size and the variability of the sample statistic $\bar{X}$. As the sample size n increases, the variability of $\bar{X}$ decreases since $\sigma_{\bar{x}}$ = $\sigma/\sqrt{n}$ falls. Suppose the population standard deviation is $\sigma = 100$. Then the following relations always hold:

n	1	4	16	25	100
σ_X	100	50	25	20	10

For $n = 1$, $\sigma_{\bar{x}} = 100$. This is intuitively appealing. If we were to randomly select one element of the population, we would expect our choice to exactly mirror the underlying population. Larger sample sizes give smaller standard errors. Notice that as n approaches infinity, the standard error approaches zero. What does this imply? It tells us that as n increases, the sampling distribution of $\bar{X}$ converges to μ, the mean value of the population mean. This is also intuitively obvious. We would expect the value of $\bar{X}$ to be closer to μ as the sample size becomes larger. This feature is illustrated in Figure 7-8.

It is necessary to distinguish between situations where the random variable X is normally distributed and where it is not. Where X is *not* normally distributed, the theorem states that $\bar{X}$ is only approximately normal. As we have hinted, the approximation improves very rapidly with increasing sample size. This feature can be illustrated by a simple example. Suppose the population characteristic of interest is the time between the arrival of recreation parties at a wilderness park. Parties are delivered by buses, which arrive at various intervals, but never more than 2 h apart. We will assume that the distribution of intervals is uniform. Because the interval cannot be negative, the random variable X runs from 0 to 2, with a mean of 1 h.

Using a pseudo-random random number generator, hypothetical samples of X were drawn by computer. One thousand random samples were drawn for sample sizes $n = 1, 2, 4, 16,$ and 32. The mean for each sample was computed and accumulated in the frequency distributions plotted in Figure 7-9. Note that for a samples of size $n = 1$, the mean is just the single value contained in the sample. Figure 7-9(a) shows the sample values more or less uniformly distributed, as would be expected given that X is uniform. For $n = 1$, the sample mean $\bar{X}$ is far from normally distributed, as it cannot be any different than the distribution of X. However, as the sample size increases, the central limit theorem suggests

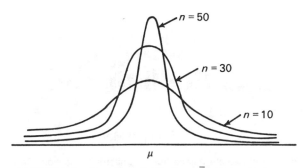

FIGURE 7-8. Sampling distributions $\bar{X}$ with increasing n.

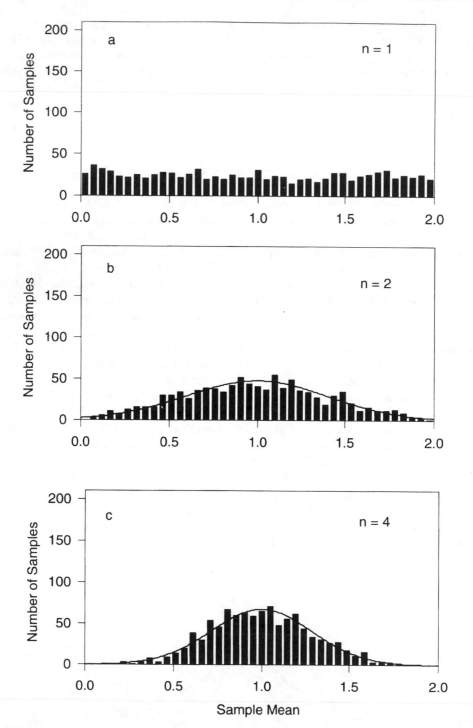

FIGURE 7-9. Distribution of sample means from a uniform distribution. Solid curve is the normal distribution given by the central limit theorem.

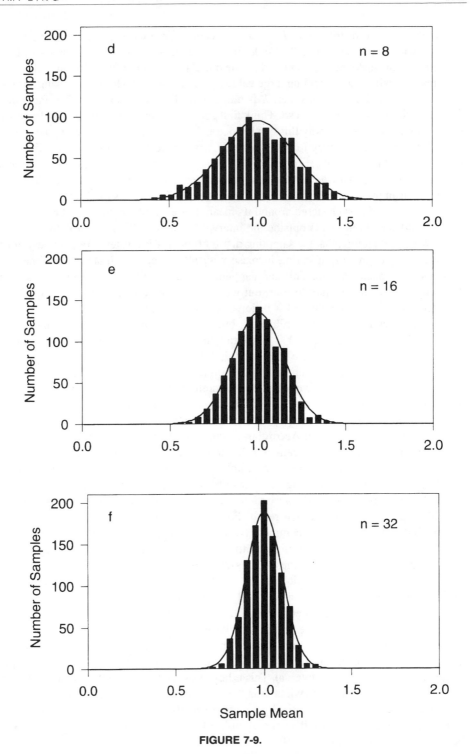

FIGURE 7-9.

that the frequency distribution of $\bar{X}$ should become progressively closer to normal, no matter what the distribution of X. We do, in fact, see this behavior in Panels (b–f). Notice how each of the histograms is centered on the population mean μ. We explained this earlier in terms of small values canceling large values. Notice also that, although X is uniform with no value preferred above any other, $\bar{X}$ is much more likely to be near μ than far away. A little thought reveals why this is so. Think first about what is needed to obtain a large value for $\bar{X}$. Obviously, the only way this can happen is for the sample to have many large values of X. Similarly, the only way to obtain a small $\bar{X}$ is for the sample to contain mainly small values. Both of these events are relatively unlikely. It will be much more common to obtain samples that contain a *mixture* of high and low values. Clearly, a *mixed* sample will produce a sample mean of *intermediate* value. Because intermediate values of $\bar{X}$ are more common, the distribution of $\bar{X}$ is clustered around its mean. For example, with an n of only 16, none of the 1000 sample means is outside the interval 0.5 to 1.5 (Figure 7-9e).

We see, therefore, that the sampling distribution of $\bar{X}$ is clustered rather than uniform. Why does the degree of clustering increase with n? To see this, imagine that one of our samples has a mean of zero. This can happen, but only if all the sample values are zero. This might occur for n equal to four, but would be extremely unlikely for n equal to 32. In other words, extreme values of $\bar{X}$ become less and less likely as n increases. Figure 7-9 shows this behavior clearly for uniform X, but we would see the same result regardless of how X is distributed. The closer X is to normal, the closer to normal will be the distribution of $\bar{X}$. In the extreme case where X is normal, $\bar{X}$ will be exactly normal.

As defined above, the central limit theorem applies to both continous and discrete random variables. We might wonder how a discrete random variable can possibly give rise to a continuous (normal) variable. To see this, consider the probability distribution in Figure 7-10(a). The random variable takes on only three values (1, 2, or 3), with the central value less likely than the other two. According to the central limit theorem, if we sample from this distribution, the sample means will be approximately normal for large n. In Figure 7-10(b), we have drawn 1000 random samples of size two and plotted a histogram of the resulting sample means. Although X takes on only three values, there are five possible values for the sample mean. Because there are three ways for a sample to have a mean of two, $\bar{x} = 2$ is the most common result. For $n = 4$, there are still more outcomes, with intermediate values still more common. Notice how the probability of extreme outcomes ($\bar{x} = 1$, $\bar{x} = 3$) decreases for this larger sample (Figure 7-10c). As the sample size increases to eight and beyond, more outcomes are possible, and they are distributed ever more closely to the central value. In the limit (infinite n), the number of outomes is infinite, and the distribution of $\bar{X}$ is continuous, as implied by normality.

The great value of the central limit theorem is that it provides a way of deducing results concerning the outcome of a sample based only on a knowledge of the population mean and standard deviation. In particular, the theorem allows us to determine the *probability* that the sample mean statistic $\bar{X}$ is larger than a given value, or smaller than a given value, or falls within a given interval. This is best illustrated by a simple numerical example. Let us return to the example in which X is the interval between arrivals at a wilderness park. This time let us suppose that arrival gaps are normally distributed with $\mu = 8$ h with $\sigma = 4$ h. What is the probability that in a sample of $n = 36$ arrival intervals the mean time between arrivals is greater than 10 h?

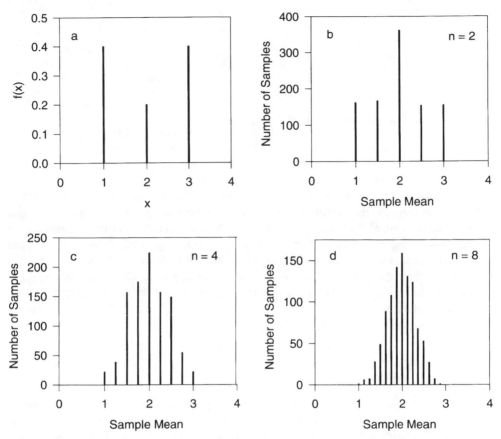

FIGURE 7-10. Sampling from a discrete probability distribution. As *n* increases, sample means become more numerous, approaching a continuous normal distribution.

Since the random variable X is normally distributed, we know the distribution of $\overline{X}$ is normal, with a mean of 8 and a standard deviation $\sigma_{\overline{x}} = 4/\sqrt{36} = 0.667$ hours. This sampling distribution is shown in Figure 7-11. We are interested in determining the probability that $\overline{X}$ is greater than 10 h, the shaded area of the sampling distribution. To find this probability, we must first standardize the variable $\overline{X}$, using

$$z = \frac{\overline{x} - \mu}{\sigma_{\overline{x}}} = \frac{10-8}{0.667} = 3.0$$

Then, using the normal table, we see the desired probability is $P(Z > 3) = 0.001$.

In Chapter 8 we will use this theorem in another way. We will take a characteristic of a sample such as $\overline{X}$ and then determine the probability that it comes from a population with hypothesized population characteristics.

The central limit theorem applies not only to the sample mean, but also in fact, to *any*

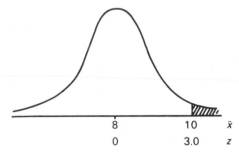

8	10	$\bar{x}$
0	3.0	z

FIGURE 7-11. Sampling distribution of arrival time intervals for $\mu = 8$, $\sigma = 4$, and $n = 36$.

statistic computed by summing the values of a sample. Because of this, we will be able to apply the theorem to other statistics, like the sample proportion p. Still other statistics, such as the sample standard deviation S, cannot be handled by the central limit theorem and have different sampling distributions. In some cases these sampling distributions are known, but only under certain restrictive assumptions about the distribution of X. By way of contrast, the central limit theorem has the great advantage of requiring almost no assumptions. Indeed, it is precisely because the central limit theorem is so widely applicable that the normal distribution is so important in statistics. The specification of other (nonnormal) sampling distributions is left to Chapter 8.

7.6. Geographic Sampling

There are at least two instances in which geographers must take spatially distributed samples. First, there is the case in which the geographer must go into the field and sample some areally distributed phenomenon. A biogeographer may wish to study the distribution of a particular plant species or even the distribution of seeds from a particular plant or set of plants. As another example, field sampling may be required for the compilation of a soil map. Second, sometimes areal sampling is necessary in direct sampling of locations on a map. This is particularly important when we are sampling from maps over which some phenomenon is continuously distributed. Land-use maps of cities, soil maps, and vegetation maps are typical examples of this form of sampling. For phenomena *discretely* distributed on a map, conventional sampling procedures can be used. Consider the problem of selecting a set of farms distributed on a map of some rural area. It is sufficient to compile a list of all farms in the area and then sample from this sampling frame in the usual way. We may even sample from all parts of the map by dividing the list into strata based on location. Such a procedure is basically independent of the map itself. For the complex patterns that exist on most continuously variable maps, such a procedure is totally inappropriate. Instead, these maps are sampled by using traverse samples (or lines), quadrats, or points. These three alternatives are illustrated in Figure 7-12.

Before examining the sampling designs used with each of these techniques, we specify the sampling frame implicit in spatial sampling. This is equivalent to the ordered list of the elements in a population used in conventional nonspatial sampling. For sampling from a map, the Cartesian coordinate system is sufficient to identify any point on the map

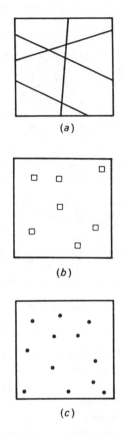

FIGURE 7-12. (a) Traverse samples, (b) quadrat samples, (c) point samples.

and hence any element of the population. This coordinate system meets all the requirements of a sampling frame. It is ordered with a unique address for each element of the population. The (x, y) coordinate pairs are mutually exclusive because no two points on the map can possibly have the same coordinates. Cartographers will recognize that we are ignoring problems of spatial distortion associated with small-scale maps. Finally, it is exhaustive of all the elements in the population because it covers the entire map.

Consider the coordinate system used to sample from the area illustrated in Figure 7-13. The two axes follow the usual north–south and east–west orientations, with the origin placed at the lower left, or southwest, corner of the map. The axes are then subdivided into units at any desired degree of accuracy. Two possibilities are shown in Figure 7-13. The simplest one-digit grid uses the numbers 0 to 9 to identify particular points on both the x and y axes. Then random numbers can be used to locate any point on the map. The first random number drawn is used to define the x coordinate, and the second number drawn the y coordinate. If one-digit random numbers are used, then only the 100 points shown as grid intersections can actually be sampled. This *available population* is quite different from the

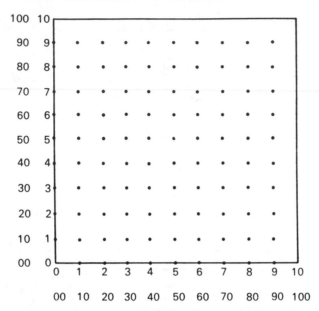

FIGURE 7-13. Coordinate systems and areal sampling.

target, or total, *population*, which is the entire map. If two-digit random numbers are used for the grid on each axis, then there are $100 \times 100 = 10,000$ possible points to be included in this sample. The grid used to define the sampling frame should be sufficiently fine that features with only small areas are included in the available population. Depending on the map in question, it may be necessary to use two-, three-, or even four-digit grid references so that the set of available points adequately covers the target map.

Quadrat Sampling and Sampling Traverses

One of the more commonly used sampling methods is quadrat sampling. Quadrat sampling differs slightly from the quadrat systems used to analyze point patterns. Variations on three different sampling designs are illustrated in Figure 7-14. First, consider the random quadrat sampling design in (a). Such a design can be developed by using three simple rules:

1. Select a quadrat size. There is some choice in size, but larger quadrats sometimes become difficult to analyze. If one is forced to calculate, say, the proportion of the quadrat in various land use categories, then this task itself becomes extremely time-consuming for a large quadrat.
2. Draw two random numbers. Use the first random number to locate the *x* coordinate of the quadrat centroid and the second to locate the y coordinate of the centroid.
3. Select as many quadrats as necessary to reach a sample size deemed sufficient.

As a variation of this algorithm, it is also possible to randomly orient the quadrats by selecting a third random number to control the orientation of the quadrats. This alternative

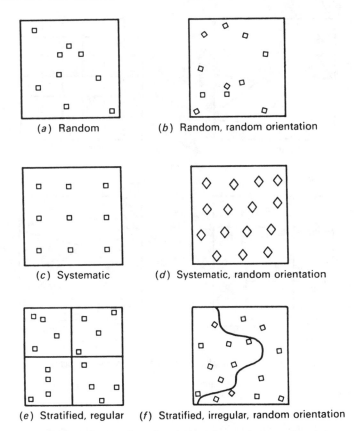

FIGURE 7-14. Sampling designs for quadrat sampling.

is shown in Figure 7-14(b). For very complex maps, the irregular distribution of the map pattern *within* a quadrat may make analysis very difficult. In such a case, point sampling may be a desirable alternative.

A systematic sampling design is illustrated in Figure 7-14(c). A pair of random numbers is used to locate one quadrat on the map, and the remaining quadrats are placed according to a prespecified spacing interval that can yield a sufficiently large sample. In Figure 7-14(d) a systematic sample with a random orientation is illustrated. In the stratified sampling designs of Figure 7-14(e) and (f), the number of quadrats in any stratum is proportional to the area of the map in that stratum. These areally defined strata may be based on either regularly or irregularly shaped portions of the map. In both designs the quadrats are randomly located within each stratum.

Four designs for sampling traverses are illustrated in Figure 7-15. To locate these traverses, the sampling frame must be slightly modified. As illustrated in Figure 7-15(a), the grid references are defined in a clockwise manner from the lower left corner. Again, there is some choice of precision depending on the number of digits used in the grid referencing system. Let us suppose we wish to sample L mi of linear traverse. For the random sampling design, this is accomplished in the following way. First, a single two-digit

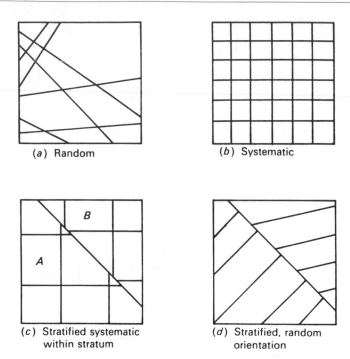

(a) Random   (b) Systematic

(c) Stratified systematic   (d) Stratified, random
within stratum   orientation

FIGURE 7-15. Sampling designs for traverse sampling.

random number is drawn. This number specifies a point on the map boundary that can be used as one end of the first traverse. A second number is used to locate a point on the boundary for the other end of the traverse. A line connecting these two points determines the first traverse. Next, we determine the length of this line. Let this length be l_1 mi. We now have l_1 mi in our sample and require only $L - l_1$ mi to complete the sample. Two more random numbers are used to locate a second traverse, the length of the sample traverse is updated to $l - l$, the remaining length to be sampled is reduced to $L - l_1 - l_2$ mi. This process is continued until *exactly* l mi, or alternatively until *at least* L mi of traverse, has been selected for sampling.

Let us consider the sequence of traverses generated in a study attempting to estimate the proportion of a 100-mi^2 map that is woodland, given a traverse of 200 mi. Figure 7-16 illustrates the sequence of traverses used to generate the sample. The first traverse is 15 mi long, of which 7 mi pass through woodland (W) areas and 8 mi through nonwoodland (NW) areas. These figures are recorded in the first three columns. A cumulative, or running, total of the traverse length, in woodland and in nonwoodland areas, is provided in the last three columns. Notice that to complete the sample of *exactly* 200 mi, the last traverse is truncated, and only 7 of its 10 mi are included in the sample. Alternatively, a sample of $L = 203$ mi could be used by accepting the complete length of the final traverse in the sample. The estimated proportion of woodland on the map is 78/200 = 0.39.

The systematic design of Figure 7-14(b) is based on a regular pattern of perpendicular

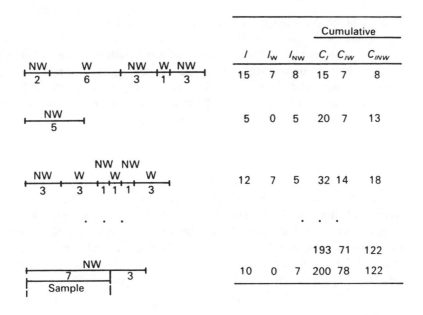

$$\text{Proportion of woodland} = \frac{78}{200} = .39$$

FIGURE 7-16. Estimating the proportion of woodland on a map by using traverses.

traverses with a prespecified interval. Suppose a sample of $L = 250$ traverse miles is to be taken from a 25-mi² area. There will be five traverses in each of the north–south and east–west directions, each 25 mi long. A spacing of 5 mi between traverses is used. A single random number can be used to locate the east–west traverses and another to locate the north–south ones. Where there are periodicities on the map, particularly linear or near-linear features, this sampling procedure can lead to biased estimates of certain map characteristics. Streets, highways, stream networks, eskers, and other similar map features can lead to significant sampling error when maps are sampled by using a systematic design.

It is possible to randomly select the orientation of traverses in a systematic sample, although this is slightly more difficult to operationalize. Additional modifications are also required to sample from an irregularly shaped area. Nevertheless, significant time efficiencies can be realized from systematic sampling, often well worth the added complexity.

An areally stratified sampling design for an irregular map is illustrated in Figure 7-15(c). The areas or strata may be defined contour intervals, geomorphic or geologic features, or even municipal boundaries in some urban area. Two strata, A and B, are depicted on the map of Figure 7-15(c), in the proportion 60 : 40. Sixty percent of the traverse length should thus be drawn from area A and 40% from B. As usual, random numbers are drawn to locate individual traverses.

As each traverse is selected and enumerated, a running total is kept of the total traverse length in each of the strata. Once the length of the traverses in area A reaches 60%

of the desired sample traverse length, only the parts of subsequent traverses in stratum B are used to complete the sample. If the traverses first exhaust the desired length in area B, then subsequent traverse lengths in B are ignored, and only lengths in A are included in the sample. This procedure can be extended to handle any number of areas or strata, of regular or irregular shape. For regular shapes, though, it may be easier to sample each stratum individually. Linear traverses for areally stratified maps can also be taken in a systematic manner and with a randomly chosen orientation, as in Figure 7-15(d).

Point Sampling

Perhaps the simplest form of spatial sampling is point sampling. Selecting a *simple random point sample* from a map is really not much more difficult than selecting a random sample from a conventional sampling frame. To identify each point in the sample, two random numbers are necessary. One is used for the X coordinate and the other for the Y coordinate. Two random numbers are drawn for each point in the sample until n points have been located. Figure 7-17(a) illustrates a typical random point pattern. There are usually a few areas on the map where no points are sampled and a few areas of relatively high point density. Since the design is purely random, there is always the possibility that a non-representative sample could be drawn.

To ensure that all parts of the map are sampled and there are no overconcentrations of sampled points, a *systematic random point sample* can be drawn. A typical systematic sample is illustrated in Figure 7-17(b). The 16 points for this particular map are regularly spaced at equal intervals in terms of both the X and Y coordinates. This design is exceptionally easy to operationalize. The interval is selected so that the number of points that fall on the map equals the desired sample size. First, the point in the southwest corner of the map is randomly chosen within the small area delimited by the dashed lines, then others are chosen at fixed intervals away. Only two random numbers are required for a systematic point sample of any size. Once the location of a single point in the design is known, the location of all the other points is known.

Because this design selects points at regular intervals, the systematic design can lead to a biased estimate of the proportion of a map covered by some phenomenon. Consider the land-use map of a large city. There are many structural regularities in the land-use map of the typical North American city. Streets tend to follow a rectangular grid with a standard distance between street intersections. Successive points taken at regular intervals might all lie on streets. Or perhaps only more of the points lie on streets. In either case, the estimate of the land use in a city devoted to streets would be biased. The bias admitted with a systematic sample is usually small, except for cases in which the map has a systematic linear or near-linear trend. Most often, the ease with which this sample can be drawn outweighs the slight bias introduced by systematization. Systematic sampling is widely used in studies of soils and vegetation.

When it is used in spatial sampling, *stratified random sampling* has the same advantage as systematic sampling, namely, more or less complete coverage of the map. But, as illustrated in Figure 7-17(c), the procedure is less likely to pick up any regularities in the map. Stratified designs have smaller sampling errors than either simple random or systematic designs. A *cluster sample* can also be taken from a map that has been areally stratified. As shown in Figure 7-17(d), the usual procedure is to randomly select certain

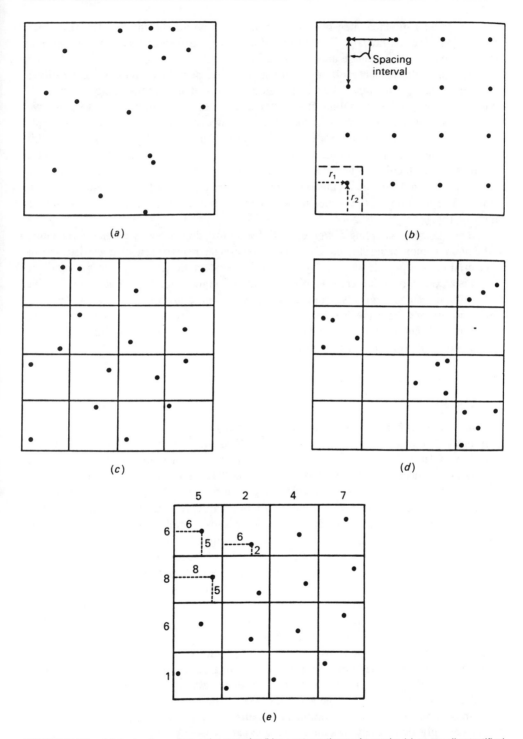

FIGURE 7-17. (a) A simple random point sample; (b) a systematic areal sample; (c) an areally stratified random sample; (d) a cluster sample; (e) a stratified, systematic, unaligned areal sample.

areas on the map and then intensively sample randomly within the clusters. The advantage of sampling ease, particularly in field sampling, must be weighed against the possible disadvantage of omitting large areas of the map.

A hybrid design incorporating elements of these simpler designs is the *stratified, systematic, unaligned sample* shown in Figure 7-17(e). It is stratified, since the map is divided into cells before sampling. It is systematic because it is unnecessary to draw two random numbers for each point in the sample. However, unlike the case of a simple systematic sample, the pattern produced by this procedure is not aligned. To develop this axis, we define a simple (x, y) grid within each cell. A single random number is drawn for each row and each column. The row random number defines the X coordinate within each cell in that row, and the column random number locates the Y coordinate for all cells in that column. The locations of the points in the three most northwesterly cells are illustrated in Figure 7-17(e).

The choice of sampling design in spatial sampling depends very much on the nature of the phenomenon being studied. Each spatial design shares the basic characteristics of its nonspatial counterpart. If the underlying map has systematic features, then a systematic design is a poor choice. If there is *no* obvious pattern on the map, then all designs will give fairly similar results. Typically, stratified samples are more precise than random samples, which in turn are more precise than cluster samples. A research experiment to test the precision of all sample designs found the stratified, systematic, unaligned sample design to be decidedly superior to all other designs.

7.7. Summary

Samples can be drawn from a population in a number of ways. They are usually taken to minimize both the time and the cost of collecting data. Four steps usually precede the actual collection of data: specification of the population, delineation of a sampling frame, selection of a sampling design, and pretest of the data collection procedures. For purposes of statistical inference, a probability sample is the required sampling procedure. Although a probability sample still can lead to the selection of an unrepresentative sample, it is *always* possible to specify the probable accuracy of the results. Simple random samples are the most commonly used method of data collection, although other variations sometimes prove necessary and desirable in practice. *For the remainder of this text, all inferential procedures are based on the assumption that the sample has been collected in this manner.*

Sample statistics are random variables. The value of any sample statistic varies from sample to sample for all samples of a fixed size drawn from the same population. For sufficiently large samples, the sampling distribution of many statistics such as $\bar{X}$ or P is known to be normal or approximately normal. By knowing the characteristics of sampling distributions, it is possible to answer probability questions about one sample. That is, we can determine the likely sampling error in our sampling experiment. In Chapter 8, we show how these sampling distributions can be used to either *estimate* population characteristics or make judgments concerning their hypothesized value.

In Section 7.6, we briefly examined the nature and problems of geographic sampling. An exceedingly rich variety of sample designs can be devised based on linear, point, or quadrat sampling. All methods find considerable application in the geographical literature.

FURTHER READING

Textbooks devoted to advanced topics in sampling theory include Mendenhall, Ott, and Schaeffer (1971) and Cochran (1977). Many textbooks of this type, and in particular the text by Cochran, require considerable background. A more practical approach to sampling questions is taken by Sudman (1976). There are literally hundreds of journal articles on practical techniques of sampling. Two very useful journals to consult are the *Journal of Marketing Research and Public Opinion Quarterly*. For an analysis of sampling techniques in geography, see Dixon and Leach (1978) and Berry and Baker (1968).

B. J. L. Berry and A. Baker, "Geographic Sampling," in B. J. L. Berry and D. F. Marble (eds.), *Spatial Analysis* (Englewood Cliffs, N.J.: Prentice-Hall, 1968, pp. 91–100).

W. Cochran, *Sampling Techniques* (New York: Wiley, 1977).

C. J. Dixon and B. Leach, *Sampling Methods for Geographical Research* (Norwich, England: Geo Abstracts, 1978).

W. Mendenhall, L. Ott, and R. L. Schaeffer, *Elementary Survey Sampling* (Belmont. Calif.: Duxbury, 1971).

S. Sudman. *Applied Sampling* (New York: Academic, 1976).

PROBLEMS

1. Explain the meaning of the following terms:

   a. Sample
   b. Population
   c. Sampling error
   d. Sampling bias
   e. Nonresponse bias
   f. Sampling frame
   g. Target population
   i. Sample design
   j. Pilot sample or pretest

   k. Random number table
   l. Simple random sample
   m. Systematic sample
   n. Stratified random sample
   o. Sample statistic
   p. Sampling distribution
   q. Finite-population correction factor
   r. Standard error of a sampling distribution
   s. Stratified, systematic, unaligned point sample.

2. Explain why, in general, it is advisable to sample rather than to attempt a complete census of a population.

3. Give three examples of finite populations and three examples of infinite populations.

4. Differentiate among (a) probability samples, (b) quota samples, (c) convenience samples, (d) judgmental samples, and (e) volunteer samples.

5. Why does a simple random sample from an infinite population differ from one from a finite population?

6. In stratified sampling, it is desirable to define strata that are internally homogeneous but different from other strata. In cluster sampling, both internal heterogeneity and between- cluster homogeneity are desirable. Why?

7. Why is a sample statistic a random variable?

8. Select a hypothetical population of 100 members. Draw the frequency distribution of some variable of interest. Illustrate the operation of the central limit theorem by drawing repeated samples of sizes $n = 4$ and $n = 10$ and drawing the sampling distributions of sample means. How do these sampling distributions illustrate the implications of the central limit theorem?

9. Distinguish among (a) the population mean of a variable μ, (b) the sample mean $\bar{X}$, and (c) the mean of the sampling distribution of the random variable $\bar{X}$, or $\mu_{\bar{X}}$.

10. A population consists of the following seven observations:

Observation	A	B	C	D	E	F	G
Value	4.0	3.0	5.0	7.0	4.0	8.0	11.0

   a. Find the population mean and standard deviation.
   b. Create a table listing all samples of size $n = 2$. For each sample, find the sample mean $\bar{x}$. Draw a histogram of the sampling distribution.
   c. Find the mean and standard deviation of the sampling distribution of means generated in part (b).
   d. Repeat parts (b) and (c) for samples of size $n = 3$.

11. The map below classifies the land use in a square city into four categories: residential, commercial, industrial, and open space. The percentage of the city land use in each class is as follows: residential (R), 69%; commercial (C), 8%; industrial (I), 10%; open space (0), 13%. Estimate the proportion of the map in each land-use category, using a point sample of 25 observations. Draw one sample each, using the following designs: simple random, stratified random, systematic, and systematic stratified unaligned.

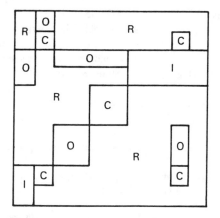

8

Parametric Statistical Inference:
Estimation

The principal objective of statistical inference is to draw conclusions about a population based on a subset of that population, the sample. To be perfectly correct, statistical inference includes all procedures based on using a *sample statistic* to draw inferences about a *population parameter*. In Chapter 1, we distinguished between two ways of drawing inferences about population parameters. The first is *estimation*, in which we guess the value of a population parameter based on sample information. This process is illustrated in Figure 8-1 for the population parameter μ. From some statistical population, we draw a sample of size n and elect to use the sample statistic $\bar{X}$ in making inferences about μ. Once the sample is taken, the values of the sample variables $X_1, X_2, \ldots, X_n$ are known to be $x_1, x_2, \ldots, x_n$, and the *value* of the test statistic $\bar{x}$ is used to estimate the population parameter μ.

The second way of drawing inferences about the value of a population parameter is called hypothesis testing. When we test a hypothesis, we make a reasonable assumption about the value of μ (or some other population parameter) and then use sample information to decide whether or not this hypothesis is supported by the data. For example, we may be interested in a population parameter π, the proportion of residents of a city who moved in the last year. In particular, we wish to determine whether the city has some abnormally high rate of residential mobility, say greater than $\pi = .20$. A sample of residents is taken, and the percentage of movers within the last year is determined as p. We then use this value of p to test whether $\pi > .20$ or $\pi < .20$ is the more reasonable conclusion.

Both *estimation* and *hypothesis* testing are based on statistical relationships between samples and populations. It should not be surprising, therefore, that these two methods of statistical inference are intimately related. The relationship is fully explained in Chapter 9. In this chapter, we focus solely on estimation. Section 8.1 explains the distinction between two main categories of statistical estimation, *point* and *interval* estimation. This section also develops criteria differentiating good estimators from bad ones. In Section 8.2, point estimators for the population parameters μ, π, and σ are considered. Section 8.3 discusses interval estimation of the first two parameters and introduces the notion of a *confidence interval*. Finally, in Section 8.4 we examine how the estimation methods can be applied "in reverse" to determine the sample size needed to estimate a parameter with a given degree of confidence.

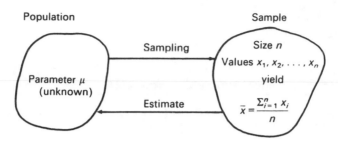

FIGURE 8-1. Estimating the value of the population parameter μ.

8.1. Statistical Estimation Procedures

The procedures used in statistical estimation can be conveniently divided into two major classes: point estimation and interval estimation.

DEFINITION: POINT ESTIMATION

In making a point estimate, a single number is calculated from the sample and is used as the best estimate of some unknown population parameter.

This number is called a *point estimate* because the population parameter is estimated by one point on the number line. For example, we might say that based on our sample of some size n, the best estimate of the proportion of movers in a city is $\pi = .26$. Note that we do not get the actual value of π from the sample, but rather compute some estimate of the true value.

DEFINITION: INTERVAL ESTIMATION

In interval estimation, the sample is used to identify a range or interval within which the population parameter is thought to lie, with a certain probability.

We might say, for example, that we are 95% certain that the proportion of people who moved in the city last year is in the interval [.20, .32]. Let us consider each type of estimation in more detail.

Point Estimation

Point estimation of some population parameter is illustrated in Figure 8-2. The population parameter can be *any* population parameter such as the mean, variance, standard deviation, or proportion. The random sample is composed of n random variables $X_1, X_2, \ldots, X_n$. The estimator of the population parameter is some function of the sample random variables.

DEFINITION: STATISTICAL ESTIMATOR AND STATISTICAL ESTIMATE

A statistical estimator is a function of the n random variables $X_1, X_2, \ldots, X_n$ in a sample. An estimator is, therefore, also a random variable. Once the sample is taken, the values of the sample random variables are known and are denoted $x_1, x_2, \ldots, x_n$. The *value* of the estimator $\hat{\theta}$, can thus be calculated. It is known as the *statistical estimate* of the population parameter θ.

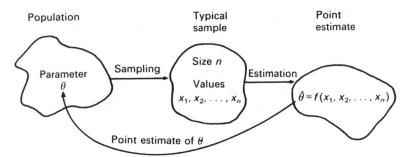

FIGURE 8-2. Point estimation of a population parameter θ.

As an example, consider the population mean parameter. There are several possible point estimators for μ. The sample mean is the obvious choice, but why should it be chosen over the sample median, or even the sample mode? All are simple functions of values x_1, $x_2, \ldots, x_n$, and can therefore serve as estimators.

There are two principal criteria for selecting an estimator for a population parameter θ: *bias* and *efficiency*. These can be understood most easily by thinking about the error associated with an estimate. Obviously, when we use a sample estimate to guess at the value of a parameter, we do not expect the estimate to exactly equal the population value—we expect some estimation error $\hat{\theta} - \theta$. (Here a positive error represents an overestimate, whereas a negative value implies too low an estimate.) Moreover, we know different samples will give different estimates, so the error will vary from sample to sample. Both bias and efficiency relate to error of estimation $\hat{\theta} - \theta$. In the case of bias, we are identifying the mean, or expected error. If the estimator on average gives values that are too high, we would say it is biased high. Similarly, if the estimator tends to underestimate θ, it is biased low. To arrive at the bias, we need to know something about the sampling distribution of the estimator; in particular, we need to know its mean value. Figure 8-3 shows the sampling distributions for two estimators, $\hat{\theta}_1$ and $\hat{\theta}_2$. The first is biased low, as its mean is less than the true value. We would very likely prefer $\hat{\theta}_2$, because on average it neither underestimates nor overestimates θ. In other words, $\hat{\theta}_2$ is *unbiased*. Bias is related to error in the following way:

$$\text{Bias} = \text{Mean Error} = E(\hat{\theta} - \theta) = E(\hat{\theta}) - \theta$$

DEFINITION: UNBIASED ESTIMATOR
An estimator $\hat{\theta}$ of a population parameter is said to be unbiased if the expected value of the estimator is equal to the population parameter. That is, $\hat{\theta}$ is unbiased if $E(\hat{\theta}) = \theta$.

The sample $\bar{X}$ mean is a good example of an unbiased estimator. Given a random sample, the expected value of $\bar{X}$ is μ, the same value we are trying to estimate. In other words, we would compute the bias of $\bar{X}$ as

$$\text{Bias of } \bar{X} = E(\bar{X}) - \mu = \mu - \mu = 0$$

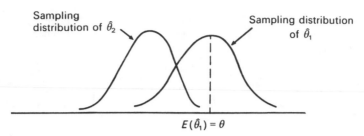

FIGURE 8-3. Sampling distributions for a biased and an unbiased estimator of θ.

Notice that using an unbiased estimator does not imply that an *individual* estimate will equal the true value, or even that it will be close to the true value. We know only that the errors are zero *on average*. Small individual errors are expected only if the distribution of $\hat\theta$ is tightly clustered around θ. Figure 8-4 shows this graphically for two estimators, both of which are unbiased. In both cases the mean estimation error is zero, but $\hat\theta_3$ is more tightly clustered. If we use $\hat\theta_3$, we are more likely to obtain an estimate that is near the true value. We would say that $\hat\theta_3$ is more *efficient* than $\hat\theta_4$.

To make the concept of efficiency more precise, we need to consider estimation errors more carefully. We have seen that bias contributes to error and is clearly undesirable. Does this mean we always choose unbiased estimators? The answer is no, because bias is not the only source of estimation error. Rather than minimize mean error (bias), we probably want individual errors to be small. After all, we have only a single sample in hand; thus, it isn't necessarily helpful to know that the mean error is small. We would prefer to know that the error expected from a single (individual) sample is small. Consider, for example, Figure 8-5. Even though it is biased, estimator $\hat\theta_5$ would probably be preferred to $\hat\theta_6$. The degree of bias introduced by using this estimator is more than offset by its high efficiency. Clearly, to evaluate an estimator, we need a measure that tracks more than just bias. Bias is inadequate because positive and negative errors cancel one another, giving low mean error even when individual errors are large. What we are after is some index that measures the departure between $\hat\theta$ and θ without regard to sign.

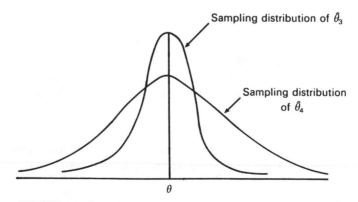

FIGURE 8-4. Sampling distributions for two unbiased estimators of θ.

One approach is to use squared differences. Rather than calculate the expected error, we will calculate the expected *squared error*. This seems like a subtle difference, but can have profound effects on the choice of estimator. The expected squared error is known as the *mean squared error*, or MSE, and is given by:

$$MSE = E(\hat{\theta} - \theta)^2$$

This is appealing because, although individual errors can be positive or negative, all are squared; thus there is no cancellation. After a little algebra, it is easy to arrive at an alternative, more useful expression for MSE:

$$MSE = E(\hat{\theta} - \theta)^2 = [E(\hat{\theta}) - \theta]^2 + E[\hat{\theta} - E(\hat{\theta})]^2 = BIAS^2 + VARIANCE$$

In other words, MSE is the sum of bias squared plus the variance of $\hat{\theta}$. Bias tells us where the sampling distribution is centered, whereas variance indicates the amount of dispersion around the central (mean) value. If both are small, the sampling distribution is both centered near the true value *and* tightly clustered around the true value. In short, the estimator will be efficient when MSE is small. This suggests a definition of efficiency in terms of MSE:

DEFINITION: EFFICIENCY OF AN ESTIMATOR
The efficiency of an estimator is inversely proportional to its mean square estimation error, MSE. As MSE increases, efficiency decreases. For an estimator to be efficient, both its bias and variance must be small.

For example, consider the sample mean. We know that it is unbiased, and that its variance is σ^2/n. Thus we can say $MSE = 0 + \sigma^2/n$. As the sample size increases, MSE decreases, and the efficiency improves. (This is expected, of course, as larger samples contain more information, and thus they ought to give better estimates.) Looking again at Figure 8-4, we might think of the curves as representing $\bar{X}$ based on different sample sizes, with $\hat{\theta}_3$ coming from the larger sample. Alternatively, the curves might represent two

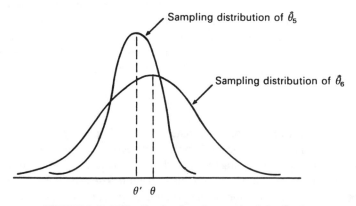

FIGURE 8-5. Difficulties in choosing a potential estimator.

completely different estimators. For example, $\hat{\theta}_4$ could be the sample median, and $\hat{\theta}_3$ the sample mean. Both of these are unbiased; thus, we do not prefer one over the other on the basis of bias. If X has a normal distribution, however, it can be shown that the sample mean has a smaller variance than any other unbiased estimator of μ, including the median. It is therefore more efficient than the median and would be preferred over the median.

Notice that bias and sampling variability both affect MSE. If we choose an estimator with a small MSE, we are saying we don't care how the errors arise, we simply want them to be small. If this means using a biased estimator, so be it. Looking again at Figure 8-5, the preferred estimator has smaller MSE, in spite of nonzero bias. We expect, therefore, that an individual estimate based on $\hat{\theta}_5$ will be closer to the true value than one based on $\hat{\theta}_6$.

MSE obviously provides useful information about an estimator, but also suffers from an important shortcoming. Because it is the average squared error, MSE has dimensions of θ^2 and is sometimes hard to interpret. For example, if we are estimating mean age in years, MSE has units of "square years," hardly a meaningful dimension. To overcome the interpretation difficulty, a related efficiency measure is more commonly used, the *root mean square error*, or RMSE. This is simply the square root of MSE:

$$\text{RMSE} = \sqrt{\text{MSE}} = \sqrt{\text{BIAS}^2 + \text{VARIANCE}}$$

Now we have two measures of overall error, MSE and RMSE. Should we choose an estimator that minimizes the former or the latter? Happily, this is never an issue. Because RMSE is a monotonic function of MSE, minimizing MSE is equivalent to minimizing RMSE. In other words, if an estimator is the minimum MSE estimator, it is also the minimum RMSE estimator. Choosing to minimize RMSE instead of MSE does not give a different estimator, it simply gives a different numerical measure of error, one that happens to be easier to interpret than MSE. We emphasize that RMSE always has the same dimensions as the parameter being estimated. It is interpreted as the average estimation error, where the averaging process involves squared, not absolute, errors. We will see RMSE again in Chapters 14 and 17, where it will be used as a way of summarizing errors of prediction.

As a final general comment about point estimation, we should mention that a number of commonplace statistical methods are based on minimum-variance unbiased estimates. That is, we first restrict the search to unbiased estimators, then choose the one with smallest MSE. Such estimators are often called best unbiased estimators, where the word "best" indicates minimum variance. For example, assuming X is normal, the sample mean is the best unbiased estimator of μ. No other unbiased function of the sample values has a smaller variance; thus, no other unbiased estimator has a smaller MSE.

Interval Estimation

Interval estimates improve on point estimates by providing a range of values for θ. Although we might know that in general some point estimator gives the best guess as to the true value of the population parameter, we have no idea about how good a particular estimate is. In statistical estimation, we obviously want to know how far the point estimate is likely to be from the actual value of the population parameter. This is the province of interval estimation, and is illustrated in Figure 8-6. Sampling from a population, we utilize our sample values $x_1, x_2, \ldots, x_n$ to arrive at two points that together define an interval, or range of values for the parameter θ. The two points represent a lower limit $\hat{\theta}_L$ and an upper limit $\hat{\theta}_U$ for the interval. In making an interval estimate, we claim *with a known degree of confidence* that the interval contains the value of the population parameter. For this reason, interval estimates are often termed *confidence intervals*. A typical conclusion might be, "Based on a sample of 100 light bulbs, we estimate the population mean lifetime of bulbs to be between 1000 and 1200 hours, with 95% confidence." The word "confidence," when it is used in this context, has a particular meaning, thoroughly explained in Section 8.3. The usefulness of our inference is based on the *probability* that the specified confidence interval contains the actual value of the population parameter θ. In the example cited above, we would attach a probability of .95 to the interval [1000,1200] based on a sample size of n = 100. In doing so we recognize that because of sampling variability, the interval constructed might be in error; that is, it might not contain the actual mean lifetime. The confidence level, .95, indicates that the probability of error is known to be .05, so that the chances of error are 5 out of 100.

Just as there are good point estimators, so are there good interval estimators. The principal requirement of a good interval estimator is that the specified interval $[\hat{\theta}_L, \hat{\theta}_U]$ be

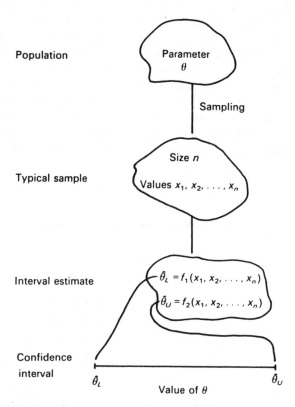

FIGURE 8-6. Interval estimation of a population parameter θ.

as small as possible for a given sample size and confidence level. Interval estimators for π and μ are covered in Section 8.3.

8.2. Point Estimation

The point estimators for the population parameters μ, π, and σ^2 are given in Table 8-1. Let us consider each in turn.

Population Mean θ

Without question, the most common estimator of the population mean μ is the sample mean $\bar{X}$. Its popularity is in large part explained by the fact that it combines high efficiency with no bias. In fact, one can show that if X is normal, the sample mean is more efficient than all other unbiased estimators. For example, its variance is about 56% smaller than the variance of the sample median. Though both are unbiased estimates of μ, the sample mean is preferred because the expected estimation error (RMSE) is smaller. For distributions that depart from normality, even symmetric distributions, the sample mean is not necessarily best. Consider, for example, a distribution that has more probability mass in its tails than the normal distribution. A sample from such a distribution is more likely to contain very high or very low values than would be true for the normal distribution. In Chapter 2 we noted that $\bar{X}$ is sensitive to extreme values, thus it can be pulled away from μ by a single unusually large or small observation. Here, then, is a case where the sample mean is not very satisfactory, and we seek an estimator that is more robust.

An obvious alternative is the sample median. Because it responds only to the central value in the sample, it is not affected by the extremes. Compared to $\bar{X}$, the median therefore has a smaller RMSE for some nonnormal distributions. Yet another alternative is to use a *trimmed mean*. The sample values are placed in rank order, and an equal number of observations are discarded from the bottom and top of the sample. The trimmed mean is

TABLE 8-1
Point Estimators of μ, π, and σ^2

Population parameters	Point estimator	Formula for point estimate
μ	$\bar{X}$	Mean of sample values: $\bar{x} = \dfrac{1}{n} \sum_{i=1}^{n} x_i$
	Median	Middle value in sample (50%-ile)
	25% Trimmed Mean	Mean of middle 50% of values in sample
	10% Trimmed Mean	Mean of middle 80% of values in sample
π	P	$p = x/n$ where x = number of successes in n trials
σ^2	S^2	$s^2 = \dfrac{1}{n-1} \sum_{i=1}^{n} (x_i - \bar{x})^2$

just the mean of the sample values that remain. For example, the 25% trimmed mean discards one-quarter from both ends and is the average of the middle half of the sample. The procedure effectively removes (trims) the extreme values in the sample, so that they cannot influence the estimate. The amount of data discarded can be as much or little as seems appropriate, though typically the trim fraction is between 20% and 50%.

In the limit, when all except one observation have been trimmed away, the trimmed mean converges to the sample median. In other words, we can think of the median as an extreme example of a trimmed mean. Likewise, we can think of $\bar{X}$ as a trimmed mean with a trim fraction of zero. In addition to the median and trimmed mean, there are other robust estimators that can outperform the sample mean. Regardless of which is used, if we want to claim that an estimator is better than $\bar{X}$, we must know the probability distribution of X. Given that we do not even know the mean of X, it's unlikely that we will have detailed knowledge about the distribution of X. In actual practice, therefore, it will be hard to claim with certainty that a particular estimator is better than the sample mean. Nevertheless, one should not blindly assume $\bar{X}$ is always the appropriate choice. Prudence suggests that several estimators be computed and compared with one another. When they give widely divergent values, one should be cautious in using the sample mean.

We note in passing that there is another, and very compelling, reason to use $\bar{X}$ as an estimator—its sampling distribution is known! Although we might have no information about the underlying distribution of X, we know from the central limit theorem that $\bar{X}$ is approximately normal. This allows us to assign probabilities to various values of $\bar{X}$, and, as will be seen below, provides a way of attaching uncertainty to the estimate of μ. That is, by using $\bar{X}$, we are able to say something about how good the estimate is. If some other estimator is used, we will have obtained a point estimate, but will have no idea about how close it is to the true value of μ.

Population Proportion π

The recommended estimator of the population proportion is the sample statistic $P = X/n$ where X is the number of "successes" found in the sample of n observations. The parameter π is one of the two parameters in the binomial distribution and corresponds to the proportion of "successes" in the population at large. To estimate π, we simply calculate the observed number of successes x in a sample of n trials and estimate π by x/n. Suppose, for example, we are interested in the proportion of residents in a metropolitan area who choose public transit for their journey to work. A sample of 100 residents finds 26 took public transit. Our best estimate of π is $p = 26/100 = .26$.

Population Variance σ^2 and Standard Deviation σ^2

For the population variance σ^2, the suggested estimator is S^2. Given a sample $x_1, x_2, \ldots, x_n$ the sample value of this random variable is

$$S^2 = \frac{\sum_{i=1}^{n}(x_i - \bar{x})^2}{n - 1}$$

Recall that $\sigma^2 = E(X - \mu)^2$, the mean square departure from the population mean. It therefore makes perfect sense that our estimate of σ^2 is found by averaging the squared departures from the sample mean. Not so obvious is why we should divide by $n - 1$ rather than n. The answer is that doing so gives an unbiased estimator. If we took many samples of size n from a population, the values of S^2 computed for each sample would average to σ^2. On the other hand, if we were to divide by n, the estimator would be biased low and tend to underestimate σ^2. The formal proof of this is beyond an introductory text [see, for example, Pindyck and Rubinfeld (1981)], but it is not hard to understand the problem in general terms. In estimating σ^2, we are replacing the unknown population mean μ with the sample mean $\bar{X}$. That is, we are computing squared deviations around the sample mean, not the population mean. From Chapter 2, we know that the squared deviations around the sample mean are smaller than the deviations around any other number. This implies that the deviations around $\bar{x}$ are *smaller* than the deviations around μ. But it is deviations around μ that we are trying to estimate! Clearly, if the mean deviation around $\bar{x}$ is smaller than the mean deviation around μ, it must underestimate σ^2. In order to compensate for this bias, we inflate the sum of squares by dividing by a number smaller than n, namely $n - 1$. In addition to being unbiased, S^2 is also the most efficient (minimum MSE) unbiased estimator of σ^2, providing X is normally distributed.

The estimator for σ, the population standard deviation, is the square root of S^2, or S. Although it is most commonly used, it is actually a biased estimator of σ. However, the degree of bias is very small for reasonable sample sizes, and S is much more convenient than alternative unbiased estimators. We will use it exclusively.

8.3. Interval Estimation

Statistical inference typically includes both an inference and some measure about the reliability of the inference. In point estimation, we select a single *value* that is used as an estimate of the population parameter in question. Though we might know the estimator is BUE, this tells us nothing about how good a particular estimate is. Ideally, along with the estimate, we should provide information about how far the estimate is likely to be from the true value. What would be useful is some measure of the probability that the population parameter of interest lies within a certain range. Specifying these ranges, and the probabilities attached to them, is the goal of interval estimation. In this section we show how to develop interval estimates for two population parameters, μ and π, based on a knowledge of the sampling distributions of the estimators, $\bar{X}$ and P.

Interval Estimator for μ When X Is Normally Distributed

Interval estimates for μ are based on $\bar{X}$ and its sampling distribution. With regard to the sampling distribution of $\bar{X}$, remember that it is normal with mean and standard deviation $\sigma/\sqrt{n}$. Knowing this, we can find the probability of $\bar{X}$ falling into an interval, or range of values. This is most easily done if we convert to the standard normal using the Z transformation. Consider the sampling distribution shown in Figure 8-7. Since 95% of the area under the standard normal curve is between -1.96 and $+1.96$, we know 95% of the values of $\bar{X}$ lie between -1.96 and $+1.96$ standard deviations of μ. That is, we can say

$$P\left(-1.96 \leq\leq \frac{\bar{X} - \mu}{\sigma_X} \geq 1.96\right) = 0.95 \tag{8-1}$$

Multiplying all parts of the inequality by σ_X gives

$$P(-1.96\sigma_X \leq \bar{X} - \mu \leq 1.96\sigma_X) = 0.95$$

Now substitute for σ_X to get

$$P\left(-1.96\frac{\sigma}{\sqrt{n}} \leq \bar{X} - \mu \leq 1.96 \frac{\sigma}{\sqrt{n}}\right) = 0.095$$

Subtracting $\bar{X}$ and multiplying by -1 expresses Equation (8-1) in terms of the unknown μ as

$$P\left(\bar{X} -1.96\frac{\sigma}{\sqrt{n}} \leq \mu \leq \bar{X} + 1.96 \frac{\sigma}{\sqrt{n}}\right) = 0.95 \tag{8-2}$$

How can this probability statement be interpreted? Let us suppose repeated samples of size n are taken from a population with a mean of μ and a standard deviation σ. (See Figure 8-8.) Using the data in each of these samples, we calculate a value for $\bar{X}$, using

$$\bar{x} = \frac{x_1 + x_2 + \ldots + x_n}{n}$$

Each time we find a new interval with upper and lower bounds $\bar{x} + 1.96\sigma/\sqrt{n}$ and $\bar{x} - 1.96\sigma/\sqrt{n}$. Then, as is illustrated in Figure 8-8, 95% of the samples yield an interval containing the true value of μ. The other 5% of the samples give intervals that do not contain μ.

Intervals specified in this way are known as *confidence intervals*. In practice, a confidence interval is constructed from a single sample. Obviously, this interval either contains μ or it does not. Because of the randomness inherent in sampling, we cannot be completely sure which of these two possibilities occurs. However, we do know from the central limit theorem that there is a .95 chance of selecting a *single sample* that produces an interval containing μ; thus, we are 95% confident in the interval obtained from our

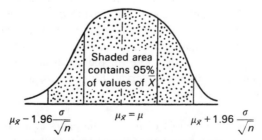

FIGURE 8-7. Sampling distribution of $\bar{X}$.

Population   Repeated samples of size *n*   Confidence intervals

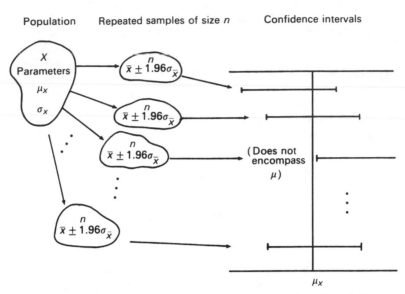

FIGURE 8-8. Interval estimates constructed from repeated samples of size *n*.

particular sample. This is a classical interpretation of a confidence interval, which states that the interval has a 95% chance of containing μ. In other words, the classical interpretation is that 95 times out of 100, intervals constructed in this manner will contain the true population mean.

The subjectivist interpretation of a confidence interval is different, and somewhat less cumbersome. It states that there is a 95% chance that μ is within the interval. Equivalently, we could say that the probability of μ being within the interval is .95. Note that for an objectivist, this is nonsense. For them μ is a constant, not a random variable, and there are no probabilities that can be assigned to μ. Only the subjectivist view of probability as a degree of certainty will admit the alternative interpretation. We are 95% sure about the location of μ, thus can attach a probability to μ being in the interval.

Confidence intervals can be constructed for any desired degree of confidence. Depending on the problem at hand, we may wish to be 90, 95, or 99% or even more confident in our result. Let $1 - \alpha$ be the desired degree of confidence specified as a proportion. For example, if α is .05, the degree of confidence is .95, representing 95% confidence. Similarly, α values of .1 and .01 correspond to 90 and 99% confidence. Let us define $z_{\alpha/2}$ as the value of the standard normal table such that the area to the *right* of $z_{\alpha/2}$ is $\alpha/2$. For a 95% confidence interval, we wish $\alpha/2 = .05/2 = .025$ of the area of the curve to the right of $z_{\alpha/2}$ (See Figure 8-9.) From the standard normal table, $z_{.025} = 1.96$. Different confidence levels imply different values for $z_{\alpha/2}$ and therefore for the upper and lower bounds of the confidence interval. In general, confidence intervals for μ are written $\bar{x} \pm z_{\alpha/2}(\sigma/\sqrt{n})$. A simple example will explain this procedure.

EXAMPLE 8-1. A random sample of $n = 25$ shoppers at a supermarket reveals that patrons travel, on the average, 16 min from their homes to the supermarket. Furthermore,

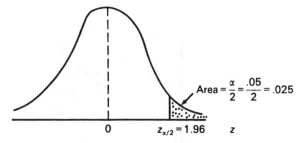

Sampling distribution of $\bar{X}$

FIGURE 8-9. Calculation of $z_{\alpha/2}$ for a 95% confidence interval.

we are told that the population standard deviation for this variable is $\sigma = 4$, and these travel times are normally distributed. The problem is to construct the 99% confidence interval for the average travel time to this supermarket. The desired interval for .99 confidence is

$$\bar{x} - z_{\alpha/2} \frac{\sigma}{\sqrt{n}} \leq \mu \leq \bar{x} + z_{\alpha/2} \frac{\sigma}{\sqrt{n}}$$

which, by substitution of $\bar{x} = 16$, $\sigma = 4$, $n = 25$, and $\alpha = .01$, can be rewritten as

$$16 - z_{.005} \frac{4}{\sqrt{25}} \leq \mu \leq 16 \, z_{.005} \frac{4}{\sqrt{25}}$$

From the normal table, $z_{.005} = 2.58$, so that the 99% confidence interval is given by

$$16 - 2.58 \frac{4}{\sqrt{25}} \leq \mu \leq 16 + 2.58 \frac{4}{\sqrt{25}} = 16 \pm 2.064 = [13.936, \ 18.064]$$

The value of μ may or may not be in this range. Our hope is that the random sample used to construct the interval is one of the 99% of all samples of size $n = 25$ that do contain the true value for μ.

The width of a confidence interval is influenced by two principal factors under control of the researcher: the desired degree of confidence $(1 - \alpha)$ and sample size n. Let us examine the impact of both factors on confidence intervals for travel time. First, note how varying degrees of confidence affect interval width (Table 8-2). As our confidence level increases, $z_{\alpha/2}$ increases, and the width of the interval increases. Figure 8-10 shows the same behavior in graphical form over a wider range of confidence levels. Notice that as the level of confidence approaches 100%, the width approaches infinity, representing the entire number line. This is because we can never be 100% sure that an interval contains μ unless the interval covers the entire range of possible values of the population variable X. If X is normal, the range of possible values runs from $-\infty$ to ∞. Notice also that the interval width increases very rapidly for confidence levels above 95%. Increasing the confidence level from .7 to .95 costs relatively little in terms of increasing width, but diminishing returns set

TABLE 8-2
Confidence Intervals for μ with Varying Degrees of Confidence

Confidence level	α	$\alpha/2$	$z_{\alpha/2}$	Lower bound $\bar{x} - z_{\alpha/2}(\sigma/n)$	Upper bound $\bar{x} + z_{\alpha/2}(\sigma/n)$	Values of bound for travel time problem		Interval Width
						Lower	Upper	
0.90	0.10	0.050	1.65	$\bar{x} + 1.65\ \sigma/n$	$\bar{x} + 1.65\ \sigma/n$	14.688	17.312	2.624
0.95	0.05	0.025	1.96	$\bar{x} + 1.96\ \sigma/n$	$\bar{x} + 1.96\ \sigma/n$	14.432	17.568	3.136
0.99	0.01	0.005	2.58	$\bar{x} + 2.58\ \sigma/n$	$\bar{x} + 2.58\ \sigma/n$	13.936	18.064	4.128

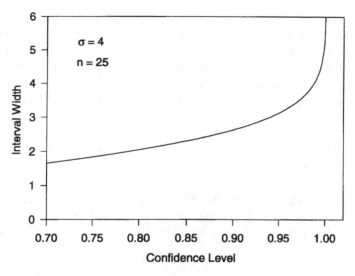

FIGURE 8-10. Effect of confidence level on interval width.

in at higher levels. We can demand any arbitrarily high level of certainty, but only at the expense of greatly increasing interval width. This behavior is consistent with the idea that a sample of fixed size contains a finite amount of information. By increasing the confidence level, we are demanding greater certainty that the interval contains μ. We have not added any new information; thus, this increase in certainty can only come at the expense of a wider interval. That is, we become increasingly certain, but make progressively less precise statements regarding the value of μ.

Table 8-3 summarizes the impact of sample size on interval width. Interval widths are calculated for sample sizes of 25, 100, and 400. As the sample size increases, the width of the confidence interval decreases, everything else being equal. The same relation is shown in Figure 8-11, where it can be seen that the width approaches zero as the sample size approaches infinity. The center of the interval is always the sample mean $\bar{X}$, but this approaches μ as n approaches infinity. With the interval narrowing and at the same time becoming more nearly centered on μ, we see the *confidence interval* for μ converging on the *value* of μ with larger sample sizes. This has great intuitive appeal. As our sample

TABLE 8-3
Confidence Intervals for μ with Varying Sample Sizes

Sample size n	Confidence level	$z_{\alpha/2}$	Lower bound $\bar{x} - z_{\alpha/2}(\sigma/\sqrt{n})$	Upper bound $\bar{x} + z_{\alpha/2}\sigma/\sqrt{n}$	Interval width
25	0.95	1.960	14.432	17.568	3.136
100	0.95	1.960	15.216	16.784	1.568
400	0.95	1.960	15.608	16.392	0.784

includes more and more members of the population, we become more and more knowledgable about the population. If we have included the entire population, there should be no doubt about the value of the population parameter; it can simply be calculated from known information. The figure shows how gathering more information (increasing n) leads to increasingly precise statements about μ. Once again, however, we see diminishing returns. For example, doubling the sample size decreases the width, but by less than a factor of two.

There are thus two ways of decreasing the width of confidence intervals, that is, of making more precise statements about μ. First, we can decrease the level of confidence. The second, and usually more worthwhile alternative, is increasing the sample size. The formula for the confidence interval of the population mean can now be summarized.

DEFINITION: INTERVAL ESTIMATE OF POPULATION MEAN μ

Let $x_1, x_2, \ldots, x_n$ be a sample of size n randomly drawn from a normally distributed random variable X. The confidence interval with confidence level $1 - \alpha$ is

$$\bar{x} \pm z_{\alpha/2}\frac{\sigma}{\sqrt{n}} \tag{8-3}$$

where $\bar{x}$ is the sample mean, σ is the population standard deviation, and $z_{\alpha/2}$ is the value of the standard normal distribution for which the area to the right of $z_{\alpha/2}$ is $\alpha/2$.

Interval Estimation of μ with X Not Normally Distributed, σ Known

If the population random variable X is not normally distributed, then we cannot appeal to the strict form of the central limit theorem to justify the normality of the sampling

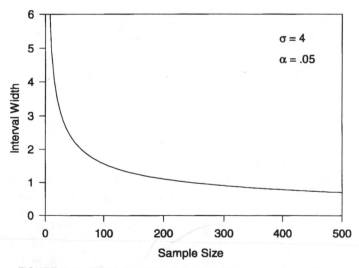

FIGURE 8-11. Effect of sample size on confidence interval width.

distribution of $\bar{X}$. All we can say is that the sampling distribution is *approximately* normal. It follows that the confidence interval given in Equation (8-3) can also be considered *approximately* correct. As a rule of thumb, the approximation can be considered sufficiently close when sample size n exceeds 30.

Interval Estimation of μ with σ Unknown

An extremely common problem in developing a confidence interval for is that the population standard deviation, σ, is unknown. After all, it does seem that μ would be known along with σ, given that is needed to compute σ. When σ is unknown, there is no way to find the standard error of the mean $\sigma_{\bar{x}} = \sigma/\sqrt{n}$; thus, we cannot use (8-3) to find the confidence interval. Luckily, there is a simple way around this problem. First, the values of the sample $x_1, x_2, \ldots, x_n$. are used to estimate σ by

$$s = \sqrt{\frac{\sum_{i=1}^{n}(x_i - \bar{x})^2}{n - 1}} \qquad (8\text{-}4)$$

Recall that this is the most commonly used estimate of σ identified in Section 8.2. With s, we can replace $\sigma_{\bar{x}} = \sigma/\sqrt{n}$ by the estimate $s/\sqrt{n}$. When this is done, however, we can no longer use the normal distribution to calculate the confidence interval. To see why, recall that confidence intervals were based on probability statements involving the random variable $\bar{X}$, having the form

$$P\left(-k \le \frac{\bar{X} - \mu}{\sigma/\sqrt{n}} \le k\right) = 1 - \alpha$$

(Equation 8-1). Knowing the middle quantity is Z, we could find k for any confidence level desired. For example, for $\alpha = .05$, k is 1.96. When σ is unknown, the analogous statement is

$$P\left(-k \le \frac{\bar{X} - \mu}{s/\sqrt{n}} \le k\right) = 1 - \alpha \qquad (8\text{-}5)$$

The middle quantity now contains two random variables ($\bar{X}$ and s) and is not a standard normal variable. Fortunately, the distribution is known, *providing X is normal*. In particular, assuming X is normal, the ratio

$$T = \frac{\bar{X} - \mu}{s/\sqrt{n}}$$

follows the so-called t distribution. This is also called *Student's t distribution*, because the inventor wrote under the pseudonym Student. The t distribution is symmetric and somewhat similar to the standard normal. Not surprisingly, its variance is larger than Z. Because the population parameter σ is replaced by an estimate s, more uncertainty is introduced, and there is greater variability in the distribution.

Obviously, the bigger the sample, the better the estimate s will be. Therefore, the bigger the sample, the more the t distribution resembles the standard normal. We see that the form of the t distribution depends on sample size n. Because we use $n - 1$ in the denominator for s in Equation (8-4), the distribution of t is tabulated according to its *degrees of freedom* (df) rather than the sample size n. The degrees of freedom are always equal to the sample size minus 1, or $n - 1$. Figure 8-12 compares the standard normal with the t distributions for degrees of freedom 2, 10, and 30. At df = 30, note the similarity of the standard normal and t distributions.

One way of quantifying the convergence of the standard normal and the t distributions with increasing degrees of freedom is to examine the value of t corresponding to a certain tail area of standard normal. For example, we know that a value of $z = 1.96$ for the standard normal leaves .025 of the area under the curve in the tail. Now, if we compare this to the values of t for several members of the family of t distributions, we find

Degrees of freedom (df)	2.5% Value of t
1	12.706
2	4.303
3	3.182
10	2.228
20	2.086
30	2.042
40	2.021
60	2.000
120	1.980
Infinity	1.960

As the sample size and degrees of freedom increase, the value of t converges to 1.96. Moreover, by the time df = 30 or so, there is a close correspondence between z and t. Table A-4 has t *values* for several other confidence levels. In all cases t converges to z with increasing df. Just as the normal table was used to find k in Equation (8-1), the t *table* can be used to find k in Equation (8-5).

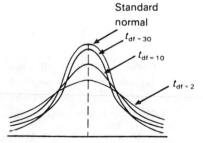

FIGURE 8-12. Normal distribution and the t distribution for 2, 10, and 30 df.

$$P\left(-k \le \frac{\bar{X} - \mu}{\sigma/\sqrt{n}} \le k\right) = 1 - \alpha$$

$$P\left(-k \le \frac{\bar{X} - \mu}{s/\sqrt{n}} \le k\right) = 1 - \alpha \qquad (8.5)$$

$$T = \frac{\bar{X} - \mu}{s/\sqrt{n}}$$

Where before we used $z_{\alpha/2}$ for k, now we use $t_{\alpha/2,n-1}$. The notation is similar to $z_{\alpha/2}$, in that we seek an upper-tail probability of $\alpha/2$. The notation for t includes the degrees of freedom $n - 1$, indicating that the appropriate value of t depends on sample size.

The use of the t distribution in constructing confidence intervals for μ with σ unknown is summarized as follows:

DEFINITION: CONFIDENCE INTERVAL ESTIMATION FOR μ WITH UNKNOWN POPULATION STANDARD DEVIATION σ

Let $x_1, x_2, \ldots, x_n$ be a sample of size n drawn from a normal population. The confidence interval for μ at level of confidence $1 - \alpha$ is

$$\bar{x} \pm t_{\alpha/2,n-1} \frac{s}{\sqrt{n}} \qquad (8\text{-}6)$$

where $\bar{x}$ is the sample mean, s is the sample standard deviation estimated using (8-4), and $t_{\alpha/2,n-1}$ is the value of the t distribution with degrees of freedom equal to $n - 1$, such that the area to the right of t is $\alpha/2$.

The confidence interval specified by Equation (8-6) differs from that of Equation (8-4) in two ways. First, because σ is unknown, it is replaced by the estimate s. Second, a value from the t distribution is used rather than one from the z distribution. As an example of confidence interval formula (8-6), let us return to the travel time example. The sample size is $n = 25$, and the mean is $\bar{x} = 16$. Now, let us change the situation slightly by assuming that the population standard deviation is unknown but has been estimated as $s = 4$. The confidence interval formula from the t distribution in (8-6) is

$$\bar{x} \pm t_{\alpha/2,n-1} \frac{s}{\sqrt{n}}$$

which for $1 - \alpha = .95$, or 95% confidence is equal to

$$16 \pm 2.064 \frac{4}{\sqrt{25}} = 16 \pm 1.651 = [14.349, 17.651]$$

The value of $t = 2.064$ is found in the t table for $t_{.025,24}$ by using the column labeled .025 and the row (for degrees of freedom) corresponding to 24. This interval is slightly wider than the interval found when σ was known, [14.432, 17.568]. Because we are in a situation of greater uncertainty with σ unknown, this is expected. Because the t and the z values give very similar confidence intervals in large samples, the normal distribution with z is often used, even in cases where σ is unknown. For small samples, t should always be used. It is important to remember that (8-6) requires that X be normally distributed. If this is not true, the confidence intervals will be approximate, with the approximation improving as sample size increases.

Interval Estimation for π

For binomial random variables, the sample proportion of successes can be estimated by using

$$P = \frac{X}{n} \tag{8-7}$$

where X represents the number of successes in n trials. To define the confidence interval for π, it is necessary to specify the sampling distribution of the statistic P. This is not too difficult, because P is an average of sorts and therefore falls under the central limit theorem. To see why P is an average, think of the sample as a collection of 0s and 1s, where 1 corresponds to a "success." The number of successes [X in (8-7)] is just the sum of the sample values Σx_i, because only the 1s contribute to the sum. Dividing by n gives the proportion of successes P, which also happens to be the mean of the sample. As a result, the sample mean P is approximately normally distributed for large sample sizes. To determine the parameters of the sampling distribution μ_p and σ_p, we can use what we know about the parameters of the binomial random variable X. First, note the mean of X, $E(X) = n\pi$ [see formula (6-11)], and therefore

$$\mu_p = E(P) = E\left(\frac{x}{n}\right) = \frac{1}{n} E(X)$$

$$= \frac{1}{n}(n\pi) = \pi$$

As expected, the distribution of P is centered on π, the population proportion. This simply proves a point made in Section 8.2: P is an unbiased estimator of π. The standard deviation of the sampling distribution of P can be found in a similar way by using the formula for the variance (V) of a binomial random variable. Recall that if X is binomial, its variance is $V(X) = n\pi(1 - \pi)$ [see formula (6-12)]. With this, the variance of P can be found as:

$$\sigma_p^2 = V\left(\frac{X}{n}\right) = V\left(\frac{1}{n} X\right)$$

$$= \frac{1}{n^2} V(X) = \frac{n\pi(1 - \pi)}{n^2}$$

$$= \frac{\pi(1 - \pi)}{n}$$

and thus

$$\sigma_P = \sqrt{\frac{\pi(1 - \pi)}{n}}$$

The principal difficulty here is that the standard deviation of P, or σ_P, itself requires a knowledge of π for estimation. This problem is solved by substituting our sample proportion p for π. The confidence interval for π can now be summarized.

DEFINITION: CONFIDENCE INTERVAL FOR π

Let X be a binomial random variable representing the number of successes in n trials where the probability of success in each trial is π. The confidence interval for π at a level of confidence $1 - \alpha$ is

$$p - z_{\alpha/2}\sqrt{\frac{p(1 - p)}{n}} \leq \pi \leq p + z_{\alpha/2}\sqrt{\frac{p(1 - p)}{n}} \tag{8-8}$$

where p is the proportion of successes in a sample of size n and $z_{\alpha/2}$ is the value of the standard normal distribution with an upper-tail probability of $\alpha/2$.

The confidence interval for in Equation (8-8) should be used only for large samples, say $n > 100$. For smaller sample sizes, the exact sampling distribution of the statistic P is the discrete binomial distribution. The binomial mass function (6-11) can be used to generate the exact sampling distribution, or else tables can be consulted.

EXAMPLE 8-2. An urban geographer interested in the proportion of residents in a city who moved within the last year conducts a survey. Summary statistics reveal 27 of the 270 people surveyed changed residences. What are the 95% limits on the true proportion of people who moved in the city?

Because the sample size of 270 is greater than 100, we can utilize formula (8-8) with the estimate $p = 27/270 = .10$. The limits are therefore

$$.10 \pm 1.96\sqrt{\frac{(.10)(.90)}{270}}$$

$$.10 \pm .036$$

$$[.064, .136]$$

EXAMPLE 8-3. Before a referendum on a bylaw requiring mandatory recycling city, a random sample of $n = 100$ voters is taken, and their voting intentions obtained. Of those surveyed, 46 favor the bylaw, and 54 are opposed. What are the 90% confidence limits on the population proportion in favor of this bylaw?

Once again the sample is large, so formula (8-8) can be used. With $p = .46$, the interval is

$$.46 \pm 1.65 \sqrt{\frac{(.46)(.54)}{100}}$$

$$.46 \pm .082$$

$$[.378, .542]$$

Notice that the interval includes $\pi = 0.5$. Despite the fact that only 46% in the sample favor the bylaw, it is still possible that the true proportion in favor is a majority.

8.4. Sample Size Determination

One common question asked of statisticians is, "How large a sample should I take in my research?" This is an extremely complicated question involving many different issues. In certain restricted conditions, however, we can use our knowledge of sampling distributions and confidence intervals to determine appropriate sampling sizes. To determine the sample size necessary in a study, we must know three things:

1. Which population parameter is under study? Is it a mean μ, proportion π, or some other parameter?
2. How close do we wish the estimate made from our sample to be to the actual (or true) value of the population parameter?
3. How certain (or, in statistical jargon, confident) do we wish to be that our estimate is within the tolerance specified in item 2?

The answer to item 1 tells us which sampling distribution governs the statistic of interest. In this chapter we limit ourselves to the population mean and proportion π.

The procedure is illustrated within the context of sample size determination for estimating μ. Suppose we wish to take a sample that is capable of yielding a point estimate of μ that is no more than E units away from the actual value of it. Short for *error, E* is a statement of the level of precision we desire. If E is small, we are saying we can tolerate only a small error in the estimate; if it is large, our required precision is less. Let us also assume that we wish to be 95% confident. In other words, a 5% chance of drawing a sample of size n with an estimate $\bar{x}$ for μ that is more than E units away from is acceptable.

To see how we can determine the sample size n meeting these requirements, let us examine the sampling distribution of $\bar{X}$ and the 95% confidence interval illustrated in Figure 8-13. We know that the interval $\mu \pm 1.96\sigma/\sqrt{n}$ contains exactly 95% of the values of the statistic $\bar{X}$. On this basis, the 95% confidence interval for μ with an estimate $\bar{x}$ is

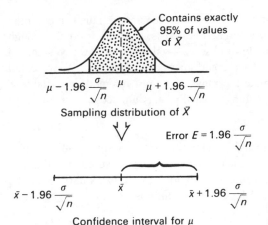

Sampling distribution of $\bar{X}$

Error $E = 1.96 \dfrac{\sigma}{\sqrt{n}}$

Confidence interval for μ

FIGURE 8-13. Relationship between sampling distribution, confidence limit, and error.

$\bar{x} \pm 1.96\sigma/\sqrt{n}$. If we wish to be no more than E units from μ with the estimate $\bar{x}$ then we can set

$$E = 1.96 \frac{\sigma}{\sqrt{n}}$$

which is one-half the interval width. In general, for a level of confidence $1 - \alpha$, we can set

$$E = z_{\alpha/2} \frac{\sigma}{\sqrt{n}} \qquad (8.9)$$

We now must solve (8-9) for n. With a few straightforward algebraic manipulations we find

$$n = \left(\frac{z_{\alpha/2}\sigma}{E} \right)^2 \qquad (8\text{-}10)$$

This formula indicates that the required sample size depends on the desired level of confidence (the higher the confidence, the larger $z_{\alpha/2}$, and hence the larger n), the desired precision (the higher the precision, the smaller E, and hence the larger n), and the variability of the underlying population (the greater the variability in the population, the larger σ, hence the larger n). All these relations have great appeal. It makes perfect sense that increased precision implies a greater sample size.

The single difficulty with Equation (8-10) is that seldom will σ be known beforehand, except where a population census has been completed. If this were the case, μ would be known and not require estimation! Usually, then, we must use an estimate of σ to determine n. The obvious candidate is s. If a rough estimate of s is not available from a recent survey, then a small pilot study can be undertaken to generate s. This is a common procedure in social science research. If no estimate of s exists, then the following rough rule of thumb can be used: It is very unlikely to observe a value more than two standard deviations above or below the mean. So, if we could guess at the range of X, we could take that as 4σ. For example, if we estimate the range of values for random variable X to be 100, we could use 25 as an estimate for σ. Usually, it is unnecessary to resort to this crude procedure.

EXAMPLE 8-5. A soils scientist is interested in determining the number of soil samples in a specific soil horizon necessary to estimate the extractable P_2O_5 (measured in milligrams per 100 g). Previous research indicates the standard deviation of P_2O_5 is about 0.7 mg. The desired precision is $E = .2$, and a confidence level for the study is set at $\alpha = 0.5$, or 95%. Using formula (8-10) gives

$$ n = \left[\frac{1.96(0.7)}{.2} \right]^2 = \left(\frac{1.372}{.2} \right)^2 = 47.06 $$

So, roughly 47 samples should be taken. Note that the entire procedure is based on the assumption that the random variable X is normally distributed, or that the required sample size is large enough ($n > 30$), and the central limit theorem applies.

An exactly analogous argument can be used to develop the following sample size formula for the population proportion π:

$$ n = \left[\frac{z_{\alpha/2}\sqrt{p(1 - p)}}{E} \right]^2 \tag{8-11} $$

This formula uses the estimate p for π in the confidence interval formula. Evidently some knowledge of the likely value of π, say from a previous study, can be used to make the necessary sample size decision. If no estimate of p is available, then a useful upper bound on n can be found by assuming $p = .5$. With $p = .5$, the value of $p(1 - p)$ is maximized, as shown in the following table:

p	0.30	0.40	0.50	0.60	0.70
$\sqrt{p(1 - p)}$	0.46	0.49	0.50	0.49	0.46

Because $\sqrt{p(1 - p)}$ appears in the numerator of (8-11), the value of n is maximized by maximizing this expression. We see that using .5 as an estimate is conservative in the sense that the value of n will be, if anything, larger than actually needed.

EXAMPLE 8-6. An urban geographer is interested in determining the sample size necessary to estimate the proportion of households in a large metropolitan area who moved in the last year. Past experience indicates that approximately 20% of all residents in North American cities move each year. How many households should be sampled? Assume a

TABLE 8-5
Summary of Point Estimators and Confidence Intervals for π and μ

Population parameter	Point estimator	Formula for confidence intervel	Appropriate conditions
μ	$\bar{X}$	$\bar{x} \pm z_{\alpha/2}(\sigma/\sqrt{n})$	Exact for any sample size when population standard deviation is known and X normally distributed Approximate when X is not normally distributed but $n > 30$.
μ	$\bar{X}$	$\bar{x} \pm t_{\alpha/2,n-1f}(s/\sqrt{n})$	Exact when population standard deviation is unknown and X normally distributed Approximate when X is not normal but $n > 30$.
π	P	$p \pm z_{\alpha/2}\sqrt{p(1-p)/n}$	Approximate when $n > 100$.

confidence level of 95% and that the geographer will tolerate an error of $E = .03$. Using Equation (8-11) with an estimate of $p = .2$ yields

$$n = \left[\frac{1.96\sqrt{(0.2)(0.8)}}{.03}\right] = 682.95$$

or approximately 683 households. It might also be interesting to determine an upper bound on the sample size by assuming $p = 0.5$:

$$n = \left[\frac{1.96\sqrt{(0.5)(0.5)}}{.03}\right]^2 = 1067.11$$

Under these conditions, a very large sample of 1067 is required. If the researcher feels the percentage of people in the city who moved is likely to be larger than the 20% characteristic of North American cities, it would be wise to enlarge the sample.

8.5. Summary

This chapter has developed the ideas in statistical inference used to estimate the value of some population parameter. We distinguished between two types of statistical estimation: point estimation and interval estimation. As is suggested by its name, a point estimate is a single value or point used to predict some population parameter, and it is based on some function of the sample values $(x_1, x_2, \ldots, x_n)$. To indicate the reliability of an estimate, it is common to construct a confidence interval—a range of values known to contain the true population value with a prescribed level of confidence. The more commonly used confidence levels are 90, 95, and 99%, but the technique is general enough to produce intervals for any desired confidence level. One can increase the confidence level to an arbitrarily high value, but only at the expense of a wider interval. That is, every confidence interval represents a trade-off between precision (interval width) and degree of certainty (confidence level). The formulas for point and interval estimates for and are summarized Table 8-5. Although other point estimates are available, their sampling distributions are in general unknown; thus, they do not easily lead to confidence intervals.

REFERENCES

R. S. Pindyck and D. L. Rubinfeld, *Econometric Models and Economic Forecasts*, 2d ed. (New York: McGraw-Hill, 1981).

FURTHER READING

Most introductory textbooks on statistical methods for geographers also contain the material covered in this chapter. Usually, these methods are discussed in the context of representative applications in various fields. The problems at the end of this chapter also indicate several possible applications of these techniques in geographical research.

PROBLEMS

1. Explain the meaning of the following concepts:
   a. Point estimation
   b. Interval estimation
   c. Unbiased estimator
   d. Statistical estimator
   e. Efficient estimator
   f. Statistical estimate
   g. Confidence interval

2. Discuss the two principal types of statistical estimation.

3. When should the T distribution be used in place of the Z distribution in the construction of confidence intervals for the population mean?

4. The yields in metric tons per hectare of potatoes in a randomly selected sample of 10 farms in a small region are 32.1, 34.4, 34.9, 30.6, 38.4, 29.4, 28.9, 32.6, 32.9, and 44.9. Assuming that these yields are normally distributed, determine a 99% confidence interval on the population mean yield.

5. Past experience shows that the standard deviation of the distances traveled by consumers to patronize Chinese restaurants is 2 km. How large a sample is needed to estimate the population mean distance traveled within 0.5 km? The probability of being correct should be set at .95.

6. The proportion of automobile commuters in a given neighborhood of a large city is unknown. A random sample of 50 households is taken, and 38 automobile commuters are found. Determine the 95% confidence interval of the proportion of commuters by automobile in the neighborhood.

7. A historical geographer is interested in the average number of children of households in a certain city in 1800. Rather than spend the time analyzing each entry in the city directory, she decides to sample randomly from the directory and estimate the size from this sample. In a sample of 56 households, she finds the average number of children to be 4.46 with a standard deviation of 2.06.

   a. Find the 95% confidence interval, using the t distribution.
   b. Find the 95% confidence interval, using the z distribution.
   c. Which interval is more appropriate? Why?

8. A geographer is asked to determine the sample size necessary to estimate the proportion of residents of a city who are in favor of declaring the city a nuclear-free zone. The estimate must not differ from the true proportion by more than .05 with a 95% confidence level. How large a sample should be taken? At 99%?

9. Suppose we double the size of a sample taken when trying to estimate the confidence interval for a mean. What will the effect be on the width of the confidence interval, assuming all other parameters (significance level and standard deviation) are held constant?

9

Parametric Statistical Inference:
Hypothesis Testing

Chapter 8 presented the form of statistical inference known as estimation and covered the techniques used for point and interval estimation for a few representative statistics. In this chapter, the form of statistical inference known as *hypothesis testing* is introduced. The methods used in hypothesis testing are closely related to confidence interval estimation. They are simply two ways of looking at the same problem, and both are based on the same theory. In Section 9.1, a highly structured method for testing hypotheses known as the *classical test of hypothesis* is outlined. Section 9.2 presents a newer and widely used variant of this procedure, the PROB-VALUE, or *p*-VALUE method. Hypothesis tests for population parameters μ and π are detailed in Sections 9.3 and 9.4. The link between hypothesis testing and confidence intervals is explained in Section 9.5. In many practical problems, the principal concern is not to test a hypothesis about the value of some parameter for a *single* population. Rather, we may want to compare *two* populations with regard to some quantitative characteristic or parameter. For example, suppose we have measures of carbon monoxide concentration from different residential neighborhoods in some city. We wish to compare the two samples. The statistical question being asked is whether these two samples come from a single population or from two different populations. So-called two-sample tests of hypotheses for means and proportions are explained in Chapter 10. These tests also utilize the basic testing method developed in Sections 9.1 and 9.2.

9.1. Key Steps in Classical Hypothesis Testing

A general conceptual framework for hypothesis testing is illustrated in Figure 9-1. We are interested in making inferences about the value of a population parameter θ. We hypothesize a value for this unknown population parameter, say $\theta = \theta_0$. A random sample of a given size n is collected having values $x_1, x_2, \ldots, x_n$. From these sample data a point estimator $\hat{\theta}$ is calculated. On the basis of $\hat{\theta}$, we evaluate our hypothesis by determining whether the sample value $\hat{\theta}$ does or does not support the contention that $\theta = \theta_0$. If $\hat{\theta}$ is close to θ_0, we might be led to believe the hypothesis is true. If $\hat{\theta}$ is very much different from θ_0, it is less likely that the hypothesis $\theta = \theta_0$ can be sustained. Recall that there are two reasons why $\hat{\theta}$ might differ from $\theta = \theta_0$. First, the hypothesis $\theta = \theta_0$ may actually be untrue. Second,

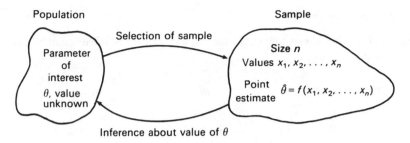

FIGURE 9-1. Sampling and statistical hypothesis testing.

$\hat{\theta}$ might differ from θ_0 simply because of sampling error. The framework shown in Figure 9-1 can be illustrated with a simple example involving residential mobility.

The average residential mobility in a North American city can be measured by the probability that a typical household in the city changes residence in a given year. Most studies have found that about 20% of the population of the city moves in any one year, or $\pi = .2$. On this basis, we would expect a typical *neighborhood* to have this overall rate of residential mobility. Let us now consider a specific neighborhood in a specific city. How could we determine whether this neighborhood has the citywide mobility rate? We first *hypothesize* a value of $\pi = \pi_0 = .2$. From a random sample of households in the neighborhood, we then determine the proportion which has moved into the neighborhood in the past year. Since this is a sample proportion, we call this value p, a specific value of the estimator P. Now, on the basis of p, we must decide whether this is a typical neighborhood (in terms of residential mobility). If p were close to our hypothesized value of .2, we would probably decide that the hypothesis is true and conclude it is a typical neighborhood. If, however, the sample proportion were $p = .8$, we would be less likely to believe the hypothesis $\pi = .2$. In rejecting our hypothesis, we must always consider the possibility this is an incorrect judgment. Our judgments are always based on samples and are therefore subject to sampling error.

The purpose of this section is to outline a rigorous procedure for testing any hypothesis of this type. A classical test of hypothesis follows the six steps shown in Table 9-1. We now consider each of these six steps, in turn, and then illustrate the procedure using a straightforward example.

TABLE 9-1
The Six Steps of Classical Hypothesis Testing

Step 1. Formulation of hypothesis.
Step 2. Specification of sample statistic and its sampling distribution.
Step 3. Selection of a level of significance.
Step 4. Construction of a decision rule.
Step 5. Compute value of the test statistic.
Step 6. Decision.

Formulation of Hypotheses

Statements of the hypothesis to be evaluated in a classical test *always* can be expressed in one of these three ways:

(A)	(B)	(C)
H_0: $\theta = \theta_0$	H_0: $\theta \geq \theta_0$	H_0: $\theta \leq \theta_0$
H_A: $\theta \neq \theta_0$	H_A: $\theta < \theta_0$	H_A: $\theta > \theta_0$

The symbol H stands for hypothesis. There are two parts to any hypothesis, labeled H_0 and H_A. In H_0 a statement is made about the value of some population parameter θ. The value θ_0 is the hypothesized, or conjectured, value for θ. In the first form (A), the statement H_0: $\theta = \theta_0$ means we are interested in deciding whether the value of the population parameter θ is equal to θ_0. In H_0, we assert that θ does equal θ_0. This is called the *null hypothesis*. We can either reject this null hypothesis or accept it, depending on the information we collect from the sample. If H_0 is not true, something else must be. As an alternative to H_0, we offer the statement H_A: $\theta \neq \theta_0$. Obviously, if we reject the statement $\theta = \theta_0$, then the only remaining possibility is $\theta \neq \theta_0$. Thus H_A is called the *alternate hypothesis*. In our residential mobility example, the parameter of interest is π, the population proportion, so the null and alternate hypotheses are

$$H_0\text{: } \pi = .2$$
$$H_A\text{: } \pi \neq .2 \qquad\qquad (9\text{-}1)$$

This short form is an especially useful way of summarizing the hypotheses in a classical test. The population parameter of interest is clearly identified. There are two other possible formats for H_0 and H_A, shown as hypothesis sets *B* and *C*. Sometimes we are interested in asserting, not an *exact* value for the population parameter θ, but a *range* of values. In format *B*, for example, we wish to test whether our sample is consistent with the statement $\theta \geq \theta_0$. That is, can we say the true population parameter is *at least* as large as θ_0? The alternative is that θ is definitely *less than* θ_0, as is expressed in H_A. In format *C*, we are interested in whether our sample supports the statement $\theta \leq \theta_0$; that is, the true value of the population parameter is *no larger than* θ_0. In this case, the alternate hypothesis must be that θ is *larger than* θ_0.

Which of these three forms should we choose? The correct format depends on the question we wish to answer. To see this, let us return to our example of neighborhood residential mobility. If we express our hypotheses concerning residential mobility in the form of (9-1), we are interested in determining whether the neighborhood being studied is different than the city, either more *or* less mobile. Were we to express the hypotheses as in format *B*

$$H_0\text{: } \pi \geq .2 \qquad H_A\text{: } \pi < .2 \qquad\qquad (9\text{-}2)$$

we would be testing a hypothesis about whether the neighborhood has a lower than average residential mobility. If we can reject H_0, then our alternate hypothesis tells us that the neighborhood is more stable than the average. If the neighborhood under study is an older suburb where there is predominantly single-family housing, this may be a hypothesis well worth evaluating. Notice that by rejecting the null hypothesis H_0, we are accepting H_A, that the neighborhood is less mobile than the city at large. Unlike format (A), this form of H_A does not allow for the neighborhood to be more mobile.

By casting our hypotheses in format C

$$H_0: \pi \le .2 \qquad H_A: \pi > .2 \qquad\qquad (9\text{-}3)$$

we could test the hypothesis of whether the neighborhood has a higher turnover than average. If we were studying the residential neighborhood around a major university, a high rate of mobility might be expected. In all three formats A, B, and C, what we are really interested in saying is stated in the alternate hypothesis H_A, which we prove by evaluating H_0 and then rejecting it, if possible. This apparently peculiar way of operating is actually the only way of proceeding. The reasons why will become clear when we discuss the third step in the hypothesis testing procedure.

Hypothesis format A is sometimes referred to as a *two-sided*, or *nondirectional*, test. This is because the alternate hypothesis does not say on which *side*, or *direction*, the true population parameter θ lies in relation to the hypothesized value θ_0. Formats B and C are usually called *directional hypotheses* because the alternate hypothesis H_A does specify on which side of θ_0 the true parameter lies. Sometimes format B is termed a *lower-tail test* and format C an *upper-tail test*, on the basis of the direction specified in H_A. The reasons for these names and the choice of formats among A, B, and C are explained further in the examples that follow.

In setting up H_0 and H_A, it is essential that they be *mutually exclusive* and *exhaustive*. We are trying to choose between the two hypotheses, so there must be no chance that both are true. H_0 must exclude the possiblity of H_A, and vice versa. In addition, we want to be able to say that if H_0 is false, H_A is definitely true. There must be no possibility that some other (unspecified) hypothesis is true instead of H_A. In order to meet this requirement, H_0 and H_A must contain (exhaust) all possible values of θ.

Whenever sample data are used to choose between H_0 and H_A, there are risks of making an incorrect decision. These errors are known as *inferential errors*, because they arise when we make an incorrect inference about the value of the population parameter. Table 9.2 summarizes the way errors arise in hypothesis testing. Note that there are four possible outcomes resulting from any decision about the null hypothesis. In two cases, the

TABLE 9-2
Decisions and Outcomes in a Classical Test of Hypothesis

Decision	True state of nature	
	H_0 is true	H_0 is false
Reject H_0	Type I error	No error
Accept H_0	No error	Type II error

decision is the correct one. This occurs when we rightfully reject a false H_0 or accept a true H_0. In the other two cases we commit *either* a Type I or a Type II error. If we reject H_0 when it is true, we commit a Type I error; if we fail to reject H_0 when it is false, we commit a Type II error.

DEFINITION: TYPE I ERROR

A Type I error occurs when one rejects a null hypothesis that is actually true. The probability of committing a Type I error is denoted α (alpha).

DEFINITION: TYPE II ERROR

A Type II error occurs when one accepts a null hypothesis that is actually false. The probability of making a Type II error is denoted β (beta).

In adopting a method for choosing between H_0 and H_A, it is important to evaluate the risks of making both types of error. The chances of committing Type I and Type II errors are denoted α and β, respectively; that is, $\alpha = P$ (Type I error) and $\beta = P$ (Type II error). The relationship of α and β to the various choices about H_0 is indicated in Table 9-3. Moving down the table, we select one of the two alternatives—either we reject or accept H_0. Our two choices are complementary; thus, the total probability of each column sums to 1. Notice that regardless of our decision, there is some risk that it is wrong. After all, there is no way to be completely certain of our conclusion short of taking a complete census of the population. Although there is no way of avoiding risk completely, we can *limit* the level of risk associated with our decision. The way this is done is explained in step 3 of the classical test of hypothesis.

Selection of Sample Statistic and Its Sampling Distribution

The second step of the classical procedure is to select the appropriate *sample statistic* upon which we base our decision to reject or not to reject H_0. From Chapters 7 and 8, we know that the appropriate statistic to use is the minimum error estimator of the population parameter under study. When used in this way, the sample statistic is called the *test statistic*. Test statistics for μ, π, and σ are summarized in Table 8-1. By now it should make sense

TABLE 9-3
Probabilities of Making a Correct or Incorrect Decision
in a Test of Hypothesis

	True state of nature	
Decision	H_0 is true	H_0 is false
Reject H_0	α	$1 - \beta$
Accept H_0	$1 - \alpha$	β
Total Probability	1	1

that tests concerning μ are based on the sample value of the estimator $\bar{X}$, and, similarly, tests of π are based on the estimator P.

Returning to the example of residential mobility, we see that the decision about the *proportion* of households in the neighborhood changing residence should be based on the sample proportion statistic P. What is the sampling distribution of P? Recall that if the sample is large, we can appeal to the central limit theorem for the approximate sampling distribution of P: for large n, the sampling distribution of P is approximately normal with mean $\mu_p = \pi$ and $\sigma_p = \sqrt{\pi(1 - \pi)/n}$. In this example, therefore, we can say the sampling distribution of P is approximately normal with a mean of .2 and a standard deviation of $\sigma_p = \sqrt{.2(.8)/100} = .04$.

Selection of a Level of Significance

In this third step of the classical hypothesis testing procedure, we specify the probability of error associated with our ultimate decision. Obviously, we want to be confident about whatever decision we reach; thus, we would prefer to have α and β both be small. Unfortunately, it is almost always impossible to simultaneously minimize the probability of both types of error; we can limit α, or β, but not both. Faced with this problem, classical hypothesis testing adopts the strategy of controlling only α. That is, the usual procedure is to select a very low value for α, the probability of making a Type I error. Typically α is chosen to be .10, .05, or .01. The α value selected is known as its *significance level*.

This posture has some very important implications for the way we frame H_0 and H_A. In making α small, we are saying that if we reject H_0, we will do so with only a small probability of error. In other words, if we reject H_0 we will be confident in our decision. On the other hand, because β is not controlled, we cannot be confident that a decision to accept H_0 is correct. In light of this, *the null hypothesis should be something we want to reject, rather than something we want to confirm.* Rejection of H_0 needs to be the scientifically interesting result, because rejecting H_0 is the only action we can take with low probability of error.

DEFINITION: LEVEL OF SIGNIFICANCE

The level of significance of a classical test of hypothesis is the value chosen for α, the probability of making a Type I error.

Whenever we report a decision about the null hypothesis, we always state the level of significance of our result. We say, for example, that a null hypothesis is rejected at the .05 level or that it is *statistically significant* at the .05 level. This means that the probability we are making an error is low, no higher than 5%.

The usual analogy in explaining the concept of a level of significance is based on the nature of justice in a court of law. In North American law, a defendant is innocent until proven guilty beyond a reasonable doubt. The burden is placed entirely on the prosecution to prove the defendant guilty. Why? Because society prefers to free a guilty person as opposed to imprisoning an innocent one. This is tantamount to saying that a Type I error is more serious than a Type II error. The reasonable doubt in a statistical test is the α level. If we are going to reject H_0, we must be very sure we are correct. This is why the usual levels of significance are so low. When we report a result as statistically significant, we are actually saying that it is *probably* correct. Using the word *probably* reflects our knowledge that we may be incorrect owing to random sampling.

The level of significance chosen for a particular problem depends entirely on the nature of the problem and the uses to which the results will be put. We are likely to be very demanding if we are testing toxicity levels of drugs, given the consequences of marketing a drug with serious side effects. In that instance, an α level of .0001 may be appropriate. For most social and physical science purposes, α is likely to be set equal to .10, .05, or .01. The consequences of an error, for example, in the study of neighborhood mobility, cannot be considered to be very serious. For our example of residential mobility, we choose a significance level $\alpha = .05$.

Construction of a Decision Rule

To this point, we can summarize the test of hypothesis as follows:

Step 1. H_0: $\pi = .2$ and H_A: $\pi \neq .2$.
Step 2. P is chosen as the sample or test statistic.
Step 3. $\alpha = .05$.

To construct a decision rule for H_0, it is now necessary to specify which values of the test statistic will lead us to reject H_0 in favor of H_A. First, we assume H_0: $\pi = .2$ is true and generate the sampling distribution of P, as in Figure 9-2. Note that it is centered over a value of $P = .2 = \pi$. The actual value p generated in our sample will most likely be near .2 if H_0 is true and far away from .2 if H_0 is false. For a given level of significance, we can divide the sampling distribution into two parts—those outcomes consistent with H_0 and those leading to a rejection of H_0. Since we have specified a two-tailed test, we reject H_0 only if the sample outcome is in the most extreme parts of the sampling distribution. The most extreme outcomes occur in either tail of the distribution, so we place $\alpha/2$ of our possible

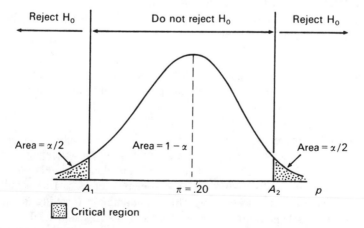

FIGURE 9-2. Sampling distribution of P, centered on hypothesized value $\pi = .2$.

outcomes in each tail. If the value of p is in the remaining $1 - \alpha$ fraction of outcomes, we cannot reject the null hypothesis. The values of p that lead us to reject H_0 are known as the *critical region* of the test. When we construct a decision rule for our hypothesis, we are specifying the values for A_1 and A_2 in Figure 9-2. These are sometimes called the *action limits*, or *critical values*, for the hypothesis.

DEFINITION: CRITICAL REGION AND CRITICAL VALUES
The critical region corresponds to those values of the test statistic for which the null hypothesis is rejected. The less extreme (more central) limits of the critical region are the critical values.

It is easy to see how the decision rule is constructed by considering the residential mobility example. Suppose we randomly sample $n = 100$ households in the neighborhood and calculate p, the proportion of people who moved in the last year. What values of p will lead us to conclude the neighborhood is less mobile than the city average (i.e., what is A_1?), and what values of p will lead us to conclude the neighborhood is more mobile than the city average (i.e., what is A_2)? Let us solve this problem by generating the sampling distribution of P for $n = 100$, $\pi = .2$, and $\alpha = .05$. The appropriate sampling distribution is illustrated in Figure 9-3. First, note that the sampling distribution in this figure is drawn to be normal (or approximately so) since we know from step 2 that this is the appropriate sampling distribution for the test statistic P. Also, note that we have placed $\alpha/2 = .025$ of the outcomes for P in each tail of the distribution.

To determine the values of the critical limits A_1 and A_2, we use our knowledge of the normal distribution. First, we know this sampling distribution has a mean centered on $\pi = .2$. The standard deviation of this sampling distribution is

$$\sigma_P = \sqrt{\frac{\pi(1 - \pi)}{n}} = \sqrt{\frac{.2(.8)}{100}} = .04$$

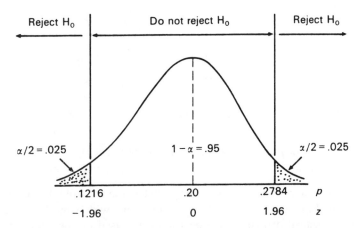

| Reject H_0 | Do not reject H_0 | Reject H_0 |

$\alpha/2 = .025$ $\quad 1 - \alpha = .95 \quad$ $\alpha/2 = .025$

.1216 .20 .2784 p

-1.96 0 1.96 z

FIGURE 9-3. Determining critical region limits for residential mobility example, $\alpha = .05$.

The critical limits on the sampling distribution are determined by the conditions

$$P(P < A_1) = \frac{\alpha}{2} \quad \text{and} \quad P(P > A_2) = \frac{\alpha}{2} \tag{9-4}$$

According to (9-4), we must choose A_1 such that such that the probability of the sample statistic P being less than A_1 is $\alpha/2$. Because the sampling distribution of P is normal, A_1 and A_2 are symmetric about $\pi_0 = .2$. To find A_1, we use the standardizing formula for the normal distribution:

$$z_{A_1} = \frac{A_1 - \pi_0}{\sigma_P} = z_{\alpha/2}$$

From the significance level α, we can say z_{A_1} is the value of z that has $\alpha/2$ in the lower tail of the distribution. In other words, $z_{A_1} = z_{\alpha/2}$. Similarly,

$$z_{A_2} = \frac{A_2 - \pi_0}{\sigma_P} = z_{1-\alpha/2}$$

can be used to determine z_{A_2}. Substituting the known values of π, α, and σ_P yields

$$z_{.025} = \frac{A_1 - .2}{.04} \quad \text{and} \quad z_{1-.025} = \frac{A_2 - .2}{.04} = z_{.975}$$

We find $z_{.025}$ by looking in the body of the normal table to find the z-score that has .025 of the area under the curve in the left tail. Using column 4 of Table A-3 gives a z value of -1.96. The other z-score, $z_{.975}$, calls for .025 of the area in the right tail, corresponding to $z = 1.96$ (Table A-3, column 3). Substituting the z values into our equations, we find the critical limits for our decision rule:

$$\frac{A_1 - .2}{.04} = -1.96 \qquad \frac{A_2 - .2}{.04} = 1.96$$

$$A_1 - .2 = -.0784 \qquad A_2 - .2 = .0784$$

$$A_1 = .1216 \qquad A_2 = .2784 \tag{9-5}$$

The decision rule can now be summarized as follows:

Reject H_0 if $p \le .1216$ or $p > .2784$. Otherwise, do not reject H_0.

This is a very compact way of expressing what action we will take depending on the value of p generated in the sample. There is an equivalent, widely used way of expressing the decision rule. Instead of making the decision rule based on the value of the sample statistic p, we express it in terms of z.

Reject H_0 if $z \leq -1.96$ or $z > 1.96$. Otherwise, do not reject H_0.

In this case, we determine the value of z for the decision rule on the basis of the standardizing formula

$$z = \frac{p - \pi_0}{\sigma_P} \tag{9-6}$$

What we are doing in this transformation is locating the observed value of sample statistic p within the standard normal distribution. In using Equation 9-5 to find the critical limits A_1 and A_2, we are actually reworking (9-6) to solve for p.

Let us now explain the reasoning behind the use of these decision rules in greater detail. If the null hypothesis $\pi = .2$ were true, we would expect the sample value p to be fairly close to .2. For $\alpha = .05$, we would expect p to be in the range [.1216, .2784] with probability $1 - .05 = .95$. If the sample gives a value for p outside that range, we reject H_0. This decision rule is thus based on the sampling distribution of the test statistic P, obtained under the assumption that H_0 is true.

At first thought, this procedure seems wrong: we have said H_0 should be a hypothesis we want to reject, yet we assume it is true in performing the test. Actually, this is not strange at all. Consider the courtroom example. As a prosecutor, we want to reject the assumption of innocence. In order to do so, we evaluate evidence under the assumption of innocence. If things have occurred that are unlikely under that assumption, we reject innocence in favor of the alternative hypothesis. This is expressed in a slightly different way in Figure 9-4.

Compute the Value of the Test Statistic

At this point, we collect our sample and calculate the value, p, of the sample statistic P. Suppose that 26 of 100 randomly selected households in the neighborhood have moved in the past year. Then the value of the test statistic is $p = .26/100 = .26$. We can use this value directly with the decision rule constructed in the previous step. Alternatively, if we wish to express our result in terms of z, we calculate

$$z = \frac{.26 - .2}{.04} = \frac{.06}{.04} = 1.50$$

and use this value in conjunction with the decision rule based on z.

Decision

The last step is to formally evaluate the null hypothesis based on the sample. This is simply a matter of examining the calculated value of the test statistic in relation to the decision rule. Because $p = .26$ is between .1216 and .2784, it *does not* lie in the critical

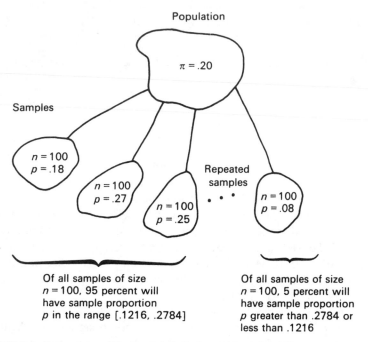

FIGURE 9-4. Determining critical region limits for residential mobility example, α = .05.

region. We therefore cannot reject H_0. If we were to reject H_0 on the basis of this sample, the probability of making a Type I error would exceed .05. In adopting a significance level of α = .05, we have said a 5% error is the largest we can tolerate, so we must not reject H_0. Notice that even if the computed value of p were .2783, we still could not reject H_0. As a matter of fact, this rigidity in our decision rule is one problem with the classical test of hypothesis. Because the decision to reject or not is binary, it does not admit shades of certainty regarding H_0. But surely, would not a sample with $p = .26$ and a sample with $p = .2$ give us varying degrees of belief in H_0? In Section 9.2 we introduce the PROB-VALUE method, which gets around this problem. Instead of simply rejecting H_0 in favor of H_A, we determine our degree of belief in H_A; that is, we compute the value of α implicit in our sample.

What if $p = .31$ is the calculated sample value of p? In this instance, the decision is different. We reject H_0 because p falls in the critical region (.31 > .2784). Rejecting H_0 leads to two possibilities. Either we made a correct decision and $\pi \neq .2$, or we obtained a sample that did come from a population with $\pi = .2$. In the latter case, we made a wrong decision and committed a Type I error. However, we should feel relatively secure, knowing the probability of such an occurrence is necessarily less than .05.

To evaluate the hypothesis in terms of a decision rule based on z, we note $z = 1.5$ calculated in the previous step is in the range $[-1.96, 1.96]$, and thus we must not reject H_0. Of course, any decision based on z is equivalent to one made by using p, and vice versa. Some texts in statistics always work with z, others with the value of sample statistic p, and still others both. The advantage of using p (or $\bar{x}$ or s, as the case may be) is that the decision is being made using the units in which the variable of interest is measured. In the current example, we could say that we will reject the hypothesis that the neighborhood has a typical level of residential mobility if the proportion of people who moved recently is less than .1216 or greater than .2784. To say the same thing in terms of some standardized normal variable removes the decision from the context of the problem at hand. Nevertheless, it is the more common approach. The complete text of hypothesis testing for the example is summarized in Table 9-4. It provides a convenient way for beginning students to organize classical tests in inferential statistics, and we will use it throughout the remainder of the text. In subsequent examples in this chapter, many of the variations in the classical test are introduced. In particular, we will cover the directional tests depicted in formats B and C, also called one-tailed tests.

There is one final, very important issue regarding decisions in hypothesis testing. In our example, the sample value did not lie in the critical region. It would seem that we should therefore accept H_0. Strangely enough, we did not do so, saying instead "We therefore cannot reject H_0." In other words, rather than accept H_0, we make no decision. The reason for this has to do with our choice to make α small. If we were to accept H_0, we would do so with an error probability of β, a value we have made no attempt to control. How can we calculate β? It seems clear that because β applies when H_A is true, we need to know how likely it is to obtain $p = .26$ when H_A is true. Here we have a problem, because H_A is vague, simply saying $\pi \neq .2$. H_A admits many values for π, actually an infinite number. If we are going to accept H_0, we must be convinced that *none* of these possibilities is true. On the basis of $p = .26$, are we prepared to say π is definitely .2, rather than .25, or .27? Clearly, if we conclude H_0 is true, we run a large, unknown risk of error. Though we could make this decision, we would could not attach an uncertainty value to our decision, and would therefore accomplish nothing.

It is often said that anything can be proved with statistics. Our experience with hypothesis testing thus far suggests nothing could be more wrong! Instead of proving things with statistics, we find that we are only sometimes able to *disprove* some things (H_0) and can do that only with some chance of error (α). Fortunately, the level of error is known and is under our control.

TABLE 9-4
Summary of Test of Hypotheses for Residential Mobility Example

---

Step 1. H_0: $\pi = .2$ and H_A: $\pi \neq .2$.

Step 2. P is chosen as the sample statistic.

Step 3. $\alpha = .05$.

Step 4. Reject H_0 if $p < .1216$ or $p > .2784$. (See Figure 9-3.)

Step 5. From random sample, $p = .26$.

Step 6. Because $.1216 < p < .2784$, do not reject H_0.

9.2. PROB-VALUE Method of Hypothesis Testing

In Section 9.1 we outlined a strict procedure for hypothesis testing known as the classical method. This procedure is useful as a means of understanding the issues that lie at the heart of all hypothesis testing, such as the distinction between a Type I and a Type II error. However, the classical procedure is less often used in actual practice than an alternative procedure known as the PROB-VALUE, or p-VALUE, method. (We will call it *PROB-VALUE testing* rather than p-VALUE since the letter p is used in so many other contexts.) What is wrong with hypothesis testing done in the classical way? First, it requires us to choose a value for α, the level of significance for the test. Unfortunately, there are few instances where this value can be rationally chosen. There is no disputing that it should always be small, but exactly how small? We can easily think of cases where we would reject H_0 at $\alpha = .05$ but not at $\alpha = .01$. If two researchers happen to choose different α levels, they might reach different conclusions from the same data. This is unsettling, given that convention rather than theory drives the choice of α.

A second shortcoming is that our findings are dichotomous. We choose α, and report only whether or not H_0 is rejected for that α. Another researcher, who prefers a different α, cannot tell from our report if she/he would arrive at the same decision. The classical method can therefore be criticized as not providing enough information. Along these same lines, suppose we subscribe to the subjectivist view of probability. If so, we might be willing to think of H_0 as having some probability of falsehood. (The objectivist interpretation doesn't allow this—H_0 is either true or it isn't, and no probability can be assigned.) The classical method of testing offers no information regarding how confident we are about a decision to reject. But surely, if the test statistic is far in the tail of the sampling distribution, we will be more confident about rejecting H_0 than if the statistic is barely within the critical region. From a subjectivist viewpoint, we have various degrees of certainty about H_0's falsehood depending on where in the critical region the statistic lands. If the degree of certainty is available, it should be reported along with our decision to reject H_0.

The PROB-VALUE method is preferred because it addresses both of the deficiencies mentioned above. As it is best explained by example, we will reconsider the test about residential mobility developed in Section 9.1. After using the example to show how the PROB-VALUE method differs from the classical method, we will turn to more general applications.

PROB-VALUE Method for a Two-Sided Test

Figure 9-3 illustrates the decision rule developed for the hypothesis set

$$H_0: \pi = .2 \qquad H_A: \pi \neq .2$$

for the residential mobility problem. Selecting a significance level of $\alpha = .05$ led to this decision rule:

1. Reject H_0 if $p < .1216$ or $p > .2784$ (or $z < -1.96$ or $z > 1.96$).
2. Do not reject H_0 if $.1216 \leq p \leq .2784$ (or $-1.96 \leq z \leq 1.96$).

Here p is the value of the sample proportion statistic P, and z is the value of the standardized variable

$$Z = \frac{P - \pi_0}{\sqrt{\pi_0(1 - \pi_0)/n}}$$

Because our sample proportion is .26, we do not reject H_0. Now, the principal difficulty with the classical method is that it provides no information as to *exactly* how likely it is that the null hypothesis is false. *Any* value of p in the critical region leads us to reject H_0. But a sample proportion $p = .9$ suggests that H_0 is less likely to be true than a sample proportion $p = .28$ (just on the edge of the critical region). It is just this problem that the PROB-VALUE method is designed to solve.

In the example, the actual sample proportion $p = .26$ leads us not to reject H_0. But because it is close to the critical limit of $p = .2784$, we might wonder, "If we were to reject H_0 on the basis of $p = .26$, how much faith would we have in this decision?" To see this, look at Figure 9-5. On this diagram we have established the critical region at the value of $p = .26$ on the upper side of $\pi_0 = .2$ and at $p = .14$ on the lower side (.2 − .06). Together, these two values define another of critical region. Associated with this new critical region is a new, different, α-value. If we happened to have used the new α, the test result would have been just on the margin of rejection. The probability of a Type I error would have been *exactly* equal to the new α.

Simply put, this new α is the "probability" in the PROB-VALUE method. The method calls for us to find and report the value of α for which the sample would just barely permit us to reject H_0. The phrase "just barely" means that the sample statistic is on the edge of the critical region. Notice how the PROB-VALUE method differs from the classical method. There we chose the probability α, which determined the critical region. Here we are using the test statistic to determine a critical region, and are finding a probability from the area of that region. How do we interpret this probability? It should be clear that the

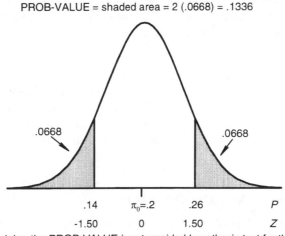

PROB-VALUE = shaded area = 2 (.0668) = .1336

.0668   .0668

| .14 | π_0=.2 | .26 | P |
| -1.50 | 0 | 1.50 | Z |

FIGURE 9-5. Determining the PROB-VALUE in a two-sided hypothesis test for the residential mobility example.

PROB-VALUE is the probability of making a Type I error. In other words, should we elect to reject H_0, the PROB-VALUE tells us how likely it is that we are wrong.

Returning to Figure 9-5, the PROB-VALUE, the area in the shaded region, can be easily determined knowing that P is normally distributed. First, convert the sample value of $p = .26$ to a standardized value z, using

$$z = \frac{p - \pi_0}{\sqrt{\pi_0(1 - \pi_0)/n}} = \frac{.26 - .2}{\sqrt{.2(1 - 2)/100}} = 1.5$$

Next, determine the area to the right of $z = 1.5$. Using the normal table gives .0668. Because the distribution is symmetric, the area in both tails is $2(.0668) = .1336$. This is the PROB-VALUE attached to H_0 for this problem. The probability of .0668 is doubled because we are evaluating a two-sided alternative. Recall that in the classical test of a two-sided hypothesis the significance level α is halved, and $\alpha/2$ is placed in the two rejection areas at the tails of the sampling distribution.

The PROB-VALUE of .1336 can be interpreted in the following way: *it is the lowest value at which we could set the significance of the test and still reject H_0*. If we reject H_0 on the basis of $p = .26$, we are saying that the probability of committing a Type I error of $\alpha = .1336$ is acceptable to us. In short, the PROB-VALUE indicates the *degree of belief* that H_0 is false. As the PROB-VALUE approaches 0, we are ever more sure that rejecting H_0 is the correct decision. Calculating a PROB-VALUE is clearly a superior alternative to the classical test. It avoids the thorny problem of needing to dismiss an alternate hypothesis simply because the sample statistic lands just outside the critical region. It also allows other researchers to interpret our sample results using whatever level of uncertainty (α) they choose.

DEFINITION: PROB-VALUE

The PROB-VALUE associated with a null hypothesis is equal to the probability of obtaining a value of the sample statistic as extreme as the value observed, if the null hypothesis is true. If a decision is made to reject H_0, the PROB-VALUE gives the associated probability of a Type I error.

Virtually all statistical packages now compute PROB-VALUEs. These values report what the sample data tell us about the credibility of a null hypothesis, rather than forcing a decision about H_0 on the basis of a possibly arbitrarily defined level of significance. Of course, determining a PROB-VALUE does not preclude performing a classical test. Rejecting a null hypothesis with a PROB-VALUE of .05 or less is equivalent to a classical test of hypothesis with $\alpha = .05$.

PROB-VALUE Method for a One-Sided Test

Determining the PROB-VALUE in a one-sided test of hypothesis, either lower or upper tail, differs in only one respect from the method discussed above. Suppose the hypothesis being evaluated in the residential mobility example is

$$H_0: \pi \le .2 \qquad H_A: \pi > .2$$

This is test format C, in which we are examining whether the neighborhood has a *higher* mobility rate than average. The PROB-VALUE associated with the sample value $p = .26$ is .0668, as illustrated in Figure 9-6. There is no need to double the probability in a one-sided test because *the critical region is always placed in one tail or the other, depending on whether it is a lower- or an upper-tail test.*

Does it make sense that the PROB-VALUE is lower for the one-tailed test than for the two-tailed test? Consider the fact that in doing a one-tailed test, we are saying the only alternative to H_0 is $\pi > .2$. In the two-sided test, we allowed for the additional possibility $\pi < .2$. In other words, we come to the one-sided test with more knowledge, having restricted H_A by one-half. Because we ask a narrower question, the sample is able to give a more precise answer to the question we ask. Here the value of $p = .26$ indicates the neighborhood is experiencing higher mobility than average. A PROB-VALUE of .0668 (as opposed to .1336) reflects our improved ability to evaluate H_0. The lesson is that in a one-tailed test, a less extreme sample result is needed to reject H_0. Whenever one has good reason for using a one-tailed test, it should be used. Doing so will maximize use of sample information.

In Figure 9-6, we have drawn the sampling distribution of P assuming $\pi = .2$. Likewise, we computed the PROB-VALUE assuming $\pi = .2$. Looking at the null hypothesis, we see that it allows for many values for π. In fact, *any* value is allowed so long as it is less than or equal to 0.2! Clearly, had we used some other value, say $\pi = .1$, the sampling distribution and resulting PROB-VALUE would be different. Why, then, did we choose $\pi = .2$?

Imagine that we use one of the allowable (smaller) values. Obviously, the sampling distribution in Figure 9-6 will move to the left. The sample result $p = .26$ stays the same, so that the shaded area and PROB-VALUE must *decrease*. This shows that the PROB-VALUE value we obtained using $\pi = .2$ is the *largest* that can be obtained under this null hypothesis. We can be sure, therefore, that no matter which value of π is true, our PROB-VALUE is, if anything, too large. If the computed PROB-VALUE leads us to reject

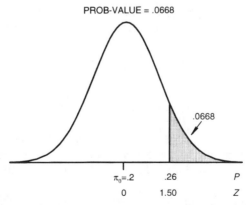

FIGURE 9-6. Determining the PROB-VALUE in a one-sided hypothesis test for the residential mobility example.

H_0, we can be positive that H_0 would be rejected for all the other values of π admitted under H_0. In other words, using $\pi = .2$ is a conservative approach.

9.3. Hypothesis Tests Concerning the Population Mean μ and π

In the remainder of this chapter, a series of hypothesis tests are described for the population parameters μ and π. Examples illustrate typical situations where these tests may be used and revisit many of the theoretical issues discussed in Sections 9.1 and 9.2. Because it is so common, all examples employ the PROB-VALUE method. However, readers are reminded that the PROB-VALUE method permits classical testing. In particular, if the PROB-VALUE obtained is lower than some α level chosen a priori, the null hypothesis should be rejected.

The first test concerns hypotheses related to the population mean μ. The sample mean statistic $\bar{X}$ is the appropriate test statistic in this case, and its sampling distribution is given in the central limit theorem. Just as we distinguished between two cases for specifying a confidence interval for μ, hypothesis tests of μ must be based what is known about the population standard deviation σ:

Case 1. Population standard deviation σ is known.
Case 2. Population standard deviation σ is unknown.

Case 1: Population Standard Deviation σ Known

In Chapter 7 we showed that the sample mean statistic $\bar{X}$ is approximately normally distributed with mean $E(\bar{X}) = \mu_{\bar{x}} = \mu$ and standard deviation $\sigma_{\bar{x}} = \sigma/\sqrt{n}$. Tests of hypothesis concerning μ can therefore be evaluated by using the standard normal statistic

$$Z = \frac{\bar{X} - \mu}{\sigma_{\bar{x}}} = \frac{\bar{X} - \mu}{\sigma/\sqrt{n}} \tag{9-7}$$

This standard normal statistic finds extensive application in hypothesis testing. As we will see below, it can be used to generate decision rules for tests of sample proportions. In fact, for many different sample statistics $\hat{\theta}$, the general form of the standard normal statistic

$$Z = \frac{\hat{\theta} - \theta_0}{\sigma_{\hat{\theta}}} \tag{9-8}$$

can be used to construct confidence intervals or test hypotheses. In (9-8), θ_0 is the hypothesized value of the population parameter and the mean of the sampling distribution of the sample statistic $\hat{\theta}$. The denominator or $\sigma_{\hat{\theta}}$ is the standard deviation of the sampling distribution of $\hat{\theta}$, sometimes called the *standard error* of the sample statistic. The purpose of the standardized statistic Z is to locate the value of the sample statistic within the standard normal distribution.

The value of z obtained from Equation (9-8) for a particular sample tells us how many standard deviations (or standard errors) is $\hat{\theta}$ from the mean of the sampling distribution. The higher the absolute value of z, the more unusual is the sample, and the less likely that H_0

is true. To test H_0, we simply need to assign a probability to the observed value under the assumption that H_0 is true. The PROB-VALUE method for any sample statistic that is normally distributed will always be as follows for a *two-sided test*:

1. Compute the sample value z from (9-8) using the observed $\hat{\theta}$ and hypothesized value θ_0.
2. Find the PROB-VALUE: $P(Z < -z) + P(Z > z) = 2\,P(Z > z)$

Let us return to hypothesis testing for μ by considering the following example.

EXAMPLE 9-1. The mean household size in a certain city is 3.2 persons with a standard deviation of $\sigma = 1.6$. A firm interested in estimating weekly household expenditures on food takes a random sample of $n = 100$ households. To check whether the sample is truly representative, the firm calculates the mean household size of the sample to be 3.6 persons. Test the hypothesis that the firm's sample is representative of the city with respect to household size. What PROB-VALUE can be attached to this hypothesis?

Solution. The values of $\mu = 3.2$ and $\sigma = 1.6$ are given, and we know $\bar{X}$ is approximately normal for $n = 100$. The null and alternate hypotheses are

$$H_0: \mu = 3.2 \qquad H_A: \mu \neq 3.2$$

A two-sided test is needed because we want to know if the sample families are different from the general population, either smaller or larger. The sample has mean $\bar{x}$ of 3.6, from which we compute z:

$$z = \frac{3.6 - 3.2}{1.6/\sqrt{100}} = \frac{.4}{.16} = 2.5$$

Using column 7 from Table A-3, we find a PROB-VALUE of 0.012. We therefore conclude it is highly unlikely that the sample is representative of the larger population. Figure 9-7 shows the results graphically. Note that if we were testing this hypothesis at the .05 level, we would reject H_0. Testing at the .01 level, we would not reject H_0.

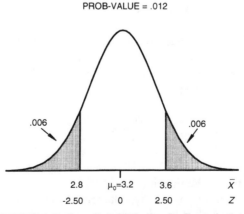

FIGURE 9-7. Sampling distribution for Example 9-1.

This example illustrates an important point about tests for μ. Household size is a discrete variable, and one with a small range of values (few families have more than five children, none less than zero). We can therefore be sure that X is not normally distributed. However, with the large sample size used here, $\overline{X}$ will be very close to normal—so close that the error is negligible. Note that if we were to repeat the test with a small n, we would have to consider the PROB-VALUE approximate, knowing that we are working with a crude approximation of the sampling distribution of $\overline{X}$.

Case 2: Population Standard Deviation σ Unknown

If σ is unknown, it must be estimated from the sample using the estimator S. From Chapter 8, we know the standardized random variable

$$T = \frac{\overline{X} - \mu}{S/\sqrt{n}} \tag{9-9}$$

is t-distributed with $n - 1$ degrees of freedom, providing X is itself normal. Thus, for a normally distributed random variable X, Equation 9-9 can be used to test hypotheses and determine PROB-VALUEs. When X is not normally distributed but n is large, the use of (9-9) may be considered to give approximate results.

When we used the t-distribution for confidence intervals, we chose a probability (α) and used a Table A-4 to find the corresponding value of t. Here we need to do just the reverse. In the PROB-VALUE method we compute a value of t, and use a table to arrive at the corresponding probability. Tables A-5 and A-6 give one- and two-tailed probabilities for the t-distribution. Notice that the rightmost columns (infinite df) are equal to standard normal probabilities.

EXAMPLE 9-2. A commuter train advertisement claims that the time on the train between a certain station and the downtown terminal is 23 min. A random sample of $n = 30$ trains in a single month yields an average time of 26 min with a standard deviation of 6 min. Test the assertion that the trains are actually slower than advertised, using $\alpha = .05$.

Solution. Viewing this as a hypothesis testing problem, we can put

$$H_0: \mu \leq 23 \qquad H_A: \mu > 23$$

We choose a one-sided test because we are asking only if the trains are slower than advertised. If we reject H_0, we can make the assertion we would like: Average times are longer than advertised. Based on a sample of size $n = 30$, and the estimates $\overline{x} = 26$, $s = 6$, we compute the observed t-value:

$$t = \frac{26 - 23}{6/\sqrt{30}} = 2.74$$

From Table A-6, we obtain a PROB-VALUE of about .005, 10 times smaller than the .05 level we have chosen for α. We therefore reject H_0.

Notice that we used 23 for the hypothesized value of μ. Once again, this is a conservative approach. Though H_0 admits many values for μ, each with a different PROB-

VALUE, we can be sure that none is larger than the PROB-VALUE we get using $\mu = 23$.

The tests of hypothesis concerning the population parameter μ are summarized in Table 9-5. The table distinguishes between two cases depending on whether σ is known beforehand. Both cases give exact PROB-VALUEs when X is normal, and approximate values otherwise. The approximations improve with increasing sample size and will give excellent results for $n > 30$ or so.

Hypothesis Tests Concerning the Population Proportion π

This test was used as the example in presenting the classical test of hypothesis in Section 9.1. We cover it again here, describing in more detail how the sampling distribution is obtained. As we shall see, there are two variations depending on the sample size.

Consider first the situation when n is large. As was seen in Chapter 6, the random variable X is a Bernoulli variable, taking on only two values, zero or unity. To compute the sample proportion p, we divide the number of 1s in the sample by n. This is equivalent to computing the sample mean $\bar{X}$, which we know to be approximately normally distributed for large n. We can say therefore, that the random variable P is approximately normal. As in tests for μ, we will compare the sample value for the test statistic with the value hypothesized under H_0. If the sample value diverges greatly from the hypothesis, we will reject H_0 in favor of H_A.

EXAMPLE 9-3. Household surveys of residents in a suburban neighborhood have *consistently* found the proportion of residents taking public transit to be .16. The city council, at the urging of neighborhood residents, increased the bus service to this neighborhood. However, the council vowed to discontinue the service if there was not an increase in patronage. Three months after the introduction of the new service, the council hired a consultant to survey $n = 200$ residents and determine their current modes of transportation. Forty-two residents indicated they took public transit. Should the council continue the service?

Solution. Since the proportion of residents taking public transit has consistently been .16 and the council has specified that the proportion must *increase*, we can specify the two hypotheses as follows:

$$H_0\!: \pi \le .16 \qquad H_A\!: \pi > .16$$

This is an upper-tail test. The corresponding sampling distribution of P is shown in Figure 9-8. With $\pi = .16$, the standard deviation of the sampling distribution for a sample size of $n = 200$ is $\sigma_P = \sqrt{.16(.84)/200} = .026$. The sample value for p is $42/200 = .21$. Using the hypothesized value for π, we compute z as

$$z = \frac{.21 - .16}{.026} = 1.9$$

According to Table A-3, the PROB-VALUE is $P(Z > 1.9) = .029$. The null hypothesis can therefore be rejected with an error probability just less than 3%. If the city council deems this an acceptable level of error, it will continue the new bus schedule. Alternatively, a

TABLE 9-5
Single-Sample Tests for μ

Background
A single random sample of size n is drawn from a population, and the sample mean $\bar{x}$ is calculated. The sample value will be compared to the hypothesized value to determine if H_0 should be rejected. The sampling distribution of $\bar{X}$ is known exactly if X is normal, which permits one to attach an exact PROB-VALUE to the sample result. If X is approximately normal, the PROB-VALUE is approximate.

Hypotheses

H_0: $\mu = \mu_0$	H_0: $\mu \geq \theta_0$	H_0: $\mu \leq \theta_0$
H_A: $\mu \neq \mu_0$	H_A: $\mu < \mu_0$	H_A: $\mu > \mu_0$
(A)	(B)	(B)
(two-tailed)	(one-tailed)	(one-tailed)

Test statistic
Case 1: σ known. The test statistic Z is normally distributed. Using the sample mean $\bar{x}$, the observed value is

$$Z = \frac{\bar{x} - \mu_0}{\sigma/\sqrt{n}}$$

Case 2: σ unknown. The test statistic T is t-distributed with $n - 1$ degrees of freedom. Using the sample mean and standard deviation $(\bar{x}, s)$, the observed value is

$$T = \frac{\bar{x} - \mu_0}{s/\sqrt{n}}$$

PROB-VALUE (PV) and Decision Rules
Two-tailed test (format A):

$$PV = P(|Z| > z) = P(Z < -z) + P(Z > z) \qquad PV = P(|T| > t) = P(T < -t) + P(T > t)$$

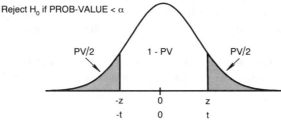

One-tailed test (format B):

$$PV = P(Z < -z) \qquad PV = P(T < -t)$$

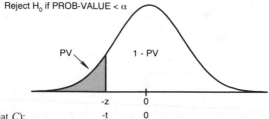

One-tailed test (format C):

$$PV = P(Z > z) \qquad PV = P(T > t)$$

(continued)

Reject H_0 if PROB-VALUE < α

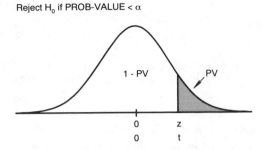

council requiring stronger evidence might authorize another survey, or perhaps even pro-nounce the new service a failure.

The procedure described above will not work for small n, because the sample proportion P will not be close to a normal distribution. However, the tools needed for small n have already been presented. Instead of working with the *proportion* of the sample taking on a value of unity, we can work with the *number* of 1s in the sample. That is, rather than use the fraction of the sample that is unity, we will count the number that are unity. If we let X be the number of 1s in the sample, we know that X follows the binomial distribution. Knowing the distribution of X lets us find the probability associated with any observed value.

EXAMPLE 9-4. In unpolluted waters, abnormal growths are known to occur in 15% of all sturgeon. A sample of 10 fish obtained near a chemical plant yields three fish with tumors. Is there reason to believe fish living in waters near the plant have an unusually high tumor rate? Use a 10% confidence level for α.

Solution. The question calls for a one-tailed test, with null and alternate hypotheses:

$$H_0: \pi \le .15 \qquad H_A: \pi > .15$$

We let X be the number of successes in 10 trials, where an abnormal growth is called a "success." The sample has a 30% success rate, which seems quite large if the population rate π is only 15%. To assess the probability, we make use of the fact that X is binomial, assuming the value for π indicated in H_0. We need to find the probability of obtaining three or more successes in 10 trials. If three or more successes occur often purely by chance, we will not reject H_0. On the other hand, if the sample value turns out to be unusual, we will embrace the alternate hypothesis.

PROB-VALUE = .029

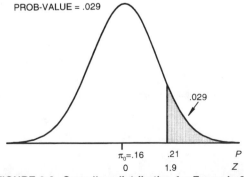

FIGURE 9-8. Sampling distribution for Example 9-3.

We need $P(X \geq 3)$. Binomial probabilities are given in Table A-2 for $\pi = .15$ and $n = 10$. Rather than sum the probabilities over $X = 3, 4, \ldots, 10$, we can use

$$P(X \geq 3) = 1 - P(X = 0) - P(X = 1) - P(X = 2)$$
$$= 1 - .1969 - .3474 - .2759 = .1970$$

If we reject H_0, we run a 20% chance of a Type I error. Because this is greater than α, we do not reject H_0. This sample does not allow us to conclude fish living near the chemical plant are experiencing an unusually high rate of abnormal growths.

One-sample tests for π are summarized in Table 9-6.

9.4. Relationship between Hypothesis Testing and Confidence Interval Estimation

Hypothesis testing and confidence interval estimation are closely related methods of statistical inference. In fact, the *significance level* of a test is actually the complement of the

TABLE 9-6.
Single-Sample Tests for π

Background
A single random sample of size n is drawn from a population, and we observe the number of successes in the sample, x. The random variable X follows the binomial distribution. If n is small, we will assign a probability to the sample value directly using the binomial distribution. For large n, we will divide by the sample size to obtain the sample proportion p. Using the central limit theorem, we know its associated random variable P is approximately normal.

Hypotheses

$H_0: \pi = \pi_0$	$H_0: \pi \geq \pi_0$	$H_0: \pi \leq \pi_0$
$H_A: \pi \neq \pi_0$	$H_A: \pi < \pi_0$	$H_A: \pi > \pi_0$
(A)	(B)	(B)
(two-tailed)	(one-tailed)	(one-tailed)

Test Statistic
Case 1: Small n. The test statistic is the number of successes in the sample X. This is a binomial variable:

$$P(X = x) = C(n, x)\pi_0^x(1 - \pi_0)^{n-x} \qquad x = 0, 1, \ldots, n$$

Case 2: Large n. The test statistic Z is based on the proportion of successes in the sample, and is approximately normal. Using the sample proportion p, the observed value of the test statistic is

$$Z = \frac{P - \pi_0}{\sqrt{\pi_0(1 - \pi_0)/n}}$$

PROB-VALUE (PV) and Decision Rules

In the following, the large sample probabilities are approximate, whereas the binomial PROB-VALUEs are exact.

Two-tailed tests (format A)

$$PV = P(|Z| > z) = P(Z < -z) + P(Z > z)$$

(continued)

Reject H$_0$ if PROB-VALUE < α

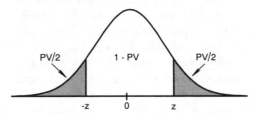

One-tailed tests (format *B*):

$$PV = P(Z < -z) \qquad PV = P(X \leq x)$$

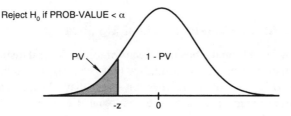

One-tailed tests (format *C*):

$$PV = P(Z > z) \qquad PV = P(X \geq x)$$

Reject H$_0$ if PROB-VALUE < α

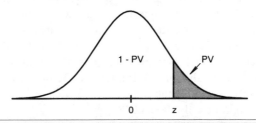

confidence in a confidence interval. (The fact that we denote the level of significance as α and the confidence level as $1 - α$ is no coincidence.) The relationship between these two procedures for statistical inference is best explained by a simple example. For this purpose, let us reconsider Example 9-1, which dealt with household sizes. Based on the sample value for *z*, we showed that the PROB-VALUE for the two-tail test is .0124.

Now, for comparative purposes, let us construct both the 95 and the 99% confidence intervals for μ, using the sample $\bar{x} = 3.6$ and $σ = 1.6$:

$$95\% \text{ interval: } \bar{x} \pm z_{.025} \frac{σ}{\sqrt{n}} = 3.6 \pm 1.96 \frac{1.6}{\sqrt{100}} = [3.28, 3.91]$$

$$99\% \text{ interval: } \bar{x} \pm z_{.005} \frac{σ}{\sqrt{n}} = 3.6 \pm 2.85 \frac{1.6}{\sqrt{100}} = [3.19, 4.01]$$

Notice that the 95% confidence interval does not contain the hypothesized value of $\mu_0 = 3.2$, but the 99% interval does. Because the PROB-VALUE was found to be .0124, this is the expected result. That is, the PROB-VALUE says we are between 95% and 99% sure that $\mu \neq 3.2$, as do the two confidence intervals (see Figure 9-9).

This interpretation can now be extended as follows. If the $(1 - \alpha)$ confidence interval *does not contain* the hypothesized value θ_0, we can *reject* the hypothesis $\theta = \theta_0$ at the α level of significance. On the other hand, if the $(1 - \alpha)$ confidence interval includes θ_0, testing the hypothesis does not allow H_0 to be rejected. We see that associated with every confidence interval is an implicit two-sided hypothesis test. The α level of the test is equal to $(1 -$ confidence level).

9.5. Statistical Significance Versus Practical Significance

Hypothesis testing provides a formal, objective means of evaluating a statistical hypothesis. When we obtain a statistically significant result, we know that H_0 should be rejected with no more than a small (α) chance of error. Does this mean that all statistically significant results are important from a practical or scientific standpoint? Certainly not! Whether or not we obtain important results depends on whether or not we test important hypotheses. Although this truth seems obvious, and is universally acknowledged, it is often overlooked.

To see how easy it is to go wrong on this point, think back to the public transit example. Let us suppose that everyone concerned is satisfied with the 95% confidence level employed in the test. If that is so, then we certainly proved that ridership increased following the bus-schedule change. The question is, did we obtain a result of any practical significance? Well, what are we prepared to say about the increase? Can we tell the city council that ridership is now .21 (the sample p)? No, we did not test that hypothesis—we tested for an unspecified increase, however small. Thus, all we can say from the test is that ridership is more than the old value of .16. It might be .21 or even higher, or it might be much lower, say .1600001. Is it likely the council would want to fund the new bus schedule knowing the increase might be only .0000001?

Unfortunately, situations similar to the above occur all too often, where a null hypothesis of no difference is adopted. There are instances where rejecting such a hypothesis is important, but not many. One example might be testing the efficacy of a drug, where

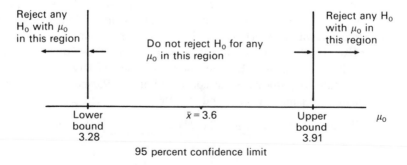

FIGURE 9-9. Relationship between confidence interval estimation and hypothesis testing for Example 9-1.

even a marginal mortality decrease translates into lives saved, and is therefore important. Most often, however, we are not interested in marginal differences. Showing that state income is below the national average is not important, unless the difference is large enough to make for a different standard of living or has other economic importance. Likewise, showing that a new agricultural practice decreases erosion is not worthwhile, unless erosion declines by a scientifically or environmentally significant amount. Practical and scientific importance usually require more than just the existence of a difference.

There are two solutions to this problem of scientific significance. One approach is to test only important hypotheses. Instead of settling for any difference, members of the city council should have asked whether or not ridership increased by some practically significant amount. For example, they might have looked for an increase large enough to recover the marginal cost of the improved service. Finding that, they could be confident that the neighborhood is paying for the extra service. Of course, the council needn't look for complete cost recovery; they might ask for only a 50% payback. For that matter, they could ask that the service more than pay for itself in order to be continued. The point is, we should have been looking for a difference large enough to be of practical importance. Having defined scientific importance, we could adjust H_0 and H_A accordingly. For example, if the council deems 20% ridership as the minimum needed to justify the new service, we could test $\pi \leq .2$ against $\pi > .2$.

The other alternative is to use a confidence interval rather than a hypothesis test. If the confidence interval includes only scientifically significant values, then we have obtained an important result.

Notice that both alternatives call for us to define what is "scientifically significant" or "practically significant." Just what these terms mean is not a statistical issue, but rather one that depends on the subject being investigated. However difficult it is to arrive at these definitions, it is necessary to do so. Although it might seem that testing for no difference avoids the problem, in point of fact it simply defines "significant" to include immeasurably small differences. Hardly ever is this an acceptable definition.

9.6. Summary

In the form of statistical inference known as hypothesis testing, a value of a population parameter is assumed or hypothesized, and sample information is used to see whether or not this hypothesis is tenable. One of three forms of a null hypothesis is chosen, depending on the nature of the problem under study. A six-step procedure known as the classical test of hypothesis has been outlined and applied to several single-sample tests for μ and π. A significance level α is chosen by the researcher and represents the acceptable level of risk for a Type I error. The probability of committing a Type II error, β, is usually unknown and not controlled. Because of this, one will normally either reject H_0 or make no decision. The risks of both types of error decrease with increasing sample size.

A variant of the classical test known as PROB-VALUE method has been explained, as well as the relationships between the classical test, the PROB-VALUE method, and confidence interval estimation. By using these procedures, it is possible to evaluate hypotheses and determine their statistical significance. Any evaluation must include a statement of this significance level, although it can be expressed as a PROB-VALUE or made implicit in a confidence interval.

In any application of hypothesis testing, one must be careful to differentiate between statistical significance and practical significance. Showing some result to be statistically significant does not necessarily impart any practical or scientific significance. Proving the statistical significance of a meaningless hypothesis is of no value. Only a thorough understanding of the problem under study can lead to the specification of hypotheses worthy of statistical evaluation.

FURTHER READING

The material covered in this chapter is contained in most elementary textbooks of statistical methods. Additional examples of situations in which hypothesis testing can be undertaken are described in the texts listed at the end of Chapter 2 and in the problems.

PROBLEMS

1. Explain the meaning of the following terms or concepts:
   a. Classical test of hypothesis   f. Test statistic
   b. Null hypothesis   g. Critical, or rejection, region
   c. Alternate hypothesis   h. Action limits (or critical values)
   d. Directional hypothesis   i. Decision rule
   e. Significance level   j. PROB-VALUE method

2. Differentiate between (a) a one-sided (or one-tail) and a two-sided test, (b) a Type I and a Type II error, and (c) statistical and practical significance.

3. List the steps in the classical test of hypothesis.

4. When should one use the T-distribution rather than the Z-distribution in a hypothesis test of population mean?

5. For a given t, the probabilities in Table A-6 are half those in Table A-5. Why?

In each of the following questions, follow the steps in the classical test of hypothesis outlined in this chapter. Where necessary, modify this procedure to incorporate the requirements of the PROB-VALUE method.

6. Department of Agriculture and Livestock Development researchers in Kenya estimate that yields in a certain district should approach the following amounts, in metric tons per hectare: groundnuts, .50; cassava, 3.70; beans, .30. A survey undertaken by a district agricultural officer on farm holdings headed by women reveals the following results:

	$\bar{x}$	s
Groundnuts	0.40	.03
Cassava	2.10	.83
Beans	0.40	.30

There are 100 farm holdings in the sample.

a. Test the hypothesis that these farm holdings are producing at the target levels of this district for each crop.

b. Test the hypothesis that the farm holdings are producing at lower than target levels.

c. Why would you want to be extremely careful in interpreting the results of this study, even if great care were taken in selecting a truly random sample?

7. A stream has been monitored weekly for a number of years, and the total dissolved solids in the stream averages 40 parts per million and is constant throughout the year. Following a recent change in land use in the drainage basins of the stream, a fluvial geomorphologist finds that the mean parts per million of dissolved solids in a 25-week sample to be 52 with a standard deviation of 32. Has there been a change in the average level of dissolved solids in this stream?

8. Mean annual water consumption per household in a certain city is 6800 L. The variance is 1,440,000. A random sample of 40 households in one neighborhood reveals a mean of 8000.

a. Test the hypothesis that these households have (i) different and (ii) higher consumption levels than the average city household.

b. Comment on the practical significance of this result.

c. What is the PROB-VALUE for each of the hypotheses in part (a)?

9. An exhaustive survey of all users of a wilderness park taken in 1960 revealed that the average number of persons per party was 2.6. In a random sample of 25 parties in 1985, the average was 3.2 persons with a standard deviation of 1.08.

a. Test the hypothesis that the number of persons per party has changed in the last 25 yr.

b. Determine the corresponding PROB-VALUE.

10. A manufacturer of barometers believes that the production process yields 2.2 percent defectives. In a random sample of 250 barometers, 7 are defective.

a. To test whether the manufacturer's belief is confirmed by this sample, calculate the PROB-VALUE for a test of hypothesis of H_0: $\pi = .022$.

b. At which of the following significance levels will the null hypothesis be rejected? (i) .20, (ii) .10, (iii) .05, (iv) .01, (v) .001.

11. The average score of geography majors on a standard graduate school entrance examination is hypothesized to be 600; that is, $H_0 : \mu = 600$. A random sample of $n = 75$ students is selected. The decision rule used to reject H_0 is

$$\text{Reject } H_0: \qquad \bar{x} < 575 \text{ or } \bar{x} > 625$$
$$\text{Do not reject } H_0: \qquad 575 \leq \bar{x} \leq 625$$

Assume that the standard deviation is 100 and X is normally distributed. Find α for this test.

12. A shipper declares that the probability that one of the shipments is delayed is .05. A customer notes that in her first $n = 200$ shipments, 12 have been delayed.

a. Test the assertion of the shipper that $\pi = .05$ at $\alpha = .05$.

b. Calculate the corresponding PROB-VALUE.

10

Parametric Statistical Inference: Two Sample Tests

In a large number of inferential problems, the goal is to compare two populations with one another. It might be that we want to estimate the mean difference between two populations, or more commonly, perhaps we want to test whether or not two populations have the same mean. For example, suppose we want to compare the average distances traveled to purchase clothing by males and by females. To do this, we would take a random sample of male shoppers and an independent random sample of female shoppers and try to draw some conclusion about the population means based on the sample means. Because we are comparing statistics from two samples, these problems are known as *two-sample tests*. They could equally well be labeled *two-population tests*, since the fundamental question being asked is whether or not the samples come from the same population. For the shopping behavior example, the question we are trying to answer is, Do women and men travel, on the average, the same distance for purchases (are they from the same population?), or do they travel different distances (are they from two populations?)?

In Section 10.1 we consider a widely used test, the two-sample difference-of-means test. The following section extends the theory to include confidence intervals for the difference in population means. In in Section 10.3 we introduce a difference-of-means test for paired observations, followed by the two-sample difference-of-proportions test in Section 10.4. As will be seen, the difference-of-means tests require that we know (or assume) something about the variances of the two populations. In particular, we need to know whether or not they are equal. It will therefore be useful to have a way of testing for equality of variances. This is covered in Section 10.5. As will be seen, much of theory behind these procedures has been presented already in Chapter 9, so that what follows is mainly an extension of one-sample hypothesis testing. The PROB-VALUE method is used throughout.

10.1. Two-Sample Difference-of-Means Test for $\mu_1 - \mu_2$

For two-sample tests, it is necessary to use additional subscripts to distinguish between the two samples and populations. Thus, μ_1 and μ_2 are the means of population 1 and population

2, respectively. Similarly, there are n_1 units sampled from the first population and n_2 units drawn from the second population. $\bar{X}_1$ and $\bar{X}_2$ are the random variables corresponding to the sample mean statistics for sample 1 and sample 2, respectively, and $\bar{x}_1$ and $\bar{x}_2$ are the values of these random variables calculated from the two samples. To complete the notation, we use a double subscript to label the sample values x_{ij}. The first subscript, i, is the population from which the sample value has been drawn ($i = 1, 2$), and subscript j designates the jth sample unit drawn from the ith population. The necessary data for a two-sample difference-of-means test are of the following form:

Sample from population 1	Sample from population 2
x_{11}	x_{21}
x_{12}	x_{22}
$\vdots$	$\vdots$
x_{1n_1}	x_{2n_2}

From these we calculate the values of sample means statistics:

$$\bar{x}_1 = \frac{\sum_{j=1}^{n_1} x_{1j}}{n_1} \qquad \bar{x}_2 = \frac{\sum_{j=1}^{n_1} x_{2j}}{n_2}$$

The methods described in this section are based on the assumption that the population random variables X_1 and X_2 are normally distributed. If they are not normal, then the tests will still be approximately correct as long as n_1 and n_2 are large, say $n_1, n_2 \geq 30$.

Hypotheses about μ_1 and μ_2

There are a number of questions we can ask about the relationship between the two population means μ_1 and μ_2. The most basic test is the one alluded to above, namely, are the means the same? In this test we are looking for *any* difference between the two population means, however small. This is the most common test, and the one most computer programs report by default. There are one-sided and two-sided versions of this test, giving the two test formats A and B:

(A)	(B)
$H_0: \mu_1 = \mu_2$	$H_0: \mu_1 = \mu_2$
$H_A: \mu_1 \neq \mu_2$	$H_A: \mu_1 > \mu_2$

As was true in one-sample testing, the formats differ only with respect to the alternate hypothesis. Format A, the two-sided version, is used when there is no prior information about the direction of the difference. In other words, it is used when we want to show the means are different, but don't care which one is larger. Format B, on the other hand, is used to show that μ_1 is the larger of the two. For example, perhaps we want to prove that years before a volcanic eruption are warmer than years immediately following an eruption; i.e., we want to prove that volcanic aerosols lead to cooling. We would use a one-sided test if there is no chance that volcanoes cause warming, or if there is no interest in showing they do so. As discussed in Section 9.1, a one-tailed test is preferable, because it has greater discriminating ability than the corresponding two-tailed test. It does, however, have the disadvantage of looking only for $\mu_1 > \mu_2$; thus, it will mislead the researcher whenever the opposite is true.

Notice that neither format seems to test for $\mu_2 > \mu_1$. Actually, this is easily accommodated in Format B. When making a one-tailed test, we must know *before* the test which mean we want to prove is larger. (If we don't know this, then we should be doing a two-tailed test.) The subscripts 1 and 2 are arbitrary, so we are free to label the two populations however we like. If we are doing a one-tailed test, we will always let population 1 be the one whose mean is supposedly larger. Doing so lets us always cast the alternate as H_A: $\mu_1 > \mu_2$. Of course, when doing a two-sided test there is no *a priori* relationship between the means, and we are free to label the populations either way.

Other hypotheses arise when we want to show the difference in the means *exceeds* some value. For example, maybe we want to show that volcanoes cause a cooling of *at least* 1°F. It might be that temperature changes smaller than 1°F are not of any climatic importance, and therefore are not worth reporting. Or, to take another example, perhaps tax equalization policies are employed only when district incomes differ by $10,000 or more. We need to show not that two districts have different incomes, but rather that the income difference is $10,000 or greater. Both of these examples call for testing that the difference in the means exceeds some threshold difference, D_0. The procedure is to hypothesize that the mean difference is less than or equal to D_0. If the sample means differ by much more than the hypothesized difference, we will reject H_0 in favor of the alternate hypothesis.

Again there are one- and two-sided versions of the test:

(C)	(D)
H_0: $\lvert \mu_1 - \mu_2 \rvert \leq D_0$	H_0: $\mu_1 - \mu_2 \leq D_0$
H_A: $\lvert \mu_1 - \mu_2 \rvert > D_0$	H_A: $\mu_1 - \mu_2 > D_0$

Format C is the two-sided version. By considering the absolute value of the difference in means, we are ignoring which of the two is the larger one. That is, we focus on only the magnitude of the difference without regard to whether it is positive ($\mu_1 > \mu_2$) or negative ($\mu_1 < \mu_2$). In format D, we are trying to show that μ_1 is larger than μ_2 by an amount that exceeds D_0. It is a one-sided test because the direction of difference is predicted beforehand, and only that direction is considered in the test. Like before, population 1 is always the population we want to show as having the larger mean.

In the hypotheses above, D_0 is much like θ_0 in Chapter 9. It is a value assigned before the test, something that reflects the research question being addressed. It constitutes what the researcher deems to be the smallest difference of any scientific or practical importance. If we discover a difference larger than D_0, we have by definition obtained a significant result. Notice that if D_0 is zero, we are saying that *any* difference is worth knowing about, regardless of size. Not surprisingly, using $D_0 = 0$ in sets C and D reduces them to sets A and B. We see, therefore, that the tests for no difference (A, B) are a special case of the tests for differences greater than D_0, (formats C and D). Because they are more general, we will use the latter formats. When we need to test for equality of the means, we will set D_0 equal to zero.

Sampling Distributions

The basic inferential framework for comparing two population means is illustrated in Figure 10-1. There are two populations with means μ_1 and μ_2 and variances σ_1^2 and σ_2^2, respectively. From each population a sample has been drawn, and the values of the sample mean and sample variance are calculated. For sample 1 these values are $\bar{x}_1$ and s_1^2, and for sample 2 the values are $\bar{x}_2$ and s_2^2. Naturally, these sample values form the basis of decisions about the population means. To decide about the difference in the population means, we examine the difference between the sample means $\bar{x}_1 - \bar{x}_2$. If the sample difference is large, we are likely to conclude the population means are different. Of course, in order arrive at a decision rule, we need information about the sampling distribution of the random variable $\bar{X}_1 - \bar{X}_2$. Knowing how this difference varies because of sampling error, we can assign a probability to what we find from the samples in hand.

To arrive at the sampling distribution of $\bar{X}_1 - \bar{X}_2$, we need to know something about the population variances σ_1^2 and σ_2^2. Are we likely to know these values? Hardly ever. After all, how would we come to know the population variances but not the population means? Though a test exists for situations where the variances are known, it is hardly ever used and will not be described. Instead, we will consider two other situations, corresponding to whether or not we believe the variances are equal. (Note that we need not know the value of two things to assert they differ.) Both cases are based on a T statistic in the form

$$T = \frac{(\bar{X}_1 - \bar{X}_2) - D_0}{\hat{\sigma}_{\bar{X}_1 - \bar{X}_2}} \tag{10-1}$$

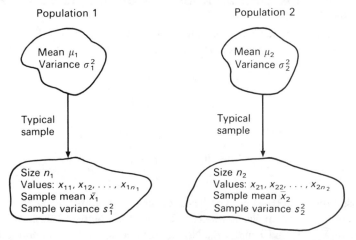

FIGURE 10-1. Inferential framework for comparing population means.

This looks complicated, but is simply another version of Equation (9-8). The numerator compares the difference in sample means to the hypothesized difference D_0. If the population means really do differ by an amount D_0, we expect the numerator to be near zero. The denominator is an estimate of the standard deviation of the difference in sample means. If we meet certain assumptions, and choose the right estimator $\hat{\sigma}_{\bar{X}_1-\bar{X}_2}$, then T in Equation (10-1) will follow the t-distribution. Note that the random variable of interest is $\bar{X}_1 - \bar{X}_2$. The proper standard error estimate depends on population variances of X_1 and X_2:

Case 1. Population variances equal: $\sigma_1^2 = \sigma_2^2$

In this case there is a single (unknown) population variance $\sigma^2 = \sigma_1^2 = \sigma_2^2$. We have two sample variances s_1^2 and s_2^2, each of which provides an estimate of σ^2. Rather than use just one or the other, it stands to reason that we ought to combine the two sample variances into the *pooled variance estimate*

$$s_p^2 = \frac{(n_1 - 1)s_1^2 + (n_2 - 1)s_2^2}{(n_1 - 1) + (n_2 - 1)}$$

Notice that this is a weighted average of the two sample variances. If n_1 is larger than n_2, then the estimate from sample 1 will be more important in s_p^2. If the sample sizes are the same, the formula weights the two sample variances equally. Recall that Equation (10-1) calls for an estimate $\hat{\sigma}_{\bar{X}_1-\bar{X}_1}$. In this equal-variance case, the appropriate estimate is

$$\hat{\sigma}_{\bar{X}_1-\bar{X}_2} = s_p\sqrt{\frac{1}{n_1} + \frac{1}{n_2}} \tag{10-2}$$

With this as background, we can state the sampling distribution of $\bar{X}_1 - \bar{X}_2$ in the form of a theorem.

DEFINITION: SAMPLING DISTRIBUTION OF $\bar{X}_1 - \bar{X}_2$, EQUAL POPULATION VARIANCES

Assume X_1 and X_2 are normal with a difference in means $\mu_1 - \mu_2 = D_0$. If the variance σ^2 is the same for both populations, then the following has a t-distribution:

$$T = \frac{\bar{X}_1 - \bar{X}_2 - D_0}{\hat{\sigma}_{\bar{X}_1-\bar{X}_2}} = \frac{\bar{X}_1 - \bar{X}_2 - D_0}{S_p\sqrt{1/n_1 + 1/n_2}} \tag{10-3}$$

with degrees of freedom

$$\text{df} = n_1 + n_2 - 2$$

To use Equation (10-3) in hypothesis testing, we would substitute sample values $\bar{x}_1$, $\bar{x}_2$ and s_p for the random variables $\bar{X}_1$, $\bar{X}_2$, and S_p. Figure 10-2 shows the sampling distribution of $\bar{X}_1 - \bar{X}_2$ and T under H_0. Notice that because we are assuming H_0 is true, the distributions are centered on D_0 and zero, respectively.

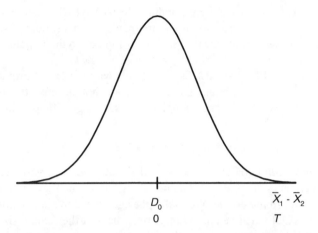

FIGURE 10-2. Sampling distribution for a difference-of-means.

EXAMPLE 10-1. A geographer interested in comparing the shopping patterns of men and women interviewed 95 men and 115 women and determined the distances traveled by each respondent to the store at which the last major clothing purchase was made. (A major purchase was defined as any purchase greater than \$50.) The average distance traveled by men was 6.2 mi and by women 14.8 mi. The standard deviations were 17.5 and 23.2 mi for the male and female samples, respectively. Assuming that the variances of the two populations are equal, test the hypothesis that women on average travel at least 2 mi farther than men.

Solution. We need to make a one-sided test, as we are testing only for women traveling farther than men. To use format *B*, we must therefore name the women sample 1 and the men sample 2. With this settled, we are given $n_1 = 115$, $n_2 = 95$, $\bar{x}_1 = 14.8$, $\bar{x}_2 = 6.2$, $s_1 = 23.2$, and $s_2 = 17.5$. The hypothesized mean difference is $D_0 = 2$, so the hypothesis set is

$$H_0: \mu_1 - \mu_2 \leq 2 \qquad H_A: \mu_1 - \mu_2 > 2$$

If trip lengths are normally distributed, then we can use Equation (10-3) to find an observed *t*. First, we form the pooled variance estimate:

$$s_p^2 = \frac{(115 - 1)23.2^2 + (95 - 1)17.5^2}{(115 - 1) + (95 - 1)} = \frac{90147}{208} = 433$$

Should we want to estimate σ, we could use $s_p = \sqrt{433} = 20.8$. The observed *t* is then

$$t = \frac{14.8 - 6.2 - 2}{20.8 \sqrt{1/115 + 1/95}} = \frac{8.6 - 2}{20.8\sqrt{.0192}} = 2.29$$

In this example we have $115 + 95 - 2 = 208$ degrees of freedom. According to the one-tailed T-table (Table A-5), the corresponding PROB-VALUE is about .012. If we reject H_0, we run a small chance of error, slightly more than 1%. This is not too surprising, given the large difference in the sample means (8.6 mi), compared to the hypothesized difference (2 mi).

The probabilities (PROB-VALUEs) found this way are exact, providing the underlying random variables are normal. If X_1 and X_2 are not normal, the results are only approximate, and one will need large samples to ensure a good approximation.

Case 2. Population variances not equal: $\sigma_1^2 \neq \sigma_2^2$

Suppose we are not willing to believe that X_1 and X_2 have the same variance. We might not know what the variances are, but might be unwilling to assume they are equal. It turns out that we can use an approximate test. We shall once again need to assume normality, but this time the PROB-VALUEs will be approximate regardless of the sample size. As in Case 1, the essential idea is best expressed as a theorem.

DEFINITION: SAMPLING DISTRIBUTION OF $\bar{X}_1 - \bar{X}_2$, POPULATION VARIANCES UNEQUAL

Assume X_1 and X_2 are normal with a difference in means $\mu_1 - \mu_2 = D_0$ and variances $\sigma_1^2 \neq \sigma_2^2$. Then the following has an approximate t-distribution:

$$T = \frac{\bar{X}_1 - \bar{X}_2 - D_0}{\hat{\sigma}_{\bar{X}_1 - \bar{X}_1}} = \frac{\bar{X}_1 - \bar{X}_2 - D_0}{\sqrt{S_1^2/n_1 + S_2^2/n_2}} \tag{10-4}$$

with degrees of freedom given by

$$\text{df} = \frac{(S_1^2/n_1 + S_2^2/n_2)^2}{(S_1^2/n_1)^2/(n_1 - 1) + (S_2^2/n_2)^2/(n_2 - 1)} \tag{10-5}$$

Alternatively, the (approximate) degrees of freedom can be found from

$$\text{df} = \min(n_1 - 1, n_2 - 1)$$

Here, we do not pool the variances, but keep them separate from one another. We still have an estimate $\hat{\sigma}_{\bar{X}_1 - \bar{X}_1}$, but the estimate has changed from Equation (10-2) to

$$\hat{\sigma}_{\bar{X}_1 - \bar{X}_1} = \sqrt{S_1^2/n_1 + S_2^2/n_2} \tag{10-6}$$

Because it does not require the estimate s_p, (10-6) is simpler than the previous formula for $\hat{\sigma}_{\bar{X}_1 - \bar{X}_2}$. Unfortunately, this comes at the price of a much more complicated expression for the degrees of freedom. When working a problem by hand, one can use the degrees of freedom associated with the smaller of the two samples. That is, for df one can use $n_1 - 1$ or $n_2 - 1$, whichever is smaller. Doing so is usually conservative, in the sense that it understates df. Computed PROB-VALUEs will therefore be larger than those given by

(10-5), making it *less* likely that H_0 is rejected. If we are able to reject H_0 using the simple version of df, we can be sure it would be rejected using Equation (10-5).

EXAMPLE 10-2. Repeat Example 9-1, this time assuming $\sigma_1^2 \neq \sigma_2^2$.
Solution. We use the simple rule for degrees of freedom, which gives df = 95 − 1 = 94. The sample *t*-value is

$$t = \frac{14.8 - 6.2 - 2}{\sqrt{23.2^2/115 + 17.5^2/95}} = \frac{8.6 - 2}{\sqrt{7.90}} = 2.35$$

Again using the one-tailed table, the PROB-VALUE is found to be .010. We can therefore conclude with confidence that the population means differ by more than 2 mi. Notice that the PROB-VALUE here is very similar to .012, the value obtained when we pooled the sample variances. We see that for this particular example, it hardly matters whether or not the sample variances are pooled. This is a consequence of the sample variances being so similar to one another, causing Equations (10-3) and (10-4) to give almost the same observed *t*. In fact, if the samples happen to have exactly the same variance, the equations give identical results. In situations where s_1 and s_2 are dissimilar, the pooled and separate variance methods will give very different results. We defer the question of how to choose between the two methods until Section 10.4. Both versions of the difference-of-means test are summarized in Table 10-1.

10.2. Confidence Intervals for $\mu_1 - \mu_2$

Suppose we are interested in estimating the difference in two population means. Assuming independent random sampling, it is easy to show that the best unbiased point estimator for the difference is just $\overline{X}_1 - \overline{X}_2$. If we want to establish a confidence interval around the estimated difference, we can use the theory outlined in the previous section. As before, we will need to assume normality for the random variables X_1 and X_2. The confidence intervals will rely on the *t*-distribution and will have the form

$$\overline{x}_1 - \overline{x}_2 \pm t_{\alpha/2} \hat{\sigma}_{\overline{x}_1 - \overline{x}_1} \tag{10-7}$$

where the confidence level is $1 - \alpha$. As we discussed in Chapter 8 regarding one-sample intervals, we need a *t*-value that leaves $\alpha/2$ of the area in each tail. The required *t* is denoted $t_{\alpha/2}$.

The *t*-value is multiplied by $\hat{\sigma}_{\overline{x}_1 - \overline{x}_2}$, our estimate of the standard error of the difference in sample means. Like before, the formula used for $\hat{\sigma}_{\overline{x}_1 - \overline{x}_1}$ depends on whether or not we can assume the populations have the same variance. If so, we use Equation (10-2); if not, we use Equation (10-6).

EXAMPLE 10-3. Assuming the population variances are different, set up a 90% confidence interval for the difference in mean travel distance for men and women.
Solution. We take the degrees of freedom to be 94, based on the smaller sample. A

TABLE 10-1.
Tests and Confidnce Intervals for Difference-of-Means.

Background
Two random samples of size n_1, n_2 are drawn with sample means $\bar{x}_1, \bar{x}_2$ and sample variances s_1^2, s_1^2. It is assumed that the underlying random variables X_1, X_2 are normal. Sampling distributions for the difference in means $\bar{X}_1 - \bar{X}_2$ depend on whether the population variances are equal or unequal.

Hypotheses
For the one-tailed test, let μ_1 be the larger of the two population means.

$\quad$ $H_0: |\mu_1 - \mu_2| \le D_0$ $\qquad$ $H_0: \mu_1 - \mu_2 \le D_0$
$\quad$ $H_A: |\mu_1 - \mu_2| > D_0$ $\qquad$ $H_A: \mu_1 - \mu_2 > D_0$
$\qquad$ (two-tailed) $\qquad\qquad$ (one-tailed)

Test Statistic
Case 1: $\sigma_1^2 = \sigma_2^2$ known. The standard error of the difference has $n_1 + n_2 - 2$ degrees of freedom, and is found from

$$\hat{\sigma}_{\bar{X}_1 - \bar{X}_1} = \sqrt{\frac{(n_1 - 1)s_1^2 + (n_2 - 1)s_2^2}{(n_1 - 1) + (n_2 - 1)}} \; \sqrt{\frac{1}{n_1} + \frac{1}{n_2}}$$

Case 2: $\sigma_1^2 \ne \sigma_2^2$. The standard error estimate has degrees of freedom shown in Equation 10-5. It is computed from

$$\hat{\sigma}_{\bar{X}_1 - \bar{X}_1} = \sqrt{s_1^2/n_1 + s_2^2/n_2}$$

In both cases the test statistic is

$$t = \frac{\bar{X}_1 - \bar{X}_2 - D_0}{\hat{\sigma}_{\bar{X}_1 - \bar{X}_1}}$$

PROB-VALUE (PV) and Decision Rules
Two-tailed test $\qquad\qquad\qquad$ $PV = P(|T| > t) = P(T < -t) + P(T > t)$

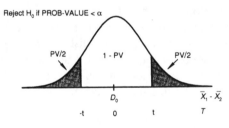

One-tailed test $\qquad\qquad\qquad\qquad\qquad$ $PV \; P(T > t)$

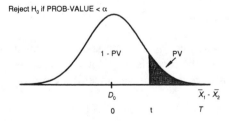

Confidence Intervals
For a confidence level $1 - \alpha$, the interval is

$$\bar{x}_1 - \bar{x}_2 \pm t_{\alpha/2} \hat{\sigma}_{\bar{X}_1 - \bar{X}_1}$$

90% confidence level calls for $t_{.05}$. According to Table A-4, the value is about 1.66. Referring back to Example 10-2, the estimate for $\hat{\sigma}_{\bar{x}_1 - \bar{x}_1}$ is $\sqrt{7.90} = 2.82$. Using these values with the sample means, the confidence interval is

$$\bar{x}_1 - \bar{x}_2 \pm t_{.05} \, \hat{\sigma}_{\bar{x}_1 - \bar{x}_2}$$

$$14.8 - 6.2 \pm 1.66 \, (2.82) = 8.6 \pm 4.67 = [3.93, 13.2]$$

Note the implied hypothesis tests in the confidence interval. Because it does *not* include zero, we can reject the hypothesis that the mean difference is zero (at the 10% level). In other words, we are sure the population means are different. In addition, the interval does not contain 2; thus, we are sure the difference is greater than 2 mi. As expected, this result is consistent with our hypothesis test for a difference exceeding 2 mi. Recall, however, that tests implied in confidence intervals are two-tailed, whereas we considered only a one-tailed alternative in Example 10-2.

10.3. The *t* Test for Paired Observations

The difference-of-means test presented above assumes that the random samples are independent of one another. In other words, it's assumed that there is no relation between individuals in one sample and individuals in the other sample. We could re-order the observations in either sample and not affect the test at all. Or, if we found that sample 2 was somehow contaminated with measurement error, we could collect a new sample 2 without regard to sample 1, because the samples are independent.

In many instances, particularly in experimental situations, the observations do not have this type of independence. For example, rather than collect separate samples of men and women in the shopping-trip study, we might have interviewed only household partners (e.g., married couples). This would give two equal-sized groups, one female, the other male, but the groups would not be independent samples. Instead, corresponding members of the groups would be associated with one another and likely share traits that are important to the study. For example, knowing that female 17 lives 10 miles from her nearest store, we know that male 17 also lives 10 miles from his nearest shopping opportunity. We would say that we have a sample of *paired observations*, also called a sample of *matched pairs*.

This section presents a method for dealing with samples like this. Special methods are needed because the observations violate the assumption of independent samples. As we will see, the presence of paired observations is not a liability, but actually improves our ability to detect differences in the populations. The hypotheses examined in paired comparisons are the same as in two sample tests for the mean. That is, we can test for a difference in the means $\mu_1 \neq \mu_2$, or can test for the difference exceeding some value D_0. In other words, no change in hypothesis format is needed. There is, however, a change in the random variable used to examine the hypotheses.

In the two-sample tests, we used the difference in sample means $\bar{X}_1 - \bar{X}_2$ to draw inferences about the population means. With paired observations, the random variable used is the *difference in X* measured on each member of the pair. That is, we work with the difference between the first members of each sample, the second members of each sample, and so on through the nth members of each sample. Let us denote the pairs of values in the experiment as (x_{1j}, x_{2j}) where x_{1j} is the jth observation in sample 1, and x_{2j} is the corresponding (paired) observation in sample 2. In matched-pairs testing, we examine the differences between members in the pairs

$$d_j = x_{1j} - x_{2j} \qquad j = 1, 2, \ldots n.$$

Obviously, the differences d_j can be positive or negative. A positive value means the observation in sample 1 is larger, whereas a negative d_j means the jth observation in sample 2 is larger. As usual, we can estimate the standard deviation of the differences from

$$s_d = \sqrt{\frac{\sum\limits_{j=1}^{n}(d_j - \bar{d})^2}{n - 1}} \tag{10-8}$$

where $\bar{d}$ is the average difference

$$\bar{d} = \frac{\sum\limits_{j=1}^{n} d_j}{n} \tag{10-9}$$

Notice that if most of the differences d_j are positive, $\bar{d}$ will probably be positive, and we will conclude that $\mu_1 > \mu_2$. On the other hand, if the positive values roughly balance the negative values, $\bar{d}$ will be near zero, and we would not have any evidence for a difference in the population means. This suggests that we should examine the average difference to test hypotheses and to set up confidence intervals.

Equation (10-9) gives $\bar{d}$, a value for the sample mean difference. Following our custom of using an upper case letter for random variables, we would say $\bar{d}$ is a particular value of the random variable $\bar{D}$. If we knew the sampling distribution of $\bar{D}$, we could attach probabilities to sample values $\bar{d}$, and thereby test hypotheses. Fortunately, the sampling distribution is known, as given in the theorem below.

DEFINITION: SAMPLING DISTRIBUTION OF PAIRED-OBSERVATION MEAN $\bar{D}$

Assume X_1 and X_2 are normal with a difference in means $\mu_1 - \mu_2 = D_0$. Given a random sample of n paired observations, the following has an approximate t-distribution:

$$T = \frac{\bar{D} - D_0}{S_d/\sqrt{n}} \tag{10-10}$$

with $n - 1$ degrees of freedom.

Equation (10-10) has what is by now a familiar form: departure from a hypothesized mean value divided by an estimated standard error. When (10-10) is used in practice, sample values $\bar{d}$ and s_d are used for the random variables $\bar{D}$ and S_d. Figure 10-3 illustrates the sampling distribution of $\bar{D}$, transformed to T by (10-10). If H_0 is true, the distribution of $\bar{D}$ will be centered on D_0, as is shown in the figure. (See Table 10-3 for a summary of the test.)

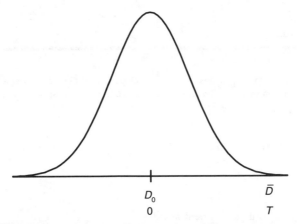

FIGURE 10-3. Sampling distribution for paired-comparison test.

EXAMPLE 10-4. A cartographer tests the time taken by introductory students to perform a set of tasks involving the extraction of information from a set of maps. At the end of the course, the same students are given an identical test, and the times are recorded. Have the students learned how to use maps more effectively? Has the time on the task decreased, because they now take less time to extract information from maps? Or, alternatively, do students now take more time on the test because they have become more careful and are extracting more information than at the start of the course? Test for a change in the mean time at the $\alpha = .1$ level.

Solution. It is clear that the two sets of sample times are not independent. Because the time values in sample 2 are measured on the same persons appearing in sample 1, a paired-comparison is needed. We are going to perform a two-tailed test, since the question

TABLE 10.2
Data for Example 10-4

Student	Exam 1	Exam 2	Change
1	11	9	2
2	23	21	2
3	17	13	4
4	14	16	-2
5	16	13	3
6	21	17	4
7	9	10	-1
8	18	15	3
9	26	21	5
10	19	20	-1
Mean	17.40	15.50	1.90
	$\bar{x}_1$	$\bar{x}_2$	$\bar{d}$
Std. Dev.	5.23	4.37	2.42
	s_1	s_2	s_d

TABLE 10-3
Paired-Comparison Tests

Background
A sample of paired observations (x_{1j}, x_{2j}) is drawn. For each pair, the difference $d_j = x_{1j} - x_{2j}$ is computed. The mean and standard deviation of the differences are computed, $\bar{d}$ and s_d, respectively. It is assumed that the underlying variables X_1, X_2 are normal.

Hypotheses

$H_0: |\mu_1 - \mu_2| \leq D_0$ $H_0: \mu_1 - \mu_2 \leq D_0$
$H_A: |\mu_1 - \mu_2| > D_0$ $H_A: \mu_1 - \mu_2 > D_0$
    (two-tailed)   (one-tailed)

Test Statistic
The the test statistic has $n - 1$ degrees of freedom and is

$$t = \frac{\bar{d} - D_0}{s_d}$$

PROB-VALUE (PV) and Decision Rules
Two-tailed test

$$PV = P(|T| > t) = P(T < t) + P(T > -t)$$

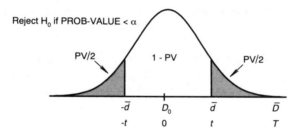

One-tailed test

$$PV = P(T > t)$$

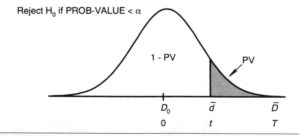

asked calls for a change in either direction (decrease *or* increase in exam time). In testing for the existence of a difference, we set the hypothesized difference $D_0 = 0$. The hypotheses are therefore

$$H_0: \mu_1 - \mu_2 = 0 \qquad H_A: \mu_1 - \mu_2 \neq 0$$

Suppose the times, in minutes, are as in Table 10-2. First, we compute $\bar{d} = 1.9$ and $s_d = 2.42$. This gives an observed t

$$t = \frac{\bar{d} - D_0}{s_d/\sqrt{n}} = \frac{1.9 - 0}{2.42/\sqrt{10}} = \frac{1.9}{.765} = 2.48$$

With 9 degrees of freedom, the corresponding PROB-VALUE is .035 (Table A-6). We can therefore (easily) reject H_0 at the 10% level.

At the beginning of this section, we claimed that paired observations are an advantage. To see why, imagine that we had treated this as a two-sample test. That is, let's pretend we have two independent samples of size 10 and perform the same test. Skipping the intermediate computations, the results are

$$t = \frac{\bar{X}_1 - \bar{X}_2 - 0}{\hat{\sigma}_{\bar{x}_1 - \bar{x}_2}} = \frac{17.4 - 15.5}{2.16} = .88$$

The PROB-VALUE is .38, so we are far from being able to reject H_0. Using these two samples, we cannot find any evidence that the mean time on task has decreased or increased. The paired comparison test finds a difference in the population means, but the two-sample test does not. Does this make sense? Are paired observations better able to discriminate between populations? As a matter of fact, they are superior, everything else being equal.

Consider first the workings of the two-sample test. It detects a population difference when the difference *between* sample means is larger than variability *within* the samples. That is, $\bar{x}_1 - \bar{x}_2$ must overwhelm $\hat{\sigma}_{\bar{x}_1 - \bar{x}_1}$. Looking at the data, it seems clear that most of the sample variance arises from differences between individuals. For example, person 1 takes about 10 min on the exams, whereas person 2 is inherently slower, requiring more than 20 min. For the two-sample test to find a difference in population means, the difference must stand out against this background variation.

On the other hand, the paired comparison test controls for (removes) this background variation. It operates on the *change* in time on task, the difference column. The only variability it finds is variability in the right-hand column, so it never sees that persons 1 and 2 are different. In fact, because both improved by the same amount, it considers them to have the same value. We see that the paired-comparison test ignores person-to-person variability. This is good, because that variability is of no interest and only confuses things. We don't care if person 1 is slow or fast, we just want to know if she/he got faster or slower on exam 2. By focusing on the each pair's change, the paired-comparison test identifies a difference missed by the two-sample test.

This is a general result, not limited to the example above. The lesson learned is that if we can obtain paired observations, we should do so. The resulting test will better detect differences between populations. However, care must be taken that in working with paired data, the original aims of the study are not compromised. For example, we began the shopping trip study wanting to make statements about male/female differences in travel time. If we sample couples to achieve paired data, we are restricting the ourselves to persons living with a member of opposite sex. Obviously, single people and same-sex couples will not appear in our sample. This raises a question as to whether the results will apply to the general population of men and women, or just to a subset of the population.

10.4. Two-Sample Difference-of-Proportion Tests for $\pi_1 - \pi_2$

This section considers methods for comparing two population proportions π_1 and π_2. Just as the two-sample difference-of-means test examines the difference between two population means $\mu_1 - \mu_2$, here we are concerned with the difference $\pi_1 - \pi_2$. Inferences concerning population difference are based on the difference in sample proportions, $P_1 - P_2$. In hypothesis testing, we assume under H_0 that the population difference is no larger than D_0, and compare the sample difference to the hypothesized difference. If we find a sample difference departs from the hypothesized value, H_0 is rejected. Once again, there are two hypothesis frameworks, representing one- and two-sided tests:

(A)	(B)
$H_0: \lvert \pi_1 - \pi_2 \rvert \le D_0$	$H_0: \pi_1 - \pi_2 \le D_0$
$H_A: \lvert \pi_1 - \pi_2 \rvert > D_0$	$H_A: \pi_1 - \pi_2 > D_0$

To test for proportions differing by *any* amount, D_0 is set equal to zero. In the one-sided format (*B*), population 1 always refers to the one whose proportion is supposed to be larger. Hypothesis tests (and confidence intervals) are based on using the central limit theorem to arrive at the sampling distribution for the difference in sample proportions. This is stated in the form of a theorem:

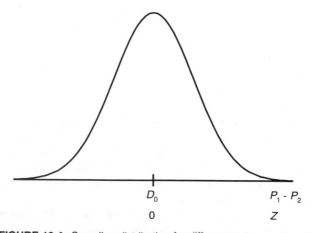

FIGURE 10-4. Sampling distribution for difference-of-proportions test.

DEFINITON: SAMPLING DISTRIBUTION OF $P_1 - P_2$

Assume X_1 and X_2 are Bernoulli variables with population proportions which differ by $D_0 = \pi_1 - \pi_2$. Given two random samples with n_1, $n_2 > 100$, the following is approximately a standard normal variable:

$$Z = \frac{P_1 - P_2 - D_0}{\hat{\sigma}_{P_1 - P_2}} = \frac{P_1 - P_2 - D_0}{\sqrt{P_1(1 - P_1)/n_1 + P_2(1 - P_2)/n_2}} \qquad (10\text{-}11)$$

Equation (10-11) is used to find probabilities under the assumption H_0 is true. It transforms a difference in sample proportions to a z-distribution. Figure 10-4 shows the corresponding probability distributions. Notice that the difference of proportions is centered on D_0, and the z-distribution is centered on zero. To calculate an observed z, we would, of course, use sample values p_1 and p_2.

EXAMPLE 10-5. A study is conducted to determine whether an equal proportion of two groups of farmers in a particular agricultural region adopted a new innovation in fertilizer for corn. The first group of farmers is composed of Mennonites, and the second group consists of all other farmers in the region. A random sample of 100 Mennonite farmers produced a sample proportion of .68 who said they used the fertilizer in the current growing season. A random sample of 144 from the second group yielded a sample proportion of .58. Test the hypothesis that the two population proportions π_1 and π_2 are equal at the $\alpha = .05$ level of significance.

Solution. Since the two sample sizes are greater than 100, the two-sample difference-of-proportions test can be applied. We are given $p_1 = .68$, $p_2 = .58$, $n_1 = 100$, and $n_2 = 144$. The hypothesized value is no difference; thus, $D_0 = 0$, and we need a two-tailed test. The sample test statistic is

$$z = \frac{.68 - .58 - 0}{\sqrt{.68(.32)/100 + .58(.42)/144}}$$

$$= \frac{.10}{\sqrt{.0022 + .0017}} = \frac{.10}{.0622} = 1.61$$

The PROB-VALUE is .110, larger than the 5% level acceptable here. On the basis of these samples, we cannot conclude there is any difference in the adoption rate between the two groups.

10.5. Confidence Intervals for the Difference of Proportions

Using the theorem in the previous section, we can easily construct confidence intervals for $\pi_1 - \pi_2$. The denominator in Equation (10-11) is $\hat{\sigma}_{P_1 - P_2}$, an estimate of the standard deviation of the difference of the sample proportions. So, we know that $P_1 - P_2$ is approximately normal, and we know its standard deviation. We can therefore build confidence intervals of the form

TABLE 10-4
Tests and Confidence Intervals for Difference-of-Proportions

Background
Two random samples of size n_1, $n_2 \geq 100$ are drawn; sample proportions are p_1, p_2.

Hypotheses

$H_0: |\pi_1 - \pi_2| \leq D_0$ $H_0: \pi_1 - \pi_2 \leq D_0$
$H_A: |\pi_1 - \pi_2| > D_0$ $H_A: \pi_1 - \pi_2 > D_0$
     (two-tailed)   (one-tailed)

Test Statistic
The test statistic is

$$z = \frac{p_1 - p_2 - D_0}{\sqrt{p_1(1 - p_1)/n_1 + p_2(1 - p_2)/n_2}}$$

PROB-VALUE (PV) and Decision Rules
Two-tailed tests

$$PV = P(|Z| > z) = P(Z < -z) + P(Z > z)$$

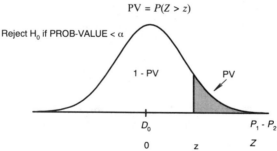

One-tailed tests

$$PV = P(Z > z)$$

Confidence intervals
For a confidence level $1 - \alpha$, the interval is

$$p_1 - p_2 \pm z_{\alpha/2} \sqrt{p_1(1 - p_1)/n_1 + p_2(1 - p_2)/n_2}$$

$$p_1 - p_2 \pm z_{\alpha/2}\hat{\sigma}_{P_1-P_2}$$

EXAMPLE 10-6. A climatologist is studying the probability of precipitation associated with two different summer air masses. In a sample of 160 days of air mass type A, it rains on 84 days. It rains 60 days out of 130 days dominated by air mass B. Construct a 95% confidence interval for the difference in the probability of precipitation.

Solution. We have large samples, so can use the central limit theorem. The sample proportions are $p_1 = 84/160 = .53$ and $p_2 = 52/130 = .40$. The standard error of the difference is

$$\hat{\sigma}_{P_1-P_2} = \sqrt{.53(47)/160 + .40(.60)/130} = \sqrt{.0034} = .0583$$

and the confidence interval is therefore

$$.53 - .40 \pm z_{.025}(.0583) = .13 \pm 1.96\,(.0583) = .13 \pm .114 = [.016, .244]$$

Looking at the limits of this confidence interval, we see one value greater than zero, and the other less than .25. On this basis, we can say the two air masses almost certainly differ in rainfall probability, but the difference is probably less than 25%.

10.6. Test for Equality of Variances $\sigma_1^2 = \sigma_2^2$

The two-sample difference-of-means test requires that we make a decision about whether or not to pool the sample variances. As we discussed, if there is reason to believe the population variances are different, we should not use the pooled variance estimate. In this section we describe a test that can sometimes help in this decision. Though we motivate the test as an aid in difference-of-mean tests, it is useful whenever one needs to compare population variances. The null hypothesis is that the variances are equal, $\sigma_1^2 = \sigma_2^2$. As would be expected, H_0 is evaluated on the basis of the sample variances. If they are very different, H_0 is rejected. The test differs from what we have seen so far in that it is based on the *ratio* of the sample variances, s_1^2/s_2^2, rather than their difference $s_1^2 - s_2^2$. Whereas previously test

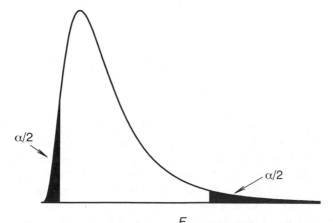

FIGURE 10-5. *F* distribution, showing equal-probability tails.

values near zero indicated no difference, here test values near 1 will suggest no difference. Conversely, whenever the observed ratio is much larger or smaller than 1, we will be inclined to reject H_0.

To arrive at a firm decision rule, we need to know the sampling distribution of the test statistic. It so happens that under certain conditions, the distribution of the ratio of the sample variances (S_1^2/S_2^2) is known. (This is why the ratio rather than the difference of the variances is used as the test statistic.) In particular, if X_1 and X_2 are normally distributed, then the ratio of the variances follows something called the F distribution.

A graph of F is shown in Figure 10-5. It takes on values from 0 to $+\infty$, but is never negative. This makes sense, because variances are never negative; thus, there is no way for the ratio of two variances to be negative. Although F is always skewed, the details of its shape are controlled by two parameters, called the degrees of freedom r_1 and r_2. The first (r_1) is associated with the numerator S_1^2, whereas the second, r_2, is the degrees of freedom associated with the denominator S_2^2. Because both the numerator and denominator are variances, the degrees of freedom are $n_1 - 1$ and $n_2 - 1$. Notice that we can think of F as a family of distributions, one member for each pair of values (r_1, r_2).

DEFINITION: SAMPLING DISTRIBUTION OF THE RATIO OF VARIANCES
Assume X_1 and X_2 are both normally distributed, with variances σ_1^2 and σ_2^2. Given independent random samples of size n_1 and n_2, then the statistic

$$F = \frac{S_1^2/\sigma_1^2}{S_2^2/\sigma_2^2}$$

will follow an F distribution with $n_1 - 1$ and $n_2 - 1$ degrees of freedom.

As mentioned above, when testing for equality of variances, we take $\sigma_1^2 = \sigma_2^2$ as the null hypothesis. The hypothesis set is therefore

$$H_0: \frac{\sigma_1^2}{\sigma_2^2} = 1 \qquad H_A: \frac{\sigma_1^2}{\sigma_2^2} \neq 1$$

If H_0 is true, we can write

$$F = \frac{S_1^2/\sigma_1^2}{S_2^2/\sigma_2^2} = \frac{S_1^2}{S_2^2} \frac{\sigma_2^2}{\sigma_1^2} = \frac{S_1^2}{S_2^2} \tag{10-12}$$

Equation (10-12) says that under H_0, the ratio of the sample variances is distributed like F. If we observe a ratio that has low probability of occurrence, we will doubt H_0. Knowing the distribution of F lets us assign PROB-VALUEs to any observed ratio. Most computer programs perform an F test for equality of variances by reporting the PROB-VALUE directly. If we are working by hand, we will need to use tabled values of F. Unfortunately, because F is not symmetric, and because it depends on two parameters, tables of F are somewhat complicated. One such table is presented as Table A-7. This table gives two-tailed values of F, which is what we want if we have no preconceived idea about which variance is larger. That is, this table gives the probability of landing in either tail of the distribution. Notice that all the F-values are greater than 1, which implies that s_1^2/s_2^2 is greater than 1. The table assumes that we have divided the larger sample variance by the smaller one, so that the ratio cannot be less than unity. In other words, in this table s_1 is always the larger of the two sample variances. To use the table, we enter with r_1 and r_2, the degrees of freedom associated with the numerator and denominator. Looking in the body of the table, we find something near our observed F, and read the corresponding PROB-VALUE under the α heading.

EXAMPLE 10-7. Samples of size 25 and 16 have standard deviations 37.5 and 72.6. Is there reason to believe that the population variances are different?

Solution. We take $s_1 = 72.6$, $n_1 = 16$, $s_2 = 37.5$, $n_2 = 25$. To compute the F statisic, the standard deviations are squared:

$$F = \frac{72.6^2}{37.5^2} = \frac{5271}{1406} = 3.75$$

Using $r_1 = 15$, $r_2 = 24$, the relevant portion of Table A-7 is

				r_1						
r_2	α	10	12	15	20	24	30	60	120	∞
24	.2	:	:	1.78	:	:	:	:	:	:
	.1	:	:	2.11	:	:	:	:	:	:
	.05	:	:	2.44	:	:	:	:	:	:
	.02	:	:	2.89	:	:	:	:	:	:
	.01	:	:	3.25	:	:	:	:	:	:
	.001	:	:	4.56	:	:	:	:	:	:

The observed F is between the 3.25 and 4.56; thus, the two-tailed PROB-VALUE is between .01 and .001. If we reject H_0, there is less than a 1% chance of error. We can therefore conclude that the population variances are different.

Our interest in this test is that it provides a way of deciding whether or not to pool sample variances when drawing inferences about population means. Most likely, we would use the F test before setting up a confidence interval for $\mu_1 - \mu_2$, or before testing some hypothesis about μ_1 and μ_2. If the F test suggests the variances differ, we use the separate variance formula, Equation (10-4). What happens if we fail to reject the hypothesis of equal variances? Normally, it isn't wise to accept H_0, because the probability of a Type II error β is not controlled. What should we do? We can't reject H_0, but are reluctant to accept H_0. About all one can do is to use both the pooled and separate variance estimates. That is, perform the difference-of-means test using both variance estimates, computing PROB-VALUEs both ways. If the PROB-VALUEs differ by a large amount, or lead to divergent conclusions, that should be reported. We might note that failure to reject the assumption of equal variances is more likely when the sample variances are similar to one another. Fortunately, as the sample variances become more similar, the difference between pooling and not pooling diminishes.

We should emphasize that the F test is quite sensitive to departures from normality.

TABLE 10-5
Test for Equal Variances

Background
Two random samples of size n_1, n_2 are drawn with sample variances s_1^2, s_2^2. It is assumed that the underlying random variables X_1, X_2 are normal.

Hypotheses
H_0: $\sigma_1 = \sigma_2$
H_A: $\sigma_1 \neq \sigma_2$
(two-sided)

Test Statistic
Let s_1^2 be the larger of the two sample variances. The test statistic is

$$f = \frac{s_1^2}{s_2^2}$$

PROB-VALUE (PV)
If F is an F-distribution with $n_1 - 1$ and $n_2 - 1$ degrees of freedom, then

$$PV = P(F > f)/2$$

When the random variables X_1 and X_2 are highly skewed, or depart in other ways from normal, the test is not reliable. This is not true for the t-test, nor for confidence intervals based on the t-distribution. Not only is t robust under moderate departures from normality, but it can be used even in strongly nonnormal situations if the sample sizes are large.

10.7. Summary

In this chapter, statistical hypothesis testing procedures have been developed for testing the difference between two population means $\mu_1 - \mu_2$, the difference between two population proportions $\pi_1 - \pi_2$, and the ratio of two population variances σ_1^2/σ_2^2. For the difference-of-means test, two cases are differentiated on the basis of the most appropriate assumption regarding the population variances. The suitability of these assumptions can be examined by using the test for the equality of variances outlined in Section 10.6. We also discussed a difference-of-means test for paired, or matched, observations. Paired observations do not constitute independent samples; thus, they require special treatment. In addition to hypothesis testing, we also covered interval estimation for the difference in population means and proportions.

We close this chapter by remarking that in both one- and two-sample hypothesis testing, it is important to distinguish between the practical and the statistical significance of results. These two types of significance are not equivalent, nor do they always occur simultaneously. For example, suppose a random sample of police emergency response times was taken in 1990. Then another random sample of these times was taken in 1995. A two-sample difference-of-means test shows that response times have increased. Suppose the result is highly significant, say a PROB-VALUE of .0001. Is this of any concern?

Whether the conclusion has any practical significance depends on the exact null hypothesis being examined. If, as usual, the null hypothesis is expressed as $\mu_1 - \mu_2 = 0$ (or $\mu_1 = \mu_2$), the result may have no practical significance. The test says nothing about how much response times have increased, only that they have increased. An increase of 10 sec is of no practical concern, but an increase of 10 min might be. As we noted in the previous chapter, proving the statistical significance of meaningless hypotheses is of no value.

PROBLEMS

1. Explain the meaning of (a) pooled variance and (b) paired samples.
2. Fertilizer XLC is applied to eight identical plots of land planted with corn. The same corn seed is also applied to eight other identical plots of land for a control treatment. At the end of the growing season, the total yield for each plot of land is measured. The data are shown in the following table:

	Treatment	
Plot No.	Control	Fertilizer XLC
1	75	86
2	66	80
3	68	83
4	71	87
5	73	82
6	74	79
7	70	81
8	71	78

   a. Use a t test to determine whether yields on XLC-treated plots differ from those of control plots.
   b. Test the hypothesis that using fertilizer XLC improves corn yield.

3. Hourly traffic volumes are monitored over the evening rush hour on a sample of 10 residential streets. Shortly after the survey, a series of turning restrictions are put in place in an attempt to lower traffic volumes. Hourly traffic volumes are again monitored on the same sample of streets, on the same day of the week, during identical weather conditions. The volumes are summarized as follows:

	Hourly traffic volume	
Street	Before	After
Fisher Street	650	600
Pine Street	240	300
Foul Bay Road	860	750
First Avenue	350	320
Magnolia Street	500	500
Renfrew Road	600	420
Hartley Boulevard	570	400
Fort Street	420	400
Pearson Street	480	450
Snedecor Road	530	560

   a. Have the turning restrictions decreased traffic volumes?

   b. Is this a fair test? Think of reasons why it might be considered to be invalid. (Hint: Might it violate any assumptions of statistical testing?)

4. The minimum daily temperature recorded at two locations in the first week of February is determined for a random sample of $n = 60$ different days over a 100-yr climate record. Based on these samples, the following statistics are computed:

	Sample size	Mean minimum temperature (C)	Standard deviation
Location A	$n = 60$	$\bar{x} = 7.3$	$s = 4.2$
Location B	$n = 60$	$\bar{x} = 6.8$	$s = 3.7$

   a. Assuming that the population variances at the two locations are equal, find a 95% confidence interval on the mean temperature difference between the two locations, $\mu_A - \mu_B$. Is one location colder than the other, on average, during this period? Explain.

   b. Is the assumption of equal variances reasonable? Test the equality of these two variances at $\alpha = .05$. Determine an appropriate PROB-VALUE for this test.

5. In several empirical studies of the patronage of consumers in rural areas, the assumption of perfect rationality is examined by comparing the actual distance traveled by a consumer to purchase a given good or service to the distance of the closest location offering that service to the consumer. Consider the actual and closest distances for the following set of rural residents:

	Distance (km)	
Consumer	Actual	Closest
1	7	4
2	13	8
3	3	3
4	4	3
5	5	2
6	5	4
7	9	8
8	7	6
9	7	5
10	8	3
11	13	2
12	4	4

   a. Test the hypothesis that consumers tend to patronize the closest outlet for this service.

   b. Should it be a one- or two-tailed test? Does it matter?

   c. Is this a reasonable way to evaluate this assumption of rationality in consumer behavior? Can you think of others?

6. Why is there no paired comparison test for population proportions?

11

Nonparametric Statistics

All the methods of statistical inference discussed so far have required specific assumptions about the nature of the population from which the samples were obtained. We select a population parameter of interest (μ or σ or π), collect a random sample, calculate a point estimator ($\bar{x}$ or s or p), construct a sampling distribution, and then employ hypothesis testing decision rules or confidence interval formulas. The most common assumption is that the population distribution function of the random variable is normal with mean μ and variance σ^2. Because these methods require a number of assumptions about one or more population parameters, they are termed *parametric statistical methods*. The tests described in this chapter are referred to as *nonparametric*, or *distribution-free, tests*. Compared with parametric tests, nonparametric tests require less stringent assumptions about the nature of the underlying population distribution function of the random variable being analyzed.

There are two situations in which nonparametric methods can be applied. When the level of measurement of the variable is nominal or ordinal, most parametric methods cannot be applied. It is impossible for a categorical or ordinal variable to be normally distributed. This is the first situation where nonparametric statistical methods prove useful. The second situation occurs when the nature of the underlying population distribution is unknown or left unspecified. Usually, this occurs when we wish to use a parametric test but we cannot meet its exacting requirements. For example, we might find that a variable has a highly skewed or even bimodal distribution and suspect the underlying population distribution to be nonnormal. In fact, one class of nonparametric tests, *goodness-of-fit tests,* can be used to test the assumption of normality. An observed frequency distribution can be compared to the distribution we might expect according to some theory or by some assumption. Quite frequently, one of the first steps in parametric testing is to use a goodness-of-fit test to see whether the assumption of normality is warranted. When the assumption of normality is rejected by a goodness-of-fit test, the alternative is to employ a nonparametric procedure.

In Section 11.1 comparisons are made between parametric and nonparametric procedures. The advantages and disadvantages of each type of method are described. Which type of test should be used in a specific case is shown to depend on a number of conditions. Sections 11.2 to 11.5 describe several nonparametric procedures useful in four different situations. Section 11.2 discusses the nonparametric equivalents to several parametric tests presented in Chapter 9. Goodness-of-fit tests are presented in Section 11.3. In Section 11.4, tests dealing with contingency tables or cross-classified count data are described. A common assumption in all statistical methods is randomness. One specific test of this assumption is explained in Section 11.5. The chapter is summarized in Section 11.6.

11.1. Comparison of Parametric and Nonparametric Tests

Both parametric and nonparametric tests have points of superiority. The choice of statistical test to be undertaken depends on several factors. Some advantages and disadvantages are outlined in the following paragraphs.

Statistical Efficiency

When a data set satisfies all the assumptions of a given parametric test, the parametric test should be applied. Any nonparametric alternative is almost always significantly *less powerful* than the comparable parametric procedure. The probability of making a Type II error is usually appreciably lower with parametric tests. However, violations of a parametric test's assumptions can change the sampling distribution of the test statistic and thus the power of the test.

Robustness

A test is said to be *robust* against the violation of a certain assumption if the power of the test is not significantly affected by the violation. For example, the one-sample t test described in Chapter 9 is still powerful when the population is nonnormal. In fact, because of the central limit theorem, the larger the sample, the less critical the assumption of normality. The power of the test, in turn, depends on the extent to which the violation of the assumption alters the sampling distribution of the test statistic. Robustness is a matter of degree. Violations of some assumptions, such as normality, may be less critical than a violation of, say, the homogeneity of variance. When a specific assumption is violated, it is important to check, first, how large the violation is and, second, the sensitivity of the test to the departure. Where the violation is significant, it is often advantageous to revert to a nonparametric test which does not require that particular assumption. Or it may be possible to *transform* the variable so that the assumption is not violated. Applying a transformation can sometimes correct for nonnormality. An example of this procedure is illustrated in Section 11.3.

Scope of Application

Since they are generally based on fewer or less restrictive conditions than parametric tests, nonparametric tests can be legitimately applied to a much larger set of populations. Even when it is known that the populations are skewed or bimodal or rectangular, most nonparametric tests can still be applied. Also, statistics calculated from variables measured at the nominal or ordinal scale can be tested by using nonparametric methods. This is a distinct advantage in survey design. It can be difficult to extract interval-scale responses from mail questionnaires or personal interviews. Preference ratings or approve/disapprove questions typically yield ordinal and/or nominal responses. As an example, consider the problem of getting respondents in a mail questionnaire to accurately answer a question about their annual income. For reasons of confidentiality, many respondents are willing to answer a question concerning income only if it allows them to check a box corresponding to a range of income, such as "From $10,000 to $15,000."

Simplicity of Derivation

The derivation of the sampling distribution of the test statistics utilized in parametric tests is based on quite complicated theorems such as the central limit theorem. Usually, the form of the sampling distribution is a function of the form of the population probability distribution. Given the mathematical complexity of the normal distribution [recall the equation for the normal density function, (5-30)], many users of parametric statistical methods cannot comprehend the derivation of the key results upon which the tests are based. Some proponents of nonparametric statistics suggest that many who employ parametric statistics must therefore apply them in a "cookbook" manner.

In contrast, the sampling distributions of the test statistics used in nonparametric tests are derived from quite basic combinatorial arguments and are easily understood. Several examples which illustrate the simplicity of the derivations are given in this chapter. For this reason, nonparametric tests are an inviting alternative to many geographers and other social scientists. The claim of superiority based on simplicity must be tempered by the fact that nonparametric tests are also misused in some social science applications. At times, they are inappropriately applied or else interpreted incorrectly.

Sample Sizes

The size of the available sample is an important consideration in determining whether to use a parametric or nonparametric test. When sample size is extremely small, say n is less than 10 or 15, violations of the assumptions of parametric tests are particularly hard to detect (see Section 11.3 for an example). Yet a violation of an assumption can have extremely deleterious effects under these conditions. The use of nonparametric tests in these situations is often advisable. In general, the larger the sample, the more advantageous a parametric test becomes. The implications of the central limit theorem support this assertion. Where sample size is large, the presence of nonnormality is often insignificant.

As a final point concerning the relation between parametric and nonparametric methods, one should realize that "nonparametric" means only that one does not make parametric assumptions about the random variable being observed. It does *not* mean that parametric distributions have no place in the application of those methods. As a matter of fact, familiar distributions (normal, χ^2, etc.) appear frequently in nonparametric methods. Though this might seem counter-intuitive, the explanation is really quite simple.

Typically, one makes very weak (nonparametric) assumptions about a random variable. The random variable is sampled, and a test statistic is computed. Very often the test statistic will have follow a normal distribution, or some other distribution we think of as "parametric." This is especially common for large sample sizes. When sample sizes are small, the sampling distribution typically follows a less familiar, but nevertheless known distribution. In the small-sample case, most tests yield PROB-VALUES directly, and a classical test of hypothesis can only be approximated. In the large-sample case, classical and PROB-VALUE tests can be performed with equal ease. To emphasize this point, we will use PROB-VALUE tests in the small-sample examples, and classical tests for large samples.

In sum, parametric tests have a significant advantage over nonparametric tests when their assumptions are met. Violations of these assumptions introduces inexactness into the test. The seriousness of the violation depends on the size of the sample, the specific assumption violated, and the robustness of the test to the particular violation.

11.2. One-, Two-, and *k*-Sample Tests

The first set of nonparametric procedures that we discuss are the one- and two-sample nonparametric equivalents to the *t* tests described in Chapter 9. They are used whenever an interval-scaled data set violates one of the assumptions of the appropriate parametric test or when the variables of the data set are at an ordinal level of measurement.

Sign Test for the Median

The simplest parametric test discussed in Chapter 9 compares the sample mean $\bar{x}$ and some hypothesized (or known) value of the population mean μ, specified in the null hypothesis. The *sign test* is used to test the value of the sample median against some hypothesized value of the population median η_I (Greek letter eta), specified in H_0. Let us denote the true population median as η. The sample median M is an unbiased consistent estimator of η, just as the sample mean $\bar{x}$ estimates μ. The null hypothesis in the sign test for the median is that the hypothesized value of the median η_I is equal to the true median η, that is, $H_0: \eta_I = \eta$. If H_0 is true, then in any sample of n observations we are just as likely to get an observation above the value η_I as below it. If an observation x_i lies above η_I, then the difference $x_i - \eta_I$ has a positive sign. If it lies below η_I, then the difference has a negative sign. Thus, the null hypothesis suggests we should get about an equal number of plus and minus signs. Put another way, the probability of getting a negative sign is $p = P(X < \eta_I) = .5$. Or, the probability of getting a negative sign is $p = P(X_i < \rho_1) \neq .5$. In any random sample we will not necessarily get an equal number of positive and negative signs. The question boils down to this: How few minus signs (or, equivalently, how few plus signs) must there be before we can reject the null hypothesis that η_I is the true median?

The test statistic is simply the number of observations in the sample with values below the value of the population median specified in the null hypothesis. Call this number B where B is a *binomial random variable* with $\pi = .5$. There are only two outcomes, and there are n observations or trials. Thus, the sign test turns out to be equivalent to a test of the population proportion of $\pi = .5$. The test is summarized in Table 11-1.

EXAMPLE 11-1. The travel time of an individual's journey to work is randomly sampled on 20 different workdays over 3 months. Each day, the total door-to-door travel

TABLE 11-1
Sign Test for the Median

Specification
The n observations are compared with the value of the median hypothesized in H_0. The number of observations below the hypothesized median is counted. Call this number B. Equivalently, estimate $p = B/n$.

Hypotheses
H_0: $\eta = \eta_I$, or equivalently $\pi = P(X_i < \eta_I) = .5$, the true median equals the hypothesized median.
H_A: (a) $\eta \neq \eta_I$, or equivalently $\pi = P(X_i < \eta_I) \neq .5$ (two-tail).
　　(b) $\eta < \eta_I$, or equivalently $\pi = P(X_i < \eta_I) < .5$ (one-tail)
　　(c) $\eta > \eta_I$, or equivalently $\pi = P(X_i < \eta_I) > .5$ (one-tail).

Test statistic
The proportion of the n observations below the median is p, and B = the number of observations below the median.

Decision rule
For small samples, the binomial distribution can be used to test this hypothesis. The exact PROB-VALUE for B can be determined from binomial tables. For large samples, the normal approximation to the binomial is used, and a test for P is utilized.

Large-sample test
Reject H_0 if $p < .5 - z_{\alpha/2}\sqrt{(.5)(.5)/n}$ or $p > .5 + z_{\alpha/2}\sqrt{(.5)(.5)/n}$ for a two-tail test. For one-tail, reject H_0 if $p < .5 + z_\alpha\sqrt{(.5)(.5)/n}$ for H_A(b), and reject H_0 if $p > .5 - z_\alpha\sqrt{(.5)(.5)/n}$ for H_A(c).

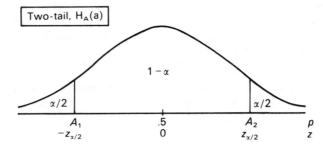

where $A_1 = .5 - z_{\alpha/2}\sqrt{(.5)(.5)/n}$ 　　$A_2 = .5 - z_{\alpha/2}\sqrt{(.5)(.5)/n}$

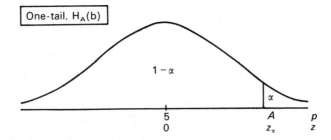

where $A = .5 + z_\alpha\sqrt{(.5)(.5)/n}$ 　　　　　　　　　　　　　　　　*(continued)*

TABLE 11-1 (continued)

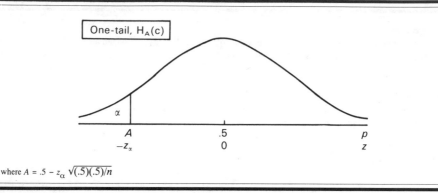

where $A = .5 - z_\alpha \sqrt{(.5)(.5)/n}$

time is recorded, and the following values are obtained: 22, 23, 25, 27, 27, 28, 28, 29, 29, 31, 33, 37, 37, 38, 39, 42, 43, 43, 43, 58. For purposes of simplicity, the times have been sorted in ascending order. Do these data support the null hypothesis that the true median travel time is 30 min for this trip, at the .05 level of significance?

First, it is necessary to calculate the number of minus signs, or observations below the hypothesized median of $\eta_I = 30$. Since the observations are in order, a simple count reveals $B = 9$ observations below the hypothesized median of 30. Or, equivalently, the probability of a negative sign in this sample is $p = B/n = 9/20 = .45$. Since $np = 20(.45) = 9$ is greater than 5, the sampling distribution of P is closely approximated by the normal (see Table 8-6). Since the null hypothesis makes the assumption that $\pi = .5$, we estimate the standard deviation of the sampling distribution of p as $\sqrt{(.5)(.5)/20} = \sqrt{.0125} = .112$. The standard hypothesis testing procedure for a test of p versus π, as described in Chapter 9, is followed in Table 11-2. The null hypothesis cannot be rejected.

EXAMPLE 11-2.   A recreation geographer randomly samples $n = 9$ visitors to a neighborhood park. Each visitor is asked to rate satisfaction with the park on a five-point scale with labels *completely dissatisfied, dissatisfied, neutral, satisfied,* and *completely satisfied.* The responses are then scored by assigning a value of 1 to *completely dissatisfied,* 2 to *dissatisfied,* and so on. The nine responses are 2, 3, 1, 2, 1, 5, 3, 1, and 2. Are these respondents satisfied with this neighborhood park?

A test of hypothesis for this question uses H_0: $\eta \leq 3$ versus the alternate hypothesis H_A: $\eta > 3$. We are interested in whether we can say, at $\alpha = .05$, that park users *are* satisfied and therefore use a one-tail test. The sequence of signs is $- 0 - - - + 0 - -$. There are two observations with a value of zero. By convention, these observations are omitted, and the sample size is reduced to $n = 7$ observations with $B = 6$ minus signs. The probability of obtaining a negative sign is $p = 6/7 = .86$. Given the small sample size, we use the binomial tables directly. For $\pi = .5$ and $n = 7$ these probabilities are as follows:

TABLE 11-2
Sign Test for Example 11-1

Hypotheses
H_0: $\eta_I = 30$, the median travel time is 30 min.
H_A: $\eta_I \neq 30$, the median travel time is not 30 min.

Test statistic and test to be employed
H_0 can be tested by using the sign test where the test statistic is the proportion of observations with a negative sign, p.

Level of significance
$\alpha = .05$.

Decision rule (two-tailed test)

$A_1 = .5 - z_{\alpha/2}\,\sqrt{(.5)(.5)/n}$

$A_2 = .5 + z_{\alpha/2}\sqrt{(.5)(.5)/n}$

$A_1 = .5 - 1.96\,\sqrt{.25/20} = .28$

$A_2 = .5 + 1.96\,\sqrt{.25/20} = .72$

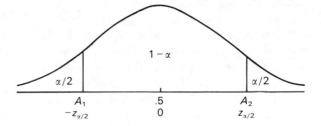

If $.28 \leq p \leq .72$, conclude H_0; if $p < .28$ or $p > .72$, conclude H_A.

Computation of test statistic
$p = 9/20 = .45$

Decision
Since $.28 \leq .45 \leq .72$, we accept the null hypothesis and conclude the median $\eta_I = 30$.

---

x	0	1	2	3	4	5	6	7
$P(x)$	.0078	.0547	.1641	.2734	.2734	.1641	.0547	.0078

To construct a decision rule for $\alpha = .05$, we must include all outcomes in the lower tail whose combined probabilities are less than or equal to .05. Since outcomes 0 and 1 taken together, $.0078 + .0547 = .0625$, exceeds .05, only the outcome of 0 can be used to reject the null hypothesis. The decision rule is therefore: If $B = 0$, conclude H_A; if $1 \leq B \leq 7$, conclude H_0. Since $B = 6$, we must accept the null hypothesis. Notice that the selection of an α level is quite difficult in this case. Unlike the normal or the t distribution, the sampling distributions of nonparametric test statistics often take large discrete jumps, particularly where the sample size is small. The use of the PROB-VALUE method is superior to classical hypothesis tests in these cases.

Mann-Whitney Test

The Mann-Whitney test is useful for comparing two populations on the basis of independent random samples. By assuming only that the two population distributions have the same shape, the test is used to determine whether the two populations have the identical location or position. The situation is depicted in Figure 11-1. The two populations are labeled X and Y and have the same shape. The exact form of the distribution need not be specified. If the two populations are identical, then the shift parameter δ (Greek letter delta) between the two medians η_X and η_Y is zero. If $\delta > 0$, as in Figure 11-1, then the median (or location, or position) of population Y is to the right of population X. If $\delta < 0$, then the distribution of Y is to the left of X. This test is comparable to the two-independent-sample t test described in Chapter 9. It is possible to specify either a one- or a two-tail version of the hypotheses depending on whether the alternate hypothesis specifies the sign of δ.

Let $x_1, x_2, \ldots, x_{n_1}$ and $y_1, y_2, \ldots, y_{n_2}$ be two independent random samples of size n_1 and n_2, respectively. Combine the n_1 observations of X from the first sample with the n_2 observations of Y, and arrange them in ascending order. Then assign ranks to the observations in this combined sample. Rank 1 is assigned to the lowest-valued observation, and rank $n_1 + n_2$ is assigned to the highest-valued observation. If the variable is collected at the ordinal level, then it is important that the rankings be obtained over the combined set of observations. For example, we might have respondents rank a series of 10 neighborhoods in terms of residential desirability. After the responses are recorded, the 10 neighborhoods are divided into two categories, say inner-city and suburban, creating two different samples. It is now possible to test the preferences of the respondents to inner-city or suburban locations. This would be impossible if the rankings were obtained separately for each type of neighborhood.

The Mann-Whitney test is based on the idea that if the two independent samples are drawn from the same population, then the average ranks of the two samples should be almost equal. The test statistic is simply the sum of the ranks from the first sample of X. Denote this sum S. The smallest possible value of S is the sum of the first n_1 ranks. This occurs when all the observations in the first sample are smaller than all the observations in the second sample. So the minimum value is $S = 1 + 2 + \cdots + n_1 = n_1(n_1 + 1)/2$. The largest possible value of S is the sum of the last n_1 ranks, or $(n_2 + 1) + (n_2 + 2) + \cdots + (n_2 + n_1)$. Therefore the largest possible value of S is $n_1 n_2 + n_1(n_1 + 1)/2$. For example, suppose $n_1 = n_2 = 5$. The two extreme values of S occur in the following ways:

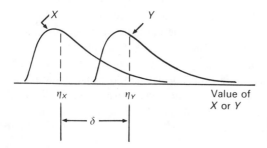

FIGURE 11-1. Two population distributions with identical shapes differing in the shift parameter δ

	Minimum S		Maximum S	
	X	Y	X	Y
	Sample 1	Sample 2	Sample 1	Sample 2
	1			1
	2			2
	3			3
	4			4
	5			5
		6	6	
		7	7	
		8	8	
		9	9	
		10	10	
	15		40	

The minimum value is $S = 1 + 2 + 3 + 4 + 5 = 5(5 + 1)/2 = 15$. The maximum value is $S = 6 + 7 + 8 + 9 + 10 = 5(5) + 5(5+ 1)/2 = 40$. If the calculated value of S is near the minimum value of S, then the distribution of X is to the left of Y and $\delta > 0$ (as in Figure 11-1). Conversely, if Y is to the right of X and $\delta < 0$, then the calculated value of S will be near the maximum value of S. If $\delta = 0$, as is hypothesized in H_0, then all possible orderings of the combined samples are equally likely and S will be between these two extremes.

To derive hypothesis testing rules or confidence intervals, it is necessary to know the sampling distribution of S when $\delta = 0$. For small samples, it is possible to work out the exact sampling distribution of S by determining the value of S for all possible orderings of the $n_1 + n_2$ observations. For example, if $n_1 = 2$ and $n_2 = 3$, the value of S for each possible permutation of the five values could be calculated in the following format:

Observation ranks		
Sample 1	Sample 2	S
1, 2	3, 4, 5	3
1, 3	2, 4, 5	4
1, 4	2, 3, 5	5
2, 3	1, 4, 5	5
⋮	⋮	⋮
4, 5	1, 2, 3	9

Tables of this sampling distribution are available in specialized texts on nonparametric statistics listed at the end of this chapter. When both n_1 and n_2 exceed 10, the sampling distribution of S is approximately normal with mean

$$E(S) = \frac{n_1(n_1 + n_2 + 1)}{2} \tag{11-1}$$

TABLE 11-3
The Mann-Whitney Test

Specification
The test compares two independent random samples of size n_1 and n_2 drawn from the two populations of X and Y, respectively. The measurement scales of X and Y are at least ordinal. Ranks are assigned to the *combined* sample; the lowest-valued observation is assigned the rank 1 and the highest observation the rank $n_1 + n_2$.

Hypotheses
H_0: $\delta = 0$, the two random samples are drawn from the same population.
H_A: (a) $\delta \neq 0$, the two random samples are drawn from different populations.
    (b) $\delta > 0$, the sample Y lies to the right of X, or $\eta_Y > \eta_X$.
    (c) $\delta < 0$, the sample Y lies to the left of X, or $\eta_Y < \eta_X$.

Test statistic
The sum of the ranks from the sample X is S.

Decision rule
For small sample sizes with $n_1, n_2 < 10$, the exact PROB-VALUE can be determined by generating the sampling distribution of S. This is accomplished by considering all possible arrangements of the $n_1 + n_2$ observations. Tables of S are available in several advanced nonparametric textbooks. For large samples, the normal approximation is used.

Large-sample test
Reject H_0 if
$$S < n_1(n_1 + n_2 + 1)/2 - z_{\alpha/2}\sqrt{n_1 n_2(n_1 + n_2 + 1)/12}$$

or if

$$S > n_1(n_1 + n_2 + 1)/2 + z_{\alpha/2}\sqrt{n_1 n_2(n_1 + n_2 + 1)/12}$$

for a two-tailed test. For one-tail test, reject H_0 if

$$S < n_1(n_1 + n_2 + 1)/2 - z_{\alpha}\sqrt{n_1 n_2(n_1 + n_2 + 1)/12}$$

for $H_A(b)$, and reject H_0 if

$$S > n_1(n_1 + n_2 + 1)/2 - z_{\alpha}\sqrt{n_1 n_2(n_1 + n_2 + 1)/12}$$

for $H_A(c)$.

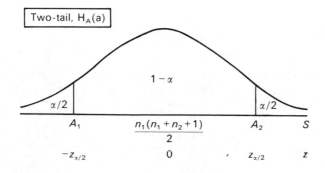

TABLE 11-3 (continued)

where $A_1 = n_1(n_1 + n_2 + 1)/2 - z_{\alpha/2}\sqrt{n_1 n_2 (n_1 + n_2 + 1)/12}$

$A_2 = n_1(n_1 + n_2 + 1)/2 + z_{\alpha/2}\sqrt{n_1 n_2 (n_1 + n_2 + 1)/12}$

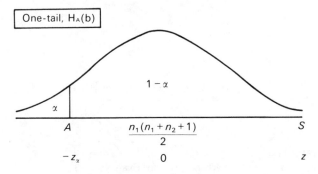

where $A = n_1(n_1 + n_2 + 1)/2 - z_\alpha\sqrt{n_1 n_2 (n_1 + n_2 + 1)/12}$

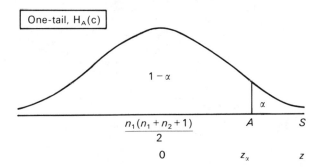

where $A = n_1(n_1 + n_2 + 1)/2 + z_\alpha\sqrt{n_1 n_2 (n_1 + n_2 + 1)/12}$

$$\sigma^2(S) = \frac{n_1 n_2 (n_1 + n_2 + 1)}{12} \tag{11-2}$$

The hypothesis that $\delta = 0$ can then be tested by using the conventional format described in Chapter 9. This procedure is summarized in Table 11-3. The use of the test is illustrated in Example 11-3.

EXAMPLE 11-3. Carbon monoxide concentrations are sampled at 20 intersections in the residential areas of a city. All intersections have approximately equal average daily traffic volumes. Ten of the intersections are controlled by yield signs, and 10 are controlled by stop signs. The concentrations of carbon monoxide (in parts per million) are as follows:

Intersections with yield signs	Intersections with stop signs
10	31
15	9
16	21
17	28
22	14
27	12
11	58
12	29
18	22
20	29

TABLE 11-4
Calculations for Mann-Whitney Test for Example 11-3

Combined ordered observations		Combined ranks	
X	Y	X	Y
	9		1
10		2	
11		3	
12	12	4.5	4.5
	14		6
15		7	
16		8	
17		9	
18		10	
20		11	
	21		12
22	22	13.5	13.5
27		15	
	28		16
	29, 29		17.5, 17.5
	31		19
	58		20
$n_1 = 10$	$n_2 = 10$	$\Sigma = 83$	$\Sigma = 127$

$$S = 83$$

$$E(S) = \frac{10(10 + 10 + 1)}{2} = 105$$

$$\sigma^2(S) = \frac{10(10)(10 + 10 + 10 + 1)}{12} = 175$$

$$\sigma(S) = 13.23$$

Because the stop signs cause more cars to travel at lower speeds, encourage periods of engine idling, and create more acceleration-deceleration phases, it is hypothesized that these intersections will have higher concentrations than those with yield signs. Is this hypothesis supported by the data?

The calculations required for the Mann-Whitney test of the carbon monoxide pollution levels are shown in Table 11-4. In the first two columns, the data are ordered within the combined set. In the last two columns, these observation values are replaced by the appropriate ranks. Notice the convention used when ties are encountered. Two observations have carbon monoxide concentrations of 12 parts per million. These two observations are ranked 4th and 5th. Each is assigned the average rank of $(4 + 5)/2 = 4.5$. The convention of assigning tied observations the average rank can be used for two, three, or more tied observations. For example, if three observations were tied for what would be the 5th, 6th, and 7th ranks, all three would be assigned a rank of $(5 + 6 + 7)/3 = 6$. However, the test assumes there are *no* ties in the data and is really only approximately correct when ties exist. The decision rule for a hypothesis test in a case where numerous ties exist requires modification. These modifications are discussed in the nonparametric statistics texts cited under Further Reading.

TABLE 11-5
Test of Hypothesis for Example 11-3

Hypotheses
H_0: $\delta = 0$, the two samples have the same carbon monoxide concentration.
H_A: $\delta < 0$, intersections with yield signs have lower concentrations of carbon monoxide.

Test statistic and test to be employed
H_0 can be tested by using the Mann-Whitney test. Since the stop sign observations have a much longer variation and a possible extreme value (58 ppm), it is preferred to the parametric alternative. The test statistic is S, the sum of ranks of yield sign observations.

Level of significance
$\alpha = .05$.

Decision rule

$A = n_1(n_1 + n_2 + 1)/2 - z_\alpha\sqrt{n_1 n_2(n_1 + n_2 + 1)/12}$

$\quad = 105 + (-1.96)(13.23)$

$\quad = 79.07$

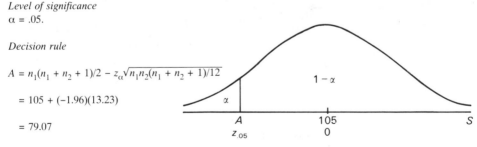

If $S < 79.07$, reject H_0; if $S \geq 79.07$, do not reject H_0.

Computation of test statistic
From Table 11-4, $S = 83$.

Decision
Since $S = 83 > 79.07$, we reject the null hypothesis and conclude that intersections controlled by yield signs cannot have lower carbon monoxide concentrations.

The sum of the ranks of X yields $S = 83$. Since n_1 and n_2 are greater than or equal to 10, the normal approximation to the sampling distribution of S is utilized: $E(S) = 105$ and $\sigma(S) = 13.23$. The formal test of hypothesis is detailed in Table 11-5. Since $S = 83$ lies above the critical value of 79.07, the null hypothesis is not rejected, and we find no evidence that concentration of carbon monoxide at intersections controlled by stop signs is any different from that at intersections controlled by yield signs.

Wilcoxon Signed-Ranks Test

The *Wilcoxon signed-ranks test* is the nonparametric equivalent to the matched-pairs t test presented in Chapter 9. The sample data consist of n matched pairs (x_i, y_i) of an interval-scaled variable. Like the matched-pairs t test, this nonparametric test utilizes the differences $D_i = y_i - x_i$. These differences are related to the individual populations of X and Y in the manner illustrated in Figure 11-2. When the y_i's are larger than the x_i's, as in Figure 11-2, the differences tend to be positive and the median difference η_D is also positive. Or, if the y_i's are generally smaller than the x_i's, then the differences are mostly negative and $\eta_D < 0$. The median difference may be thought of as a parameter which measures how far apart the distributions of X and Y are located. When the median difference is zero, the two distributions are in the same location. The sign of the median indicates the relative location of population X relative to population Y. If the distribution of difference is *symmetric*, then $\eta_D = \mu_D$ and the Wilcoxon signed-ranks test is equivalent to the matched-pairs t test for the difference of population means.

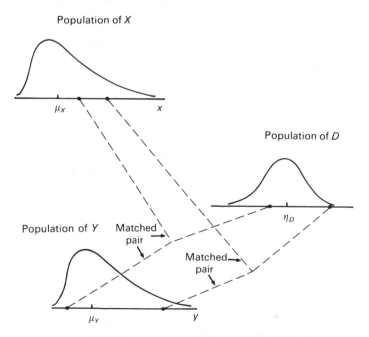

FIGURE 11-2. Wilcoxon signed-ranks test.

The calculation of the test statistic is quite simple. First, calculate the differences D_i, $i = 1, 2, \ldots, n$. Eliminate any observations with $D_i = 0$, and decrease the sample size appropriately. Consider now the absolute value of the differences $|D_i|$. These absolute values are ranked, with the lowest rank assigned to the smallest difference. Ties are handled in the manner described in the Mann-Whitney test. To each rank we also attach a sign of plus or minus according to whether the difference is positive or negative. The test statistic T is simply the sum of the signed ranks.

Under the null hypothesis of $\eta_D = 0$, we are just as likely to get a positive difference as a negative difference. Therefore, the rank associated with any absolute difference has an equal probability of being positive or negative. The expected value of T is therefore zero. The maximum value of T occurs when all the differences are positive. In this instance T is the sum of the ranks $1, 2, \ldots, n$. The sum of the first n integers is $n(n + 1)/2$, and this is the maximum value of T. The minimum value of T is $-n(n + 1)/2$ and occurs when all differences are negative. For a two-sided test, values of T close to zero lead us to conclude H_0: $\eta_0 = 0$. Values of T close to $n(n + 1)/2$ or $-n(n + 1)/2$ lead to the conclusion H_A: $\eta_D \neq 0$. A one-tail version of the test can also be employed.

The sampling distribution of T under H_0 has mean and variance

$$E(T) = 0 \tag{11-3}$$

$$\sigma^2(T) = \frac{n(n + 1)(2n + 1)}{6} \tag{11-4}$$

The sampling distribution depends only on the number of differences n. The exact sampling distribution of T can be easily generated for small sample sizes. For example, for $n = 3$ the sampling distribution is as follows:

Signed ranks			T
1	2	3	6
-1	2	3	4
1	-2	3	2
1	2	-3	0
-1	-2	3	0
-1	2	-3	-2
1	-2	-3	-4
-1	-2	-3	-6

Tables of T are available but are generally not needed unless n is very small. For $n \geq 10$, the sampling distribution of T is approximately normal with mean and variance given by Equations (11-3) and (11-4). The Wilcoxon signed-ranks test is completely specified in Table 11-6.

EXAMPLE 11-4. Ten water samples are taken at different sites on a river just downstream from the location of a proposed cannery. All samples are taken on the same day during summer drought conditions. Water quality measurements of the samples are re-

TABLE 11-6
Wilcoxon Signed-Ranks Test

Specification
The n matched pairs (X_i, Y_i) are interval-scaled variables. Let $D_i = Y_i - X_i$. The absolute values of the differences D_i are then marked. A sign is attached to each rank, depending on whether it is positive or negative.

Hypotheses
H_0: $\eta_D = 0$, the population median difference equals zero.
H_A: (a) $\eta_D \neq 0$, the population median difference is not zero.
    (b) $\eta_D > 0$, there is a positive difference, and distribution Y lies to the right of X.
    (c) $\eta_D < 0$, there is a negative difference, and distribution X lies to the right of Y.

Test statistic
$T = \min (T^+, T^-)$

Decision rule
Tables of the exact sampling distribution of T are available in many statistics textbooks. However, these tables are rarely needed since the normal approximation is applicable for sample sizes as small as $n = 10$. Reject H_0 if

$$T < -z_{\alpha/2}\sqrt{(n(n + 1)(2n + 1)/6}$$

or if

$$T > z_{\alpha/2}\sqrt{n(n + 1)(2n + 1)/6}$$

for a two-tail test. For H_A(b), reject H_0 if

$$T > z_\alpha \sqrt{n(n + 1)(2n + 1)/6}$$

and for H_A(c) reject H_0 if

$$T < -z_\alpha\sqrt{n(n + 1)(2n + 1)/6}$$

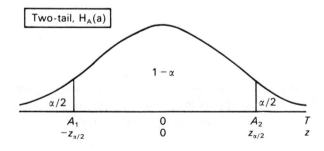

where $A_1 = -z_{\alpha/2} \sqrt{n(n + 1)(2n + 1)/6}$

$\quad\quad A_2 = z_{\alpha/2}\sqrt{n(n + 1)(2n + 1)/6}$

TABLE 11-6 (continued)

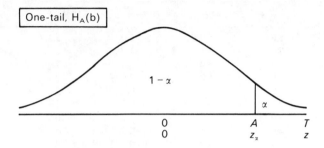

One-tail, $H_A(b)$

where $A = z_\alpha \sqrt{n(n + 1)(2n + 1)/6}$

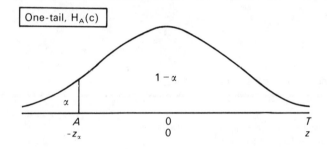

One-tail, $H_A(c)$

where $A = -z_\alpha \sqrt{n(n + 1)(2n + 1)/6}$

TABLE 11-7
Dissolved Oxygen Levels at 10 Selected Sites,
Summer Drought Conditions

Site	Dissolved oxygen, mg/L	
	Before	After
1	5.0	6.0
2	7.2	5.5
3	4.5	3.0
4	7.0	5.0
5	7.0	6.2
6	2.4	2.5
7	3.0	0
8	4.0	3.1
9	2.8	2.2
10	4.2	3.0

TABLE 11-8
Calculations for Wilcoxon Signed-Ranks Test for Example 11-4

Site i	Before x_i	After y_i	Difference D_i	Rank	Signed rank
1	5.0	6.0	+1.0	5	+5
2	7.2	5.5	-1.7	8	-8
3	4.5	3.0	-1.5	7	-7
4	7.0	5.0	-2.0	9	-9
5	7.0	6.2	-0.8	3	-3
6	2.4	2.5	+0.1	1	+1
7	3.0	0	-3.0	10	-10
8	4.0	3.1	-0.9	4	-4
9	2.8	2.2	-0.6	2	-2
10	4.2	3.0	-1.2	6	-6
					$T = -43$

TABLE 11-9
Wilcoxon Signed-Ranks Test for Example 11-4

Hypotheses
H_0: $\eta_D = 0$, there is no change in the median DO levels.
H_A: $\eta_D \neq 0$, the median DO has changed.

Test statistic and test to be employed
Given the small sample size and the apparent nonnormality of the data, the Wilcoxon signed-ranks test is employed. Test statistic is T, the sum of the signed ranks.

Level of significance
$\alpha = .05$.

Decision rule
$A_1 = -z_{\alpha/2} \sqrt{n(n + 1)(2n + 1)/6}$

$A_2 = +z_{\alpha/2} \sqrt{n(n + 1)(2n + 1)/6}$

$A_1 = -1.96 \sqrt{10(11)(21)/6} = -38.46$

$A_2 = +1.96 \sqrt{10(11)(21)/6} = +38.46$

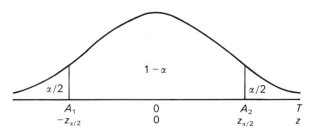

If $T < -38.46$ or $T > +38.46$, conclude H_A.

Computation of test statistic
From Table 11-7, $T = -43$.

Decision
Since $T < -38.46$, reject H_0; conclude there has been a change in DO Levels.

corded. The measure of water quality employed is the dissolved oxygen (DO) content in micrograms per liter. Label these 10 samples *before*. Shortly after the opening of the cannery and its waste treatment system, new DO measurements are taken at the same location. Label these samples *after*. The water quality measurements are listed in Table 11-7. Has there been a change in DO levels?

The calculations for the Wilcoxon signed-ranks test of these data are illustrated in Table 11-8. The first three columns contain the (x_i, y_i) pairs and the differences D_i. The next column contains the ranks of the absolute differences, the lowest (rank 1) for observation 6 and the highest (rank 10) for observation 7. Eight of the differences have negative signs, and two have positive signs. The signed ranks are shown in the last column. The test statistic T is the sum of the signed ranks, or −43. It is near the lower extreme of T which ranges from −55 to +55 for sample size $n = 10$. Since there are 10 differences, the normal approximation can be used. A complete test of hypothesis is given in Table 11-9. The calculated test statistic of $T = -43$ leads to the rejection of H_0 at $\alpha = .05$. It is concluded that DO levels in the river decreased with the opening of the cannery. To calculate the two-sided PROB-VALUE for this problem, we convert T to Z, using $Z = (T - 0)/\sigma(T)$. Substituting $T = -43$ and $\sigma(T) = 19.62$ yields $Z = -2.19$. From the normal table we see that the two-sided PROB-VALUE equals $2(.0143) = .0286$.

The Wilcoxon signed-ranks test can also be used to test whether the median of a set of numbers equals some specified value, say $\eta = \eta_l$. If x_i, $i = 1, 2, \ldots, n$, represents the observations in a random sample from a population, then define the differences $D_i = x_i - \eta_l$. These differences are then used to test whether $\eta_D = 0$. If the hypothesis that $\eta_D = 0$ cannot be rejected, it follows that the median of the x_i's must be equal to η_l. Used in this way, the Wilcoxon signed-ranks test is an alternative to the sign test for the median.

Kruskal-Wallis Test

The *Kruskal-Wallis test* extends the Mann-Whitney test to the case when there are more than two populations. It is the nonparametric equivalent to one version of a powerful parametric test known as the *analysis of variance*, called ANOVA. The data for this test can be arranged to the following format:

Sample 1	Sample 2	. . .	Sample k
x_{11}	x_{21}	. . .	x_{k1}
x_{12}	x_{22}		x_{k2}
. . . .			
x_{1n_1}	x_{2n_2}	. . .	x_{kn_k}

where x_{ij} is the value of jth observation in the ith sample. In the k samples there are $n = n_1 + n_2 + \cdots + n_k = \Sigma_{i=1}^{k} n_i$ observations. The test utilizes the ranks of the individual observations within the *combined* set. If the measurement scale is ordinal, then the ranks must be obtained within the combined set, just as in the Mann-Whitney test.

The Kruskal-Wallis test is also very simple. It is based on a comparison of the sums of the ranks for each of the k independent samples. The first step is to rank the observations. Let $r(X_{ij})$ be the rank of the jth observation of the ith sample within the combined set of n

observations. Ties are handled in the usual way. Define r_i as the sum of the rankings in the ith sample:

$$r_i = \sum_{j=1}^{n_i} r(x_{ij})$$ (11-5)

Under the null hypothesis that the k samples come from the same population, the average ranks in each sample should be approximately equal. If the average ranks differ, then this supports the alternative hypothesis that *at least one* of the samples comes from a different population than the other samples. If only the locations of the population distributions differ, and not the shapes, then the Kruskal-Wallis test is equivalent to testing the equality of the k population means $\mu_1 = \mu_2 = \cdots = \mu_k$.

The test statistic for the Kruskal-Wallis test is

$$H = \frac{12}{n(n+1)} \sum_{i=1}^{k} \frac{r_i^2}{n_i} - 3(n+1)$$ (11-6)

if there are no ties in the k samples. If there are numerous ties in the data, then a correction factor must be used. The sampling distribution of H depends primarily on the sum of the ranks in each sample r_i. For small sample sizes, the sampling distribution of H can be generated by considering all distinguishable combinations of the n rankings for the k groups. For example, consider the smallest possible problem, a $k = 3$ sample problem with $n_1 = 2, n_2 = n_3 = 1$, and $n = 4$ observations. The possible combinations of rankings are shown in the following table:

	Sample		
1	2	3	H
1, 2	3	4	2.7
1, 2	4	3	2.7
3, 4	1	2	2.7
3, 4	2	1	2.7
2, 3	1	4	2.7
2, 3	4	1	2.7
1, 3	2	4	1.8
1, 3	4	2	1.8
2, 4	1	3	1.8
2, 4	3	1	1.8
1, 4	2	3	0.3
1, 4	3	2	0.3

Notice that arrangements of the rankings *within* a sample are ignored. For example, no distinction can be made between these two combinations

Sample 1	Sample 2	Sample 3	Sample 1	Sample 2	Sample 3
1, 2	3	4	2, 1	3	4

TABLE 11-10
Kruskal-Wallis Test

Specification

There are k independent random samples with n_i $i = 1, 2, \ldots, k$, observations. In total there are $n = \sum_{i=1}^{k} n_i$ $n_1 + n_2 + \cdots + n_k$ observations. The individual observations are ranked within the complete data set of n observations. Calculate the sum of the ranks r_i for each of k samples.

Hypotheses

H_0: $\eta_1 = \eta_2 = \eta = \eta_k$, all samples are drawn from the same population. If the population distributions have the same shape and variance, then this is equivalent to a test· of the k population means $\mu_1 = \mu_2 = \cdots = \mu_k$.

H_A: At least one median differs from the rest.

Test statistic

$$H = \frac{12}{n(n + 1)} \sum_{i=1}^{k} \frac{r_i^2}{n_i} - 3(n + 1)$$

Decision rule

Tables of the critical values of H for $k = 3$ samples where all $n_i \leq 5$ are available in most nonparametric textbooks. For all other cases the distribution of H is approximately χ^2 with $k - 1$ degrees of freedom. Reject H_0 if $H > \chi^2(1 - \alpha, k - 1)$. This is a one-tail test.

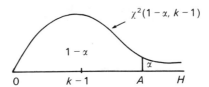

where $A = \chi^2(1 - \alpha, k - 1)$.

since in both cases the two lowest-ranked observations are in the first sample. The value of H is equal in these cases. Under the null hypothesis that the three samples come from the same population, each of these combinations of rankings is equally likely. If the actual sample contained the rankings 1, 2; 3; 4 with a calculated value of $H = 2.7$, then we see from the enumeration table that it has a 6/12, or 50 percent, chance of occurring. Thus, if we reject the null hypothesis of equality, we do so at an $\alpha = .50$ level of significance.

However, it is not always necessary to consult detailed tables of H for a Kruskal-Wallis test. Whenever $k > 3$ and/or $n_i > 5$ for at least one sample, the sampling distribution of H is χ^2 with $k - 1$ degrees of freedom. The Kruskal-Wallis test is summarized in Table 11-10.

EXAMPLE 11-5. Suppose the sample of carbon monoxide concentrations utilized in the Mann-Whitney test example is augmented by 10 more observations. These 10 observations are from intersections with no traffic controls. This augmented data set is now of the form of the Kruskal-Wallis input:

Intersections with yield signs	Intersections with stop signs	Intersections with no control
10	31	17
15	9	14
16	21	12
17	28	8
22	14	6
27	12	19
11	58	13
12	29	23
18	22	5
20	29	7

TABLE 11-11
Calculations for the Kruskal-Wallis Test for Example 11-5

Intersections with yield signs		Intersections with stop signs		Intersections with no controls	
Observation value	Rank	Observation value	Rank	Observation value	Rank
				5	1
				6	2
				7	3
				8	4
		9	5		
10	6				
11	7				
12	9	12	9	12	9
				13	11
		14	12.5	14	12.5
15	14				
16	15				
17	16.5			17	16.5
18	18				
				19	19
20	20				
		21	21		
22	22.5	22	22.5		
				23	24
27	25	28	26		
		29, 29	27.5, 27.5		
		31	29		
		58	30		
$r_1 = 153$		$r_2 = 210$		$r_3 = 102$	

$$H = \frac{12}{30(31)} \left(\frac{153^2}{10} + \frac{210^2}{10} + \frac{102^2}{10} \right) - 3(31) = 7.53$$

TABLE 11-12
Kruskal-Wallis Test for Example 11-5

Hypotheses
H_0: All three samples come from the same population. Carbon monoxide concentrations are the same at intersections with stop signs, yield signs, and no controls.
H_A: At least one of the samples differs from the rest.

Test statistic and test to be employed
The test statistic of the Kruskal-Wallis test H.

Level of significance
$\alpha = .05$.

Decision rule
$A = \chi^2(1 - \alpha, k - 1)$
  $= \chi^2(0.95, 2) = 5.99$

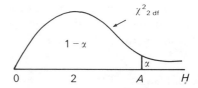

Reject H_0 if $H > 5.99$; if $0 \le H \le 5.99$, accept H_0.

Computation of test statistic
From Table 11-11, $H = 7.53$.

Decision
Since $H = 7.53 > 5.99$, reject H_0 and conclude that carbon monoxide concentrations tend to vary with intersection controls.

Suppose that we have not yet analyzed any part of this data set and do not know the results of the Mann-Whitney test. In fact, the best procedure for analyzing k-sample data is to test the k-sample data for significance and then to perform two-sample tests on pairs of samples if it can be shown that more than one population is involved.

The necessary calculations are illustrated in Table 11-11. Notice how ties are handled. For the carbon monoxide value of 12 parts per million, there are three tied observations, one in each sample. The rank assigned to each is $(8 + 9 + 10)/3 = 9$. The sum of the ranks in the three samples of $r_1 = 153$, $r_2 = 210$, and $r_3 = 102$ is 465. For problems calculated by hand, a useful computational check is to compare the sum of the ranks of the k samples $\Sigma_{i=1}^{k} r_i$ to the sum of the first n ranks $n(n + 1)/2$. These two should be equal. In this example $\Sigma_{i=1}^{k} r_i = 465 = 30(31)/2$. The complete test of hypothesis is given in Table 11-12. The calculated value of $H = 37.53$ leads to a rejection of the null hypothesis at $\alpha = .05$. We now know that at least one of these samples differs from the other two. The obvious questions are: Does only one sample differ from the other two, and if so, which one? and Are all three samples from different populations? We already know that intersections with yield signs have lower concentrations of carbon monoxide than intersections with stop signs. To answer this question completely, however, Mann-Whitney tests between each pair of samples are required.

11.3. Goodness-of-Fit Tests

It is frequently necessary in empirical research to test the hypothesis that a random variable has a specified theoretical probability distribution. This hypothesis may be of interest for two reasons: to confirm a theory or a model or to determine whether the assumptions of a particular statistical inference procedure are satisfied. As an example of the first case, we may wish to test the randomness of a particular point pattern by using the Poisson probability distribution. Because virtually all classical or parametric statistical inference procedures make the assumption of normality, it is often necessary to check this assumption before one proceeds. This is an example of the second reason for testing hypotheses about the distribution of a random variable. Goodness-of-fit tests are used for these purposes. The label *goodness of fit* is used because the purpose of the test is to see whether a particular probability distribution is a good model for the sampled population. This judgment is made according to whether the probability distribution specified in H_0 provides a good fit for the sample data. The two most commonly used tests are the χ^2 and Kolmogorov-Smirnov tests.

Chi-Square Test

The χ^2 test is widely used goodness-of-fit test. The basic structure of the test is shown in Table 11-13. The sample data are nominal. Counts are obtained for each of the k classes. Denote the observed number or count of observation of the ith class f_i. The sum of the observed frequency counts over the k classes equals the total sample size n:

$$\sum_{i=1}^{k} f_i = n \tag{11-7}$$

By using the probability distribution specified in the null hypothesis, a set of expected frequencies are determined. Let the expected number of observations in the ith class be denoted F_i. Of course,

$$\sum_{i=1}^{k} F_i = n \tag{11-8}$$

The χ^2 test is based on a comparison of the observed frequencies f_i and the expected frequencies F_i. The closer the set of F_i to f_i, the more likely it is that the distribution of the sample reflects the probability distribution specified in the null hypothesis. The farther apart the observed and expected frequencies, the less likely H_0 is true. The test statistic is the sum of the relative squared differences:

TABLE 11-13
Structure of a χ^2 Goodness-of-Fit Test

Class	Observed frequency	Expected frequency with H_0	Relative squared difference
1	f_1	F_1	$(f_1 - F_1)^2/F_1$
2	f_2	F_2	$(f_2 - F_2)^2/F_2$
k	f_k	F_k	$(f_k - F_k)^2/F_k$
Total:	n	n	$\chi^2 = \sum_{i=1}^{k} (f_i - F_i)^2/F_i$

$$X^2 = \sum_{i=1}^{k} \frac{(f_i - F_i)^2}{F_i} \tag{11-9}$$

First, note that when $f_i = F_i$ for any class, the contribution of that class to X^2 is zero. Second, note that larger differences are penalized more severely than smaller differences since the difference is squared in the numerator of (11-9). The denominator is used to standardize these errors. Suppose, for example, we expect a count of 40 in some class and we observe 38. This difference of 2 leads to a $(38 - 40)^2/40 = .1$ contribution to X^2. However, a difference of 2 from an observed count of 6 to an expected count of 4 leads to a $(6 - 4)^2/4 = 1$ contribution to X^2. Even though the difference of 2 is the same in each case, the latter situation causes an increase in X^2 ten times as large.

When n is large, the test statistic X^2 is distributed approximately as a χ^2 random variable with $k - m - 1$ degrees of freedom. One degree of freedom is lost because the observed and expected counts both sum to the sample size n. Also, 1 degree of freedom is lost for each parameter of the probability distribution specified in H_0 which must be estimated from sample data. The number of parameters to be estimated m varies with the distribution specified under H_0. For the Poisson distribution, $m = 1$ since only the parameter λ need be estimated. For the normal, two parameters must be estimated, μ and ρ. If the null hypothesis includes the values of the required parameters, then no parameters need be estimated from the sample. Thus, there are two cases for the relevant hypotheses.

1. H_0: The distribution is normal. H_A: The distribution is not normal. Degrees of freedom are $k - 2 - 1$.
2. H_0: The distribution is normal with mean $\mu = a$ and variance $\sigma^2 = b$. H_A: The distribution is not normal with mean a and variance b. Degrees of freedom are $k - 1$.

Case 1 is more common. Our interest lies only in whether we can safely assume the distribution is normal. In case 2, however, we are interested in whether the distribution is normal *and* in the particular parameters for the mean and variance. If we rèject the null hypothesis in the second case, we are *not* necessarily saying the distribution is not normal. We are saying only that it is not normal with the given parameters. The distribution may be normal with mean $= c$ and variance $= d$.

The sample size required for the χ^2 approximation depends on the expected frequencies under H_0, or F_i. As a general rule, the expected frequency in each class should exceed 2, and most should exceed 5. If more than 20 percent of the classes have expected frequencies less than 5, the χ^2 approximation is not valid. If there are several classes with $F_i < 5$ or even one with $F_i < 2$, it is usual to group adjacent categories so that the combined expected frequency exceeds 5. For example, if the expected frequencies of the first three classes in a problem are

i	1	2	3
F_i	1.27	3.88	5.99

they can be combined to

i	1	2
F_i	5.15	5.99

so that the required expected frequency in each class exceeds 5. Ordinarily, only adjacent classes are grouped in this way. In almost all goodness-of-fit tests, there is an explicit ordering of the categories according to the value of the variable. It would thus make little sense to group categories 3 and 9 to satisfy the minimum frequency requirement. The principal reason for combining categories to achieve a minimum expected frequency of 5 is the extreme sensitivity of the X^2 test statistic to small differences when F_i is small. For example, a difference of $f_i - F_i = 4$ when $F_i = 1$ leads to a contribution of 16 to the statistic X^2. The combination of categories has one drawback. Each time two categories are combined, there is a loss of a degree of freedom.

TABLE 11-14
Chi-Square Goodness-of-Fit Test

Specification
The χ^2 test compares an observed set of *frequencies* in k classes f_i, $i = 1, 2, \ldots, k$ to a set of expected frequencies F_i, $i = 1, 2, \ldots, k$. It is required that $\sum_{i=1}^{k} f_i = \sum_{i=1}^{k} F_i = n$. It is assumed that the sample is random and reasonably large.

Hypotheses
H_0: The population has a specific form, for example, normal or normal with mean μ and variance σ^2.
H_A: The population distribution is not of the form specified in H_0.

Note: In some instances H_0 can specify the specific parameters of the population. In other cases, only the family of the distribution is specified, and the parameters are estimated from the sample data.

Test statistic

$$X^2 = \sum_{i=1}^{k} \frac{(f_i - F_i)^2}{F_i}$$

With $k - m - 1$ degrees of freedom, m is the number of parameters estimated from sample data.

Decision rule
If $X^2 > \chi^2(1 - \alpha, k - m - 1)$, conclude H_A.

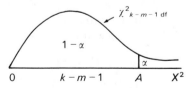

where $A = \chi^2(1 - \alpha, k - m - 1)$

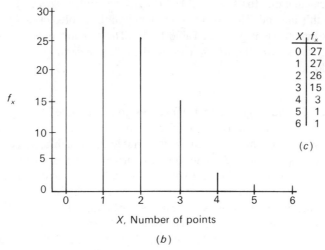

0	1	2	2	0	1	2	0	1	1
3	2	2	0	0	2	1	2	0	1
0	1	0	0	0	2	3	1	3	1
4	2	3	1	3	1	0	2	3	0
1	1	0	1	2	2	2	0	2	1
1	4	2	3	1	5	2	2	0	3
1	2	0	2	0	0	3	1	4	0
0	2	1	2	1	2	3	3	3	1
1	3	6	0	3	3	0	0	0	0
1	0	2	1	2	2	0	1	1	2

(*a*)

f_x

30
25
20
15
10
5
0

0   1   2   3   4   5   6

X, Number of points

(*b*)

X	f_x
0	27
1	27
2	26
3	15
4	3
5	1
6	1

(*c*)

FIGURE 11-3. Maps for Example 11-6.

The χ^2 goodness-of-fit test is summarized in Table 11-14. Two examples illustrate the application of this test. Example 11-6 utilizes a *discrete* probability distribution, the Poisson, in a test of point-pattern randomness. It also illustrates the procedure for pooling classes to yield $F_i > 5$. Example 11-7 represents one of the most common uses of the χ^2 test, testing a variable for normality prior to the application of a parametric statistical test. Also, this problem illustrates a useful method of applying a goodness-of-fit test to a continuous probability distribution.

EXAMPLE 11-6.   Let us consider a problem in point-pattern analysis. Specifically, let us test whether the distribution of points in the map of Figure 11-3 is random. To do this, we must see whether the frequency distribution of points by quadrat can be considered to be Poisson with $\lambda = 1.5$. The observed and expected numbers of points per quadrat are as follows:

Points per quadrat	Observed number of quadrats	Expected number of quadrats
0	27	22.31
1	27	33.47
2	26	25.10
3	15	12.55
4	3	4.71
5	1	1.41
6+	1	0.45

To apply the χ^2 test, we require frequencies greater than 2. To satisfy this assumption, the final three categories can be pooled into a single class of 4+ points. This pooling is illustrated in the calculations for X^2 of Table 11-15. The original seven classes are now reduced to five by this method. The expected frequency of the pooled class 4+ is now 6.57. The formal test of hypothesis is given in Table 11-16. The calculated test statistic of $X^2 = 3.103$ is not significant at $\alpha = .05$, which leads to the acceptance of H_0. The point pattern is apparently random.

EXAMPLE 11-7. Consider again the 50 DO values given in Table 2-1. Suppose we now wish to test whether this variable is a normal random variable. First, it is necessary to specify a set of classes. Since this is a continuous variable, the specification of appropriate classes is not as easy as it is for a discrete random variable. One efficient way to define classes for continuous random variables is to divide the area under the density function into k intervals, each with a probability of occurrence of $1/k$. Figure 11-4 illustrates this procedure for a normal distribution and $k = 6$ classes. The cumulative z scores defining the class bounds are shown at each class limit. The first class includes all observations with a standardized value less than $z_{.167} = -.97$. The second class includes all observations between $-.97$ and $z_{.333} = -.43$, and so on.

It is easy to convert these z scores to values of any variable X by using the identity

TABLE 11-15
Calculations for χ^2 Test for Example 11-6

Class	Number of points X	Observed frequency f_i	Probability under H_0	Expected frequency F_i	$f_i - F_i$	$(f_i - F_i)^2/F_i$
1	0	27	.2231	22.31	4.69	0.99
2	1	27	.3347	33.47	−6.47	1.25
3	2	26	.2510	25.10	0.90	0.03
4	3	15	.1255	12.55	2.45	0.48
5	4 } 5 } ≥ 4	3 } 1 } 5	.0657	6.57	−1.57	0.38
	6 }	1 }	1.000	100.00	6.000	$X^2 = 3.13$

TABLE 11-16
Chi-Square Test for Example 11-6

Hypotheses
H_0: The distribution of points in the map of Figure 11-3 is Poisson; that is, it is a random point pattern.
H_A: The distribution of points is not Poisson.

Test statistic
A χ^2 goodness-of-fit test is used. Test statistic is X^2 with $k - 2 = 5 - 2 = 3$ degrees of freedom. The parameter $\lambda = 1.5$ is estimated from the map.

Level of significance
$\alpha = .05$.

Decision rule
$A = \chi^2(.95, 3) = 7.82$
If $X^2 \geq 7.82$, accept H_A.

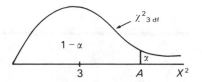

Computation of test statistic
From Table 11-14, $X^2 = 3.13$.

Decision
Since $X^2 < 7.82$, do not reject H_0.

$z = (x - \bar{x})/s$. For the DO lake data we know (from Chapter 2) that $\bar{x} = 5.58$ and $s = .6392$. Therefore, the value of X (or DO value) corresponding to $z = -.97$ can be determined by substitution:

$$-.97 = \frac{x - 5.58}{.6392} \qquad \text{or} \qquad x = 4.96 \qquad (11\text{-}10)$$

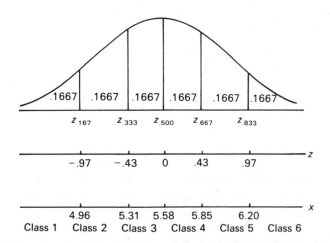

FIGURE 11-4. Dividing the standard normal distribution into six intervals of equal probability.

TABLE 11-17
Testing Lake DO Values for Normality by Using χ^2

Class	Bounds in X	f_i, observed frequency	Probability under H_0	F_i, expected frequency	$f_i - F_i$	$(f_i - F_i)^2/F_i$
1	<4.96	9	.1667	8.333	0.667	.05
2	4.96–5.31	6	.1667	8.333	−2.333	.65
3	5.32–5.57	6	.1667	8.333	−2.333	.65
4	5.58–5.85	14	.1667	8.333	5.667	3.85
5	5.86–6.20	7	.1667	8.333	−1.333	.21
6	>6.20	8	.1667	8.333	−0.333	.01
		50	≈1.000	≈50.0	0.00	$\chi^2 = 5.43$

All other class limits can be found in the same way. The classes defined by this method are illustrated in Figure 11-4 and Table 11-17. The observed frequencies shown in Table 11-17 are generated by using the DO values of Table 2-1. The probabilities for each class are $1/k$ = 1/6 = .1667, and the expected frequencies are .1667 times the sample size of n = 50, or 8.333. The calculations for the test statistic X^2 are given in the last two columns.

The procedure for dividing the area under a density function into k classes of equal probability can be utilized for any value of k and any probability density function. How should the value of k be chosen? The greater the value of k, the more degrees of freedom for the statistical test, so it is best to maximize the number of classes. The only constraint is that the expected frequencies exceed 5. The easiest way of meeting the minimum frequency requirement is to choose k so that n/k exceeds 5 by as small an amount as possible. For n = 20 observations, it is possible to use 4 classes; for n = 63, 12 classes; for n = 100, 20 classes; and so on. For the lake DO data, the maximum value of k is 50/5 = 10 (see Problem 10).

Kolmogorov-Smirnov Test

An alternative, and frequently more powerful, goodness-of-fit test is the *Kolmogorov-Smirnov test*. This test can be used for *continuous* random variables. For discrete random variables, the test gives only approximate results. Let us begin by reviewing the concept of a *cumulative probability distribution function* of a random variable. As defined in Chapter 5, the cumulative probability distribution function of a random variable X is defined as

$$F(x) = P(X \leq x) \qquad -\infty < x < +\infty \qquad (11\text{-}11)$$

Figure 11-5 illustrates the cumulative probability distribution functions for the uniform continuous and standard normal distributions. The Kolmogorov-Smirnov test is based on examining the difference between this cumulative probability function $F(x)$ and the *sample cumulative distribution function* $S(x)$. The sample cumulative distribution function $S(x)$ specifies the proportion (or relative frequency) of *sample* values less than or equal to x for each value of x in the sample. It is an unsmoothed ogive. The method for constructing

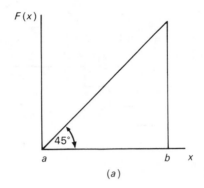

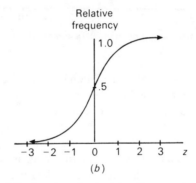

FIGURE 11-5. Cumulative probability functions for the (a) uniform continuous and (b) standard normal.

the sample cumulative distribution function for any variable is best illustrated with a small sample. Suppose the sample consists of the following values: 18, 10, 5, 2, 4, 20, 2, 8, 1, 12. First, order the observations: 1, 2, 2, 4, 5, 8, 10, 12, 18, 20. then construct a cumulative relative frequency table of the form

x	1	2	4	5	8	10	12	18	20
$S(x)$	.1	.3	.4	.5	.6	.7	.8	.9	.10

Of the 10 observations, only one is less than or equal to one, so that $S(1) = .1$. There are three observations in the sample less than or equal to 2, so that $S(2) = .3$. The remainder of the table is constructed in this way. The step-function form of $S(x)$ is clearly seen in Figure 11-6.

The key question in the Kolomogorov-Smirnov test is, Are the differences between the sample $S(x)$ and the cumulative probability distribution function $F(x)$ sufficiently large that they are unlikely to have occurred because of chance fluctuations? The test statistic is

$$D = \max_{x} |S(x) - F(x)| \qquad (11\text{-}12)$$

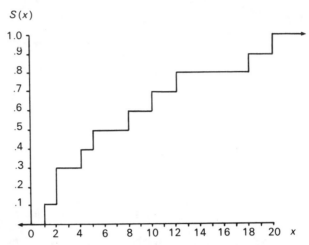

FIGURE 11-6. Graph of cumulative sample distribution function.

In other words, D is the largest absolute deviation between the cumulative sample function and the cumulative probability function. Tables of D for α = .01, .05, and .10 levels of significance are given in the Appendix. The Kolomogorov-Smirnov test is summarized in Table 11-18.

EXAMPLE 11-8. A cartographer records the times taken by 10 subjects to complete an experiment in which subjects are asked to estimate data values illustrated as proportional circles. The times in seconds are

65	62	50	80	83	64	67	62	66	68

TABLE 11-18
Kolmogorov-Smirnov Test

Specification
A cumulative distribution function $S(x)$ is generated from the random sample of n observations. This cumulative distribution function is compared to the cumulative probability distribution function of some hypothesized distribution, $F(x)$, with known parameters.

Hypotheses
H_0: $S(x) = F(x)$, the sample is from the hypothesized probability distribution with known parameters.
H_A: $S(x) \neq F(x)$, it is not.

Test statistic
$D = \max_{x} |S(x) - F(x)|$ sample size = n

Decision rule
The table of critical values for D is given in the Appendix. Reject H_0 if D exceeds the tabulated value at the chosen level of significance. Otherwise, do not reject H_0. There are n degrees of freedom.

TABLE 11-19
Calculation of D for Example 11-8

z	$S(z)$	$F(z)$	$|S(z) - F(z)|$
−1.0	.1000	.1587	.0587
0.2	.3000	.5793	.2793
0.4	.4000	.6554	.2254
0.5	.5000	.6915	.1915
0.6	.6000	.7257	.1257
0.7	.7000	.7580	.0580
0.8	.8000	.7881	.0119
2.0	.9000	.9772	.0772
2.3	1.0000	.9893	.0107

$$D = \max_z |S(z) - F(z)| = .2793$$

Past research indicates that these times tend to be normally distributed with a mean of 60 s and a standard deviation of 10 s. Use the Kolmogorov-Smirnov test to determine whether it is likely that these times could be a random sample from a normal distribution with these parameters.

First, standardize the values, using $z_i = (x_i - \mu)/\sigma$. The 10 z scores are

| 0.5 | 0.2 | −1.0 | 2.0 | 2.3 | 0.4 | 0.7 | 0.2 | 0.6 | 0.8 |

and when they are reordered by value, they are

| −1.0 | 0.2 | 0.2 | 0.4 | 0.5 | 0.6 | 0.7 | 0.8 | 2.0 | 2.3 |

The cumulative sample distribution function is therefore

z	−1.0	0.2	0.4	0.5	0.6	0.7	0.8	2.0	2.3
$S(z)$	.1	.3	.4	.5	.6	.7	.8	.9	.10

The calculations required for the test statistic D are given in Table 11-19. The values for $F(z)$ are extracted from the table of the standard normal distribution. The value of $D = .2793$ for $n = 10$ degrees of freedom is not significant at $\alpha = .05$. From the table of the Kolmogorov-Smirnov test statistic in the Appendix, we note that a value of $D = .409$ is required to reject H_0 at this level of significance. We conclude that the times taken by this sample of 10 subjects could be a random sample from a normal distribution with mean 60 and standard deviation 10. See Figure 11-7.

The Kolmogorov-Smirnov test can also be used with grouped data from any cumulative frequency table of a continuous variable. The values of $S(x)$ are the relative cumulative frequency values at the endpoints of the intervals, and the value of x is simply the upper limit of the interval. The principal disadvantage of using grouped data for a continuous variable is the loss of degrees of freedom. There is only 1 degree of freedom

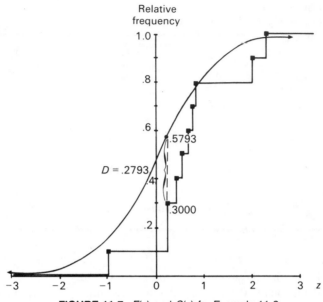

FIGURE 11-7. $F(z)$ and $S(z)$ for Example 11-8.

for each class or category rather than one for each observation. Of course, this must be weighed against the reduced computational load necessary to calculate D. Except perhaps when the computations are being completed by hand, it is advisable to use the original ungrouped data.

11.4. Contingency Tables

A *contingency table* is a cross-classified set of frequency counts for two nominally or ordinally scaled variables, each having two or more classes. In Chapter 5, we introduced the use of contingency tables within a discussion of *bivariate random variables* and *bivariate probability functions*. Let us reconsider the relationship between household size and car ownership used as an example in Section 6.5. The contingency table and bivariate probability function relating these two variables are repeated in Table 11-20(a) and (b). The *test for statistical independence*, or *contingency table test*, is a nonparametric test utilizing the χ^2 distribution to determine whether two variables are statistically independent or a relationship exists between them. Recall that two random variables X and Y are said to be statistically independent if

$$P(x, y) = P(x)P(y) \tag{11-13}$$

holds for all (x, y) pairs. For the variables called car ownership and household size, this equation is violated for at least one (x, y) pair. Note that

$$P(x = 3) = .26 \qquad P(y = 1) = .25 \qquad P(x = 3, y = 1) = .10$$

TABLE 11-20
(a) Contingency Table and (b) Bivariate Probability Function for Household Size and Car Ownership Data

		Car ownership				
		0	1	2	3	Total
Household size	2	10	8	3	2	23
	3	7	10	6	3	26
	4	4	5	12	6	27
	5	1	2	6	15	24
	Total	22	25	27	26	100

(a)

		Car ownership				
		0	1	2	3	Marginal
Household size	2	.10	.08	.03	.02	.23
	3	.07	.10	.06	.03	.26
	4	.04	.05	.12	.06	.27
	5	.01	.02	.06	.15	.24
	Marginal	.22	.25	.27	.26	1.00

(b)

Therefore $P(x = 3, y = 1) \neq P(x = 3)P(y = 1)$ since $.10 \neq (.26)(.25)$. This result presupposes that Table 11-20(b) is the bivariate probability function for a *population*. Suppose we consider the data to be a sample from a *bivariate multinomial population*. It is bivariate since it concerns two different variables. It is multinomial since each nominal variable has more than two possible classes or values. The test for statistical independence of two variables in a bivariate multinomial population uses the χ^2 distribution. It is a simple extention of the goodness-of-fit test described in Section 11.3.

To fix ideas, consider Table 11-21(a). Let π_{ij} denote the population proportion in the ijth cell, and let π_i^r and π_j^c be the marginal probability distributions for the random variables X and Y, respectively. The null hypothesis in this test is that the two variables X and Y are statistically independent, or

$$H_0: \pi_{ij} = \pi_i^r \pi_j^c \qquad i = 1, 2, \ldots, r; j = 1, 2, \ldots, c \qquad (11\text{-}14)$$

where the variable X has r categories or classes and the variable Y has c categories. The alternate hypothesis is

$$H_A: \pi_{ij} \neq \pi_i^r \pi_j^c \qquad \text{for at least one } i, j \text{ pair} \qquad (11\text{-}15)$$

TABLE 11-21
(a) Bivariate Multinomial Population and (b) Sample

Y \ X	1	2	$\cdots$	C	Total
1	π_{11}	π_{12}	$\cdots$	π_{1c}	π_1^r
2	π_{21}	π_{22}	$\cdots$	π_{2c}	π_2^r
. . .					
r	π_{r1}	π_{r2}	$\cdots$	π_{rc}	π^r
Total	π_1^c	π_2^c	$\cdots$	π_c^c	1.00

(a)

Y \ X	1	2	$\cdots$	C	Total
1	p_{11}	p_{12}	$\cdots$	p_{1c}	p_1^r
2	p_{21}	p_{22}	$\cdots$	p_{2c}	p_2^r
. . .					
r	p_{r1}	p_{r2}	$\cdots$	p_{rc}	p_4^r
Total	p_1^c	p_2^c	$\cdots$	p_c^c	1.00

(b)

As usual, this population is unobserved. From a sample, such as the contingency table shown in Table 11-20(a), the *sample* proportions p_{ij} for each cell are determined as well as the row and column marginal probability functions p_i^r and p_j^c. The form of the sample tableau is illustrated in the sample by comparing the value p_{ij} to $p_i^r p_j^c$. Any discrepancies between these values suggests that the two categories may not be independent in the population. But since this is only a sample and so is subject to random or chance fluctuations, the equation may not hold *exactly* for all i, j pairs in the sample. The inferential question boils down to this: How much deviation from Equation (11-14) can there be in the sample before we can say, at a certain level of significance, that the two variables are not statistically independent?

The χ^2 test for contingency tables does not directly utilize the sample proportions given in Table 11-21(b) in the calculation of the test statistic X^2. Instead, the test is based on a comparison of the *observed frequencies* [as in Table 11-20 (a)] and a set of *expected frequencies* which should occur if the null hypothesis is true. The underlying idea behind the test is the same as in the goodness-of-fit test. If H_0 is true, then the observed and expected frequencies will be quite similar; if it is not true, then there will be substantial differences between the sets of frequencies.

The data used in the contingency table test are the raw frequencies in a sample or observed contingency table:

		Variable Y class			
		1	2	$\cdots$	c	Total
	1	f_{11}	f_{12}	$\cdots$	f_{1c}	R_1
	2	f_{21}	f_{22}	$\cdots$	f_{2c}	R_2
Variable X class	$\cdots$					
	r	f_{r1}	f_{r2}	$\cdots$	f_{rc}	R_r
	Total	C_1	C_2	$\cdots$	C_c	n

Here f_{ij} is the observed frequency in cell i, j, R_i is the sum of the frequencies in row i, C_j is the sum of the frequencies in column j, and n is the number of observations. The sum of the row frequencies $\sum_{i=1}^{r} R_i$ equals the sum of the column frequencies $\sum_{j=1}^{c} C_j$ which equals the total number of observations n.

To compute the expected frequencies F_{ij} under H_0, use the relation

$$F_{ij} = \frac{R_i C_j}{n} \tag{11-16}$$

This shortcut formula uses the respective row and column totals for cell i, j and the total number of observations n. For example $F_{11} = R_1 C_1 / n$. For cell 1, 1 of Table 11-20(a), we compute $F_{11} = 23(22)/100 = 5.06$. It is instructive to see how this simple formula relates to the null hypothesis of statistical independence. First, note that

$$p_{ij} = \frac{f_{ij}}{n} \qquad p_i^r = \frac{R_i}{n} \qquad p_j^c = \frac{C_j}{n}$$

Therefore, if the null hypothesis is true and $p_{ij} = p_i^r p_j^c$, then

$$p_{ij} = \frac{R_i}{n} \frac{C_j}{n} \tag{11-17}$$

To compute the *expected frequency* F_{ij}, we multiply the joint probability by the sample size n:

$$F_{ij} = n \frac{R_i}{n} \frac{C_j}{n} = \frac{R_i C_j}{n} \tag{11-18}$$

This is the same as Equation (11-16).

The test statistic is

$$X^2 = \sum_{i=1}^{r} \sum_{j=1}^{c} \frac{(f_{ij} - F_{ij})^2}{F_{ij}} \tag{11-19}$$

which is obtained by summing over *all* classes. The double summation is required since each class is identified by a row and column subscript. When H_0 is true, the sampling distribution of X^2 is approximately χ^2 with $(r - 1)(c - 1)$ degrees of freedom. The degrees

TABLE 11-22
Chi-Square Test for Independence in Contingency Tables

Specification
A total of n observations of two nominal/ordinal variables are cross-classified in a contingency table of dimensions $r \times c$. The observed frequency in the ijth cell is denoted f_{ij}. Expected frequencies for these cells are calculated from

$$F_{ij} = \frac{R_i C_j}{n}$$

Here R_i is the observed frequency count of the ith row, and C_j is the observed frequency count of the jth column.

Hypotheses
H_0: $\pi_{ij} = \pi_i^r \pi_j^c$ $i = 1, 2, \ldots, r; j = 1, 2, \ldots, c$
The variables are statistically independent.
H_A: $\pi_{ij} \neq \pi_i^r \pi_j^c$ for at least one ij pair
The variables are not statistically independent.

Test statistic

$$X^2 = \sum_{i=1}^{r} \sum_{j=1}^{c} \frac{(f_{ij} - F_{ij})^2}{F_{ij}} \qquad \text{with } (r - 1)(c - 1) \text{ degrees of freedom}$$

Decision rule
Reject H_0 if $X^2 > A$, where $A = \chi^2[1 - \alpha, (r - 1)(c - 1)]$.

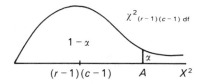

of freedom of the distribution are obtained as follows. There are rc classes. The number of estimated parameters required to compute F_{ij} is $r - 1 + c - 1$. Since both sets of marginal probabilities must sum to 1.00, only $r - 1$ of the row probabilities and $c - 1$ of the column probabilities are needed to compute F_{ij}. The available degrees of freedom are $rc -$ (number of estimated parameters) $- 1$, or $rc - (r - 1 + c - 1) - 1 = (r - 1)(c - 1)$. Just as in the goodness-of-fit test, n must be reasonably large and all $F_{ij} > 2$ and most $F_{ij} \geq 5$. The test is summarized in Table 11-22.

EXAMPLE 11-9.   Let us complete the example of car ownership and household size. First, compute the expected frequencies F_{ij}, by using Equation (11-16). These results, along with the calculation of the test statistic X^2, are given in Table 11-23. Note that all expected frequencies exceed 5, so that the χ^2 approximation is valid for these data. If this condition is violated, then it is necessary to collapse categories so that the approximation can be used. Since $r = 4$ and $c = 4$, there are $(r - 1)(c - 1) = (4 - 1)(4 - 1) = 9$ df for χ^2 when H_0 holds. For

$\alpha = .05$ we require $\chi^2(.95, 9) = 16.92$ in order to reject H_0. Because $X^2 = 37.17$ exceeds 16.92, we conclude H_A—that household size and car ownership are statistically dependent.

EXAMPLE 11-10.   Four candidates A, B, C, and D are running for office in a city divided into four major electoral districts, North (N), South (S), Ease (E), and West (W). A pollster randomly samples 100 voters in the city and asks them to express their first preference among the candidates. The results are as follows:

		Candidate				
		A	B	C	D	Total
	N	5	9	5	10	29
	S	8	6	7	1	22
Electoral district	E	9	11	10	4	25
	W	4	8	3	9	24
	Total	17	34	25	24	100

Do these data suggest that support for the candidates depends on electoral district? The expected frequencies are calculated by using Equation (11-16):

		Candidate			
		A	B	C	D
	N	4.93	9.86	7.25	6.96
	S	3.74	7.48	5.50	5.28
Electoral district	E	4.25	8.50	6.25	6.00
	W	4.08	8.16	6.00	5.76

TABLE 11-23
Calculation of Expected Frequencies and Test Statistic for Example 11-9

Expected frequency			
$\dfrac{23(22)}{100} = 5.06$	$\dfrac{23(25)}{100} = 5.75$	$\dfrac{23(27)}{100} = 6.21$	$\dfrac{23(26)}{100} = 5.98$
$\dfrac{26(22)}{100} = 5.72$	$\dfrac{26(25)}{100} = 6.50$	$\dfrac{26(27)}{100} = 7.02$	$\dfrac{26(26)}{100} = 6.76$
$\dfrac{27(22)}{100} = 5.94$	$\dfrac{27(25)}{100} = 6.75$	$\dfrac{27(27)}{100} = 7.29$	$\dfrac{27(26)}{100} = 7.02$
$\dfrac{24(22)}{100} = 5.28$	$\dfrac{24(25)}{100} = 6.00$	$\dfrac{24(27)}{100} = 6.48$	$\dfrac{24(26)}{100} = 6.24$

$$X^2 = \frac{(10 - 5.06)^2}{5.06} + \frac{(8 - 5.75)^2}{5.75} + \cdots + \frac{(6 - 6.48)^2}{6.48} + \frac{(15 - 6.24)^2}{6.24}$$

$$= 4.82 + 0.88 + \cdots + 0.04 + 12.30 = 37.17$$

Unfortunately, 4/16 or 25 percent, of the cells in the table have expected frequencies less than 5. There are three options. First, one is tempted to ignore the problem. Cell 1, 1 has an expected frequency of almost 5, and 3/16 of the cells represents only 18.75 percent of the total. This is inadvisable and, as we shall see, could lead to a false rejection of H_0. The second option is to collapse the categories so that the requirement of $F_{ij} \geq 5$ is met. It makes little sense to group any of the candidates, although we might be able to group them on the basis of their perceived location along some political spectrum such as left wing–right wing. Or we could group the electoral districts, say into two categories of north–south and east–west. Although this would solve the technical problem, it is a poor alternative since we know geographical aggregation often masks data variation.

The last and probably best alternative is to drop candidate A. It is not a perfect solution, but it solves the technical problem without any arbitrary judgmental grouping. The new observed and expected frequencies are as follows:

	Observed						Expected			
	B	C	D	Total			B	C	D	Total
N	9	5	10	24		N	9.83	7.22	6.94	24
S	6	7	1	14		S	5.73	4.22	4.05	14
E	11	10	4	25		E	10.24	7.53	7.23	25
W	8	3	9	20		W	8.19	6.02	5.78	20
Total	34	25	24	83		Total	34	25	24	83

Only two-twelfths of the cells have expected frequencies less than 5. To complete the test, we compute the test statistic X^2 from (11-19):

$$X^2 = \frac{(9 - 9.83)^2}{9.83} + \frac{(5 - 7.22)^2}{7.22} + \cdots + \frac{(9 - 5.78)^2}{7.78}$$

$$= .07 + .68 + \cdots + 1.79 = 11.93$$

Since $X^2 = 11.93$ does not exceed $\chi^2(.96, 6) = 12.59$, we might conclude H_0—voter preference for candidates B, C, and D is statistically independent of electoral districts. Interestingly, this conclusion is at odds with what would have been concluded had we ignored the minimum frequency requirement for the complete data set with candidates A, B, C, and D.

11.5 Tests for Randomness

Randomness is a key assumption for all the statistical tests we have discussed. Whether this assumption is reasonable for any set of data may be very difficult to decide—particularly when a random sampling procedure has not been used to generate the data. For example, suppose we wish to ascertain whether the sequence of annual rainfall totals over the past 30 years at some location has been generated by a random process or whether the sequence contains a trend. Even if the location was chosen randomly, the data may still not be random since the individual rainfall values are not selected from a random number table or similar device. Nonparametric methods can be used to test for randomness of a sample, even after

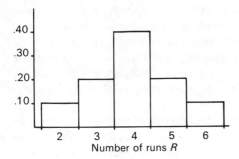

FIGURE 11-8. Probability distribution for number of runs R for $n = 6$, $n_1 = n_2 = 3$.

it has been collected and even if it has not been collected by using a randomization device. As an example of a test for randomness we present the number-of-runs test.

Number-of-Runs Test

The number-of-runs test is based on examining the number of *runs* in a sequence of items. A *run* is defined as an unbroken sequence of like items surrounded by unlike items. Let us use the convention of denoting the two types of items in the sequence by + and −. The sequence of items [+ + + + − − − −] contains two runs. The first run is the first four pluses,

TABLE 11-24
Arrangements of Signs for $n_1 = n_2 = 3$

Sequence	Number of runs	Number of realizations
+ + + − − − − − − + + +	2	2
+ − − − + + + + − − − + − + + + − − − − + + + −	3	4
+ − + + − − − + − − + + − + + − − + + − − + + − − − + + − + + + − − + − − − + − + + + + − + − −	4	8
+ − + − − + + − − + − + − + + − + − − + − + + −	5	4
+ − + − + − − + − + − +	6	2

and the second run is the four minuses. As another example, the sequence [+ − + − − − +−] contains six runs. Each of the first three signs is a run of length 1, the middle three minuses represent a fourth run, and the last two signs are the fifth and sixth runs. The total number of runs in such a sequence is a good indicator of randomness. If there are too few runs, it is possible there is some clustering pattern and the series is probably not random. If, however, there are too many runs, then there may be a repeating or alternating pattern uncharacteristic of random series.

The principle behind the number-of-runs test is best illustrated by using a short series as an example. Suppose the entire series consists of $n = 6$ observations of which $n_1 = 3$ are pluses and $n_2 = 3$ are minuses. Table 11-24 lists the 20 different arrangements of these pluses and minuses. Four runs is the most likely outcome, occurring in 8 of the 20 sequences. Suppose the observed sequence is [+ + + − − −]. This series has 2 runs. Is the number of runs in this sequence indicative of nonrandomness—or can the small number of runs be attributed to chance? The sampling distribution for the number of runs for a series of $n = 6$ observations with $n_1 = n_2 = 3$ is illustrated in Figure 11-8. This sampling distribution is compiled from Table 11-24. Although the number of runs is really a discrete distribution, it has been illustrated as a conventional histogram. From this distribution, we see there is a 10 percent chance (2 out of 20) of two runs occurring in a random sequence under these conditions.

For small values of n_1, n_2, and n it is feasible to construct the sampling distribution of the number of runs in this way. For longer series this is impractical. The sampling distribution of the number of runs R has mean and variance given by, respectively,

$$E(R) = \frac{2n_1 n_2}{n_1 + n_2} + 1 \tag{11-20}$$

and

$$\sigma^2(R) = \frac{2n_1 n_2 (2n_1 n_2 - n_1 - n_2)}{(n_1 + n_2)^2 (n_1 + n_2 - 1)} \tag{11-21}$$

Moreover, this sampling distribution is closely approximated by the normal distribution, provided n_1, $n_2 \geq 10$. The various hypotheses are summarized in Table 11-25. A one-tail test can be used if we assert there is clustering in the pattern (for example, [+ + + + − − − −]) or if there is a repeating pattern ([+ − + − + − + −]). Each of these patterns if found at one extreme of the sampling distribution. A two-tail test can also be used.

As an example, consider the 30-year sequence of annual rainfall at some hypothetical climatic station. The annual totals are listed in Table 11-26. Does the process underlying annual rainfall have a decreasing trend? The first step is to divide the observations into two categories, plus and minus. For an interval-scaled variable, it is easiest to define the two classes with respect to the median rainfall. The median rainfall over the 30-yr period is 58 in. All observations above 58 are given a plus and all those below a minus. What is to be done with the observations with value 58? It turns out that the number of runs in this example is not affected by the sign assigned to these three observations. The runs in the 30-yr sequence are illustrated in Table 11-27. Consider first the observation of year 1961

TABLE 11-25
Number-of-Runs Test

Specification
The observations are taken in the order in which they are generated. Each observation below the median is assigned a minus sign, and each observation above the median is assigned a plus sign. This procedure generates a new series of n signs, n_1 of which are positive and n_2 of which are negative. A run is defined as an unbroken series of plus or minus signs.

Hypotheses
H_0: The series of observations is generated by a random process.
H_A: (a) The series is not generated by a random process (two-tail).
   (b) The series is increasing (decreasing) (one-tail).
   (c) The series has a repeating cyclical pattern (one-tail).

Test statistic
The total number of runs of like items in the observation series is R.

Decision rule
Reject H_0 if

$$R > \left(\frac{2n_1 n_2}{n_1 + n_2} + 1 \right) + z_{\alpha/2} \sqrt{\frac{2n_1 n_2 (2n_1 n_2 - n_1 - n_2)}{(n_1 + n_2)^2 (n_1 + n_2 - 1)}}$$

or

$$R < \left(\frac{2n_1 n_2}{n_1 + n_2} + 1 \right) - z_{\alpha/2} \sqrt{\frac{2n_1 n_2 (2n_1 n_2 - n_1 - n_2)}{(n_1 + n_2)^2 (n_1 + n_2 - 1)}}$$

for a two-tail test. For a one-tail test, reject H_0 if

$$R < \left(\frac{2n_1 n_2}{n_1 + n_2} + 1 \right) - z_{\alpha} \sqrt{\frac{2n_1 n_2 (2n_1 n_2 - n_1 - n_2)}{(n_1 + n_2)^2 (n_1 + n_2 - 1)}}$$

for $H_A(b)$ and reject H_0 if

$$R > \left(\frac{2n_1 n_2}{n_1 + n_2} + 1 \right) + z_{\alpha} \sqrt{\frac{2n_1 n_2 (2n_1 n_2 - n_1 - n_2)}{(n_1 + n_2)^2 (n_1 + n_2 - 1)}}$$

for $H_A(c)$.

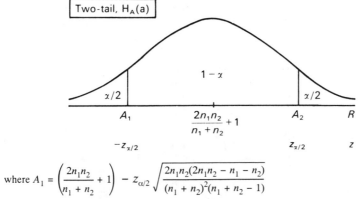

$$\text{where } A_1 = \left(\frac{2n_1 n_2}{n_1 + n_2} + 1 \right) - z_{\alpha/2} \sqrt{\frac{2n_1 n_2 (2n_1 n_2 - n_1 - n_2)}{(n_1 + n_2)^2 (n_1 + n_2 - 1)}}$$

$$A_2 = \left(\frac{2n_1 n_2}{n_1 + n_2} + 1 \right) + z_{\alpha/2} \sqrt{\frac{2n_1 n_2 (2n_1 n_2 - n_1 - n_2)}{(n_1 + n_2)^2 (n_1 + n_2 - 1)}}$$

(continued)

TABLE 10-15 (continued)

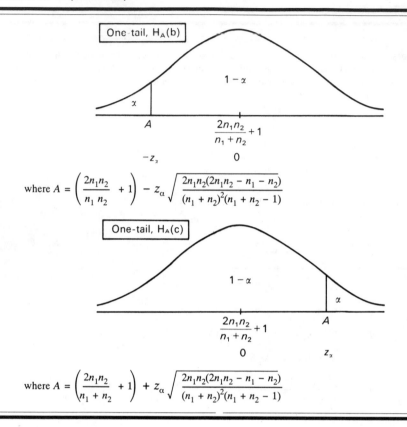

One tail, $H_A(b)$

$$1 - \alpha$$

$$\frac{2n_1 n_2}{n_1 + n_2} + 1$$

$$-z_\alpha \qquad 0$$

where $A = \left(\dfrac{2n_1 n_2}{n_1 \, n_2} + 1 \right) - z_\alpha \sqrt{\dfrac{2n_1 n_2 (2n_1 n_2 - n_1 - n_2)}{(n_1 + n_2)^2 (n_1 + n_2 - 1)}}$

One-tail, $H_A(c)$

$$1 - \alpha$$

$$\frac{2n_1 n_2}{n_1 + n_2} + 1$$

$$0 \qquad z_\alpha$$

where $A = \left(\dfrac{2n_1 n_2}{n_1 + n_2} + 1 \right) + z_\alpha \sqrt{\dfrac{2n_1 n_2 (2n_1 n_2 - n_1 - n_2)}{(n_1 + n_2)^2 (n_1 + n_2 - 1)}}$

TABLE 11-26
Annual Rainfall Levels at a Climatic Station

Year	Annual rainfall, in.	Year	Annual rainfall, in.
1951	66	1966	59
1952	65	1967	58
1953	64	1968	56
1954	65	1969	55
1955	66	1970	53
1956	64	1971	54
1957	63	1972	52
1958	62	1973	51
1959	60	1974	50
1960	59	1975	51
1961	58	1976	49
1962	57	1977	48
1964	58	1978	49
1964	59	1979	47
1965	60	1980	48

TABLE 11-27
Number-of-Runs Test for Climatic Data of Table 11-26

Year	Rainfall, in.	Run		Year	Rainfall, in.	Run	
1951	66	+		1966	59	+	Run 3
1952	65	+		1967	58	0	
1953	64	+		1968	56	−	
1954	65	+		1969	55	−	
1955	66	+	Run 1	1970	53	−	
1956	64	+		1971	54	−	
1957	63	+		1972	52	−	
1958	62	+		1973	51	−	
1959	60	+		1974	50	−	Run 4
1960	59	+		1975	51	−	
1961	58	0		1976	49	−	
1962	57	−	Run 2	1977	48	−	
1963	58	0		1978	49	−	
1964	59	+		1979	47	−	
1965	60	+		1980	48	−	

$$R = 4 \qquad n = 30$$
$$n_1 = 14 \qquad n_2 = 16$$

$$E(R) = \frac{2(14)(16)}{14 + 16} + 1 = 15.93$$

$$\sigma^2(R) = \frac{2(14)(16)[2(14)(16) - 14 - 16]}{(14 + 16)^2(14 + 16 - 1)}$$

$$\sigma(R) = 2.68$$

with a rainfall of 58 in. If we assign it a plus sign, the net effect is to make run 1 longer and run 2 shorter. There is no change in the number of runs. Any "tied" value which is surrounded by two observations of opposite sign is *noncritical*. No matter which sign is applied to the observation, the number of runs remains the same.

Suppose, however, that observation 1954 had a rainfall of 58 in. If it is assigned a plus sign, then run 1 remains as it is and the entire series has 4 runs. If it is assigned a minus sign, then run 1 is broken into three separate runs and the entire series has 6 runs! This may be a significant change. The best method of handling such *critical* ties is to calculate the test statistic R twice: once by assigning the sign most conducive to rejection of H_0 and the second time by assigning the sign least conducive to rejection of H_0. The PROB-VALUES obtained by these two cases must bound the PROB-VALUE for H_0.

The calculations for the number-of-runs test for the rainfall data are given in Table 11-27. The series has $R = 4$ runs with $n_1 = 14$, $n_2 = 16$, and $n = 30$. For this size problem we calculate $E(R) = 15.93$ and $\sigma(R) = 2.68$, using Equations (11-20) and (11-21). To test the hypothesis that rainfall has been decreasing, a one-tail test is applied. A small value of R sustains this hypothesis. For $\alpha = .05$, $z_{.05} = -1.645$. Hence $A = 15.93 + (-1.645)(2.68) =$

11.52. Since $R = 4 < 11.52$, we conclude $H_A(b)$, that rainfall has been decreasing over the 1951–1980 period at this climatic station.

11.6. Summary

When the standard parametric procedures are not appropriate for analyzing a set of data, nonparametric tests can often be used. These tests are most useful when (1) the level of measurement of the variables concerned is nominal or ordinal and (2) the assumptions of the equivalent parametric procedure are appreciably violated. Violating a parametric test's assumptions causes the test to be inexact. For most, but not all, parametric tests there are some conditions for which even a fairly large violation of an assumption causes only a minor disturbance in the sampling distribution of the test statistic. However, for other assumptions, even a small violation may lead to a significant degree of error in test results. Nonparametric tests are at a significant disadvantage compared with parametric tests when a data set satisfies the assumptions of the appropriate statistical test.

Four types of nonparametric procedures have been presented in this chapter. The nonparametric analogs to the one- and two-sample parametric tests discussed in Chapter 9 are the median test, the Mann-Whitney test, and the Wilcoxon signed-ranks test. The Kruskal-Wallis test is the nonparametric equivalent to one form of the analysis of variance, a powerful parametric procedure. Goodness-of-fit tests are used to test the hypothesis that a random variable follows some specified theoretical probability distribution. The χ^2 and Kolmogorov-Smirnov tests are the most widely used goodness-of-fit tests. The χ^2 test can also be extended to the analysis of dependence between two nominal and/or ordinal variables, summarized in a contingency table.

Nonparametric tests are becoming more and more popular in geographical research. They are decidedly easier to apply than parametric tests. Moreover, these methods are applicable to the sorts of variables collected in questionnaire-based surveys. Such surveys are increasingly common in geographical research. Unfortunately, the simplicity of application of these techniques has made them available to unsophisticated users who sometimes apply them inappropriately or in an uncomprehending, cookbook way. Understanding when to use these methods also requires an understanding of when *not* to use them.

It has not been possible to describe all the nonparametric tests used by geographers in this brief overview of nonparametric methods. Other tests are described in the references cited at the end of this chapter under Further Reading.

FURTHER READING

Extensions and refinements to all the tests described in this chapter can be found in most textbooks focusing on nonparametric statistics. Examples of situations in which nonparametric statistics can be applied are described in the three textbooks below as well as the problems at the end of this chapter.

W. J. Conover, *Practical Nonparametric Statistics*, 2d ed. (New York: Wiley, 1980).
W. W. Daniel, *Applied Nonparametric Statistics* (Boston: Houghton Mifflin, 1978).
S. Siegel, *Nonparametric Statistics for the Behavioral Sciences* (New York: McGraw-Hill, 1956).

PROBLEMS

1. Explain the meaning of (a) nonparametric statistics, (b) robustness, (c) goodness-of-fit test, and (d) contingency table.

2. Briefly compare the advantages and disadvantages of parametric and nonparametric tests.

3. An urban geographer randomly samples 20 new residents of a neighborhood to determine their ratings of local shopping facilities. The measurement scale is as follows: strongly dislike, 0; dislike, 1; neutral, 2; like, 3; and strongly like, 4. The 20 responses are 0, 4, 3, 2, 2, 1, 1, 2, 1, 0, 0, 1, 2, 1, 3, 4, 2, 0, 4, 1. Use the median test to see whether the population median is 2.

4. A course in statistical methods for geographers is team-taught by two instructors, Professor Jovita Fontanez and Professor Clarence Old. In the student evaluations for the course, students were asked to indicate their preference for the two professors. Of the 13 students in the class, 8 preferred Professor Fontanez and 2 preferred Professor Old. The remaining students were unable to express a preference for either instructor. Test the hypothesis that students prefer Professor Fontanez. (*Hint*: Use the median test.)

5. Consider the data given in Table 11-11. Use two-sample Mann-Whitney tests to test whether each pair of intersection types has different carbon monoxide levels.

6. Consider the data of Problem 4 in Chapter 9. Using the Mann-Whitney test at $\alpha = .05$, test the null hypothesis that the two samples come from the same population, that is, fertilizer XLC does not increase corn yields.

7. Solid-waste generation rates measured in metric tons per household per year are collected in randomly selected areas of a city. These areas are classified as high-density, low-density, or sparsely settled. It is thought that generation rates are probably higher in the areas of highest residential density because they are likely to be served by more efficient waste collection systems and there is very little opportunity for on-site disposal or storage, particularly in comparison to sparsely settled areas. Do the following data support this hypothesis?

High density	Low density	Sparsely settled
1.84	2.04	1.07
3.06	2.28	2.31
3.62	4.01	0.91
4.91	1.86	3.28
3.49	1.42	1.31

8. Use the Wilcoxon signed-ranks test to verify the conclusions of the matched-pairs test for Problem 5 of Chapter 9.

9. The distances traveled by a random sample of 12 people to their places of work in 1970 and again in 1980 are shown in the following table:

	Distance traveled, km			Distance traveled, km	
Person	1970	1980	Person	1970	1980
1	8.6	8.8	7	7.7	6.5
2	7.7	7.1	8	9.1	9.0
3	7.7	7.6	9	8.0	7.1
4	6.8	6.4	10	8.1	8.8
5	9.6	9.1	11	8.7	7.2
6	7.2	7.2	12	7.3	6.4

Has the length of the journey to work changed over the decade?

10. Test the normality of the DO data of Table 2-1, (a) using the χ^2 test with $k = 10$ classes, (b) using the χ^2 test with $k = 6$ classes defined in Table 2-3, and (c) using the Kolmogorov-Smirnov test for the data in the form of both (a) and (b).

11. One hundred randomly sampled residents of a city subject to periodic flooding are classified according to whether they are on the floodplain of the major river bisecting the city or off the floodplain. These households are then surveyed to determine whether they currently had flood insurance of any kind. The results are shown in the following table:

	On the floodplain	Off the floodplain
Purchased insurance	50	10
Did not purchase insurance	15	25

Test a relevant hypothesis.

12. In a study researching the effects of sociocultural environments on children's evaluation of other family members, 40 subjects in Nairobi, Kenya, Oslo, Norway, and Seattle, Washington, were instructed to draw a picture of a family. The following table classifies the leading figure in the drawings made by each group of children:

	Parent	Self	Relatives
Nairobi, Kenya	5	10	25
Oslo, Norway	15	10	15
Seattle, Washington	10	25	5

Test the hypothesis that these three problems are homogeneous with respect to the relationship of the leading figure in the drawings.

13. The actual daily occurrence of sunshine in a given city over a 30-day period is calculated as the percentage of possible time the sun could have shone, had it not been for the existence of cloudy skies. If we define a sunny day as one with over 50 percent sunshine, determine whether the pattern of occurrence of sunny days is random for the following data:

Day	Percentage of sunshine	Day	Percentage of sunshine	Day	Percentage of sunshine
1	75	11	21	21	77
2	95	12	96	22	100
3	89	13	90	23	90
4	80	14	10	24	98
5	7	15	100	25	60
6	84	16	90	26	90
7	90	17	56	27	100
8	18	18	0	28	90
9	90	19	22	29	58
10	100	20	44	30	0

14. In a certain city, the mortality rate due to household fire per 100,000 inhabitants for the last 15 years is as follows: 3.65, 3.41, 3.53, 3.23, 3.50, 3.43, 3.73, 3.40, 3.70, 3.58, 3.80, 3.40, 3.38, 3.36, 3.21. Is there a trend in these rates?

III

STATISTICAL RELATIONSHIPS BETWEEN TWO VARIABLES

12

Correlation Analysis

Quite often in statistical analysis we are interested in studying the relationship *between* variables. For example, a geomorphologist might wish to explore the relationship between the annual discharge of a stream and its width downstream from the source. Also urban geographers are sometimes interested in the association between the selling price of a residential property and other structural and locational characteristics such as house size and accessibility. *Multivariate statistical techniques* are methods which can be applied in situations where the analysis involves two or more different variables. The simplest class of multivariate statistical techniques is *bivariate*, involving only two variables at a time. An investigation of the *strength* of the association between two variables is called *simple (or bivariate) correlation analysis*, and a study of the *nature* of the relationship between two variables is termed *simple regression analysis*. In simple regression analysis, we are concerned with *estimating* the value of one variable by considering its relationship with one other variable. But, as we shall see in Chapter 12, regression and correlation are intimately related. This chapter is limited to simple linear correlation. Our purpose is to illustrate the use and interpretation of various quantitative indices that express the relative strength of the *linear* relationship between two variables. Clearly this is not the only way in which two variables might be related. For example, we might find that two variables are not linearly related at all, but are perfectly or almost perfectly related in some nonlinear way. This issue is explored further in Section 12.1.

The most commonly used measure of correlation for interval or ratio variables is known as *Pearson's product-moment correlation coefficient*, or simply as *Pearson's r*. The calculation and interpretation of Pearson's r are discussed in Section 12.1. In fact, this measure of the relationship between two variables is one of the parameters of the bivariate normal distribution discussed in Section 6.4. Before beginning Section 12.1, you may find it useful to review the material in Section 6.4, particularly the presentation of the bivariate normal distribution.

It is also possible to derive indices of correlation for variables measured at the nominal or ordinal scales. These measures are discussed in Section 12.3. The last two sections of this chapter describe particular problems and applications of correlation analysis in map association studies in geography.

12.1. Pearson's Product-Moment Correlation Coefficient

In Chapter 6, the bivariate normal distribution is introduced along with the concepts of the *covariance* between two random variables $C(X, Y)$ and the correlation ρ_{xy}. All the examples utilized in Chapter 6 describe situations in which the parameters of the population distribution are known. In most practical situations, the parameters of the bivariate normal distribution, and therefore the correlation coefficient, are unknown. As is usual in statistical analysis, the correlation coefficient is then estimated from sample data. The best point estimate of rho is Pearson's r, known also, for reasons outlined below, as the product-moment correlation coefficient.

Scatter Diagrams and the Calculation of r

It is possible to calculate a correlation coefficient for any set of n paired values of two variables X and Y. Label these observations (X_i, Y_i), $i = 1, 2, \ldots, n$. Since these observations may or may not be random variables, we drop the convention of distinguishing values of the variable by a lowercase designation. Of course, if we are interested in testing the statistical significance of a correlation coefficient, it is necessary to make the additional assumption that the n observations are a true random sample for some statistical population.

First, the observations are graphed in a *scatter plot* (or *scattergram* or *scatter diagram*). Each observation or pair (X_i, Y_i) represents one dot in a scatter plot. Examples of scatter plots with corresponding values of r are illustrated in Figure 12-1. Where all points

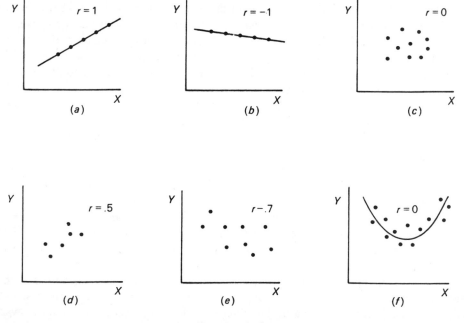

FIGURE 12-1. Scatter plots.

in the scatter plot lie on a positively sloped line, $r = 1$, as in Figure 12-1(a). If the line has a negative slope, then $r = -1$, as in (b). When $r = 0$, the scatter of points is near circular, as in (c). The general elliptical shape to the scatter plots occurs for intermediate values of r, as in (d) and (e). Note especially Figure 12-1(f). This is an example of a scatter of points with a strong *nonlinear* association, but $r = 0$. This is because Pearson's r measures only the strength of the *linear* association between variables.

The sample correlation coefficient r is obtained by substituting the appropriate point estimators for $C(X, Y)$, or σ_X and σ_Y, into Equation (6-40). The best point estimate for $C(X, Y)$ is

$$S_{XY} = \frac{1}{n-1} \sum_{i=1}^{n} (X_i - \bar{X})(Y_i - \bar{Y}) \tag{12-1}$$

The sample covariance S_{XY} measures the tendency for the two variables X and Y to covary. Consider, for example, the scatter plot of Figure 12-2. The plot suggests that the two variables X and Y are positively related. We would expect S_{XY} to be positive. The calculation of S_{XY} can be given a simple geometric explanation by using Figure 12-2. The two means $\bar{X}$ and $\bar{Y}$ divide the scatter plot into four quadrants labeled I, II, III, and IV. Let us examine the sign and magnitude of $(X_i - \bar{X})(Y_i - \bar{Y})$ in each of these four quadrants. If both X_i and Y_i are *greater* than their respective means, then the observation is in quadrant I. The product of the deviations from the mean is positive, since both deviations are positive. This holds true for all observations in quadrant I. In quadrant III, both deviations are negative since the values of X_i and Y_i are below their means. However, the product of the two deviations $(X_i - \bar{X})(Y_i - \bar{Y})$ is positive since it is the product of two negative values.

In quadrants II and IV, the deviations are of opposite signs, and therefore the product of the deviations must be negative. The outcome for S_{XY} is based on the sum of these products for all observations. Where there is a positive relation between the two variables, as in Figure 12-2, $S_{XY} > 0$. Why? Because there are many points in quadrants I and III with large positive values for $(X_i - \bar{X})(Y_i - \bar{Y})$, but only a few points in quadrants II and IV where the product is negative. If, however, the scatter suggested a negative relation between X and

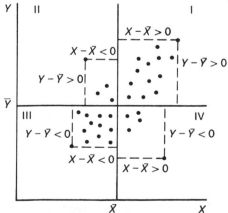

FIGURE 12-2. Geometric illustration of correlation coefficient derivation.

Y, there would be many points in quadrants II and IV and few in quadrants I and III. In this case, the covariance turns out to be negative. If the scatter reveals a virtual equal scatter in all quadrants, the covariance S_{XY} is near zero.

The sample covariance S_{XY} has the same disadvantage as $C(X, Y)$ for measuring the strength of the relation between the two variables. It is highly influenced by the units in which the two variables are measured. Just as $C(X, Y)$ is standardized by σ_X and σ_Y in the correlation coefficient ρ_{XY}, the sample covariance is standardized by S_X and S_Y, so that

$$r = \frac{S_{XY}}{S_X S_Y} = \frac{[1/(n-1)] \sum_{i=1}^{n}(X_i - \bar{X})(Y_i - \bar{Y})}{\sqrt{[1/(n-1)] \sum_{i=1}^{n}(X_i - \bar{X})^2} \sqrt{[1/(n-1)] \sum_{i=1}^{n}(X_i - \bar{X})^2}} \qquad (12\text{-}2)$$

Equation (12-2) can be simplified by canceling the term $1/(n-1)$ which appears in the numerator and the denominator. This yields

$$r = \frac{\sum_{i=1}^{n}(X_i - \bar{X})(Y_i - \bar{Y})}{\sqrt{\sum_{i=1}^{n}(X_i - \bar{X})^2} \sqrt{\sum_{i=1}^{n}(Y_i - \bar{Y})^2}} \qquad (12\text{-}3)$$

For computing purposes, especially when the calculations are performed by hand or a nonprogrammable hand calculator is used, it is helpful to rearrange (12-3) to yield

$$r = \frac{\sum_{i=1}^{n} X_i Y_i - (\sum_{i=1}^{n} X_i)(\sum_{i=1}^{n} Y_i)/n}{\sqrt{\sum_{i=1}^{n} X_i^2 - (\sum_{i=1}^{n} X_i)^2/n} \sqrt{\sum Y_i^2 - (\sum_{i=1}^{n} Y_i)^2/n}} \qquad (12\text{-}4)$$

This is the most convenient formula for calculating Pearson's r.

To illustrate the calculation of the correlation coefficient by using these alternative formulas, consider the following three examples:

EXAMPLE 12-1.   The four points $(0, 3)$, $(1, 5)$, $(2, 7)$, and $(3, 9)$ fall on the line $Y = 2X + 3$. As is evident in Figure 12-3, there is a perfectly linear relationship between X and Y

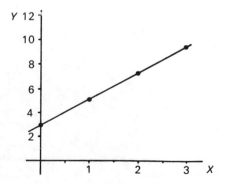

FIGURE 12-3. Graph for Example 12-1.

TABLE 12-1
Calculation of Pearson's *r* for Data of Figure 12-3

Point (X_i, Y_i)	$X_i - \bar{X}$	$(X_i - \bar{X})^2$	$Y_i - \bar{Y}$	$(Y_i - \bar{Y})^2$	$(X_i - \bar{X})(Y_i - \bar{Y})$
(0, 3)	−1.5	2.25	−3	9	4.50
(1, 5)	−0.5	0.25	−1	1	0.50
(2, 7)	0.5	0.25	1	1	0.50
(3, 9)	1.5	2.25	3	9	4.50
	0	5.00	0	20	10.00

$$r = \frac{S_{XY}}{S_X S_Y} = \frac{10/3}{\sqrt{20/3}\,\sqrt{5/3}} = \frac{10}{\sqrt{100}} = 1$$

for these four points. The sample correlation coefficient should be $r = 1$. First, calculate $\bar{X}$ = $(0 + 1 + 2 + 3)/4 = 1.5$ and $\bar{Y} = (3 + 5 + 7 + 9)/4 = 6$. The necessary calculations for Pearson's *r* obtained by using the covariance formulation in (12-3), are summarized in Table 12-1. As expected, $r = 1$.

EXAMPLE 12-2.   The six points in Figure 12-4 do not fall on a straight line. It appears that the two variables are not linearly associated. For these data let us utilize Equation (12-4). It is useful to organize the calculations for the correlation coefficient by using the tabular format of Table 12-2. As the scatter diagram in Figure 12-4 suggests, these two variables have no linear association whatever and $r = 0$.

TABLE 12-2
Tabular Format for Calculation of Pearson's *r* from Data of Figure 12-4

X_i	Y_i	X_i^2	Y_i^2	$X_i Y_i$
0	2	0	4	0
0	4	0	16	0
2	2	4	4	4
2	4	4	16	8
4	2	16	4	8
4	4	16	16	16
$\sum_i X_i = 12$	$\sum_i Y_i = 18$	$\sum_i X_i^2 = 40$	$\sum_i Y_1^2 = 60$	$\sum_i X_i Y_i = 36$

$$r = \frac{36 - (12)(18)/6}{\sqrt{40 - 12^2/6}\,\sqrt{60 - 18^2/6}} = \frac{0}{\sqrt{16}\,\sqrt{6}} = 0$$

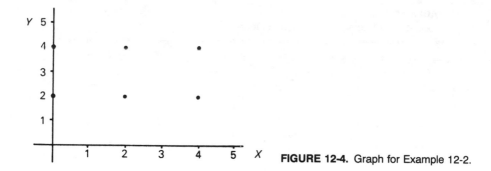

FIGURE 12-4. Graph for Example 12-2.

EXAMPLE 12-3.   In practice, most correlation coefficients lie between the two extremes of perfect correlation ($r = \pm 1$) and no linear correlation ($r = 0$). Consider the following data consisting of the selling price of a sample of 10 homes:

Selling price Y, $000	Size X, 000 ft^2
45	1.0
60	1.3
55	1.6
70	1.8
75	1.9
80	2.0
100	2.2
90	2.4
105	2.8
120	3.0

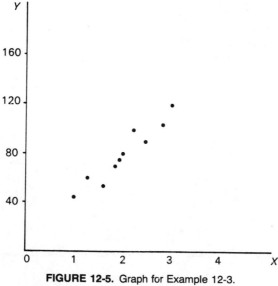

FIGURE 12-5. Graph for Example 12-3.

TABLE 12-3
Calculation of Pearson's r for Example 12-3

X_i	Y_i	X_i^2	Y_i^2	X_iY_i
45	1.0	2,025	1.00	45
60	1.3	3,600	1.69	78
55	1.6	3,025	2.56	88
70	1.8	4,900	3.24	126
75	1.9	5,625	3.61	142.5
80	2.0	6,400	4.00	160
100	2.2	10,000	4.84	220
90	2.4	8,100	5.76	216
105	2.8	11,025	7.84	294
120	3.0	14,400	9.00	360
800	20.0	69,100	43.54	1729.5

$$r = \frac{1729.5 - (800)(20.0)/10}{\sqrt{69,100 - 800^2/10} - \sqrt{43.54 - 20.0^2/10}} = \frac{129.5}{134.4} = .96$$

When illustrated in the scatter diagram of Figure 12-5, the data reveal a strong positive correlation. The necessary calculations are summarized in Table 12-3 and confirm that the selling price of a house is strongly positively associated ($r = .96$) with the square footage of the house. Put simply, larger houses tend to be more expensive.

Frequently in geographical research the data collected in a single study include observations for many different variables. In a study of house price variations in an urban area, for example, let us suppose that the following six variables are available from a sample of $n = 100$ homes:

1. Selling price, thousands of dollars (Price)
2. Size of house, thousands of square feet (Size)
3. Number of bedrooms (Bed)
4. Lot frontage, feet (Lot)
5. Age, years since constructed (Age)
6. Number of bathrooms (Bath)

Correlation coefficients can be calculated between each pair of variables in this set. These correlation coefficients are frequently summarized in a *correlation matrix* such as Table 12-4. There are several important features of such matrices. First, the correlation between a variable and itself is always 1, so that the *diagonal* elements of the matrix always contain the value $r = 1.0$. To see why this is so, examine Equation (12-4). If we substitute X for Y in this equation, the numerator and the denominator are identical; thus the ratio must be 1.0. Second, correlation matrices are symmetric. The correlation between two variables X and Y necessarily equals the correlation between variables Y and X. Note in Equation (12-4) that the identical numerical result is obtained if variable X is substituted for variable Y and vice versa.

What does Table 12-4 reveal? First, note the strong positive correlations between

TABLE 12-4
Correlation Matrix for Six Housing Variables

	Price	Size	Bed	Lot	Age	Bath
Price	1.0	.96	.83	.91	−.53	.92
Size	.96	1.0	.98	.80	.23	.91
Bed	.83	.98	1.0	.62	−.71	.94
Lot	.91	.80	.62	1.0	.07	.21
Age	−.53	.23	−.71	.07	1.0	−.77
Bath	.92	.91	.94	.21	−.77	1.0

Price and the variables Size, Bed, Lot, and Bath. For this sample of 100 homes, this indicates that higher-price homes tend to be larger and have more bedrooms, wider lot frontages, and more bathrooms. Also, these variables tend to covary. For example, larger homes generally have more bedrooms, more bathrooms, and wider lots. There are also a few strong negative correlations. Older homes are generally *cheaper* and have *fewer* bedrooms and bathrooms. Finally, there are several weak correlations. There seems to be little relation between Size and Age ($r = .23$), Age and Front ($r = .07$), and Front and Bed ($r = .21$). Besides using the correlation coefficient to describe the *degree* of association between two variables, it is also possible to test the statistical significance of any sample correlation coefficient.

Significance Testing of a Correlation Coefficient

The sample correlation coefficient r calculated by using (12-2), (12-3), or (12-4) is the best point estimator of ρ, the population correlation coefficient. Figure 12-6 illustrates the sampling procedure which underlies the significance test of r. The statistical population consists of the two random variables X and Y known to be bivariate normally distributed. The population correlation coefficient is given by

$$\rho = \frac{C(X, Y)}{\sigma_X \sigma_Y} \tag{12-5}$$

If the population cannot be observed, then ρ is unknown and the null hypothesis to be tested is H_0: $\rho = 0$. From the population, a random sample of n pairs (X, Y) is drawn, and the sample correlation coefficient r is calculated. The sampling distribution of r, under the null hypothesis that $\rho = 0$, (1) is t-distributed with $n - 2$ degrees of freedom and (2) has an *estimated* standard error of

$$s_r = \sqrt{\frac{1 - r^2}{n - 2}}$$

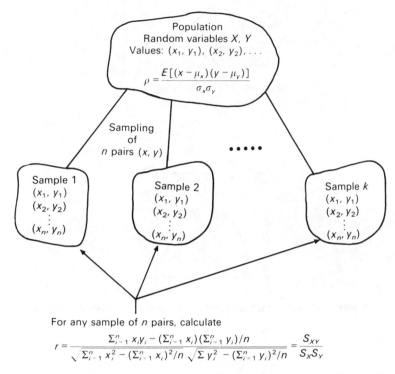

For any sample of n pairs, calculate

$$r = \frac{\sum_{i=1}^{n} x_i y_i - (\sum_{i=1}^{n} x_i)(\sum_{i=1}^{n} y_i)/n}{\sqrt{\sum_{i=1}^{n} x_i^2 - (\sum_{i=1}^{n} x_i)^2/n} \sqrt{\sum y_i^2 - (\sum_{i=1}^{n} y_i)^2/n}} = \frac{S_{XY}}{S_X S_Y}$$

FIGURE 12-6. Sampling procedure underlying significance test of correlation coefficient.

Therefore, the required test statistic is

$$t = \frac{r}{S_r} = \frac{r}{\sqrt{(1 - r^2)/(n - 2)}} = \frac{r\sqrt{n - 2}}{\sqrt{1 - r^2}} \qquad (12\text{-}6)$$

For example, consider the value of $r = .96$ obtained for the variables Price and Size for $n = 10$ houses listed in Example 12-3. Using a two-tail test at $\alpha = .05$ yields

$$t = \frac{.96\sqrt{8}}{\sqrt{1 - .96^2}} = 9.70$$

Since $t > t_{.025,\ 8\ df}$, we must reject H_0: $\rho = 0$.

To test a hypothesis where $\rho \neq 0$ or to construct a confidence interval for ρ, the procedure becomes more complicated. This is because the sampling distribution of r is

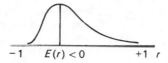

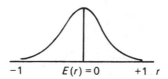

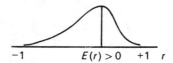

FIGURE 12-7. Sampling distributions of the sample correlation coefficient r.

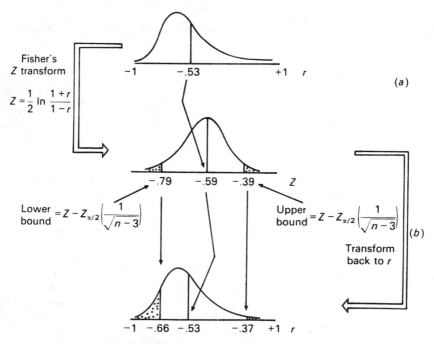

FIGURE 12-8. Confidence interval estimate of correlation coefficient ρ.

symmetric only for the case in which the sampling distribution is centered at 0. When $E(r)$ ≠ 0 (that is, we cannot assume that ρ = 0), the sampling distribution of r is skewed. It is increasingly skewed as $E(r)$ approaches ± 1. This situation is depicted in Figure 12-7. Fortunately, it is possible to use Fisher's Z transformation as an approximate method for constructing confidence intervals in this case.

The transformation $Z = \frac{1}{2} \ln [(1 + r)/(1 - r)]$ is approximately normally distributed with mean $\mu_Z = \frac{1}{2} \ln [(1 + \rho)/(1 - \rho)]$ and standard error $\sigma_Z = \sqrt{1/(n - 3)}$ for $n > 50$. With this transformation of the correlation coefficient it is possible to derive confidence intervals in the usual way. The method is best illustrated with an example.

EXAMPLE 12-4. Consider the correlation of $r = -.53$ between house selling price (Price) and the number of years since the house was constructed (Age) for $n = 100$ homes. In Figure 12-8(a) the sampling distribution of r for the case $r = -.53$ illustrates the problem in deriving a confidence interval. The transformation $Z = \frac{1}{2} \ln (.47/1.53) = -.59$ leads to the approximately normal distribution in Figure 12-8(b). From this distribution, confidence limits can be constructed in the usual way. For α = .05, or 95 percent, confidence

$$\text{Upper bound} = -.59 + 1.96\left(\frac{1}{\sqrt{97}}\right) = -.39$$

$$\text{Lower bound} = -.59 - 1.96\left(\frac{1}{\sqrt{97}}\right) = -.79$$

As usual, these bounds are symmetric around −.59. To express these confidence limits in terms of r, it is necessary to convert these values from Z to r by using the normal table in the Appendix. From this table, we see the limits on ρ are $-0.66 \le \rho \le -0.37$. The process is illustrated in Figure 12-8. Because of the asymmetry of the interval, the lower bound is $.66 - .53 = .13$ below r, and the upper bound is $.53 - .37 = .16$ unit above $r = .53$. The asymmetric nature of the interval becomes increasingly pronounced as it approaches ± 1.

12.2. Nonparametric Correlation Coefficients

Pearson's product-moment correlation coefficient is the most widely used measure of association for interval-scaled variables. Nonparametric measures of association can also be applied to pairs of variables measured at the ordinal or nominal scale. Nonparametric measures are useful in cases in which the available data are limited to these scales, or in cases in which interval variables do not meet the exacting requirements sometimes necessary for parametric measures such as Pearson's r. For example, any statistical inference concerning Pearson's r requires the assumption that the population joint probability distribution of X and Y is bivariate normal. Just as nonparametric statistical tests sometimes prove useful in univariate inferential statistics, so, too, do nonparametric measures of association in bivariate situations. Nonparametric measures can be conveniently divided into two classes: *rank correlation coefficients* for ordinal data and *nominal scale measures.*

Rank Correlation Coefficients

Ordinal measures of association, or rank correlation coefficients, are appropriate whenever the relationship between two variables X and Y is *monotonic increasing* or *decreasing*. A monotonic increasing function for Y is one that either remains constant or increases as X increases. Analogously, a monotonic decreasing function for Y is one that either remains constant or decreases as X increases. We frequently encounter such functions in many propositions in geography and the social sciences. For example, it is common to say, "The greater the X, the greater the Y" or "The greater the X, the less the Y". Both statements imply monotonocity in the relationship between X and Y. Whereas Pearson's r is a measure of the *linear* association between two variables, rank correlation coefficients cannot distinguish linear from nonlinear monotonic functions. This is entirely due to the nature of ordinal data. Recall that the ordinal level of measurement requires only that we are able to *order* the observations of a variable, and the notion of *distance* between observations is inappropriate. Although we may be able to say X_1 is greater than X_2, and X_2 is greater than X_3, we are unable to compare the differences $X_1 - X_2$ and $X_2 - X_3$. Therefore, we have no way of knowing whether the relationship between X and Y is linear, quadratic, or of some other monotonic form.

   Because it is a parametric measure of correlation based on continuous interval data, Pearson's r is the most powerful measure of association presented in this text. Rank correlation measures should be used only where (1) the level of measurement of the two variables X and Y is ordinal or (2) the assumptions required for a test of statistical significance for Pearson's r cannot be met (that is, the data are not bivariate normal). At the same time, Pearson's r is a robust measure of correlation, and the statistical test of significance is valid under modest departures from the assumption of bivariate normality.

   Two of the more commonly used measures of ordinal association are *Spearman's* ρ *(rho)* and Kendall's τ *(tau)*. They share three properties with Pearson's r:

1. Both are limited to the range $[-1, +1]$.
2. Both coefficients are positive (negative) when an increase (a decrease) in X corresponds to an increase (a decrease) in Y.
3. A value near zero indicates that the values of X are uncorrelated with the values of Y.

   Consider n observations of two variables X and Y measured on at least the ordinal scale. Assume that there are no tied values among either the values of X or those of Y. Each of the X values is replaced by its rank, the lowest rank 1 is given to the smallest value of X, and rank n is given to the highest value of X. Each of the Y values is ranked in the same way. Now, arrange the X values in ascending order with the rank of the Y values in the same sequence. Table 12-5 illustrates this procedure for the two variables Size (house size in thousands of square feet) and Price (selling price in thousands of dollars) examined in Section 12-2. The values of X are listed in column 2 and the Y values in column 3. Note that the X variable has been placed in ascending order. Let $r(X_i)$ be the rank of the ith observation X_i and $r(Y_i)$ be the rank of the ith observation Y_i. For example, observation $i = 1$ is the lowest X and lowest Y, so that $r(X_1) = r(Y_1) = 1$. Observation $i = 2$ is the second lowest X and third lowest Y, so that $r(X_2) = 2$ and $r(Y_2) = 3$. If the two variables are perfectly positively correlated, then the rank orders of the two variables will be identical. If there is

TABLE 12-5
Calculation of r_S

i	X_i	Y_i	$r(X_i)$	$r(Y_i)$	$[r(X_i) - r(Y_i)]^2$
1	1.0	45	1	1	0
2	1.3	60	2	3	1
3	1.6	55	3	2	1
4	1.8	70	4	4	0
5	1.9	75	5	5	0
6	2.0	80	6	6	0
7	2.2	100	7	8	1
8	2.4	90	8	7	1
9	2.8	105	9	9	1
10	3.0	120	10	10	0
					4

$$r_S = 1 - 6 \sum_i \frac{[r(X_i) - r(Y_i)]^2}{n(n^2 - 1)}$$

$$= 1 - 6 \frac{(1-1)^2 + (2-3)^2 + (3-2)^2 + (4-4)^2 + (5-5)^2 + (6-6)^2 + (7-8)^2 + (8-7)^2 + (9-9)^2 + (10-10)^2}{10(10^2 - 1)}$$

$$= 1 - 6(4/990) = .976$$

no correlation, then the distribution of Y ranks is random with respect to the ranks of the paired X values. If there is a perfect negative correlation, the ranks of the Y values will decrease perfectly with the increase in the ranks of X. Rank correlation coefficients assess the degree of association between the ranks. In Table 12-5 the ranks of the X's and Y's are almost perfectly correlated. Six of the ten rankings are identical, and the other four differ only by one ranking. Both Spearman's ρ and Kendall's τ are measures which tell us the precise degree of association between these two sets of ranks.

Spearman's Rank Correlation Coefficient

The idea behind Spearman's ρ is quite simple. The two sets of rankings are compared by taking the differences between the ranks, squaring these differences, summing the squares over all n rankings, and then standardizing the result so that the bounds are $[-1, +1]$. An estimate of Spearman's ρ, denoted r_S, can be computed from

$$r_S = 1 - 6 \sum_{i=1}^{n} \frac{[r(X_i) - r(Y_i)]^2}{n(n^2 - 1)} = 1 - 6 \sum_{i=1}^{n} \frac{d_i^2}{n(n^2 - 1)} \quad (12\text{-}7)$$

where $d_i = r(X_i) - r(Y_i)$. This formula can be derived directly from Pearson's r simply by substituting $r(X_i)$ for X_i and $r(Y_i)$ for Y_i. Whenever the ranks are identical, $d_i = 0$ for all i and $r_S = +1$. Whenever the ranks are in perfect reverse order, the second part of (12-7) simplifies to −2 and $r = 1 − 2 = −1$. Suppose the ranks for X are 1, 2, 3 and the corresponding ranks for Y are 3, 2, 1. Using (12-7) to calculate r_S, we find $r_S = 1 − 6(4 + 0 + 4)/[3(8)] = 1 − 48/24 = 1 − 2 = −1$. The calculation of r_S for Price and Size is summarized in Table 12-5. As is apparent from the two rankings, Price and Size are indeed highly positively correlated with $r_S = .976$. Note that this result is almost identical to the value of $r = .96$ obtained for Pearson's r. If we were to take the ranks and substitute them directly into (12-4), the result for r_S, would also be .976. It is for this reason that Spearman's r_S can be interpreted in the same way as the product-moment correlation coefficient.

If there are tied values for any of the ranks, then it is still possible to calculate Spearman's r_S by using the convention for ties suggested in Chapter 10. Each tied value is assigned the average of the ranks that would have been assigned had no ties existed. If the number of ties is small, then Equation (12-7) can be used to compute r_S. Whenever there are a substantial number of tied values, a correction factor should be used. Details are available in most advanced textbooks on nonparametric texts such as Siegel (1956). As a rule of thumb, the correction factor can be ignored when the number of pairs of ties is less than one-fourth of the sample.

If the n pairs of observations represent a random sample from a bivariate population, then it is possible to test the statistical significance of r_S. Since it is a nonparametric measure, Spearman's coefficient makes no assumption whatsoever about the nature of the underlying population distribution. For $n < 10$ the significance of r_S can be determined directly from tables. When $n \geq 10$, it is usual to test Spearman's r_S by using a standard normal deviate. In this case the sampling distribution of r_S is approximately normal with a standard error of $1/\sqrt{n − 1}$. Under the null hypothesis that there is no relationship between the two variables X and Y in the population, the test statistic

$$Z = \frac{r_S − 0}{1/\sqrt{n − 1}} = r_S \sqrt{n − 1} \tag{12-8}$$

can be used to test the significance of r_S. For Price and Size we have $r_S = .976$ and $n = 10$, so that $Z = .976\sqrt{9} = 2.93$. This is significant at $\alpha = .05$ since it exceeds $Z = 1.96$. Confidence limits on ρ_S can be computed from

$$r_S − Z_{\alpha/2}\left(\frac{1}{\sqrt{n − 1}}\right) \leq \rho_S \leq r_S + Z_{\alpha/2}\left(\frac{1}{\sqrt{n − 1}}\right) \tag{12-9}$$

The 95 percent limits on ρ_S are $.976 − 1.96(1/\sqrt{10 − 1}) \leq \rho_S \leq .976 + 1.96(1/\sqrt{10 − 1})$ which leads to $.323 \leq \rho_S \leq 1.629$ (or 1.0 since this is the upper bound on Spearman's rank correlation coefficient).

Kendall's τ

Spearman's r_S is the most widely used measure of ordinal correlation. It can be derived directly from Pearson's r and is extremely easy to calculate. Kendall's τ is an alternative

TABLE 12-6
Calculation of Kendall's τ

i	X_i	Y_i	$r(X_i)$	$r(Y_i)$	N_C	N_D
1	1.0	45	1	1	9	0
2	1.3	60	2	3	7	1
3	1.6	55	3	2	7	0
4	1.8	70	4	4	6	0
5	1.9	75	5	5	5	0
6	2.0	80	6	6	4	0
7	2.2	100	7	8	2	1
8	2.4	90	8	7	2	0
9	2.8	105	9	9	1	0
10	3.0	120	10	10	0	0
					43	2

$$\tau = \frac{43 - 2}{10(9)/2} = \frac{41}{45} = .91$$

correlation coefficient based on an entirely different computational procedure. The first step is to rank the observations in the same way as for Spearman's r_S. Table 12-6 contains the rankings for the variables Size and Price. Note that the first five columns are identical to the first five columns of Table 12-5. The calculation of Kendall's τ is based on the comparison of each pair of observations. For example, compare observations $i = 1$ and $i = 2$. Note that $r(X_1) = 1 < r(X_2) = 2$ and $r(Y_1) = 1 < r(Y_2) = 3$. This implies that the rankings of variables X and Y are in agreement or ordinally correct for this pair of observations. Pairs of observations which have this property are said to be *concordant*. Consider now observations $i = 2$ and $i = 3$. Note that $r(X_2) = 2 < r(X_3) = 3$ but $r(Y_2) = 3 > r(Y_3) = 2$. This pair is said to be *discordant* since the rankings of variable X are not in agreement with those of variable Y. To compute Kendall's τ, we must compare all pairs of observations and record the number of discordant and concordant pairs. If there are n observations, then there are $\binom{n}{2} = [n(n - 1)]/2$ comparisons to be made. Of these N_C are concordant and N_D are discordant. Kendall's τ uses the function

$$\tau = \frac{N_C - N_D}{n(n - 1)/2} \tag{12-10}$$

as the measure of association between variables X and Y. If all the pairs are concordant, then $N_C = [n(n - 1]/2$ and $N_D = 0$, so that $\tau = +1$. When $N_C = N_D$, $\tau = 0$ and there is no correlation. When $N_D = [n(n - 1]/2$ and $N_C = 0$, we know $\tau = -1$.

The most confusing part of the calculation of τ is the enumeration of the pairs of observations. This procedure is simplified by organizing the calculations in the form of Table 12-6. Consider any observation i. Since the X's are in ascending order, we know that

$r(X_i) < r(X_{i+1})$. But since the Y's are not in ascending order, we must count all the values of $r(Y_k)$ that are larger than $r(Y_i)$ but farther down the list than i. These will be the discordant values. All other comparisons must be concordant. This procedure is repeated, beginning at row $i = 1$ and continuing down the list until row $i = n$. To see this operation, let us begin in Table 12-6 at row $i = 1$. Since $r(X_1) = r(Y_1) = 1$ comparisons with all other nine observations must be concordant, they must have ranks greater than 1. In the last column we record nine concordant pairs and zero discordant pairs. Continuing with observation $i = 2$, we note $r(X_2) = 2$ and $r(Y_2) = 3$. Therefore any rank in the Y column below $i = 2$ and less than 3 must be discordant. Looking down the list of the Y ranks, we see that only one observation has a Y rank of less than 3. Note that the pair $i = 2, 3$ is discordant since $r(X_2) < r(X_3)$ but $r(Y_2) > r(Y_3)$. There are seven concordant pairs and one discordant pair. If we continue down the list in this way, all $10(9)/2 = 45$ pairs will be enumerated and no double counting occurs. In this case there are 43 concordant pairs and 2 discordant pairs. Kendall's $\tau = .91$, quite close to $r_S = .976$.

The test of significance of Kendall's τ is based on examining the difference $N_C - N_D$. Kendall has shown that for $n \geq 10$ the sampling distribution of $N_C - N_D$ is approximately normal with mean 0 and standard error $\sigma_{N_C - N_D} = \sqrt{1/18n(n - 1)(2n + 5)}$. To test the null hypothesis $H_0: \tau = 0$, the appropriate test statistic is therefore

$$Z = \frac{N_C - N_D}{\sqrt{(1/18)n(n - 1)(2n + 5)}} \tag{12-11}$$

For example, for Price and Size, $n = 10$, $\tau = .91$, $N_C = 43$, and $N_D = 2$,

$$Z = \frac{43 - 2}{\sqrt{(1/18)(10)(9)(25)}} = \frac{41}{\sqrt{125}} = 3.67$$

This is significant at $\alpha = .05$ since it exceeds 1.96.

Nominal Scale Measures of Association

In Chapter 11 we introduced the χ^2 test as a method for determining whether a statistically significant association exists between two nominally scaled variables expressed in a contingency table. Although a χ^2 test can be used to verify the existence of a *significant statistical association* between the two variables, it does not necessarily tell us much about the *strength* of the association. The level of significance is not a good indicator of strength since it depends, as we found in both parametric and nonparametric testing, on the size of the sample. So once we have used χ^2 to verify the existence of a statistically significant association, the next step is to measure the strength of this relationship. As usual, we seek single summarizing measures with bounds $[-1, +1]$ or $[0, 1]$. Such measures can be used to compare several different relationships and select the strongest or weakest relationship. There are several measures for contingency tables.

Traditional Measures Based on χ^2

The difficulty with χ^2 is that the value of χ^2 in any contingency table is directly proportional to the total sample size n. Two tables with identically proportional cell frequencies will have different χ^2 values. One way of overcoming this problem is to use the ϕ (phi) coefficient:

$$\phi = \sqrt{\frac{\chi^2}{n}} \tag{12-12}$$

This measure has many desirable properties. It can also be derived independently from a 2×2 contingency table of the form

		Variable Y		
		Category 1	Category 2	Total
	Category 1	f_{11}	f_{12}	$f_{11} + f_{12}$
Variable X	Category 2	f_{21}	f_{22}	$f_{21} + f_{22}$
	Total	$f_{11} + f_{21}$	$f_{12} + f_{22}$	$f_{11} + f_{12} + f_{21} + f_{22}$

by using the formula

$$\phi = \frac{f_{11}f_{22} - f_{21}f_{12}}{\sqrt{(f_{11} + f_{12})(f_{21} + f_{22})(f_{12} + f_{22})(f_{11} + f_{21})}} \tag{12-13}$$

where f_{ij} is the observed count in cell ij.

When Equation (12-12) is used to calculate ϕ, it has a range $[0, 1]$. By using (12-13) the sign of the coefficient is retained. If $f_{11}f_{22} > f_{21}f_{12}$, then ϕ is positive; and if $f_{21}f_{12} > f_{11}f_{22}$, then it is negative. However, the sign of the coefficient is not particularly meaningful since we could change the sign simply by interchanging the order of the rows and columns of the contingency table. Since the data are purely nominal, row and column order is of no significance. Note that if the cross-products are equal, then the numerator of ϕ is zero and so ϕ must be zero. If both the off-diagonal elements f_{21} and f_{12} are zero, then $\phi = 1$ and the relationship is perfect. To see this, simply substitute the values $f_{12} = f_{21} = 0$ into (12-13). The numerator and denominator become equal, and $\phi = 1$. If, however, f_{11} and f_{22} are both zero, then $\phi = -1$. The range of ϕ is thus $[-1, +1]$.

If the general $r \times c$ contingency table, ϕ can attain a value greater than 1. To correct for this, three other measures of χ^2 are often used for contingency tables larger than 2×2:

$$\text{Tschuprow's } T = \sqrt{\frac{\chi^2}{n \sqrt{(r - 1)(c - 1)}}} \tag{12-14}$$

$$\text{Cramer's } V = \sqrt{\frac{\chi^2}{n \min(r - 1, c - 1)}} \qquad (12\text{-}15)$$

$$\text{Pearson's } C = \sqrt{\frac{\chi^2}{\chi^2 + n}} \qquad (12\text{-}16)$$

All three measures attempt to standardize χ^2 by the maximum attainable value of χ^2 so that the coefficient has a true range $[0, 1]$. For an $r \times c$ contingency table, the maximum value of χ^2 is $n \min (r - 1, c - 1)$. This the denominator of the radical in Cramer's V. Whenever the contingency table is square and $r = c$, it is easily shown that Tschuprow's $T = $ Cramer's V. In fact, for 2×2 tables $T = V = \phi$. Another commonly used measure is Pearson's contingency coefficient C. Like ϕ, T, and V, it has a value of zero whenever there is no correlation between the two variables. The maximum value of C is always less than 1 since χ^2 can never be as large as $\chi^2 + n$. Since the maximum value of χ^2 is $n \min (r - 1, c - 1)$, the maximum value of C is, by substitution for χ^2 into (12-16),

$$\min\left(\sqrt{\frac{r - 1}{r}}, \sqrt{\frac{c - 1}{c}}\right)$$

For a 2×2 table this maximum is .707, for a 3×3 table it is .816, and so on. Since the maximum value of C depends on the dimensions of the contingency table, Pearson's C is not as easily interpretable as T or V. All three contingency coefficients are functions of χ^2, and a test of significance for χ^2 is always equivalent to a test of significance of the contingency coefficient.

EXAMPLE 12-5. A rural geographer samples 100 farmers of the rural-urban fringe of a large metropolitan area. Farmers in the sample are classified according to whether they are a hobby farmer, part-time commercial farmer, or full-time commercial farmer. Also, the principal farming activity can be classified as mixed cash crops, mixed livestock, or dairy. The results of the survey are summarized in the following table:

		Principal farm activity			
		Mixed cash crops	Mixed livestock	Dairy	Total
	Hobby	8	12	5	25
	Part-time	10	9	11	30
Ownership type	Full-time	7	14	24	45
	Total	25	35	40	100

How strongly associated are the variables ownership type and crop choice?

TABLE 12-7
Calculation of Nominal-Scale Association Measures

Observed

	Mixed cash crop	Mixed livestock	Dairy	Total
Hobby	8	12	5	25
Part-time	10	9	11	30
Full-time	7	14	24	45
Total	25	35	40	100

Expected

	Mixed cash crop	Mixed livestock	Dairy	Total
Hobby	6.25	8.75	10	25
Part-time	7.50	10.50	12	30
Full-time	11.25	15.75	18	45
Total	25	35	40	100

$$\chi^2 = \frac{(8 - 6.25)^2}{6.25} + \frac{(12 - 8.75)^2}{8.75} + \cdots + \frac{(14 - 15.75)^2}{15.75} + \frac{(24 - 18)^2}{18}$$

$$= 9.13$$

$$T = \sqrt{\frac{9.13}{100\sqrt{(2)(2)}}} = .214$$

$$V = \sqrt{\frac{9.13}{100(2)}} = .214$$

$$C = \sqrt{\frac{9.13}{9.13 + 100}} = .289$$

The strength of the association is measured by any of the three contingency coefficients T, V, or C. In Table 12-7 the calculations commence with the computation of χ^2. A value of $\chi^2 = 9.13$ has a PROB-VALUE in the range $.05 \leq$ PROB-VALUE $\leq .10$ with 4 degrees of freedom. This value is then substituted into Equations (12-14) to (12-16) to estimate T, V, and C. All three coefficients indicate a rather weak association between ownership type and crop choice. Note the equality of coefficients V and T. This is to be expected since the observed contingency table has dimensions 3×3.

Other Measures for Nominal Variables

The handicap of all coefficients based on χ^2 is the lack of a meaningful interpretation of the coefficient except at the extremes of 0 and +1. To get around this, Goodman and Kruskal have devised a coefficient which is easily interpretable. The calculation and interpretation of Goodman and Kruskal's τ_b are best explained in the context of an empirical example. Consider once again the relationship between farm ownership type and crop choice examined above. Suppose that we are asked to predict the ownership type of each of the 100 farmers as they are presented to us one by one. The only information known to us is that there are 25 hobby farmers, 30 part-time farmers, and 45 full-time farmers. That is, we know the marginal totals by ownership type. How many errors would we expect to make in classifying these 100 farmers?

First, consider the hobby farmers. We know that 75 of the 100 farmers are not hobby farmers. Therefore, of the 25 farmers we decide to place in this category, we should expect to make (in the long run) 25(75/100) = 18.75 errors or incorrect classifications. This product is simply the total number of farmers assigned to the hobby class multiplied by the probability that any one of them is not a hobby farmer. Similarly, we can expect to incorrectly assign 30(70/100) = 21 part-time farmers and 45(55/100) = 24.75 full-time farmers. In total there are 18.75 + 21 + 24.75 = 64.5 likely classification errors. Only about one-third of the farmers would be correctly classified.

Now, suppose we are given some additional information about the farmers—their principal farming activity. Can this additional information be used to reduce our classification errors? First, let us suppose the farmer is introduced as one whose principal farming activity is mixed cash crops. It is necessary to place 8 of the 25 cash crop farmers into the hobby farm category. Since 17 of the 25 mixed cash crop farmers are not hobby farmers, we can expect to make 8(17/25) = 5.44 errors. For part-time farmers the expected number of errors is 10(15/25) = 6, and for full-time farmers we expect to make 7(18/25) = 5.04 errors. Consider now the second column. These are mixed livestock farmers. Using the same logic, we would expect to make 12(23/35) = 7.89 hobby farmer misclassifications, 9(26/35) = 6.69 part-time farmer errors, and 14(21/35) = 8.40 full-time farmer errors. For the last category of dairy farmers, the respective errors are 5(35/40) = 4.375, 11(29/40) = 7.98, and 24(16/40) = 9.6. Adding these quantities leads to a total of 61.42 errors.

Goodman and Kruskal's τ_b is defined as the proportional reduction in assignment errors for the first variable (ownership type) gained by the knowledge of the second variable (principal farming activity). Specifically, define

$$\text{Goodman and Kruskal's } \tau_b = \frac{\text{Number of} \quad \overset{-}{} \quad \text{Errors with}}{\text{original errors} \quad \text{additional variables}} \qquad (12\text{-}17)$$
$$\frac{}{\text{Number of original errors}}$$

In the example above, $\tau_b = (64.5 - 61.42)/64.5 = .048$. Put simply, the assignment errors have been reduced by 4.8 percent. Suppose that the contingency table relating ownership type and principal farming activity is given by the following idealized table:

		Principal farm activity			
		Mixed cash crops	Mixed livestock	Dairy	Total
	Hobby	100	0	0	100
	Part-time	0	100	0	100
Ownership type	Full-time	0	0	100	100
	Total	100	100	100	300

If we tried to predice ownership type without utilizing crop choice, we would misclassify two-thirds of the 300 farmers. However, once we were given the crop choice of the farmer, we would make no misclassifications since the relationship between the two variables is perfect. For this table $\tau_b = 1$.

There is one problem with Goodman and Kruskal's τ_b. It is asymmetric. To calculate τ_b, we examined the error reduction in terms of the variable specified by the rows of the contingency table, using the additional information given by the variable specified in the columns of the table. If the process were reversed and the error reduction analyzed for the variable at the top of the table, a different value for τ_b would result. It is common to differentiate these coefficients by labeling the first τ_b and the second τ_a.

12.3. Correlation Coefficients and Areal Association

As we found in Chapter 2, geographers are frequently concerned with spatial distributions. The transfer of most statistical concepts to a spatial context is relatively straightforward, although it can occasionally raise some thorny conceptual issues. For correlation analysis, all that is required is that the observations refer to a set of locations expressed as areal units or as discrete points in space. Measurements of any two variables at the nominal, ordinal, or interval scales can be used in the calculation of one of the correlation coefficients described in Sections 12.1 and 12.2. As usual, the resulting coefficient indicates the direction and strength of the relationship between the pair of variables over the set of locations. When the observations refer to areal units, the correlation coefficient can be interpreted as *a standardized measure of areal association*. A correlation coefficient is a measure of the degree of similarity of the two maps of the individual variables.

Maps of nominal variables with two categories are known as *two-color maps*. Figure 12-9 illustrates two-color maps for five different nominal variables X, Y_1, Y_2, Y_3, and Y_4. Is the map of variable X similar to the maps of variables Y_1 to Y_4? The degree of similarity can be measured by any of the nominal-scale measures of association. Consider first the maps of variables X and Y_1. These two maps are identical. Each corresponding cell of the maps has an identical value and thus color. The ϕ coefficient for these two variables is +1. This can be verified by a quick examination of the contingency table for X and Y_1.

Variable Y_1

		Black = 1	White = 0	Total
	Black = 1	8	0	8
Variable X	White = 0	0	8	8
	Total	8	8	16

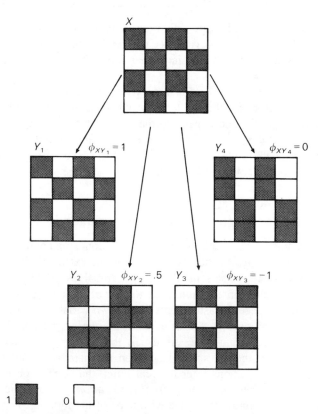

FIGURE 12-9. Five examples of two-color maps.

The zeros on the off-diagonal area indicate the perfect association between these two variables. For variables X and Y_2, the relationship is not perfect. Six of the eight black cells of the map of variable X are black in the map of variable Y_2, and two are white. Also, six of the eight white cells of map X are white in map Y_2, and two are black. The contingency table

		Variable Y_2		
		Black = 1	White = 0	Total
Variable X	Black = 1	6	2	8
	White = 0	2	6	8
	Total	8	8	16

reveals a moderately strong relationship and $\phi = .5$.

Variable X is perfectly negatively correlated with variable Y_3. Each cell that is black in one map is white in the other, and vice versa. The ϕ coefficient of -1 verifies the perfect inverse relationship between these two variables. Although the maps of variables X and Y_4 both depict alternating patterns, the ϕ coefficient of zero indicates there is no relation between these two variables. The contingency table

		Variable Y_A		
		Black = 1	White = 0	Total
Variable X	Black = 1	4	4	8
	White = 0	4	4	8
	Total	8	8	16

verifies this. Although it is easy to visualize the areal correspondence of these relatively small and simple maps, it is not as easy to visually interpret the degree of areal association between complicated maps with many cells. Quantitative measures of association are particularly useful in these cases.

Ordinal or interval areal data are usually displayed on a *k-color map*. By dividing the data of a variable into three or more classes, a *k*-color map can be used to give a visual portrait of its areal distribution. For ordinal data one can assign a single color (or pattern or shade) to each of the four quartiles of the ranks, or 10 colors to the deciles. For interval data, the data can be divided into classes in several different ways. For example, a three-color map with the classes

$$Z > 1.0 \qquad \text{(black)}$$
$$-1.0 \leq Z \leq 1.0 \qquad \text{(gray)}$$
$$Z < -1.0 \qquad \text{(white)}$$

could be used to portray the distribution of an interval scale variable sfandardized by its standard deviation through Equation (5-33). In other cases more ad hoc classes may be defined. The bounds defining color categories should be carefully chosen. It is important that the visual impression of the maps of two variables accurately reflect the degree of association between them. Just as the selection of an improper interval width can mask important characteristics of the frequency distribution of a variable, so, too, can poorly chosen classes mask the pattern of a k-color map.

Let us consider a simple example drawn from political geography. The percentage support of a particular political party is known for the 16 electoral districts of the hypothetical city illustrated in Figure 12-10. These percentages are compiled in Table 12-8. It is hypothesized that support for this particular party is closely associated with certain demographic characteristics of the individual electoral districts. In this case, let us suppose that the party appeals mainly to the age group of 25- to 45-year-olds. The percentage of each electoral district's population in this age group is also listed in Table 12-8. A quick visual check of the areal association of these two variables can be made by comparing the k-color maps for each. To simplify matters, identical class bounds are used in each map and only three colors. As is clearly illustrated in Figure 12-11, these two maps are very similar. Eleven of the 16 zones have identical colors. One of the problems with k-color maps is that it is sometimes difficult to get an accurate visual impression of the degree of similarity. Too many colors can lead to unnecessary complexity, and too few colors can mask the real variation portrayed by the data. Whenever the data are grouped into classes and then translated into color categories, there is an inevitable loss of information.

There are two important problems in the use and interpretation of correlation coeffi-

TABLE 12-8
Electoral District Data

Zone	Voter support for party, %	Electoral district population ages 25 to 45, %
1	10	20
2	50	60
3	10	10
4	60	60
5	20	10
6	50	50
7	20	20
8	50	40
9	20	30
10	40	30
11	20	40
12	40	40
13	30	30
14	40	50
15	30	40
16	30	50

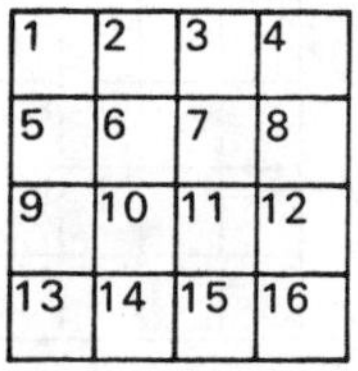

1	2	3	4
5	6	7	8
9	10	11	12
13	14	15	16

FIGURE 12-10. Sixteen electoral districts of a hypothetical city.

cients for areal data: the *scale problem* and the *modifiable areal units* problem. All correlation coefficients measure the areal association between two variables only for the particular areal units used in the calculation. As we noted in Chapter 3, when areal data are aggregated into fewer and larger areal units, there can be substantial variations in the value of most statistics. This is the *scale or aggregation problem.* And even if we use the same number, size, and shape of areal units, different partitions of space can yield different values for the correlation coefficients for the same data. This is the *modifiable areal units problem.* Neither problem invalidates the use of correlation coefficients for areal data, but each presents specific interpretation problems.

First, let us illustrate the potential problems by using the voting behavior data. Figure 12-12 illustrates four possible aggregations in the data of the original 16 districts. In each of cases A, B, C, and D the original data are aggregated to four larger equal-area zones. The values for X and Y for zone 1 of aggression A are *averages* of the four electoral districts numbered 1, 5, 9, and 13. The (X, Y) pair of $(20, 22.5) = ((10 + 20 + 20 + 30)/4, (20 + 10$

Percentage of party support

Percentage of population aged 25 to 45 years

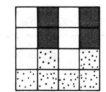

0–20%   21–40%   41–60%

FIGURE 12-11. *k*-color maps for variables in Table 12-8.

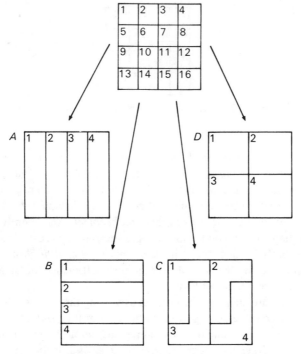

FIGURE 12-12. Alternative spatial aggregations.

TABLE 12-9
Zonal Aggregation of Electoral District Data

	Zone	Party support, %	Zone aged 25 to 45, %
Aggregation *A*	1	20	22.5
	2	45	47.5
	3	20	27.5
	4	45	47.5
Aggregation *B*	1	32.5	37.5
	2	35	30
	3	30	35
	4	32.5	42.5
Aggregation *C*	1	25	30
	2	27.5	32.5
	3	40	40
	4	37.5	42.5
Aggregation *D*	1	32.5	35
	2	35	32.5
	3	32.5	35
	4	30	42.5

+ 30 + 30)/4). The values for all zones in each of the four aggregations are computed in the same way and are summarized in Table 12-9. To see the effects of the aggregation, it is useful to compare the scatter diagrams of each of the aggregations to the original data set. As is illustrated in Figure 12-13, the data of the original 16 districts exhibit the expected strong positive correlation between these two variables. The effects of aggregation are dramatic. Notice that there is a substantial reduction in the total variation of both variables. Aggregation D is the most extreme example. The range of the variable party support declines from $60 - 10 = 50$ in the original data to $37.5 - 32.5 = 5$. A similar reduction occurs for the percentage of the population aged 25 to 45 years.

The correlation coefficients between party support and percentage of the population aged 25 to 45 years are summarized in Table 12-10. Note the large discrepancies between

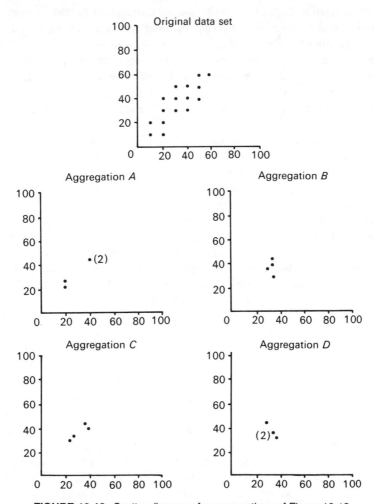

FIGURE 12-13. Scatter diagrams for aggregations of Figure 12-12.

TABLE 12-10
Correlation Coefficients for Alternative Spatial Aggregations of
Electoral District Data

Aggregation	Pearson's r
Original 16 zones	.81
Aggregation A	.99
Aggregation B	−.39
Aggregation C	.95
Aggregation D	−.94

the 16-district correlation coefficient $r = .81$ and the correlations for the four aggregations. Partly because the data have been selected in order to dramatize the possibilities, the correlations range from the almost perfect positive to the almost perfect negative. This is not uncommon when there are only a small number of observations. Even where the *scale* of the observations is held constant, there can also be large variations in the value of r. Notice that all four aggregations involve *contiguous* districts and have exactly the same areas. Despite this, the range of the correlation coefficient for these modified units has the extreme range of +.99 to −.94.

In many situations in the social sciences, the effects of data aggregation are neither as dramatic nor as random as occurs in the voting behavior case. A systematic *increase* in the value of the correlation coefficient appears to be common, particularly for cases in which contiguous grouping procedures are used. Different scales of aggregation produce different degrees of association between variables. When we attempt to interpret correlation coefficients or test hypotheses about the relationships between variables, using areal data, it is important to keep in mind what sociologists term *the ecological fallacy*. In short, it is fallacious to transfer conclusions concerning the relationship between variables obtained from data at one level of aggregation to other levels of aggregation. The most obvious fallacies occur when we attempt to make statements about the behavior of individuals, using data based on aggregates or groups of these individuals.

Consider the voting behavior data in this light. Of course, we are really interested in whether individuals aged 25 to 45 years are likely to support the party. Because of confidentiality, all that is available is data aggregated to the level of the electoral district. The fact that these two variables are strongly associated at the electoral district level does not necessarily mean that the same degree of association applies to *individual* voters. All we can legitimately say is that supporters of the party *tend to live in electoral districts with a high proportion of residents from 25 to 45 years*. The relationship holds between *areas*, not necessarily between *individuals*. In fact, it is possible that the people who support the party are of entirely different age groups. They may vote for this party in response to the 25- to 45-year-olds who populate their electoral districts! In many instances the solution to the scale problem is simply to employ individual or household data. At other times, one of the variables in which we are interested might necessarily require an areal definition. For example, individual house selling prices depend on many characteristics of the individual house and lot, but also on neighborhood variables which are based on areal aggregations of individual houses. Neighborhood stability, land-use mix, and other such variables are known to have significant effects on residential property values. Changes in scale may mask important associations between variables or overemphasize others.

Areal association studies represent an initial attempt to develop explanations for the areal variations of many phenomena. Unfortunately, the lack of geographical theory concerning the areal distribution of many phenomena hinders our ability to satisfactorily interpret the results of many correlation analyses. It is often difficult to explain observed scale and modifiable areal unit effects. Nevertheless, statistical correlation analysis represents one of the most widely used techniques of map comparison and areal association.

12.4. Spatial Autocorrelation

Correlation analysis is not limited to the comparison of the maps of two different variables. It can also be used as a measure of order or pattern in the spatial distribution of a single variable depicted on a two- or k-color choropleth map. The term *spatial autocorrelation* refers to the correlation of a variable with itself through space. If there is any systematic pattern in the spatial distribution of a variable, it is said to be spatially autocorrelated. If the pattern on the map is such that nearby or neighboring areas are more similar than more distant areas, the pattern is said to be *positively spatially autocorrelated.* This is a common occurrence in geography. The distribution of house prices or household income in a city is usually positively spatially autocorrelated. This is because wealthy households tend to live in exclusive neighborhoods, segregating themselves from lower-income households, relegating these households to other areas in the city. Many land-use activity types are agglomerated in space and also exhibit positive spatial autocorrelation. Variables collected by physical geographers can also display this pattern. Climatologists find similar temperature and precipitation levels in adjacent areas. *Negative spatial autocorrelation* describes patterns in which neighboring areas are unlike. This is not a common pattern in geography. Between the two extremes of positive and negative autocorrelation are *random* patterns which exhibit *no spatial autocorrelation.*

There are two reasons why autocorrelation is of concern to geographers. First and foremost, the search for spatial pattern is one of the dominating themes of geographical research. Areal distributions are rarely random. If there are any "laws" in geography, then surely the first is "Nearby locations are more alike than more distant ones." In statistical jargon this means the areal distributions of many phenomena are positively spatially autocorrelated. The second reason for studying spatial autocorrelation is that the inferential techniques presented in Chapters 8 to 10 are based on the assumption that the values of the observations in each sample are independent of one another. One of the ways in which spatial data might violate this assumption is if the data reveal a pattern of spatial autocorrelation. If two of the observations in a sample are from nearby locations, they are not usually independent. Our interest in spatial autocorrelation is twofold: to measure the strength of spatial autocorrelation in a map and to test the assumption of independence or randomness.

Map *A*   Map *B*

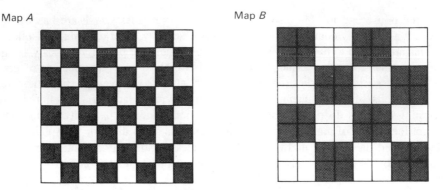

FIGURE 12-14. Map patterns.

One difficulty in devising measures of spatial autocorrelation is that many different patterns could exist in a map. Each measure and test is useful in detecting only one of these patterns. To see the problem, consider the two patterns in Figure 12-14. Both exhibit a pattern of negative spatial autocorrelation when they are examined at different scales. However, any test of the pattern is scale-dependent. Note that no black cell in map *A* is contiguous with another black cell. White cells have the same property. In map *B* each black cell is contiguous with two other black cells. A test to uncover the simple pattern in map *A* may fail to uncover the pattern in map *B* and vice versa. To illustrate the procedures of spatial autocorrelation, we restrict ourselves to simple contiguity tests which examine the correlation between a cell and those neighboring cells with which it shares a boundary. First, let us consider a simple test for nominal variables with two categories based on a two-color map. This will introduce the basic rationale behind the more complicated methods used to analyze *k*-color maps.

A Contiguity Test for Two-Color Maps

Measures of spatial autocorrelation for nominal variables mapped in two colors are based on the pattern of joins between contiguous areas or cells of the map. Each cell is colored either black (B) or white (W). The *join*, or border, between adjacent cells can be classed as white–white (WW), black–black (BB), or (BW) depending on the colors of the two cells. Only joins of nonzero length are considered, and diagonal cells are by definition noncontiguous. Different patterns of autocorrelation have different numbers of joins in each of the BB, WW, and BW categories. Compare the three maps of Figure 12-15. Map *A* is negatively autocorrelated. Adjacent cells are unlike since they are of different colors. All joins are BW. Map *C* represents the other extreme, positive autocorrelation. Most of the joins are BB or WW. Only the four along the middle boundary are BW. In between these two extremes lies map *B*. The pattern is a mix between the alternating pattern of map *A* and the clustering pattern of map *C*. There is almost an equal number of each type of join. A summary of the join counts for these three maps confirms these observations.

Map *A* 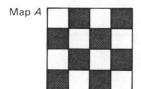   Map *B* 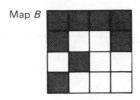   Map *C*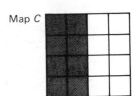

FIGURE 12-15. Two-color maps.

		Join counts			
		BB	WW	BW	Total
	A	0	0	24	24
Map	*B*	5	7	12	24
	C	10	10	4	24

The simplest tests for autocorrelation use the numbers of joins of each type as an indicator of the type of autocorrelation displayed on the map.

To illustrate the two-color test for contiguity, let us analyze map *B*. To begin, we must enumerate the number and type of joins on the map. There are $n = 16$ cells numbered consecutively, beginning in the northwest corner and proceeding by rows. There are $n_B = 8$ black cells and $n_W = 8$ white cells. Table 12-11 is a convenient format for enumerating the information required for a contiguity test. Let L_i be the number of joins of cell *i*. Cell 1 has $L_1 = 2$ joins, cell 2 has $L_2 = 3$ joins, and so on. The total number of joins on the map is $L = \Sigma_{i=1}^{n} L_i/2$. The sum of L_i is divided by 2 to avoid double counting of joins. In map *B* there are $48/2 = 24$ joins. Now, if we know a map has n_B black cells and $n_W = n - n_B$ white cells, the probability that a randomly selected cell is black is $p = n_B/n$. The probability it is white is $q = n_W/n$, and $p + q = 1$. If it is selected randomly, cell color is a *binomial* random variable.

If the areal pattern on the map is random, how many joins of each type would we expect in an *n*-cell region with *L* total joins? Begin by selecting a color for the first cell. The probability it is black is *p*, and the probability it is white is *q*. Now, consider a second cell contiguous with zone 1. The probability it is black is also *p*. The probability that the join between the two cells is BB is $p \cdot p = p^2$. To get a WW join, the required probability is $q \cdot q = q^2$. BW joins occur in two ways. Either a black cell can be followed by a white cell, or vice versa. Both probabilities are equal to *pq*, so that the probability of a BW join is $pq + pq = 2pq$. If there are *L* joins on the map, the expected number of cells of each type is simply each probability multiplied by *L*:

$$E(\text{BB}) = \mu(\text{BB}) = p^2L$$

$$E(\text{WW}) = \mu(\text{WW}) = q^2L \qquad (12\text{-}18)$$

$$E(\text{BW}) = \mu(\text{BW}) = 2pqL$$

where $\mu(\text{BB})$ is the expected or mean number of black–black joins, on a map with a random pattern. The expected number of joins adds to *L* since $p^2L + q^2L + 2pqL = (p + q)^2L = L$ (since $p + q = 1$, by definition).

TABLE 12-11
Join Counts for Map *B* of Figure 12-12

Cell	Color	Joins			L_i	$L_i(L_i - 1)$
		BB	WW	BW		
1	B	2	0	0	2	2
2	B	2	0	1	3	6
3	B	2	0	1	3	6
4	B	2	0	0	2	2
5	B	1	0	2	3	6
6	W	0	1	3	4	12
7	W	0	2	2	4	12
8	B	1	0	2	3	6
9	W	0	0	3	3	6
10	B	0	0	4	4	12
11	W	0	3	1	4	12
12	W	0	2	1	3	6
13	B	0	0	2	2	2
14	W	0	1	2	3	6
15	W	0	3	0	3	6
16	W	0	2	0	2	2
Totals	$n_B = 8$	10	14	24	48	104
	$n_W = 8$					

$$p = 8/16 = .5 \qquad L = 1/2(48) = 24$$
$$q = 8/16 = .5 \qquad K = 1/2(104) = 52$$

$$\text{BB} = 1/2(10) = 5 \qquad \text{WW} = 1/2(14) = 7 \qquad \text{BW} = 1/2(24) = 12$$

Of course, these expectations will not occur on each random two-color map, and it is necessary to specify the standard deviations of the sampling distributions of BB, WW, and BW joins. Since the details are quite complicated, the standard deviations are presented without comment:

$$\sigma(\text{BB}) = \sqrt{p^2 L + p^3 K - p^4(L + K)}$$
$$\sigma(\text{WW}) = \sqrt{q^2 L + q^3 K - q^4(L + K)} \qquad (12\text{-}19)$$
$$\sigma(\text{BW}) = \sqrt{2pqL + pqK - 4p^2 q^2(L + K)}$$

where $K = \frac{1}{2}\sum_{i=1}^{n} L_i(L_i - 1)$ is a constant for any map. It turns out that the sampling distributions of the number of joins of each type are normal under the following general conditions:

1. The number of cells is large, that is, greater than approximately 30.
2. Both $p, q > .2$.

3. The region is not elongated.
4. The join structure is not dominated by a few cells. The normal approximation is not close if one cell (or a small number of cells) has many joins and most others have only 1 or 2.

When these conditions are satisfied, the appropriate test statistics for each of the join counts are

$$Z(BB) = \frac{BB - \mu(BB)}{\sigma(BB)}$$

$$Z(WW) = \frac{WW - \mu(WW)}{\sigma(WW)} \qquad (12\text{-}20)$$

$$Z(BW) = \frac{BW - \mu(BW)}{\sigma(BW)}$$

where BB, WW, and BW are the *observed* number of joins of each type.

Since it contains only 16 cells, map B does not satisfy the requirements for normality Z_{BB}, Z_{WW}, and Z_{BW}. The computational steps required in the test can still be illustrated with this example, if we recognize the fact that it is a poor approximation in this case. Using the information contained in Table 12-11, we first calculate $p = q = .5$, $L = 24$, and $K = 52$. These values are then substituted to obtain

$$\mu(BB) = 24(.5)^2 = 6$$

$$\mu(WW) = 24(.5)^2 = 6$$

$$\mu(BW) = 2(24)(.5)(.5) = 12$$

Notice that these expected numbers of joins are $6 + 6 + 12 = 24$. The standard errors of the sampling distributions are calculated from (12-9):

$$\sigma(BB) = \sqrt{(.5)^2(24) + (.5)^3(52) - (.5)^4(24 + 52)} = 2.78$$

$$\sigma(WW) = \sqrt{(.5)^2(24) + (.5)^3(52) - (.5)^4(24 + 52)} = 2.78$$

$$\sigma(BW) = \sqrt{2(.5)(.5)(24) + (.5)(.5)(52) - 4(.5)^2(.5)^2(24 + 52)} = 2.45$$

Substituting this information into (12-20) yields the following values for $Z(BB)$, $Z(WW)$, and $Z(BW)$:

$$Z(BB) = \frac{5 - 6}{2.78} = -.36$$

$$Z(WW) = \frac{7 - 6}{2.78} = .36$$

$$Z(BW) = \frac{12 - 12}{2.45} = 0$$

Taking these values of Z to the standard normal table, we find that the PROB-VALUES for each type of join are $P(Z(BB) \geq -.36) = P(Z(WW) \geq .36) = .3594$ and $P(Z(BW) \geq 0) = .5000$. These PROB-VALUES suggest that the numbers of joins in map B frequently occur in random patterns with 24 joins. We cannot reject the null hypothesis H_0: "There is no autocorrelation of contiguous cells in map B" at any common level of significance. We must conclude that the pattern on the map is *not* spatially autocorrelated. Another example of the use of this test is provided in Chapter 13 where it is used to test one of the key assumptions of bivariate regression analysis.

This test can also be used for nominal variables with more than two classes expressed as k-color maps. The extension is quite straightforward. Details can be obtained in Cliff and Ord (1973, 1981).

A Contiguity Measure and Test for Ordinal and Interval Data

In general, it is inadvisable to group ordinal and interval data into classes and use the join-count statistical test described above to measure autocorrelation. There is a significant loss of information when the level of measurement of the variables is reduced in this way. Instead it is advisable to use a measure of spatial autocorrelation which employs the actual ordinal or interval value of a cell in its calculations. One such measure is *Geary's contiguity ratio c*. This statistic relates the difference between contiguous cells to the differences in the entire set of cells which comprise the map.

The first step is to construct a contiguity matrix. Each entry in this matrix δ_{ij} is assigned the value 1 if cells i and j are contiguous and 0 otherwise. By definition, $\delta_{ij} = 0$ for all i. For a map with n cells the contiguity matrix is square of order $n \times n$ and symmetric. The contiguity matrix for the variable percentage party support (see Table 12-8 and Figure 12-10) is given in Table 12-12. Since each of the 24 joins appears twice in the matrix, there are 48 entries of 1. Geary's contiguity ratio c uses the information in this contiguity matrix and the actual ordinal or interval cell values in the formula

$$c = \frac{(n - 1) \sum_{i=1}^{n} \sum_{j=1}^{n} \delta_{ij}(X_i - X_j)^2}{4L \sum_{i=1}^{n} (X_i - \bar{X})^2} \qquad (12\text{-}21)$$

where n is the number of cells, L is the total number of joins, and X_i is the value of cell i. This statistic takes the ratio of the sum of the squared differences between contiguous cells to the sum of the squared differences of each cell value to the mean cell value on the map. In the numerator the squared difference between cells enters into the summation

TABLE 12-12
Contiguity Matrix for Map of Figure 12-10

Cell	1	2	3	4	5	6	7	8	9	10	11	12	13	14	15	16
1		1			1											
2	1		1			1										
3		1		1			1									
4			1					1								
5	1					1			1							
6		1			1		1			1						
7			1			1		1			1					
8				1			1					1				
9					1					1			1			
10						1			1		1			1		
11							1			1		1			1	
12								1			1					1
13									1					1		
14										1			1		1	
15											1			1		1
16												1			1	

Note. Only 1s are shown.

only when the two cells are contiguous and $\delta_{ij} = 1$. The coefficient $(n - 1)/(4L)$ is used to scale the ratio, so that the expected value of c, $E(c)$, for a random map pattern is 1. Of course, we cannot expect all maps without autocorrelation to have $c = 1$. The sampling distribution of c for a map with no autocorrelation is normal with $E(c) = 1$ and a standard deviation given by

$$\sigma(c) = \sqrt{\frac{(2L + K)(n - 1) - 2L^2}{(n + 1)^2}} \tag{12-22}$$

where K is defined as $\frac{1}{2}\Sigma_{i=1}^{n} L_i(L_i - 1)$. This leads to the following test statistic:

$$Z_c = \frac{E(c) - c}{\sigma(c)} \tag{12-23}$$

The preliminary calculations for Geary's c for the party support data are given in Table 12-13. Consider cell 1. It has a value of 10 and is contiguous with two cells—cell 2 with a value of 50 and cell 5 with a value of 20. Two terms enter the numerator of c, the values of which are listed in the last column. The mean party support in the city is 32.5, so that $(X_1 - \bar{X})^2 = (10 - 32.5)^2 = 506.25$. The 16 terms in the summation for the denominator are listed in the second column, one for each cell. Table 12-13 is a convenient format for organizing the calculations for Geary's c. The values of L and K are 24 and 52, respectively. Note that these values are the same as those obtained in the two-color join-count test for map B of Figure 12-15. This is because both maps are regular 4×4 lattices.

Substituting these calculated values into (12-21) yields

$$c = \frac{15(21200)}{4(24)(3500)} = .946$$

We now calculate $\sigma(c)$:

$$\sigma(c) = \sqrt{\frac{[2(24) + 52]15 - 2(24)^2}{17^2}} = 1.097$$

The test statistic

$$Z_c = \frac{1 - .946}{1.097} = .049$$

leads us to accept the null hypothesis of no autocorrelation at $\alpha = .05$ since the calculated test statistic is less than 1.96.

TABLE 12-13
Work Table for Geary's Contiguity Ratio for Party Support Data

Cell	$(X_i - \bar{X})^2$	Joins	$X_i - X_j$	$(X_i - X_j)^2$
1	506.25	1-2	−40	1600
		1-5	−10	100
2	306.25	2-1	40	1600
		2-3	40	1600
		2-6	0	0
3	506.25	3-2	−40	1600
		3-4	−50	2500
		3-7	−10	100
4	756.25	4-3	50	2500
		4-8	10	100
5	156.25	5-1	10	100
		5-6	−30	900
		5-9	0	0
6	306.25	6-2	0	0
		6-5	30	900
		6-7	30	900
		6-10	10	100
7	156.25	7-3	10	100
		7-6	−30	900
		7-8	−30	900
		7-11	0	0
8	306.25	8-4	−10	100
		8-7	30	900
		8-12	10	100
9	156.25	9-5	0	0
		9-10	−20	400
		9-13	−10	100
10	56.25	10-6	−10	100
		10-9	20	400
		10-11	20	400
		10-14	0	0
11	156.25	11-7	0	0
		11-10	−20	400
		11-12	−20	400
		11-15	−10	100
12	56.25	12-8	−10	100
		12-11	20	400
		12-16	10	100
13	6.25	13-9	10	100
		13-14	−10	100

(continued)

TABLE 12-13 (continued)

Cell	$(X_i - \bar{X})^2$	Joins	$X_i - X_j$	$(X_i - X_j)^2$
14	56.25	14-10	0	0
		14-13	10	100
		14-15	10	100
15	6.25	15-11	10	100
		15-14	-10	100
		15-16	0	0
16	6.25	16-12	-10	100
		16-15	0	0

$$\sum_{i=1}^{n} (X_i - \bar{X})^2 = 3500.00 \qquad \sum_{i=1}^{n}\sum_{j=1}^{n} (X_i - X_j)^2 = 21{,}200$$

Limitations of Autocorrelation Measures

The join-count two-color test and Geary's contiguity ratio have two distinct limitations. First, many different partitions of space have identical join structures. Figure 12-16 illustrates three different cell patterns, each of which has the contiguity matrix

$$\begin{array}{c|cccc} & A & B & C & D \\ \hline A & 0 & 1 & 1 & 0 \\ B & 1 & 0 & 0 & 1 \\ C & 1 & 0 & 0 & 1 \\ D & 0 & 1 & 1 & 0 \end{array}$$

In (a), there are four equal sized cells, in (b) there is a single cell of dominating size, and in (c) the map does not completely encompass the area inside its exterior boundary. These

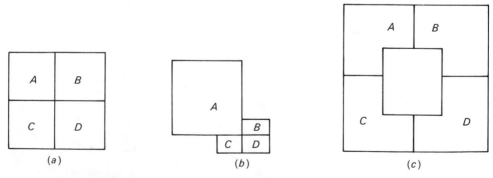

FIGURE 12-16. Different partitions of space with identical join structures.

are quite distinct patterns in themselves, and there appear to be few reasons to expect similar autocorrelation patterns.

The second problem with these measures is that they are capable of evaluating only the simplest forms of spatial autocorrelation—between contiguous zones. What about more complex patterns of areal dependency? If we use a better measure of cell separation, say intercell distance, we might be able to evaluate many more patterns than is possible with contiguity-based measures. In-depth treatment of the problem of spatial autocorrelation may be found in Cliff and Ord (1973, 1981).

12.5. Summary

This chapter has introduced the notion of relations between variables and several aspects of bivariate statistical analysis. Of particular importance is the notion of correlation, which is used to measure the degree of association between two variables measured over the same set of observations. Correlation coefficients can be defined for interval variables (Pearson's r), ordinal variables (Kendall's τ and Spearman's ρ), and for nominal variables (for example, ϕ). These measures can be used to gauge the strength of the association and as input to tests of statistical significance. Geographic applications of correlation analysis incorporate the notion of areal association. The scale and modifiable areal units problems, first described in Chapter 3, surface again. Difficulties in interpretation of correlation coefficients often arise. Correlation analysis used to uncover the spatial pattern on a single map is known as spatial autocorrelation analysis. Measures and tests of autocorrelation for two- and k-color maps are indicators of certain systematic map patterns.

REFERENCES

A. D. Cliff and J. K. Ord, *Spatial Autocorrelation* (London: Pion, 1973).
A. D. Cliff and J. K. Ord, *Spatial Processes* (London: Pion, 1981).
S. Siegel, *Nonparametric Statistics for the Behavioral Sciences* (New York: McGraw-Hill, 1956).

FURTHER READING

Examples of the application of correlation analysis in social science research are contained in virtually all introductory textbooks for statistical methods in the various disciplines. Several textbooks for geographers are cited at the end of Chapter 2. Extensions to, and applications of, nonparametric methods of correlation can be found in the textbooks cited at the end of Chapter 11. Advanced material on spatial autocorrelation can be found in the two texts by Cliff and Ord (1973, 1981) listed in the References. The problems which follow also include some examples relevant to geographers.

PROBLEMS

1. Explain the meaning of the following terms or concepts:
   a. Correlation coefficient
   b. Covariance
   c. Correlation matrix
   d. Monotonic increasing function
   e. Concordant observations
   f. Spatial autocorrelation
   g. Contiguity
   h. Ecological fallacy
   i. Two-color map
   j. k-color map

2. Explain what is meant by the concept of a correlation coefficient between (a) two ordinal variables and (b) two nominal variables.

3. What pattern would appear on a map characterized by positive spatial autocorrelation? Negative spatial autocorrelation?

4. What is the value of the correlation coefficient r between two variables X and Y when (a) variable X always takes on some constant value C and, (b) the value of variable Y is always 3 times the value of variable X? (*Hint*: Draw a graph.)

5. In each of the following cases, state whether you would expect the correlation coefficient r to be positive, negative, or near zero.
   a. X = stream discharge
      Y = stream depth
   b. X = distance from a neighborhood to the central business district
      Y = residential density
   c. X = sound level, in decibels
      Y = distance from a major urban expressway
   d. X = number of patrons of a day care center
      Y = distance traveled to the center
   e. X = percentage of neighborhood residents who are black
      Y = percentage of neighborhood residents who are Caucasian
   f. X = percentage of neighborhood residents who are Australian
      Y = percentage of neighborhood residents who are Icelandic

6. A behavioral geographer is interested in whether two different methods of estimating familiarity with a set of neighborhoods, method A and method B, give similar results. Each method uses questions concerning the knowledge of subjects of neighborhood landmarks, streets, activities, and residents. Method A uses an entirely different set of items from method B. Ten subjects are given both tests. An interval between the tests of 1 month is used to eliminate possible bias. Five subjects are given method A first, and five subjects are given method B first. These subjects are chosen randomly. The results are summarized as follows:

Subject	Scores	
	Method *A*	Method *B*
1	58	72
2	49	63
3	60	73
4	81	82
5	92	81
6	34	50
7	14	32
8	68	74
9	25	41
10	72	80

a. Plot a scattergram of these data.
b. Calculate Pearson's *r*.
c. Rank each of the variables and calculate Spearman's r_S.
d. What do the results suggest?
e. Test the significance of the correlations.

7. Calculate Kendall's τ for the data in Problem 6.

8. Consider the following two-color map:

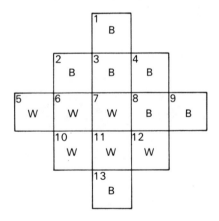

Note: Cell numbers are located in the upper left corner of each cell; B = black, W = white.

a. Use the contiguity test for two-color maps to test for possible spatial autocorrelation in this map.
b. Suppose that the values for the variable depicted on this map are, in order of cell number, 3, 4, 7, 6, 3, 5, 11, 14, 18, 21, 17, 16, 5. Calculate Geary's contiguity ratio and test it for significance.
c. Group 12 of the cells in pairs, leaving one cell ungrouped. Replace each cell with the mean of the two cells from which it has been formed. Construct a two-color map of this new

configuration with black assigned to values below the mean and white to values above the mean. Test the new map for autocorrelation.

d. Calculate Geary's contiguity ratio for the new map created in part (c).

9. Residents of a city were surveyed by telephone to determine their support for a proposal to construct a multipurpose stadium on a large tract of previously underdeveloped land on the outskirts of the city. When cross-tabulated with age of the respondent, the results of the survey are as follows:

	Age of respondent		
	<30	31–44	45+
Approve	43	13	10
Neutral or no opinion	27	7	60
Disapprove	17	72	12

a. Calculate Tschuprow's T, Cramer's V, Pearson's C, and Goodman and Kruskal's τ_b for these data.

b. Does age of respondent correlate with support for the stadium?

c. What other ways might one analyze these data?

13

Introduction to Regression Analysis

Suppose now that we wish to go beyond measuring the mere strength of association between two variables through a correlation coefficient. In particular, we might be interested in the precise mathematical form of the function linking variable X to variable Y. We might even wish to take this further and use our derived mathematical function to predict the value of variable Y from our knowledge of variable X. This is a problem in regression analysis.

Regression analysis can be broadly defined as the analysis of statistical relationships among variables. There are two types of variables in regression problems. First, there are the *dependent* variables, which are also often called *response*, or *endogenous*, variables. By convention, these variables are normally labeled $Y_1, Y_2, \ldots, Y_r$. Second, there are the *independent* or *predictor* or *exogenous* variables. These are usually labeled $X_1, X_2, \ldots, X_s$. In this chapter we confine ourselves to simple regression problems. As the name suggests, simple regression concerns itself with the case in which we have one dependent variable Y and one independent variable X. For purposes of clarity, we now suppress the subscript on these two variables since there is only one variable of each type.

Although the terms *dependent* (or response) and independent (or predictor) variables are quite conventional, no implication of causality is necessarily implied in any given case. This is true no matter how strong the statistical relationship might be. In some instances there may be strong a priori grounds for the specification of a cause-and-effect relationship and the selection of a dependent variable and one or more independent variables. In many other situations this may not be easy. Geography, and social science generally, is characterized by the recognition of relationships between variables that are unclear, ambiguous, or possibly even exhibiting two-way dependence. The difficulty in establishing clear-cut cause-and-effect relations most often derives from the nature of theory in social science. In the physical sciences, however, the specification of regression relations is often straightforward. For example, the rate at which some chemical reaction occurs can be made a function of the ambient temperature. Cause and effect are clear. The temperature is not caused by the rate of the chemical reaction! In this chapter let us assume that causal and functional dependence go hand in hand. Much of the material is easier to comprehend if we make this assumption.

Section 13.1 describes the situations in which we employ regression analysis and introduces the simple linear regression model. In Section 13.2 we discuss the procedure used to estimate the parameters of the simple regression model. Section 13.3 details various methods of assessing the goodness of fit of the regression equation to the observed data. Here we will see the intimate relationship that exists between correlation and regression. In Section 13.4 the use of a regression equation in the context of prediction is examined, as well as several issues concerning the interpretation of regression parameters and regression equations. Finally, in Section 13.5 we briefly introduce some technical and methodological issues that arise in empirical applications.

13.1. Simple Linear Regression Model

Relations between Variables

The concept of a relation between variables is a familiar one to geographers. We speak, for example, of the relationship between land rent and distance to the central business district (CBD) in cities or of the relationship between stream discharge and sediment load. However, let us now distinguish a *functional* relationship from a statistical one.

FUNCTIONAL RELATIONS

A functional relation between two variables X and Y is always expressed by a mathematical formula of the form $Y = f(X)$. Given any value of X, the function indicates the corresponding value of Y. The function can take on any form whatsoever, but let us consider one particular example involving a linear function (See Appendix 13A for a review of the elementary geometry of straight lines).

EXAMPLE 13-1. A shipper charges prospective customers according to the following schedule for a full truckload delivery:

$$Y = 50.00 + 0.40X$$

Here Y is the total shipment cost in dollars, 0.40 is the per-mile rate in dollars, and X is the distance to the destination in miles. The bills of lading for four customers provide the following data:

Shipment No.	Total cost, $	Miles
1	70.00	50
2	90.00	100
3	100.00	125
4	124.00	185

These values are plotted in Figure 13-1. Note that all points lie directly on the line. There is no error. All functional relations share this characteristic.

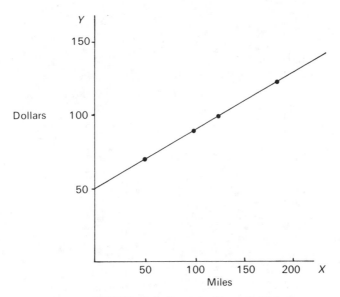

FIGURE 13-1. Graph for Example 13-1.

STATISTICAL RELATIONS

A statistical relation is not necessarily a perfect one. Except in the most unlikely cases, the observations of a statistical relation do not fall on the line or curve that expresses the functional relationship.

EXAMPLE 13-2. Scientists at an experimental agricultural station applied varying amounts of fertilizer to homogeneous 2-acre plots and recorded the yield from each plot at the end of the harvest. The graph of corn yield versus the amount of fertilizer applied is shown in Figure 13-2. The figure clearly suggests there is a relation between these two variables. As increasing amounts of fertilizer are applied, there is a corresponding increase in the yield of corn. The relation is not a perfect one, and the scatter of points suggests that some, but not all, of the variation in corn yield of these plots can be accounted for by the variations in the amount of fertilizer applied. For example, two of the plots are treated with 300 lbs of fertilizer per acre, but yields are somewhat different. Because of the scattering of points in a statistical relation, we often also call plots such as Figure 13-2 *scatter diagrams* or *scattergrams*. Each point in the scattergram represents one of our observations.

In Figure 13-2 we have also plotted a line that describes this statistical relation reasonably well. Most of the points fall close to the line, but the scatter suggests there is some variation in corn yield not accounted for by the amount of fertilizer applied. If the experimental agricultural station irrigated each plot equally, and if the underlying soil and drainage are homogeneous, we might think of this error as being random. Notice that even though we do not have an exact functional relationship, we still have what appears to be a highly useful statistical one.

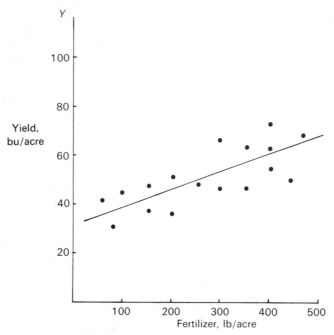

FIGURE 13-2. Graph for Example 13-2.

Linear Regression Model

A linear regression model is a formal means of expressing the two ingredients of a statistical relationship. First, the dependent variable varies systematically with the independent variable or variables. Second, a scattering of the observations occurs around this systematic component. Although the systematic component can take on any form and we can have as many variables as is necessary, we restrict ourselves to the simple two-variable regression model. We conceptualize the problem in the following way.

FIXED-X MODEL

Suppose we are able to control the value of variable X at some particular value, say at $X = X_1$. At $X = X_1$, we have a distribution of Y values and a mean of Y for this value of X. There are similar distributions of Y and mean responses at each other fixed value of X, $X = X_2, X = X_3, \ldots, X = X_n$. The regression relationship of Y on X is derived by tracing the path of the mean value of Y for these fixed values of X. Figure 13-3 illustrates a regression relation for a fixed-X model. Now the regression function that describes the systematic part of this statistical relation may be of any form whatsoever. We restrict ourselves to linear regression equations graphed in Figure 13-4(a). The function that passes through the means has the equation

$$Y_i = a + bX_i \tag{13-1}$$

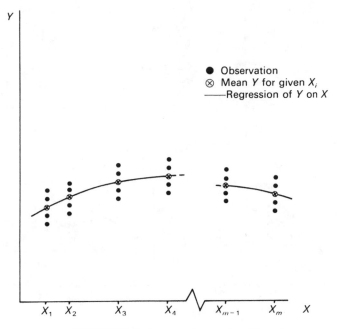

FIGURE 13-3. A regression relationship.

which is the familiar slope-intercept form of a straight line. Although we limit ourselves to linear regression equations in this chapter, regression analysis can be utilized in situations for which the path of the means follows many different functional forms. Figures 13-4(b) to (e) illustrate several potential alternatives and the appropriate functional forms.

EXAMPLE 13-3. To examine the rate of noise attenuation with distance from a major transport facility, sound recorders are placed at intervals of 50 ft over a range of 50 to 500 ft from the centerline of an expressway. Sound recordings are then analyzed, and the sound level in decibels that is exceeded 10% of the day is calculated. This measure, known as L_{10}, is a general indicator of the highest typical sound levels at a location. Recordings are then repeated for five consecutive weekdays by using the same sound recorders placed at the identical locations. A plot of the data generated from this experiment is shown in Figure 13-5.

The graph reveals the typical pattern of distance decay. The variation in decibel levels at any fixed location either is due to the effects of other variables—perhaps variations in traffic volumes—or can be thought to represent random error. Let us presume that all other variables that might have caused sound-level variations are more or less constant and that the variations do represent random error. Note that the path of the mean sound levels with distance follows the linear function superimposed on the scattergram. One would feel quite justified in fitting a linear regression equation to this scatter of points.

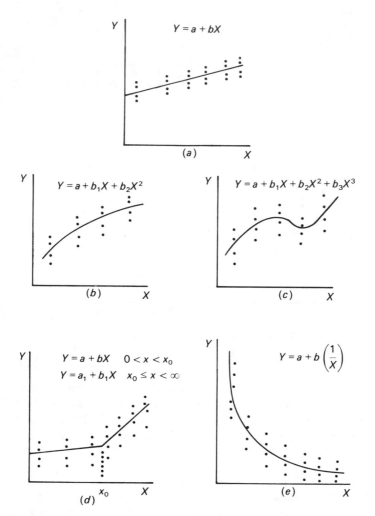

FIGURE 13-4. (a) Linear; (b) quadratic; (c) cubic; (d) piecewise linear; (e) reciprocal.

RANDOM-X MODEL

If situations in which we can control the values of X so that they are set at predetermined levels were the only conditions in which regression analysis could be applied, it would be of limited value in geographical research. Only rarely do geographers gather data of this form. They are more commonly collected in experimental situations. All the mechanical operations presented in this chapter can be performed on any set of paired observations of two variables X and Y. And, as we shall see in Chapter 14, virtually all the inferential results of regression analysis apply to data characteristic of the fixed-X model and to data in which both X and Y can be considered random variables. This greatly generalizes the applicability of the regression model. Let us consider an example of a random-X model.

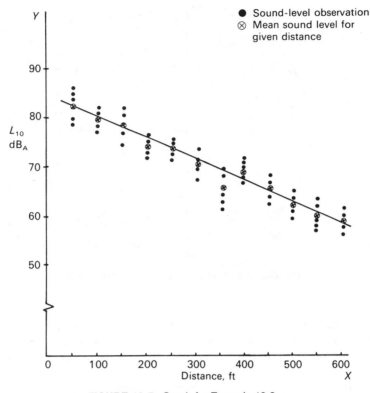

FIGURE 13-5. Graph for Example 13-3.

EXAMPLE 13-4. One of the key tasks in metropolitan area transportation planning is to make estimates of the total number of trips made by households during a typical day. Data collected from individual households are usually aggregated to the level of the *traffic zone*. Traffic zones are small subareas of the city with relatively homogeneous characteristics. The dependent variable is the number of trips made by a household per day, known as the trip generation rate. One of the most important determinants of variations in household trip generation rates is household income. We would expect the total number of trips made by a household in a day to be positively related to the income of the household.

Consider the hypothetical city of Figure 13-6 which has been divided into 12 traffic zones. Values of the average trip generation rates and household income for each of 12 zones are listed in Table 13-1. The data are typical for a North American city. Inner-city traffic zones 1 to 4 generally are composed of lower income households who make few trips per day. Suburban zones 5 to 12 are populated with higher income households with high trip generation rates. The scattergram of Figure 13-7 verifies the positive relation between trip generation and household income.

Notice how these data differ from the data of a fixed-X model. First, we do not have more than one observation of trip generation rates for any single level of household income. Second, we do not have observations of trip generation rates for systematic values of

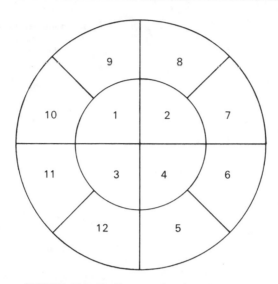

FIGURE 13-6. Traffic zones in a hypothetical city.

household income, say every $500 or $1,000 of income. Nevertheless, it appears that the simple linear regression of household trip generation on income represents a useful statistical relation. We are going to use this example throughout this chapter as well as in Chapter 14.

But first let us examine the nature of these data. In what sense can we think of variable X as being a random variable? We might think of household income as a random variable since it is derived from one particular spatial aggregation of households into traffic zones. The traffic zones of Figure 13-6 can be thought of as one random choice among all the possible ways in which we might build traffic zones from the household data. Other

TABLE 13-1
Trip Generation Rates and Household Income in a Hypothetical City

Traffic zone	Trips per household per day	Average household income $000
1	3	10
2	5	12
3	5	11
4	4	9
5	6	14
6	7	16
7	9	21
8	7	18
9	8	22
10	8	24
11	6	17
12	5	15

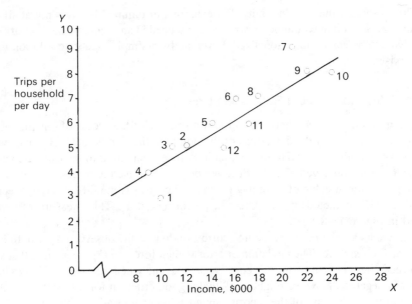

FIGURE 13-7. Scattergram for Example 13-4.

aggregations to different traffic zones would yield different average incomes and different trip generation rates. This is not a very persuasive argument. We would think of income as being a random variable if the original observations from which we generated Table 13-1 were a random sample taken from all the households in the city. If this were the case, then we might have some justification in treating income as a random variable. But what if the aggregated data represent all the households in the city? Can we make a case for treating income as a random variable? As you see, many applications of regression analysis in geography raise thorny theoretical issues. Most of these issues arise in situations in which we utilize regression analysis in an inferential mode. We discuss them further in Chapter 14.

13.2. Estimation of a Linear Regression Function

Ordinarily, we do not know the values of the regression parameters a and b, and we must estimate their values from our data. Once we know a and b, we can precisely locate our regression line within the scattergram. The method used to derive a and b from our set of observations does not depend on the type of input data at our disposal. It can be consistent with either the fixed-X or the random-X model. We will use the data from our trip generation example, given in Table 13-1.

Let us denote (X_i, Y_i) as the values of our two variables for the ith observation. There are n different observations, and thus i can take on the values 1, 2, . . . , n. Now if the scatter of points lies quite close to falling on a single line, we could probably do a reasonable job of fitting the line by eye. We could then simply estimate the values of a and b from the graph. This is seldom the case in reality. For example, in Figure 13-7, it is not at all clear where our regression line would be drawn. What we need is an objective method of fitting the line that can be easily operationalized. Fortunately, a simple algebraic solution to this problem exists.

Possible Criteria for Fitting a Regression Line

As a starting point, we must choose the criterion to assess the success of our line of best fit. An approach based on the deviations of the points to the line is ideal. Because we are interested in explaining the variation in variable Y, it seems natural to measure the vertical deviations from the observed Y to the line. We do this because the regression line should pass through the mean value of Y for any given X. Our error is the difference between the mean value of Y predicted by our line and the observed Y_i. This measure of error is illustrated in Figure 13-8.

First, we locate the point on the line corresponding to the observed X by defining $\hat{Y}_i$ (read "Y hat i") $= a + bX_i$. The deviation or error is therefore $e_i = (Y_i - \hat{Y}_i)$. Note that when the observed Y_i lies above the line, this error is positive, and when the observed Y_i lies below the line, it is negative. Now we might think that a good criterion for fitting the regression line is to minimize the sum of these errors taking into account all our observations:

$$\min \sum_{i=1}^{n} (Y_i - \hat{Y}_i) \tag{13-2}$$

Unfortunately, the positive deviations offset the negative deviations, and it is possible to draw several lines through the set of points—each with a zero sum of deviations. An example of this problem is illustrated in Figure 13-9.

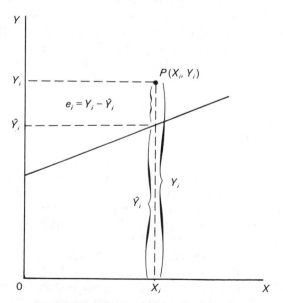

FIGURE 13-8. Error in fitting a line to a typical point.

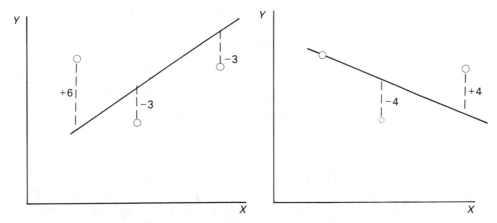

FIGURE 13-9. Both regression lines have $\Sigma_{i=1}^{3}(Y_i - \hat{Y}_i) = 0$.

To overcome this problem, we utilize the *least squares criterion:*

$$\min \sum_{i=1}^{n}(Y_i - \hat{Y}_i)^2 \tag{13-3}$$

which selects a and b to minimize the sum of squared deviations. If the line is very close to the points, then this sum is very small; as the points become increasingly spread about the line, this index increases. If all the observed points lie on the line, then the sum equals zero. The use of *squared* deviations overcomes the "sign problem" inherent in the use of Equation (13-2). Moreover, we see in Chapter 14 that there are important theoretical justifications for using this criterion.

Estimation through Least Squares

It is now time to ask, "How can we find the values of a and b that minimize the sum of squared deviations?" Let us substitute $Y_i = a + bX_i$ into Equation (13-3). The problem now becomes one of choosing a and b to

$$\min \sum_{i=1}^{n}(Y_i - a - bX_i)^2 \tag{13-4}$$

Because all the X and Y values are known, the value of this expression clearly depends on the values we select for a and b. As we change the values of these two parameters, we change the location of the line and also the value of this minimand. Although the details need not concern us, the solution requires that the following two "normal" equations be satisfied (students who have studied elementary calculus can refer to Appendix 13B for the derivation of these two normal equations):

$$na + b\sum_{i=1}^{n}X_i = \sum_{i=1}^{n}Y_i \tag{13-5}$$

$$a\sum_{i=1}^{n}X_i + b\sum_{i=1}^{n}X_i^2 = \sum_{i=1}^{n}X_iY_i \tag{13-6}$$

Let us take the first equation, (13-5). If we solve for a by dividing by n and simplifying, we get

$$a = \frac{\sum_{i=1}^{n} Y_i}{n} - b \frac{\sum_{i=1}^{n} X_i}{n} \qquad (13\text{-}7)$$

which, from the identities $\bar{X} = \sum_{i=1}^{n} X_i/n$ and $\bar{Y} = \sum_{i=1}^{n} Y_i/n$ leads to

$$a = \bar{Y} - b\bar{X} \qquad (13\text{-}8)$$

Since $\bar{X}$ and $\bar{Y}$ are calculated from the data, we can easily determine the intercept a once we know b. There is also one important implication that is derived from Equation (13-8). If we rewrite (13-8) as $\bar{Y} = a + b\bar{X}$, we see that the least squares line passes through the point $(\bar{X}, \bar{Y})$. Note that there is no error at this point.

How do we solve for b? Now that we have a, we can substitute (13-7) into our second normal equation (13-6):

$$-\sum_{i=1}^{n} X_i Y_i + \left(\frac{\sum_{i=1}^{n} Y_i}{n} - b \frac{\sum_{i=1}^{n} X_i}{n} \right) \left(\sum_{i=1}^{n} X_i \right) + b \sum_{i=1}^{n} X_i^2 = 0 \qquad (13\text{-}9)$$

This can be multiplied to yield

$$-\sum_{i=1}^{n} X_i Y_i + \left(\frac{\sum_{i=1}^{n} X_i \sum_{i=1}^{n} Y_i}{n} - b \frac{\left(\sum_{i=1}^{n} X_i \right)^2}{n} \right) + b \sum_{i=1}^{n} X_i^2 = 0 \qquad (13\text{-}10)$$

If we multiply by n and rearrange this equation, we arrive at

$$nb \sum_{i=1}^{n} X_i^2 - b \left(\sum_{i=1}^{n} X_i \right)^2 = n \sum_{i=1}^{n} X_i Y_i - \sum_{i=1}^{n} X_i \sum_{i=1}^{n} Y_i \qquad (13\text{-}11)$$

Finally, we solve for b:

$$b = \frac{n \sum_{i=1}^{n} X_i Y_i - \sum_{i=1}^{n} X_i \sum_{i=1}^{n} Y_i}{n \sum_{i=1}^{n} X_i^2 - \left(\sum_{i=1}^{n} X_i \right)^2} \qquad (13\text{-}12)$$

Equations (13-8) and (13-12) are the most straightforward computational formulas for solving linear regression problems by hand. Of course, we use computers for large-scale applications of linear regression analysis (small ones too!). Normally, the software packages use other computational formulas. By convention, a is known as the *regression intercept* or just *intercept*, and the parameter b is the *regression coefficient*.

Let us now return to our example of household trip generation. Table 13-2 contains the required calculations. It is an extremely useful tableau for summarizing the calculations required for the linear regression problem. Strictly speaking, we do not need the last column, which contains $\sum_{i=1}^{n} Y_i^2$, but we need this term to calculate the correlation coefficient

TABLE 13-2
Calculations for a Simple Linear Regression Problem

Observation No.	X	Y	XY	X^2	Y^2
1	10	3	30	100	9
2	12	5	60	144	25
3	11	5	55	121	25
4	9	4	36	81	16
5	14	6	84	196	36
6	16	7	112	256	49
7	21	9	189	441	81
8	18	7	126	324	49
9	22	8	176	484	64
10	24	8	192	576	64
11	17	6	102	289	36
12	15	5	75	225	25
$n = 13$	$\Sigma X = 189$	$\Sigma Y = 73$	$\Sigma XY = 1237$	$\Sigma X^2 = 3237$	$\Sigma Y^2 = 479$

$$\bar{X} = \frac{189}{12} = 15.75 \qquad \bar{Y} = \frac{73}{12} = 6.083$$

$$a = 6.083 - .335(15.75) \qquad b = \frac{12(1237) - (189)(73)}{12(3237) - 198^2}$$

$$= .807 \qquad\qquad\qquad = .335$$

between X and Y. As we shall see in Section 13-3, the correlation coefficient is a useful measure for assessing the goodness of fit of our regression line.

From Table 13-2 we see that the regression of household trip generation rates on household income yields

$$Y_i = .807 + .335X_i \tag{13-13}$$

where Y_i is the estimated or predicted mean number of trips generated for a given income level. Recall that a regression line follows the path of the means. It is relatively straightforward to graph our regression line within the scatter of observations. To precisely locate the line, we require two points. Although we can determine any two points simply by substituting two different values of X into (13-13), it is often simpler to use the points $(\bar{X}, \bar{Y})$ and $(0, a)$. We know $(\bar{X}, \bar{Y})$ is on the line because of (13-8), and by definition the parameter a is the intercept on the Y-axis. We simply connect these points by a straight line. If the intercept a is not close to the range of our data (this is a normal occurrence), we can utilize any other convenient point. In our example of household trip generation, the income variable is limited to range 9 to 24 ($000). We thus estimate $\hat{Y}_i = .807 + .335(20) = 7.057$ and obtain the point (20, 7.507). We use this point and our bivariate mean (15.75, 6.083) to draw the regression line in Figure 13-10.

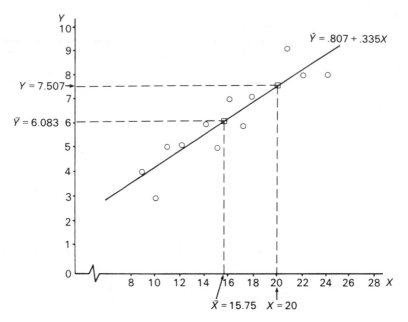

FIGURE 13-10. Drawing the regression line.

13.3. Assessing the Goodness of Fit

We now know that the least squares estimates of a and b given by Equations (13-8) and (13-12) provide the best possible fit of a straight line to the data according to criterion (13-4). Since we can calculate these two parameters from any set of observations having at least two different values of X, the question remains: How well does the line fit the original data? We cannot use the total sum of squared deviations as our index since it clearly *depends on the scale of the numbers involved as well as on the number of observations.* What we require are indices capable of assessing the goodness of fit of regression lines drawn from different sets of data. With such indices, we could compare the fit of several different regressions drawn from different data. Let us examine three different measures that can be used to assess the goodness of fit of a regression line: the *simple correlation coefficient r*, the *coefficient of determination r^2*, and the *standard error of estimate s_{yx}.*

Correlation Coefficient r

As we found in Chapter 12, the product-moment correlation coefficient called Pearson's r is a dimensionless measure of the degree of linear association between two variables X and Y. It has a range $-1 \le r \le +1$, and it can be used to compare the association between two sets of variables drawn from the same set of observations or of the same two variables drawn from different data sets. The closer r is to ± 1, the greater the strength of the linear association between the two variables. It represents a useful summary measure in regression analysis. If our two variables are positively related, then $r > 0$ and $b > 0$. If they are negatively related, then $r, b < 0$. We can thus say that a regression line with a correlation coefficient higher than that of another equation fits the data better.

As you might suspect, there is an important algebraic relation between b and r that shows the intimate relationship between correlation and regression. A quick comparison of Equations (12-4) and (13-12) should convince you of this fact. To illustrate this relationship, let us recall the definition of the correlation coefficient

$$r = \frac{\sum_{i=1}^{n}(X_i - \bar{X})(Y_i - \bar{Y})}{\sqrt{\sum_{i=1}^{n}(X_i - \bar{X})^2}\sqrt{\sum_{i=1}^{n}(Y_i - \bar{Y})^2}} \tag{13-14}$$

Let us define $x_i = X_i - \bar{X}$ and $y_i = Y_i - \bar{Y}$. With this pair of definitions we now rewrite (13-14) as

$$r = \frac{\sum_{i=1}^{n}(x_i)(y_i)}{\sqrt{\sum_{i=1}^{n}(x_i)^2}\sqrt{\sum_{i=1}^{n}(y_i)^2}} \tag{13-15}$$

This is often termed the *deviations form* of the correlation coefficient. It is used in many situations because it simplifies the notation of certain formulas such as the correlation coefficient. We can make a similar simplification of the regression coefficient:

$$b = \frac{\sum_{i=1}^{n}x_i y_i}{\sum_{i=1}^{n}x_i^2} \tag{13-16}$$

The similarity of (13-15) and (13-16) again suggests the relation between correlation and regression. To formalize this relationship, let us take the ratio of the regression coefficient to the correlation coefficient:

$$\frac{b}{r} = \frac{\sum_{i=1}^{n}x_i y_i / \sum_{i=1}^{n}x_i^2}{\sum_{i=1}^{n}x_i y_i / \left(\sqrt{\sum_{i=1}^{n}x_i^2}\sqrt{\sum_{i=1}^{n}y_i^2}\right)}$$

$$= \frac{\sqrt{\sum_{i=1}^{n}x_i^2}\sqrt{\sum_{i=1}^{n}y_i^2}}{\sum_{i=1}^{n}x_i^2} = \frac{\sqrt{\sum_{i=1}^{n}y_i^2}}{\sqrt{\sum_{i=1}^{n}x_i^2}} \tag{13-17}$$

Now if we divide both the numerator and denominator of the right-hand side of (13-17) by $1/\sqrt{n-1}$, we obtain

$$\frac{b}{r} = \frac{\sqrt{\sum_{i=1}^{n}y_i^2/(n-1)}}{\sqrt{\sum_{i=1}^{n}x_i^2/(n-1)}} = \frac{S_Y}{S_X} \tag{13-18}$$

by using the familiar formula for the standard deviation. We now express (13-18) as

$$b = r \frac{S_y}{S_X} \tag{13-19}$$

which indicates the direct algebraic link between correlation and regression analysis. This equation implies $b = 0$ if $r = 0$, and vice versa. Also, since S_Y and S_X, must always be positive, b and r must have identical signs.

Let us now return to our example of trip generation. In Table 13-3, we calculate the simple correlation coefficient between income and household trip generation as $r = .9153$. This indicates a very strong linear association between variables. Also, in this table we illustrate the numerical validation of the relation between correlation expressed by (13-19).

Coefficient of Determination r^2

The *residuals* from a regression line are defined by $e_i = (Y_i - \hat{Y}_i)$. There is one residual for each of the n observations. They provide extremely useful information on the extent to which the calculated regression line fits the data. Generally, a good regression line is one that has small residuals, and a poor regression is one with large residuals. What is needed is a measure of goodness of fit that uses these residuals and is unit-free. That is, our measure must be independent of the units in which our dependent variable Y is measured. The *coefficient of determination* is such a measure.

TABLE 13-3
Work Table for Illustrating Equation (13-19)

The simple correlation coefficient for the household trip generation data, obtained by using computational formula (12-4) with the data from Table 13-2, is

$$r = \frac{1237 - [(189)(73)]/12}{\sqrt{3237 - 189^2/12} \; \sqrt{479 - 73^2/12}}$$

$$= .9153$$

Using the computational formula for the standard deviation, we find

$$S_X = \sqrt{\frac{3237}{11} - \frac{189^2}{12(11)}} \qquad S_Y = \sqrt{\frac{479}{11} - \frac{73^2}{12(11)}}$$

$$= 4.864 \qquad\qquad\qquad = 1.7816$$

$$b = .9153 \left(\frac{1.7816}{4.864} \right)$$

$$= .335$$

This agrees with the calculation of b in Table 13-2.

Let us begin by defining the variation of variable Y about its mean as

$$\sum_{i=1}^{n}(Y_i - \bar{Y})^2 \tag{13-20}$$

We now wish to divide this total variation into two distinct parts. The first part we term the part *due to regression*, or *explained by regression*, or *explained by variable X*. The second part is the *residual variation*, which is *unexplained by the regression*, or the *error variation*.

Let us assume that we fit a regression model to our sample of observations and found the regression coefficient b to be zero. Then our best prediction for Y given any value of X is

$$\hat{Y}_i = a + 0X_i = \bar{Y} \tag{13-21}$$

and our knowledge of the value of variable X is of no use in improving our prediction of Y. Without knowing the value of X, we would predict $\bar{Y}$ since it lies in the center of the distribution. In this case, the variation of Y about its mean (13-20) represents the sum of the square of the differences between our observed Y and the predicted values $\hat{Y}_i = \bar{Y}$. Now suppose that the regression coefficient or slope of the fitted line is nonzero. We can improve the prediction of Y by accounting for the fact that the predicted values of Y_i are dependent on our observation X_i and given by $\hat{Y}_i = a + bX_i$.

To see this improvement, consider the following identity, geometrically illustrated in Figure 13-11:

$$Y_i - \bar{Y} = (Y_i - \hat{Y}_i) + (\hat{Y}_i - \bar{Y}) \tag{13-22}$$

On the left-hand side is the difference between the observed Y_i and the mean $\bar{Y}$. This is then decomposed into two parts. The first term on the right-hand side is the residual, or error e_i, and the second part represents the difference between the predicted values of Y and $\bar{Y}$. This second part represents the improvement "due to the regression of Y on variable X." We now expand this relation by summing observations and squaring both sides:

$$\sum_{i=1}^{n}(Y_i - \bar{Y})^2 = \sum_{i=1}^{n}(Y_i - \hat{Y}_i)^2 + \sum_{i=1}^{n}(\hat{Y}_i - \bar{Y})^2 + 2\sum_{i=1}^{n}(Y_i - \hat{Y}_i)(Y_i - \bar{Y}) \tag{13-23}$$

To square the right-hand side of (13-22), we have used $(a + b)^2 = a^2 + 2ab + b^2$. Although the details need not concern us, the last term on the right-hand side of (13-23) always equals zero, leaving

$$\sum_{i=1}^{n}(Y_i - \bar{Y})^2 = \sum_{i=1}^{n}(Y_i - \hat{Y}_i)^2 + \sum_{i=1}^{n}(\hat{Y}_i - \bar{Y})^2$$

or

$$\text{TSS} = \text{ESS} + \text{RSS} \tag{13-24}$$

where TSS is the total variation of Y, or total sum of squares; ESS is the error variation, or error sum of squares; and RSS is the explained variation or regression sum of squares.

We have still not developed a unit-free measure of goodness of fit, since all of our

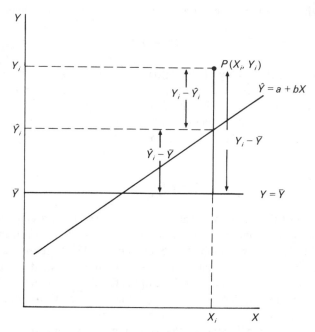

FIGURE 13-11. Decomposition of the total variation.

components of variation depend on the units in which the dependent variable Y is measured. To eliminate this difficulty, let us divide both sides of (13-24) by TSS to get

$$1 = \frac{\text{ESS}}{\text{TSS}} + \frac{\text{RSS}}{\text{TSS}} \tag{13-25}$$

We can now define the coefficient of determination.

DEFINITION: COEFFICIENT OF DETERMINATION r^2

$$r^2 = 1 - \frac{\text{ESS}}{\text{TSS}} = \frac{\text{RSS}}{\text{TSS}} \tag{13-26}$$

From this definition we see that r^2 can be interpreted as the proportion of the total variation in variable Y that is explained by the regression of variable X. Since $0 < \text{ESS} < \text{TSS}$, we see that $0 < r^2 < 1$. When ESS = 0, we have a perfect fit and $r^2 = 1$, and our statistic tells us that we have explained 100% of the variation in Y. When ESS = TSS, $r^2 = 0$, and variable X is of no value in a prediction of Y. It explains 0% of the total variation. Most often, r^2 lies between these extremes.

In general, a high value of r^2 means we have a good fit, and a low value means we have a poor fit. However, it is no accident that we define the coefficient of determination as r^2. This is because it is always numerically equal to the square of the correlation coefficient r in any simple regression problem. We can use this relationship to aid in the interpretation of the correlation coefficient. An r^2 of .8 implies that variable X "explains" 80% of the variation in variable Y. This corresponds to a correlation coefficient of approximately $r = .90$. We can thus translate r into values of r^2 and get a better "feel" for the strength of the association. The seemingly high correlation of $r = .64$ corresponds to an $r^2 = .41$ and implies that only about 40% of the variation in Y can be explained by variable X. A correlation of $r = .40$ might seem quite weak when we realize that only 16% of Y is explained by variations in X.

The coefficient of determination should also be used with caution. Geographers often place undue emphasis on obtaining a high value of r^2. Although we speak in terms of explanation, in many instances we do not have a truly causal relationship, and we should speak in terms of *statistical* explanation, not *causal* explanation. The lack of a strong theoretical base in geography means that much empirical work is not based on a sufficient theoretical foundation to be spoken of in terms of $\Sigma_{i=1}^{n}(Y_i - \bar{Y})^2 = \Sigma_{i=1}^{n}(Y_i - \hat{Y}_i)^2 + \Sigma_{i=1}^{n}(\hat{Y}_i - \bar{Y})^2$ causal explanation. Also r^2 alone is not a suitable measure to assess the utility of a regression model. We often obtain a low value of r^2 in cases in which there is a large variation across the individual units of observation. In Chapter 12 we witnessed the dependence of the correlation coefficient on the level of aggregation in the data.

Consider again the example of trip generation. If the data are based on individual households, they would probably exhibit a great deal more variation than when they were aggregated to the level of the traffic zone. We may have aggregated a number of elderly immobile households with a number of highly mobile younger households. The average trip rates and incomes of the traffic zones in this case would give a poor representation of the actual variation in these two variables in the metropolitan area. While we might be able to explain the variations in trips at the level of the traffic zone, we would probably not be able to predict or explain the variation in individual household trip generation rates with the same precision. In this case, the coefficient of determination overestimates the goodness of fit of the relationship between these two variables.

Also, it is very difficult to utilize the coefficient of determination as defined in (13-26) in a fixed-X model. A quick glance at Figure 13-5 shows why. It is impossible to predict each observed Y_i exactly, since we have different values of Y_i for one individual value of X_i. However, our equation can predict only one value of $\hat{Y}_i$ for all these observations. So $r^2 = 1$ cannot be achieved. A modified r^2 is used in this instance. In some experimental situations in physical geography, we might require this modified r^2 as a true measure of goodness of fit. Suppose we are interested in the relation between stream velocity and sediment load. We set up an experiment in the laboratory using a flume. The bottom of the flume is packed with materials typical of stream beds. We then subject the flume to a given flow velocity and measure the suspended sediment. The experiment is repeated, and measurements are recorded several times at many different flow velocities. This approach generates data for a fixed-X regression model of suspended sediment versus flow velocity. A modified r^2 [see Draper and Smith (1981, 33–43)] is required to assess goodness of fit.

Finally, it is possible to achieve unrealistically high values of r^2 in many time series studies. In time series studies, we measure one or more variables at different times. One variable growing over time is quite likely to do a good job at explaining the variation in any other variable that also is growing over time—whether or not the two variables are linked in any causal way. In sum, the coefficient of determination represents a useful summary measure of goodness of fit when it is interpreted with caution. A high value of

r^2 might, on reflection, be virtually meaningless. A low value of r^2 might be very meaningful if the dependent variable is subject to more random than systematic variation.

Let us complete our discussion by illustrating the calculation of r^2 (see Table 13-4). We might wish to keep this table separate from Table 13-2 or else just add the final four columns to Table 13-2 and make a composite table. The total variation is 34.918, the unexplained (or residual, or error) variation is 5.666, and the explained variation is 34.918 – 5.666 = 29.252. The geometric equivalents to the components of variation are illustrated in Figure 13-12. To check the calculation r^2 = .838, we compare it to the square of the correlation coefficient r from Table 13-3. Notice that .838 = .9153^2. This is a useful numerical check for your calculations.

Standard Error of Estimate $S_{Y \cdot X}$

In many applications of regression analysis, the ultimate goal is to come up with a good estimate or prediction for Y by using variable X. Our success is measured by the degree to which we can predict variable Y accurately. In this case, we are primarily interested in the numerical value of the error we are likely to make when utilizing X to predict Y. The best measure of goodness of fit in this instance is the *standard error of the estimate $S_{Y \cdot X}$*. The standard error of the estimate is defined as the standard deviation of the residuals about the regression line:

$$S_{Y \cdot X} = \sqrt{\frac{\sum_{i=1}^{n}(Y_i - \hat{Y}_i)^2}{n - 2}} \qquad (13\text{-}27)$$

TABLE 13-4
Work Table for Calculation of $S_{Y \cdot X}$ and r^2

Obs. No.	X_i	Y_i	$\hat{Y}_i$	$Y_i - \hat{Y}_i$	$(Y_i - \hat{Y}_i)^2$	$(Y_i - \bar{Y})^2$
1	10	3	4.156	−1.156	1.336	9.505
2	12	5	4.826	0.174	0.030	1.173
3	11	5	4.491	0.509	0.259	1.173
4	9	4	3.820	0.180	0.032	4.339
5	14	6	5.497	0.503	0.253	0.007
6	16	7	6.167	0.833	0.694	0.841
7	21	9	7.843	1.157	1.339	8.509
8	18	7	6.838	0.162	0.026	0.841
9	22	8	8.179	−0.179	0.032	3.675
10	24	8	8.849	−0.849	0.721	3.675
11	17	6	6.502	−0.502	0.252	0.007
12	15	5	5.832	−0.832	0.692	1.173
		$\bar{Y} = 6.083$		$\Sigma(Y_i - \hat{Y}_i) = 0.0$	$\Sigma(Y_i - \hat{Y}_i)^2 = 5.666$	$\Sigma(Y_i - \bar{Y}_i)^2 = 34.198$

$$S_{Y \cdot X} = \sqrt{\frac{5.666}{12 - 2}} \qquad r^2 = \frac{34.918 - 5.666}{34.918}$$

$$= .752 \qquad\qquad = .838$$

The subscript $Y{\cdot}X$ is used to denote the fact that we are not calculating the standard deviation of Y but the standard deviation of Y, taking into account that variable Y is regressed on variable X. The *standard error of estimate* is sometimes called the *root mean square error*, or *root MSE*. Notice that it is calculated as the square root of the average or mean squared error. Why do we divide by $n - 2$ and not n? As we shall see in Chapter 14, $S_{Y\cdot X}$ is really

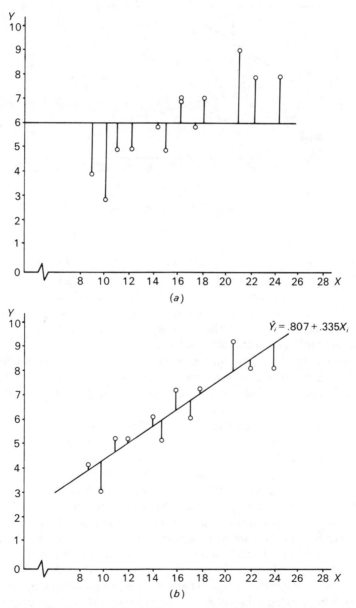

FIGURE 13-12. Geometrical representation of (a) total variation, (b) residual variation, and (c) explained variation.

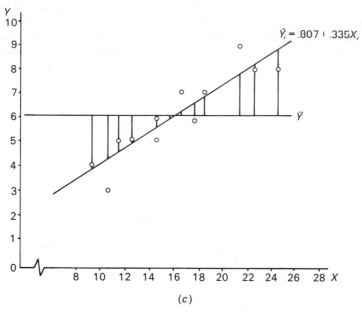

$$\hat{Y}_i = .807 + .335X_i$$

(c)

FIGURE 13-12 (continued).

an *estimate* of the standard deviation of the error term in our model. We treat our data of n observations as if they were a sample from a population. Recall that when we make an estimate in inferential statistics, we divide by the appropriate degrees of freedom, not the sample size. Most computer programs for regression calculate the standard error of estimate by using $n - 2$, not n, in the denominator. Strictly speaking, we might wish to divide by n if we consider the data a statistical population.

Why are the degrees of freedom equal to $n - 2$? This is because we lose 2 degrees of freedom in the data when we estimate the regression parameters a and b since they require $\bar{X}$ and $\bar{Y}$ in their calculations. We can also use a straightforward geometric argument. Because two points are needed to determine a line, any point in excess of $n - 2$ provides a degree of freedom in defining the location of the line. Thus $n - 2$ degrees of freedom are always associated with a simple regression based on n observations.

Once again let us return to our example of household trip generation. Table 13-4 illustrates the calculations for the standard error of estimate. We simply take the square root of the residual variation, or ESS, divided by the number of observations minus 2. Note the residual sum of squares is always generated in the computation of r^2. In this case $S_{Y \cdot X} = .752$. Since the standard error of estimate is always measured in units of the dependent variable Y, we can interpret this measure of goodness of fit as corresponding to approximately three-quarters of a trip per household per day. But does this represent a "good" fit? Two previous measures of goodness of fit suggested our relation to be extremely close: $r = .9153$ and $r^2 = .838$. Is a prediction of a household's trip generation rate within three-quarters of a trip per day as good as the values of r and r^2 seem to suggest? We can make this judgment only within the context of the problem at hand, that is, in relation to some predetermined standard of prediction.

If the standard error of estimate is .752, then 1 standard deviation on either side of the regression line roughly corresponds to 1.5 trips per household per day. Two standard deviations corresponds to three trips. If the errors about the regression line are normally distributed (and in the next chapter we make this very assumption for inferential purposes), then we expect roughly 68% of our observations to fall within 1 standard deviation about the line and 95% to fall within 2 standard deviations. So our predictive ability could be very crudely estimated as a range of about three trips per day 95% of the time. Unfortunately, for a city with 100,000 households this would mean a potential range of about 300,000 (100,000 × 3) trips in any given day! Put in this context, our equation is not nearly as useful as we might have thought—at least in the context of transportation forecasting. Thus, we should never base our judgment about goodness of fit of a regression line on one single measure.

Finally, let us consider a standardized measure of goodness of fit known as the *coefficient of variation*. Recall that in Chapter 2 we defined the coefficient of variation CV to be the ratio of the standard deviation to the mean. We use this statistic to compare the variability of two or more variables with different means and standard deviations. In the context of regression, the appropriate definition is

$$CV = \frac{S_{Y \cdot X}}{\bar{Y}} \times 100 \tag{13-28}$$

where we have multiplied by 100 to express the coefficient as a percentage. If $CV = 0$, this implies that $S_{Y \cdot X} = 0$ and we have a perfect fit. When $b = 0$ and the line of best fit reduces to $\hat{Y} = \bar{Y}$, then $CV = 100$. In most cases the CV lies between these extremes. The lower the CV, the better the relative fit of the regression line. However, we must be careful. It is possible to generate an equation with an $S_{Y \cdot X}$ that is acceptable in comparison to some given standard of prediction but that has an associated CV higher than some other equation that may be unacceptable on these grounds.

13.4. Interpreting a Regression Equation

To this point, we have described the situations in which regression can be applied, the method for calculating the line of best fit, and several measures of goodness of fit. What does all this information tell us? In any application of regression analysis, it is often useful to examine our equation within the context of the empirical problem at hand. We have just witnessed the utility of this approach in interpreting the standard error of estimate as a measure of goodness of fit. We now review the output from our regression equation in this light.

Units of Measure

Table 13-5 summarizes the units of measure of the various terms of a regression equation and describes their interpretation in the example of generation. First, we must remember that a regression equation provides a formal statistical link between two variables measured in different units. Any regression equation can be expressed as.

$$\text{Units of } Y = a + b(\text{units of } X) \tag{13-29}$$

The dimensions on both sides of this equation must be equal. It thus follows the right-hand side must be measured in units of Y. Knowing this, we can say (1) the intercept a must be measured in units of Y and (2) the slope or regression coefficient b is measured in units of Y per unit of X. Making these substitutions, we see

$$\text{Units of } Y = \text{units of } Y + [\text{units of } Y/(\text{units of } X)] \cdot (\text{units of } X)$$

$$\text{Units of } Y = \text{units of } Y \tag{13-30}$$

and we see that these units of measure ensure dimensional equivalence of both sides of the equation.

The intercept a is thus measured in units of Y and is interpreted as the value of Y when $X = 0$. For the problem in household trip generation, $a = .807$ implies that a household with no income makes, on the average, .807 trips per day. Is this plausible? The answer depends on whether the range of X over which the equation is estimated includes the value $X = 0$. If so, we might have some a priori reason to expect a to take on a positive value, a negative value, or zero. In our case we might prefer a zero intercept, since a household with no income should make no trips. However, the range of X in the data is limited to $9 < X < 24$. So we really have no knowledge of the value of Y in the vicinity of $X = 0$. Were we to have data from a broader range of income levels, we might have different expectations. It is possible that the regression model depicted in Figure 13-13 is applicable to the problem. The true regression is a nonlinear model that has a zero intercept. However, when the data are limited to the region labeled *observed range of X*, a linear equation is a reasonable

TABLE 13-5
Dimensions of Regression Model Terms

Term	Notation	Units	Units in example
Dependent variable	Y	—	Trips per household per day
Independent variable	X	—	$000s of income
Intercept	a	Y	Trips per household per day
Regression coefficient	b	Units of Y per unit of X	Trips per household per day per $000s of income
Residuals	$e_i = Y_i - \hat{Y}_i$	Y	Trips per household per day
Standard error of estimate	$S_{Y \cdot X}$	Y	Trips per household per day
Correlation coefficient	r	Dimensionless	
Coefficient of determination	r^2	Dimensionless	
Coefficient of variation	CV	Dimensionless	

estimate of the relationship in the range. In this case a nonzero intercept is meaningful. Of course, we could also argue that there is some minimum threshold income below which no observation exists, and therefore our intercept is really meaningless.

Now let us turn to the regression coefficient b. It is interpreted to be the rate of change of the dependent variable with respect to the independent variable X. We might also have some a priori grounds for expecting b to be of a certain sign and/or of a certain order of magnitude. If we are modeling a truly causal relationship, then we most certainly should know at least the sign of the regression coefficient. What if our regression calculations yield a value of b with the opposite sign to our expectations, but the goodness of fit statistics indicate high values of r and r^2? Suppose that the regression coefficient in the trip generation example turned out to be $b = -.335$, not $b = +.335$? Given our expectations that increasing incomes lead to higher trip rates—especially over the income range in the data—we would have to be skeptical of the regression results. We may have a "poor" set of data with enormous measurement errors or perhaps just a poorly chosen sample. Or the aggregation might have spoiled the data. If not, we must turn to the theory upon which we base our expectations for the sign of b. Perhaps certain preconditions or assumptions are inherent in the theory that render it inapplicable to our data. Or, we might even improve the theory by suggesting other variables that must be incorporated.

Change of Scale

What if we were to change the units of measurement of either the dependent variable Y or the independent variable X? What happens to the values of the regression equation parameters and the goodness of fit statistics? It turns out that the changes are exactly what we

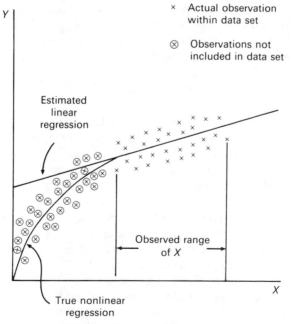

FIGURE 13-13. Regression model.

would expect and hope for. First, let us consider the independent variable X. Let us change X by measuring household income in tens of thousands dollars rather than in thousands of dollars. Instead of the sequence 10, 12, ..., 15 for X, we have $X' = 1, 1.2, ..., 1.5$. Each X' is one-tenth as large as each X so we can say $X' = 0.1\ X$. The new equation is

$$Y_i = .807 + 3.35X'_i \qquad\qquad (13\text{-}31)$$

and the new regression coefficient is 10 times as large as the old one: $b' = 10b$. But has the response of trip generation to changes in income been affected? No. Note that a \$10,000 change in income in the old regression Equation (13-13) predicts a $10 \times .335 = 3.35$ increase in trips per household per day. This is exactly the estimate given by the regression coefficient in (13-31). In general, if $X' = kX$, we find $b' = (1/k)(b)$. In the previous example $k = .1$ and thus $b' = 10b$. There is no change in intercept a. All we have done is to stretch or compress the X-axis without changing the Y axis. You may wish to prove these two results by substituting $X' = kX$ into (13-8) and (13-12).

We can also determine the effects of changes in the units of measure of variable Y. A similar analysis reveals that a change to $Y' = kY$ leads to $a' = ka$ and $b' = kb$. Whether we change X or Y or both X and Y, we do not change the magnitudes of the dimensionless measures r, r^2, or CV. All are invariant with respect to scale changes in variables X and Y. These results are very important. Regression analysis would be of limited value if we could change the goodness of fit of the equations simply by changing the scale of the numbers involved.

Interpolation and Extrapolation

One of the most common uses of regression analysis is prediction. We seek to predict or estimate the value of Y given any value of X. To illustrate the use of our regression equation, suppose we wish to predict the number of trips per household per day made by a household with an income of \$19,000. We simply substitute $X = 19$ (remember, we measure income in thousands of dollars) into the equation and obtain

$$\hat{Y} = .807 + .335(19) = 7.172 \qquad\qquad (13\text{-}32)$$

trips per day. Whenever the value of X lies within the range of our data, this process is known as *interpolation*. We can even roughly interpolate by eye, using our scattergram and regression equation. Remember, we assumed that the regression line follows the path of the means within our range of X. So we expect the regression line to pass through the mean value of Y for any value of X. Where the data contain large gaps in the value of X, this prediction may be subject to error.

Sometimes we seek to predict Y for a value of X beyond the range of the data. This is known as *extrapolation*. Generally, we are in a much poorer position because we have no knowledge of the value of Y in this range of X. We encountered the same difficulty when we tried to interpret the intercept value a when the value $X = 0$ was not within the range of the data. The problem is that the function that passes through the means may not continue to follow the linear function postulated for purposes of calibration. In Figure 13-4 notice how many different functions are linear or near linear over part of the range, but then make significant changes in slope over other parts of the range. In Chapter 14 we return to this forecasting problem and illustrate the method used to make *point and interval estimates* for $\hat{Y}$ at any desired level of confidence. Our results clearly illustrate the dangers of extrapolation.

13-5. Technical and Methodological Issues

So far we have focused on the mechanics of fitting a regression line, assessing its goodness of fit, and interpreting the parameters of the model. We close this chapter with a brief exploration of three particular problem areas. All three are relevant to both the computational operations involved in a regression model and questions of interpretation.

Data Aggregation

We have repeatedly stressed that all descriptive and inferential statistics are inextricably bound to the data from which they are derived. Any changes in these data lead to significant changes in the numerical values of the summary statistical interpretations. This issue arises in regression analysis in much the same way as we saw in correlation analysis. Let us examine the consequences of two potential changes to the data. First, suppose we aggregate the data by taking the original observations and substituting an average measure for groups of individual observations. In Chapter 12 we called the variations in the value of the correlation coefficients with increased aggregation the *scale problem.* Given the close relationship between correlation and regression analysis, it is not surprising that the same problem exists when we estimate the linear regression equation.

A simple example will illustrate. Let us aggregate the original 12 traffic zones into six *traffic districts* according to the scheme summarized in Table 13-6 and illustrated as aggregation A in Figure 13-14(a). For example, traffic *districts* 1 is formed from traffic *zones* 1 and 9. Assuming the zones have an equal number of households, we compute new values of X and Y as the mean values for the two zones:

$$\left(\frac{X_1 + X_9}{2}, \frac{Y_1 + Y_9}{2}\right) = \left(\frac{10 + 22}{2}, \frac{3 + 8}{2}\right) = (16, 5.5)$$

Table 13-7 summarizes the results of the regression analysis for the original data with $n = 12$ observations and the $n = 6$ observations of aggregation A.

First, we notice the changes in the estimates of the regression parameters a and b. The intercept *increases* to 1.299, and the coefficient *decreases* to .304. Second, there are substantial differences in the various measures of goodness of fit. The value of r^2 declines appreciably, suggesting that income explains only 72% of the variation in household trip generation rates. In this sense, our equation is worse with this new aggregation. But note the reduction in the standard error of the estimate. Does this mean we can now predict trip rates more accurately by using this equation? Unfortunately, no. The reason for this reduction in $S_{Y \cdot X}$ is related to the significant decline in the overall variation in Y as measured

TABLE 13-6
Summary of Data for Alternative Spatial Aggregations

Traffic district	Traffic zones included	X	Y
	Aggregation A		
1	1, 9	16	5.5
2	2, 8	15	6
3	3, 12	13	5
4	4, 5	11.5	5
5	6, 7	18.5	8
6	10, 11	20.5	7
	Aggregation B		
1	1, 2	11	4
2	3, 4	10	4.5
3	5, 6	15	6.5
4	7, 8	19.5	8
5	9, 10	23	8
6	11, 12	16	5.5
	Aggregation C		
1	1, 10	17	5.5
2	2, 7	16.5	7
3	3, 11	14	5.5
4	4, 6	12.5	5.5
5	5, 12	14.5	5.5
6	8, 9	20	7.5

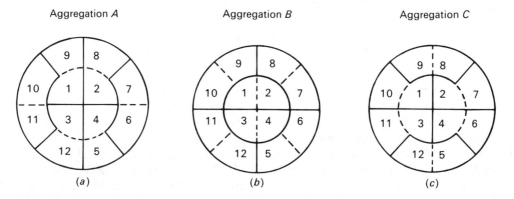

FIGURE 13-14. Alternative spatial aggregations.

TABLE 13-7
A Comparison of Regression Results for Different Aggregation Schemes

	Original data
$n = 12$	$\hat{Y}_i = .807 + .335X_i$ $r = .9153 \quad S_{Y \cdot X} = .753$ $r^2 = .838 \quad \text{TSS} = 34.918$
	Aggregation A $\hat{Y}_i = 1.299 + .304X_i$
$n = 6$	$r = .850 \quad S_{Y \cdot X} = .708$ $r^2 = .722 \quad \text{TSS} = 7.208$
	Aggregation B $\hat{Y}_i = 1.004 + .322X_i$
$n = 6$	$r = .932 \quad S_{Y \cdot X} = .695$ $r^2 = .869 \quad \text{TSS} = 14.708$
	Aggregation C $\hat{Y}_i = 1.798 + .272X_i$
$n = 6$	$r = .789 \quad S_{Y \cdot Y} = .630$ $r^2 = .622 \quad \text{TSS} = 4.208$

by TSS. Note that the total variation, or sum of squares, falls from 34.918 to 7.208. Put simply, there is simply less variation to explain. A quick examination of the data for the aggregation in Table 13-6 reveals that trip rates for the zones vary only from 5.5 to 8. In the original data in Table 13-1, household trip generation rates vary from three to nine trips per day.

To see why this equation is *not* superior to the original equation, we calculate the standard error of estimate for this new equation, *using the original 12 observations.* We substitute the values of X corresponding to the 12 traffic zones and compare the estimated trip rates to those given in Table 13-1. We find $S_{Y \cdot X} = .799$. This is clearly above the standard error of estimate from the traffic zone data of .753.

We can also illustrate variations in our equations and summary statistics if we examine different spatial aggregations for traffic districts. Even at the same scale of analysis, the use of different areal units leads to different results. Many geographers call this the *modifiable areal units problem.* Table 13-6 and Figures 13-14(a) to (c) compare three different aggregations of the original 13 traffic zones into six traffic districts. We label these aggregations A, B, and C, respectively. The regression results for each are compared in Table 13-7. Note that the regression coefficient b is always lower than in the original equation and the intercept is always higher. In fact, it is possible to define spatial aggregations that raise the slope and lower the intercept. We can never be sure of the effects of aggregation. Some aggregations lead to lower estimates of b while others lead to higher estimates. Also note the variations in the values of the correlation coefficients and the coefficients of determination. Aggregation B leads to an improvement in r^2 while the other two aggregations lead to a reduced r^2.

Finally, compare the three standard errors of estimate. All equations appear to be superior predictors of trip rates compared to the original equation. But this is really illusory. Most often the reduction in the standard error results from a significant decline in total variation TSS. Usually, these two go hand in hand. Note that aggregation C reduces the total variation in trip rates by over 83%. The sensitivity of regression equations to the size and spatial arrangement of the zones cannot be overemphasized. Our results are *valid only for the particular set of zones we use in the analysis*. When we must aggregate, it is again useful to follow the rules suggested in Chapter 12.

Specification of a Regression Equation

Although there is not a specific requirement that the dependent variable be causally related to the independent variable, we should nevertheless be extremely careful in the specification of the regression equation. Why? Because, unlike correlation analysis, regression relations are not symmetric. The regression of Y on X is not the same as the regression of X on Y unless the relation is perfect and all points lie on a straight line. If there is no relation between X and Y, and $b = 0$, then the two regression lines are perpendicular. The closer the correlation coefficient is to +1 or −1, the closer the two regression lines are. To show why the regression of Y on X and that of X and Y are different, we need only examine Equation (13-13), which defines b. For the regression of Y on X we use (13-13) to calculate b. For the regression of X on Y, we must interchange X and Y in the formula. This will not change the numerator but will change the denominator substantially since it would be a function of Y and not X.

In our example we know $r = .9153$ and expect two similar regression lines. Figure 13-15 verifies this. The difference between the lines is that the regression of Y on X is generated by minimizing the vertical deviations from the points to the line, whereas the regression of X on Y uses the horizontal ones. If we express both functions in the form $Y = f(X)$, the regression of X on Y can be re-expressed as $Y = -.220 + .400X$. (To do this yourself, simply solve $X = .549 + 2.499 Y$ for Y.) This is extremely close to the regression of Y on X which is $Y = .807 + .335X$.

Of course, this lack of symmetry is of little concern if we have a causal relation that can be specified on a priori grounds. In our example, it would be foolish to make a household's income a function of the number of trips made by the household. When we are utilizing our equation for prediction, either we must be capable of controlling the value of X (as in an experimental situation) or we must be able to predict its value independently. An equation to predict Y by using some independent variable X is of little value if we cannot provide the future values of X that we wish to use in our prediction! If we plan to use our trip generation equation to predict future household trip generation rates, we must have some way of predicting the future zonal distribution of household income.

Outliers and Extreme Values

Sometimes when we calculate the least squares regression of a relationship, we notice that one or more of the observations has an extremely large residual. Such observations are termed *outliers*—points greater than some arbitrarily defined distance from the regression line. Our attention should be drawn to any observation that is far from the systematic

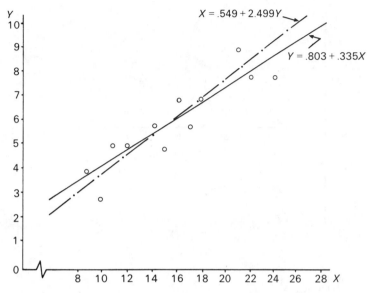

FIGURE 13-15. Comparing the regressions of Y on X and X on Y.

relation specified by the regression line. The quickest way to identify outliers is to plot the residuals versus the predicted value of Y. Table 13-8 lists the residuals for the 12 traffic zones, and Figure 13-16(a) illustrates the plot of these residuals versus Y. We notice no outliers, although traffic zones 1 and 7 have larger residuals than most others. One problem in examining residual plots such as Figure 13-16(a) is that the residuals are measured in units of the dependent variable. It is difficult to compare the residuals from two different regressions that are based on different dependent variables.

TABLE 13-8
Residuals for Trip Generation Example

Observation No.	$\hat{Y}_i$	Residual $Y_i - \hat{Y}_i$	Standardized residual $\dfrac{Y_i - \hat{Y}_i}{S_{Y \cdot X}}$
1	4.156	−1.156	−1.537
2	4.826	0.174	0.231
3	4.491	0.509	0.677
4	3.820	0.180	0.239
5	5.497	0.503	0.669
6	6.167	0.833	1.108
7	7.843	1.157	1.539
8	6.838	0.162	0.215
9	8.179	−0.179	−0.238
10	8.849	−0.849	−1.129
11	6.502	−0.502	−0.668
12	5.832	−0.832	−1.106

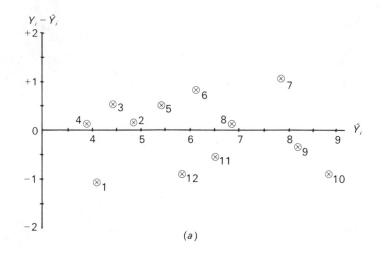

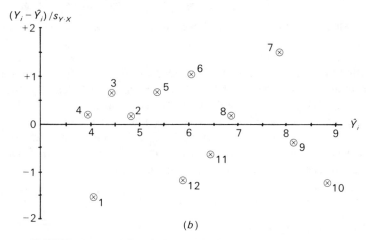

FIGURE 13-16. (a) Residual plot; (b) plot of standardized residuals.

One way of overcoming this problem is to compute and plot *standardized residuals* from

$$Z_i = \frac{Y_i - \hat{Y}_i}{S_{Y \cdot X}} \tag{13-33}$$

This expresses each residual as a standard normal deviate. We can now define an *outlier* as any observation with a standard score Z less than −3.0 *or* greater than 3.0. If the residuals are normally distributed, then an outlier would have to be far removed from the regresion line, at least 3 standard deviations away. We have selected 3 standard deviations quite arbitrarily. If most of the residuals are quite small, then a residual 2.5 standard deviations away from the line might be an outlier. Some judgment is involved here. However, since we have standardized the residuals, we can compare different regression equations, using this criterion. Figure 13-16(b) illustrates a plot of the standardized residuals for the trip generation problem. We note that all the residuals are within 2 standard deviations of the regression line, and there are not outliers. Also the standardization of the residuals does not change the *pattern* of the residuals, and Figure 13-16(a) and (b) reveals quite similar patterns.

One characteristic of the last squares procedure is that the parameter estimates *a* and *b* are extremely sensitive to the presence of outliers, because we *square the deviations*. Large residuals are penalized in the fitting procedure. This effect is similar to one we first noticed in Chapter 2. Recall the sensitivity of the mean of a set of numbers to the presence of an exteme value. For example, suppose we made a mistake in recording the trip generation rate of traffic zone 1. Rather than it being three trips per household per day, suppose is is actually 10 trips per household per day. As you can readily see in Figure 13-17, this dramatically changes the known scatter of observations. The new regression equation with this corrected information yields

$$Y_i = 3.822 + .181 \, X_i \tag{13-34}$$

Notice the changes in the two parameters. Not surprisingly, we also get very different goodness of fit statistics. The new r^2 is .2315, and the new standard error is 1.679. Figure 13-17 illustrates the two regression lines. Note the large residual of traffic zone 1 from the first regression line, and see how it is significantly reduced in the new equation. The moral is clear: Regression lines are "drawn" to outliers.

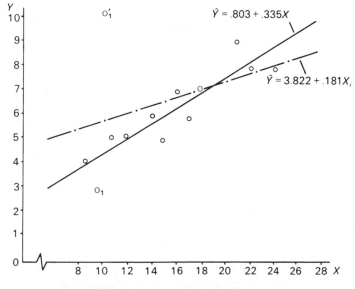

FIGURE 13-17. Effects of an outlier.

In this case we have illustrated the sensitivity of the least squares estimating procedure to outliers by using the example of a measurement error. What if the outlier is a "true"
data point and not simply the result of measurement error? Should we just eliminate the
point and proceed? It would probably lead to a better regression in that we would get a
higher value of r^2. In general, this is unwise. A more appropriate solution is to report the
results both with the outlier included in the data set and with it removed. Then we can judge
the sensitivity of the results to the nature of the underlying data. Many computer packages
for regression analysis have techniques for judging the importance of individual observations on the least squares estimates. In all cases outliers deserve careful examination. They
represent important pieces of information about the complexity of the relationship between
the two variables. Further study of the outlier might reveal an important variable that has
been omitted in the analysis, or it might indicate the existence of a nonlinear relationship.

Another potential difficulty arises when the data contain an *extreme* value of X far
removed from the other observations. This situation is depicted in Figure 13-18. A single
observation (the problem is the same if there are two or three observations) is located well
beyond the range of X that encompasses all other observations. We would not uncover this
situation by examining residual plots since the observation is really not an outlier. But the
observation is very influential in determining the location of the regression line. Notice the
lack of correlation between X and Y exhibited by the remaining observations. This problem
illustrates the importance of *basics* in regression analysis. *An initial examination of our
scattergram can tell us much about our data.* We just have to know what to look for.

Once again, we are faced with the same problem—to eliminate the observation or not.
Perhaps the best solution is to try to generate some new observations in the intervening
range of X. Perhaps the extreme value is really "typical" in this range of X. If we eliminate

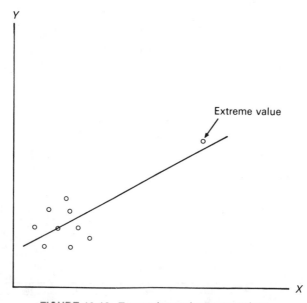

FIGURE 13-18. Regression and extreme values.

the observation, it is always wise to indicate that the results are limited to a specific range of the independent variable. Suppose we are examining the relationship between city size and in-migration in a developing country that has a primate (largest) city. The primate city may be 5 to 10 times the size of the second largest city. In terms of regression analysis, this presents a difficulty. The alternatives are limited. We cannot simply "invent" new cities that fall in the range of sizes missing from the data. Yet, to drop the primate city would be inadvisable if we really wish to understand the migration pattern of the country. Perhaps we can say just as much without regression analysis as with it. But a picture is worth a thousand words. Careful scrutiny and explanation of the scattergram would nicely complement the regression results. In general, we might say that the appropriate course of action depends on the particular problem at hand and the interests of the geographer or social scientist undertaking the study.

13.6. Summary

This chapter has presented many new ideas and a great deal of information on the mechanics of fitting a linear regression equation. Until you have experience in applying regression analysis to a variety of sets of data in widely different contexts, the material appears much more difficult than it really is. Yet we have barely scratched the surface. We have merely introduced regression analysis as a descriptive tool for studying statistical relations among variables. In Chapter 14 we extend our discussion to include the role of statistical inference in regression analysis.

Appendix 13A. Review of the Elementary Geometry of a Line

A straight line is precisely defined by the coordinates of any two points on the line since it continues forever in the same direction. Lines have a constant slope. Let us consider the points $P_1(X_1, Y_1)$ and $P_2(X_2, Y_2)$ of Figure 13-19. We define the slope of the line passing through these two points to be

$$\frac{\Delta Y}{\Delta X} = \frac{Y_2 - Y_1}{X_2 - X_1} = b \qquad (13\text{-}35)$$

where Δ means change, or difference. When $\Delta X = 1$, $\Delta Y = b$, and clearly we can interpret b as the increase in Y that results from a unit increase in X.

If we know two points on a line or one point and the slope of the line, we can determine the unique equation for that line. Since any two points on the line will suffice, we choose the two points of Figure 13-20. The coordinates are $(0, a)$ for P_0 and (X, Y) for point P. If the line is to have constant slope, then, by (13-35)

$$\frac{Y - a}{X - 0} = a \text{ and } Y - a = bX \text{ or } Y = a + bX \qquad (13\text{-}36)$$

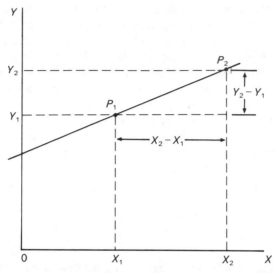

FIGURE 13-19. Slope of a line.

This the slope-intercept form of the equation for a straight line with the intercept *a* and slope *b*. The *parameters a* and *b* are unrestricted in sign or value.

Figure 13-21 illustrates several examples of straight lines and their equations. The line in Figure 13-21(b) differs from (a) in that is has a steeper slope ($1 > 0.5$). The line depicted in Figure 13-21(c) is parallel to (b) because it has the same slope ($+1$), but differs because it has a lower intercept value ($2 < 5$). In (d) we see the graph of a line with a

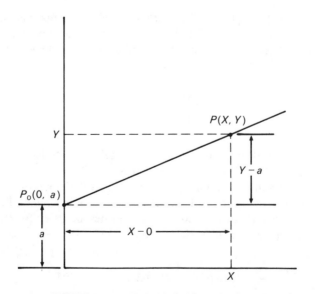

FIGURE 13-20. Derivation of equation for a line.

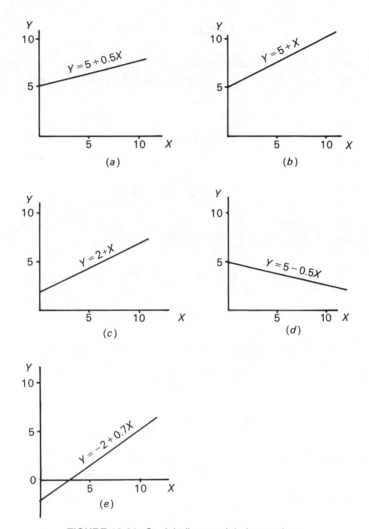

FIGURE 13-21. Straight lines and their equations.

negative slope (−0.5), and in (e) we see the graph of a line with a negative intercept (−2). Most of the interest in linear equations in the context of regression analysis is limited to lines passing through the first quadrant since variables X and Y usually take on positive values. However, no restrictions exist on the values to be taken by either the parameters of the line, a and b, or variables X and Y.

Appendix 13B. Least Squares Solution via Elementary Calculus

The normal equations (13-5) and (13-6) that lead directly to the computational formulas for a and b can be easily derived from elementary calculus. Let

$$S(a, b) = \sum_{i=1}^{n}(Y_i - a - bX_i)^2 \tag{13-37}$$

be the total sum of squared deviations from the line to the n points. The values of a and b that minimize S can be derived by differentiating (13-37) with respect to a and b, to obtain

$$\frac{\partial S}{\partial a} = -2\sum_{i=1}^{n}(Y_i - a - bX_i)$$

$$\frac{\partial S}{\partial b} = -2\sum_{i=1}^{n}X_i(Y_i - a - bX_i)$$

Setting each derivative equal to zero and simplifying yields

$$\sum_{i=1}^{n}(Y_i - a - bX_i) = 0$$

$$\sum_{i=1}^{n}X_i(Y_i - a - bX_i) = 0$$

These equations can be expanded by using simple rules of summations to yield

$$\sum_{i=1}^{n}Y_i - na - b\sum_{i=1}^{n}X_i = 0$$

$$\sum_{i=1}^{n}X_iY_i - a\sum_{i=1}^{n}X_i - b\sum_{i=1}^{n}X_i^2 = 0$$

These equations are identical to the normal equations (13-5) and (13-6) with a simple rearrangement of terms.

REFERENCE

N. Draper and H. Smith, *Applied Regression Analysis*, 2nd ed. (New York: Wiley, 1981).

PROBLEMS

1. Explain the meaning of the following terms:
   a. Dependent variable
   b. Independent variable
   c. Least squares criterion
   d. Regression intercept
   e. Regression slope
   f. Simple regression
   g. Linear regression
   h. Coefficient of determination
   i. Standard error of estimate
   j. Residual

k. Standardized residuals
l. Outliers
m. Coefficient of variation
n. Causal relation
o. Extreme values

2. Differentiate between a *functional* relationship and a *statistical* relationship.

3. What is the principal difference between the fixed-X and random-X regression model?

4. Differentiate between the use of a regression equation in interpolation and extrapolation.

5. Measurements of the L_{10} sound levels and distance from the centerline of an urban expressway are as follows:

Observation No.	Distance, ft	Sound level, dB
1	45	83
2	63	81
3	160	66
4	225	68
5	305	69
6	390	66
7	461	58
8	515	57
9	605	55
10	625	61

a. Draw a scattergram of the data with distance as the independent variable on the X-axis.
b. Using the format shown in Table 13-2, estimate the regression coefficient and intercept for the regression equation. (If you use a spreadsheet program, this is easily accomplished with a few simple formulas.)
c. Using a work table of the form of Table 13-4, calculate the coefficient of determination and standard error of estimate for this problem. (Again, a spreadsheet program is an efficient tool for this problem.)
d. Graph the regression line within the scattergram of points.
e. Interpret the meaning of the regression coefficient and intercept for this example.
f. Suppose the distance variable is measured in hundreds of feet and not feet. What would be the regression equation of this revised problem? (Hint: no new calculations are required.)
g. Generate a table of standardized residuals. Are there any outliers?
h. Which observation is most closely predicted by the equation? Least closely?
i. Construct a plot of the residuals of the equation against the predicted value of Y.
j. Construct a plot of the standardized residuals against the predicted value of Y.
k. How many decibels would you predict for L_{10} noise levels at a distance of 250 ft from the centerline? at 1000 ft? In which of these two predictions would you have more confidence? Why?

6. The following table contains the population of a set of 10 cities and the total number of serious

crimes that occurred in each city in 1990. Serious crimes are defined as murder, rape assault, robbery, and car theft.

Observation number	Population (000)	Total serious crimes
1	7,017	393,162
2	4,370	294,466
3	1,832	106,482
4	1,306	99,293
5	970	54,854
6	794	30,771
7	610	32,146
8	435	25,650
9	352	11,273
10	288	18,173

a. Convert the variable total serious crimes to a rate (i.e., occurrences per thousand population) by dividing column 3 by column 2.
b. Draw a scattergram with city size as the independent variable on the X-axis.
c. Estimate the regression equation.
d. Interpret the meaning of the regression coefficient and intercept.
e. Calculate the standard error of estimate and the coefficient of determination.
f. Generate a table of standardized residuals. Are there any outliers? Extreme values?
g. What do you think the results of the regression equation tell us about crime rates and city size?

7. A chain of fast food restaurants was interested in the relative profitability of their franchisees based on two aspects of their location: (1) the number of households in the trading area of the restaurant (in thousands), and (2) whether the franchise was located in a shopping mall or not. The following 20 observations were collected:

Obs.	1 if mall 0 if highway	Number of households	Sales $ 000s	Obs	1 if mall 0 if highway	Number of households	Sales $ 000s
1	0	155	135	11	1	62	88
2	0	93	72	12	1	280	400
3	1	178	179	13	1	220	270
4	1	215	220	14	0	110	100
5	0	129	115	15	0	127	98
6	0	158	131	16	0	185	160
7	1	172	179	17	1	89	118
8	1	95	110	18	1	147	175
9	0	183	161	19	1	99	120
10	0	202	187	20	0	130	108

a. Estimate an equation between sales and number of households for the complete data set.
b. Examine the residuals to see if location of the franchise has anything to do with the sales of the outlet.
c. Estimate an equation for two cases, one for the 10 franchises located in malls, and one for the 10 franchises located on highways.

d. Compare the slopes and intercepts of the three equations. Which locations are more profitable? Is the impact of trading area markedly different between the two types of franchises?

8. Fly-by-Night (FBN) airlines recorded the following results over the 1974–1993 period:

Year	Revenue passenger miles	Year	Revenue passenger miles
74	12.6	84	28.0
75	13.2	85	29.8
76	14.1	86	32.3
77	15.1	87	35.4
78	16.4	88	39.0
79	18.3	89	43.7
80	20.1	90	49.8
81	21.8	91	51.7
82	23.8	92	56.6
83	26.4	93	64.4

a. Enter this data into a spreadsheet program and estimate the regression equation between revenue passenger miles and time using the data as displayed.
b. Estimate the same equation using time measured as 1974, 1975, . . . , 1994.
c. Estimate the equation using time measured 1, 2, . . . , 20.
d. Compare the three equations.

9. Measurements are taken of the residential density of various areas in a city, Y, and the distance of these areas to the central business district, X. The measurements are as follows:

Observation	Distance to CBD km	Residential density person/km^2
1	1.8	1,000
2	2.0	800
3	3.0	300
4	4.3	200
5	5.4	100
6	6.8	90
7	7.3	80
8	8.6	60
9	9.4	40
10	10.9	30
11	13.7	20
12	15.5	15

a. Draw a scattergram of this data.
b. Fit a regression equation of the form $Y = Y + bX$. Draw the line within the scattergram.
c. Fit a regression equation by transforming the variable X to $\ln X$. Draw the least squares regression line within the scattergram of the data. In order to draw this second log-transformed

line within the scattergram of the original data, you will need to review your knowledge of logarithms.

d. Which equation best describes the data? Use various measures of goodness of fit to compare the two equations.

e. Suppose you are told that even numbered observations are from the west of the city and odd observations from the east. Do the results suggest this may be important, or not?

f. Suppose that distance is really measured in miles not kilometers. What effect does this have on the equation?

14

Inferential Aspects of Regression Analysis

So far, our analysis of the regression model has involved only the mechanical procedure whereby a least squares line is fit to a set of sample observations for two variables X and Y. This is but one step in the process of building a simple linear regression model. The following steps are usually components of any regression model building procedure:

1. *Specify the variables in the model and the exact form of the relationship between them.* In Chapter 13, the number of trips made per household per day is selected as the dependent variable. This variable was made a linear function of a single independent variable, household income. In many instances more than one independent variable is related to the dependent variable, or a nonlinear function is more appropriate for the systematic part of the relationship. These two issues are explored in Sections 14.4 and 14.5. Also, the model may not be completely specified until the data have been collected. For example, if we were unsure of the nature of the function linking the two variables, we might explore different functions using the available empirical data. In still other cases the structure of the regression model may be specified a priori from some given theory that links the two variables. Usually, though, the specification of the regression equation is the initial task in regression model building.

2. *Collect data.* The data collection procedure must fulfill the underlying assumptions of the simple regression model. These assumptions are detailed in Section 14.1. The exact form of the data depends on whether the fixed-X or random-X model is used as a basis for the regression.

3. *Estimate the parameters of the model.* To estimate the parameters of the regression model, the least squares procedure is used. This has been fully explained in Chapter 13.

4. *Statistically test the utility of the developed model, and check whether the assumptions of the simple linear regression model are satisfied.* This is the inferential side of regression analysis. The purpose of this step is twofold. First, significance tests are undertaken on the slope, the intercept, and the regression model as a whole. Second, hypothesis testing and interval estimation procedures are described. These two components are explained in Section 14.2. For these tests to be valid, several assumptions must be met. Section 14.3 describes methods that can be used to verify the required assumptions.

5. *Use the model for prediction.* Once it is developed, we often wish to use the regression equation to predict values of the dependent variable. In the context of trip

generation, it is common to utilize the estimated trip generation equation to predict future trip generation rates based on future values of expected household income. Normally, confidence intervals are used to establish a range within which the dependent variable can be predicted at any level of confidence $1 - \alpha$.

Whenever a regression analysis is undertaken, it may not necessarily include all these steps. Sometimes we may not be interested in prediction at all, and the ultimate goal is only to evaluate the strength of the empirical evidence in support of some prespecified theoretical model. Other times, the overriding goal may be to use the model for prediction, and interest in hypothesis testing may be secondary. For example, we might wish to develop a regression equation, using data collected for one period, and then evaluate its predictive capabilities, using data from another period. Quite often, both issues are important. The remainder of this chapter explores the computational techniques and issues involved in both hypothesis testing and prediction.

14.1. Assumptions of the Simple Linear Regression Model

In Section 13.2, two different forms of the simple linear regression model were distinguished. In the fixed-X model, it is assumed that an experiment is undertaken, and the values of the independent variable X are under the control of the researcher. Values of X—X_1, X_2, ..., X_m—are selected, and for each value we randomly sample one or more values of Y. In this representation only Y is a random variable. Alternatively, the data for the regression model can be generated by a survey in which pairs of (X, Y) values are randomly selected from some known population. This random-X model leads to two different random variables, X and Y. No matter which model is used to derive the sample data for a regression problem, the method of least squares is used to derive the two parameters of the linear equation.

To extend the analysis to include inferential statistics, it is necessary to make several assumptions about the parent populations from which the sample data are drawn. To clarify this notion, it is useful to examine Figure 14-1. The true (population) regression line of Y as a linear function of X is the heavy line. The equation of this line is

$$Y = \alpha + \beta X \tag{14-1}$$

where the convention of utilizing Greek letters for population parameters has been followed. Depending on the particular sample drawn from the population of X and Y values, we may estimate this true regression by any of a possibly infinite number of sample regression lines. Each of these equations is expressed as $Y = a + bX$, where a and b estimate (in the statistical sense) the two population parameters α and β, respectively. If we wish to make inferences about this true population regression line, then additional assumptions are required. These assumptions are identical, with one exception, for both the fixed-X and random-X models. For present purposes, the discussion of these assumptions is couched in the context of the fixed-X model.

Formally, the simple linear regression model for the fixed-X case is

$$Y_i = \alpha + \beta X_i + \varepsilon_i \qquad i = 1, 2, \ldots, M \tag{14-2}$$

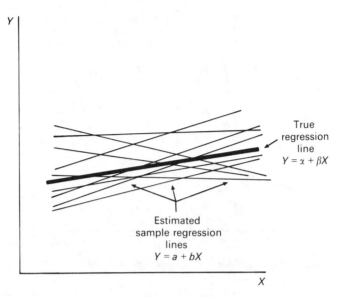

FIGURE 14-1. True regression line and several sample regression lines.

where Y_i is the value of the dependent random variable, X_i is the value of the ith level of the fixed variable X, and ε_i is the random error or disturbance. In this specification, the subscript i denotes not a particular observation but one of the levels at which the independent variable is set in an experiment. The graphical form of the fixed-X model is seen in Figure 14-2. For each level of X—$X_1, X_2, \ldots, X_M$—there is a probability distribution for the random variable Y. The mean value of the dependent variable Y for $X = X_1$ is given by $Y_1 = \alpha + \beta X_1$. To this systematic component of Y, we must add the value of the random component ε_i. It is this random error component that makes the dependent variable Y a random variable—even when the independent variable X is fixed.

What is the source of this random error term component? The error term may be conceived as being derived from two sources. First, the relationship may be imperfect owing to *measurement* error. This source of error should be minimized. The other source of error is called *stochastic* error. Suppose an experiment is conducted to predict crop yield Y based on the amount of fertilizer applied X. If several experimental plots are all treated with exactly 300 lb of fertilizer, the recorded crop yields will not be exactly equal. Why? For one thing, the soils on the various plots may not be perfectly homogeneous. If the drainage of the plots differs or if the purity of the fertilizer applied to the plots varies, then we might expect some variation in yields. In fact, a large number of other factors may affect the experiment that cannot be *completely* controlled, each having a small influence on yield. The *net* effect of these influences appears to be random. As we shall see in Section 14.5, we can incorporate several independent variables into the regression equation. However, some may be impossible to control exactly and are best treated as random or stochastic error.

Figure 14-2 is a crucial diagram for understanding the inferential aspects of regression analysis. Notice that the vertical axis is labeled *probability density* since the dependent

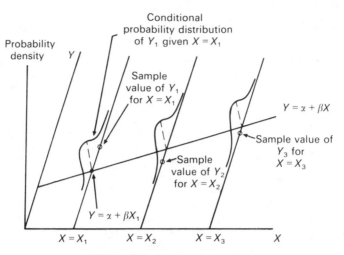

FIGURE 14-2. The fixed-X model.

variable Y is a random variable. Moreover, associated with each level of X is a *conditional probability distribution* of the random variable Y. It is important to distinguish between this unobserved population and the observed sample observations upon which the regression equation $Y_i = \alpha + \beta X$ is estimated. One such sample value is illustrated for each value of X. In fact, there may be more than one sample observation for each value of X. Normally, the experiment would be designed so that a number of different observations of Y are obtained for each value of X. However, even if there are a large number of observations for each value of X and even if the values of X are closely spaced, it would be very fortunate if the estimated regression line $Y = a + bX$ exactly coincided with this true regression equation. By making a few assumptions about the regularity of the distribution of the random error ε_i (or equivalently Y_i), it is possible to make statistical inferences about $Y_i = \alpha + \beta X$ on the basis of the sample information $Y = a + bX$. There are four essential assumptions.

ASSUMPTION 1
For the ith level of the independent variable X_i the expected value of the error component ε_i is equal to zero. This is usually expressed as $E(\varepsilon_i) = 0$ for $i = 1, 2, \ldots,$ M.

This first assumption states that, on the average, the random error component of Y_i is zero. If there are a series of sample observations of Y_i at some particular value of X, say $X = X_i$, then some of these observations should lie above the regression line and others below it. Recall that the regression line passes through the mean value of Y_i. To show this, note that $E(Y_i) = E(\alpha + \beta X_i + \varepsilon_i)$. Since α, β, and X_i are constants, $E(Y_i) = \alpha + \beta X_i + E(\varepsilon_i)$. By this first assumption $E(\varepsilon_i) = 0$, and thus $E(Y_i) = \alpha + \beta X_i$.

How can this assumption be violated? Suppose we were to fit a linear regression model to two variables that are actually related by a nonlinear function. In Figure 14-3, for example, a linear equation $Y = \alpha + \beta X$ is fit to a set of data. Note that Assumption 1 is not satisfied since the equation does not appear to follow the path of the means. In this instance, if a nonlinear equation of the form is fit to the data, it appears that Assumption 1 could be satisfied. But what if the exact form of the true regression equation that links the two variables is unknown? It may not be immediately apparent that the assumption has been violated. An examination of the residual errors from the estimated equation $Y = a + bX$ may uncover certain violations. In Figure 14-3, for example, the residuals from the line $Y = a + bX$ show a definite pattern and do not appear random at all. On the average, the Y values for $X = X_1$ and X_6 are below the line, but the Y values for X_3 and X_4 are above the line. This is definitely not random. The examination of residuals is discussed in Section 14.3. Nonlinear regression models are briefly introduced in Section 14.4.

For Assumption 1 to hold, the regression model must be correct; that is, it must represent the true underlying process. Once the correct *specification* is formally assumed, estimating the model parameters becomes a relatively straightforward mechanical procedure. In reality, however, we can never be sure that the model is correctly specified. Two types of problems exist. First, we may have omitted relevant variables from the regression equation. If so, the random error term cannot be considered random since it contains a systematic component. Second, the wrong functional form may have been chosen. Applied researchers usually examine more than one possible specification in an attempt to find the single specification that best describes the process. A detailed examination of the residuals is one of the best ways to make sound judgments concerning model specification. Relevant material is included in Sections 14.3, 14.4, and 14.5.

ASSUMPTION 2

The variance of the error component ε_i is constant for all levels of X; that is, $V(\varepsilon_i) = \sigma^2$ for $i = 1, 2, \ldots, M$.

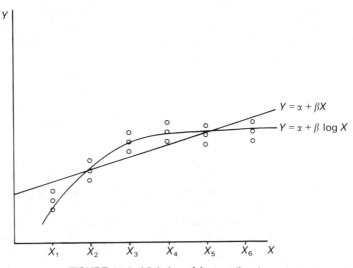

FIGURE 14-3. Violation of Assumption 1.

The second assumption also concerns the error term ε_i. The variance of the probability distribution of is equal ε_i to σ^2. Also, this is constant for all values of X. This variance is unknown, but it can be estimated from the sample residuals. As introduced in Chapter 11, this is known as the assumption of *homoscedasticity*, or simply *equal variance*. Since $E(\varepsilon_i)$ = 0 and $V(\varepsilon_i) = \sigma^2$, it is possible to specify the variance of the conditional probability distribution of Y_i or $V(Y_i)$. Note that $V(Y_i) = V(\alpha + \beta X_i + \varepsilon_i)$. Since $\alpha + \beta X_i$ is a constant, it has no variance and can be omitted. Simplifying, we see that $V(Y_i) = V(\varepsilon_i) = \sigma^2$. The probability distributions of Y_i and ε_i differ only by their means. Violations of the assumption of homoscedasticity are also usually uncovered through the analysis of residuals (see Section 14.3).

ASSUMPTION 3

The values of the error component for any two ε_i and ε_j are pairwise uncorrelated.

By this assumption, the outcome Y_i for any given value of X, say $X = X_i$ neither affects nor is affected by the outcome Y_j for any other value of X, say $X = X_j$. In terms of the covariance operator described in Chapter 12, this assumption implies $cov(\varepsilon_i, \varepsilon_j) = 0$ for all i and j. If this assumption is not satisfied, then autocorrelation is present in the error term. Autocorrelation can exist in many forms. A detailed discussion of autocorrelation is discussed in Section 14.3 within the context of the violations of the standard assumptions.

ASSUMPTION 4

The error components ε_i are normally distributed.

This last assumption completes the specification of the error term in the fixed-X regression model. It *is not required* for calculating point estimates of (a, b, and $s_{Y \cdot X}$, respectively), *nor is it required* to make point estimates of the mean value of Y for any given X_i, $\hat{Y}_i$. It *is required* for the construction of confidence intervals and/or tests of hypotheses concerning these parameters. This assumption should not be surprising since virtually all the confidence interval and hypothesis testing methods described in Chapters 9 and 10 are based on the assumption of the normality of the population distribution. The verification of this assumption is also based on an examination of the residuals from the sample regression line $Y = a + bx$. The specific techniques used are explained in Section 14.3.

For a random-X model this assumption must be augmented with an additional constraint governing the probability distribution of the random variable X. One of the most important members of the class of random-X models assumes that the distribution of X and Y is bivariate normal. Data conforming to the random-X model might be realized by a sampling procedure in which individuals, households, industrial firms, zones, regions, or cities are randomly selected as observations from a statistical population. If the underlying probability distribution of the two variables is bivariate normal (see Figure 6-20), then the standard inferential tests described in Section 14-2 are still valid. Often, however, the data used by geographers in a regression model do not conform to either the fixed-X or the random-X model. The data for the trip generation example of Chapter 13 illustrate this point. The validity of the use of the standard inferential tests in this situation is· unclear. Sometimes other justifications for their use are given.

Gauss-Markov Theorem

The principal analytical results of regression theory derive from the Gauss-Markov theorem. This theorem justifies the use of the least squares procedure for estimating the two parameters of the linear regression equation. Also, it establishes the principal properties of the least squares estimators a and b.

GAUSS-MARKOV THEOREM

Given the four assumptions of the simple linear regression model, the least squares estimators a and b are unbiased and have the minimum variance among all the linear unbiased estimators of α and β.

To examine the implications of this theorem, consider the sampling procedure depicted in Figure 14-4. From some statistical population, repeated random samples of some given size, say $n = 50$, are drawn and least squares estimators a and b calculated from (13-7) and (13-12). If the four assumptions are satisfied, then several characteristics of the

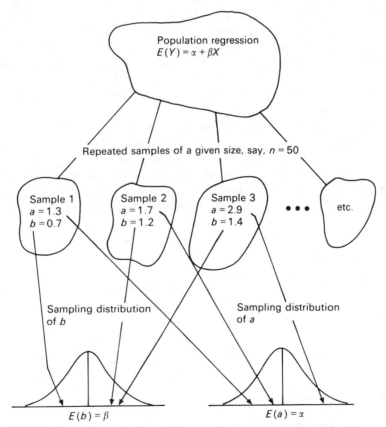

FIGURE 14-4. The sampling procedure in a regression model.

sampling distributions of a and b follow from the Gauss Markov theorem. First, estimators a and b are unbiased estimators of α and β. In Figure 14-4 the sampling distributions of a and b are centered on $E(a) = \alpha$ and $E(b) = \beta$. These sampling distributions are themselves normal. Finally, and most important, a and b are the most efficient estimators in the sense that the sampling distributions of a and b have less variability (smaller variances) than any other linear and unbiased estimators of α and β. The least squares estimators a and b are thus often called *BLUE*, an acronym for best linear unbiased estimators. Statistical tests on various aspects of the fitted regression equation $Y = a + bX$ follow directly from the Gauss-Markov theorem.

14.2. Inferences in Regression Analysis

Point Estimates

In many applications of regression analysis, it is important to estimate the value of some parameter in the regression equation. Statistical estimation in the context of regression is based on the process of statistical estimation described in Chapter 8. In Chapter 8 two types of statistical estimation were distinguished: point estimation and interval estimation. In point estimation, a single number is computed from the sample information and is used as an estimate of some population parameter. Note that all the point estimates in regression analysis *require only that the first three assumptions explained in Section 14.1 be satisfied. No assumption of normality for the error term is required.* However, the assumption of normality is *essential* for both hypothesis testing and the construction of confidence intervals. Table 14-1 summarizes the principal parameters and their estimators in regression analysis.

From the Gauss-Markov theorem, we know that a and b are the *BLUE* estimators of α and β. Similarly, $S_{Y \cdot X}$, r^2, and r^2 provide point estimates of σ, ρ, and ρ^2, respectively. If we wish to predict the mean value of Y for any level of X in the range of the sample data [or $E(Y)$], the best point estimate is clearly $E(Y) = a + bX_i$. This should not be surprising since the regression equation is defined as the path of the means if the first three assumptions outlined in Section 14.1 hold. The expressions $E(Y)$ and $\mu_{Y \cdot X}$ are interchangeable. Here

TABLE 14-1
Estimators in Regression Analysis

Regression Model Parameter	Point Estimator	Equation
α	a	(13-8)
β	b	(13-12)
$\sigma^2 [\sigma]$	$s_{Y \cdot X}^2 [s_{Y \cdot X}]$	(13-27)
ρ	r	(13-14)
ρ^2	r^2	(13-26)
$\mu_{Y \cdot X} [E(Y)]$	$a + bX_i$	(13-1)
$\hat{Y}_{new}$	$a + bX_{new}$	(13-1)

$E(Y)$ is used since Y_i is a random variable, and $\mu_{Y \cdot X}$ is used to reflect the fact that it is a population parameter that clearly depends on the value of variable X. Finally, suppose we have a single new value of X, denoted X_{new}, and wish to predict the most likely value of Y for X_{new}, or Y_{new}. The best estimator for this is again $Y_{new} = a + bX_{new}$. This is the same as $\mu_{Y \cdot X}$. However, as we shall see, the interval estimates of these two parameters are much different. This is expected. It is easier to make a prediction of the mean of several numbers than a single number. In the same sense we found the sampling distribution of X to be much more compact than the variable X itself. The standard error of the mean is equal to the standard deviation of variable X only when the sample size is 1!

Inferences Concerning the Slope of the Regression Line β

The slope of the regression line is particularly important in regression analysis. Sometimes the interest in β derives from the fact that it measures the sensitivity of variable Y to changes in variable X. For example, the slope of the trip generation equation tells a transportation planner the *magnitude* of the increase in the number of trips expected from increasing incomes. Estimating β is thus extremely important. For this purpose a confidence interval for β can be constructed by using the least squares estimator b. There is a second reason for examining β. If it is possible to infer from a sample regression line that $\beta = 0$, then it can be concluded that the equation is of no use as a predictor since $E(Y) = \alpha + (0)X = \overline{Y}$. For every value of X, the best estimate of Y is $\overline{Y}$. Whether it is possible to infer that the slope of the population regression is nonzero, therefore, has important ramifications for the use of the regression equation for predictive purposes. To establish confidence interval and hypothesis testing formulas for β, it is necessary to describe the sampling of the least squares estimator b.

The sampling distribution of b is fully characterized in the Gauss-Markov theorem (see Section 14.1 and especially Figure 14-4). It has the following features.

1. It is normal.
2. It has a mean value centered on β; that is, it is unbiased and $E(b) = \beta$.
3. It has a standard error, denoted σ_b,

$$\sigma_b = \frac{\sigma}{\sqrt{\sum_{i=1}^{n} X_i^2 - \left(\sum_{i=1}^{n} X_i\right)^2 / n}}$$

Note that the formula for the standard error of b contains one unknown, σ. However, this unknown can be estimated from the sample data by $S_{Y \cdot X}$, and an estimate of σ_b, denoted S_b, is obtained:

$$S_b = \frac{S_{Y \cdot X}}{\sqrt{\sum_{i=1}^{n} X_i^2 - \left(\sum_{i=1}^{n} X_i\right)^2 / n}} \tag{14-3}$$

It turns out that the sampling distribution of b is no longer normal once this substitution has been made. Now the sampling distribution of the statistic is t-distributed with $n - 2$ degrees of freedom. There are only $n - 2$ degrees of freedom because 2 degrees of freedom are lost in the estimation of $S_{Y \cdot X}$.

The value of the standard error S_b, depends on two separate factors. First, S_b will increase with increases in the standard error of estimate. This is expected. If the standard deviation of the conditional probability distribution of ε_i is large, one will be less confident about estimating the slope of the regression line. However, the greater the sum of squared deviations of X (that is, the greater the variability in X), the lower S_b. To see why, compare Figure 14-5(a) with (b). Suppose all the values of X used to fit the regression line are concentrated at a few values of X quite close to $\overline{X}$. This is the situation in Figure 14-5(a). The estimated slopes of various sample regression lines are likely to be quite variable. By comparison, the estimated slopes of possible regression lines in situations in which there is a considerable variability in the values of X are all quite similar, as is shown in Figure 14-5(b). In general, then, we are more confident about predicting the true value for β when the prediction is based on observations over a wide range of the independent variable X.

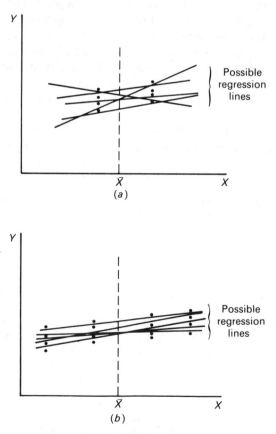

FIGURE 14-5. S_b as a function of the variability in X.

The test statistic is

$$t = \frac{b - \beta}{S_b} \tag{14-4}$$

This reduces to $t = b/S_b$ for the most common null hypothesis: H_0: $\beta = 0$.

As an alternative to a formal test of hypothesis, it is possible to construct a confidence interval for β from the point estimate b:

$$b - t_{\alpha/2, n-2} S_b \leq \beta \leq b + t_{\alpha/2, n-2} S_b \tag{14-5}$$

If this interval contains zero, then we will accept the null hypothesis that $\beta = 0$ and conclude that the regression line is of no use in prediction.

It is now possible to test the usefulness of household income variations in household trip generation rates for the regression equation of Chapter 13. First, we calculate the test statistic as

$$S_b = \frac{.752}{\sqrt{3237 - (189)^2/12}} = .0466 \text{ and } t = \frac{.335}{.0466} = 7.189$$

We see from the 2-tailed t-Table in the Appendix that this value exceeds the highest tabulated value of 5.0 for 10 degrees of freedom so that the PROB-VALUE < .001. Given the *estimate* of $b = .335$, the $100(1 - \alpha)$ confidence interval for β is

$$.335 - 2.228(.0466) \leq \beta \leq .335 + 2.228(.0466)$$

$$.2312 \leq \beta \leq .4388$$

for $\alpha = .05$. As noted in Chapter 8, 2.228 is the appropriate t-value for 10 degrees of freedom correct to 3 places for $\alpha = .05$. With 95% confidence, we can say the true slope β lies in the range [.2312, .4388]. This interval does not contain zero, and the sample regression line $Y = .807 + .335 X$ is a useful equation for predicting household trip generation rates for traffic zones with average household incomes between $9,000 and $24,000.

Inferences Concerning the Regression Intercept

Confidence intervals and hypothesis tests for the intercept are rarely used in regression model building. In many practical applications the Y-intercept does not have an important meaning. Often the range of observations of the independent variable X does not envelop the value $X = 0$. For example, the trip generation rate for a household with an annual income of $0 is not particularly meaningful. When we consider that the independent variable in this case is really the average income of all households in a traffic zone, the intercept is even less meaningful! In some cases there is an important interpretation of the intercept, and it does merit inferential consideration. Consider again the example of predicting crop yield Y as a function of the amount of fertilizer applied X to various test plots in an agricultural experiment. The intercept α represents the expected yield of the crop in the absence of fertilizer application. This has meaning as a "control" value.

The sampling distribution of a, an outcome of the Gauss-Markov theorem, has the following characteristics:

1. It is normal.
2. It has a mean value centered on a; that is, it is unbiased and $E(a) = \alpha$.
3. It has a standard error σ_a of

$$\sigma_a = \sigma \sqrt{\frac{1}{n} + \frac{\bar{X}^2}{\sum_{i=1}^{n} X_i^2 - \left(\sum_{i=1}^{n} X_i\right)^2/n}}$$

Again, since σ is unknown, it is estimated by $s_{Y \cdot x}$ and the estimated standard error of a becomes

$$S_a = S_{Y \cdot X} \sqrt{\frac{1}{n} + \frac{\bar{X}^2}{\sum_{i=1}^{n} X_i^2 - \left(\sum_{i=1}^{n} X_i\right)^2/n}} \tag{14-6}$$

The appropriate test statistic is t-distributed with $n - 2$ degrees of freedom. It is most common to test the hypothesis that $\alpha = 0$, but this test statistic can be used to test a null concerning any value of α.

$$t = \frac{a - \alpha}{S_a} \tag{14-7}$$

By using the standard t-statistic, the $100(1 - \alpha)$ confidence interval for α from a given least squares estimate a is

$$a - t_{\alpha/2, n-2} S_a \leq \alpha \leq a + t_{\alpha/2, n-2} S_a \tag{14-8}$$

As we have noted, the intercept of the household trip generation equation is not particularly meaningful. Nevertheless, a test of this intercept is useful to illustrate the necessary calculations of this test. The calculated t value is

$$t = \frac{.807}{.752 \sqrt{\frac{1}{12} + (15.75)^2[3237 - (189)^2/12]}} = \frac{.807}{.7656} = 1.054$$

leading to $0.297 \leq$ PROB-VALUE $\leq .341$. At a significance level of .05, we would thus accept the null hypothesis that $\alpha = 0$. This seems to be a reasonable conclusion. Without any income, households in a traffic zone might be expected to make zero trips. Because the intercept is well out of the range of the available data, it is of minor concern in any case. If the data were to suggest a rejection of the null hypothesis, this would also be of no real significance. Given the estimated value $a = .807$, the 95% confidence limits from Equation (14-8) are

$$.807 - 2.228(.7656) \le \alpha \le .807 + 2.228(.7656)$$

$$-.899 \le \alpha \le 2.513$$

The fact that this interval contains $\alpha = 0$ agrees with the PROB-VALUE evaluation of the null hypothesis.

Analysis of Variance for Regression

It is also possible to test the null hypothesis $\beta = 0$ by using the F distribution. The analysis of variance for regression always produces results equivalent to those obtained from the t test for the null hypothesis $\beta = 0$. In fact, the value of the F statistic computed in the analysis of variance is the exact square of the value of the sample t statistic given by (14-4). This will be demonstrated for the household trip generation equation. Although these two tests are exactly equivalent for simple linear regression, they are not equivalent for cases in which the regression equation includes more than one independent variable. In multiple regression analysis the t test is used to test the significance of a single independent variable given the null hypothesis β_j. The analysis of variance is used to test the joint hypothesis that all the independent variables have an insignificant effect on the dependent variable Y, that is, $\beta_1 = \beta_2 = \ldots = \beta_j = 0$.

The general form of the analysis of variance table for simple linear regression is given in Table 14-2. The total sum of squares (TSS) is equal to $\Sigma_{i=1}^{n}(Y_i - \bar{Y})^2$. Associated with this total variation are $n - 1$ degrees of freedom, 1 less than the number of observations. The two mean squares are calculated by apportioning these $n - 1$ degrees of freedom to two sources: the regression model and the error, or residual, variation. The degrees of freedom for the residual error (ESS) are equal to the number of observations minus the number of

TABLE 14-2
Analysis of Variance Table in Simple Regression

Source of variation	Sum of squares	Degrees of freedom	Mean square
Regression (model)	RSS	1	$\dfrac{RSS}{1}$
Error (residual)	ESS	$n - 2$	$\dfrac{ESS}{n - 2}$
Total	TSS	$n - 1$	

$$F = \frac{RSS/1}{ESS(n - 2)}$$

independent variables minus 1. In simple regression this is always equal to $n - 2$. The degrees of freedom attributable to the model are always equal to 1. Table 14-3 summarizes the results for the trip generation equation. Based on the calculated F value of 51.6, we can say PROB-VALUE < .001 for this test and conclude that household income is useful in predicting household trip generation rates. The calculated value of $F = 51.68$ is almost the exact square of the t value calculated above for the same null hypothesis:

$$(7.189)^2 \cong 51.628.$$

The difference between the figures is due solely to the number of places used in the calculation of each.

Not only are the F test for the analysis of variance and the t test identical in simple linear regression, but also they are both equivalent to the test of significance for the simple correlation coefficient r. This should not be surprising, given the relationship between correlation and regression expressed by Equation (13-19): $b = rS_y/S_x$. To see this equivalency, let us rewrite the F statistic computed for the analysis of variance in terms of the coefficient of determination r^2. Since TSS = ESS + RSS and $r^2 = $ RSS/TSS, we can rewrite the mean square due to regression, RSS/1, as r^2(TSS/1). Similarly, we rewrite the residual or error mean ESS/$(n - 2)$, as $(1 - r^2)$(TSS/$n - 2$). The F statistic can then be rewritten as

$$F = \frac{r^2(\text{TSS}/1)}{(1 - r^2)(\text{TSS})(n - 2)} = \frac{r^2(n - 2)}{1 - r^2} \qquad (14\text{-}9)$$

Now, let us compare the F statistic expressed by (14-9) to the t statistic used to test the significance of the simple correlation coefficient (12-5):

$$t = \frac{r\sqrt{n - 2}}{\sqrt{1 - r^2}}$$

Equation (14-9) is the square of (12-5).

TABLE 14-3
Analysis of Variance for Trip Generation Equation

Sources of variation	Sum of squares	Degrees of freedom	Mean square
Regression	29.552	1	29.2520
Error	5.666	10	0.5666
	34.918	11	

$$F = \frac{29.252}{.5666} = 51.6$$

Note. Students may get slightly different results depending on the number of places kept in the calculations.

Confidence Interval for $\mu_{Y \cdot x}$ of Given X

One of the most important applications of regression analysis is to use the equation $Y = a + bX$ to estimate a value for the conditional mean of Y, denoted $\mu_{Y \cdot x}$, for a given value of X. We have already noted that the point estimate of $\mu_{Y \cdot x}$ is determined by substituting the appropriate value of X, say X_0, into the equation so that $Y_0 = \mu_{Y \cdot x} = a + bX_0$. Of course, there is some error involved in such an estimate, and it is usual to construct a confidence interval around this point estimate. To develop this confidence interval, it is necessary to specify the sampling distribution of $\hat{Y}_0$. The characteristics of this sampling distribution are as follows:

1. It is normal.
2. It has a mean value of $\mu_{Y \cdot x_0}$, and it is unbiased.
3. It has a standard error $\sigma_{\hat{Y}_0}$ of

$$\sigma_{\hat{Y}_0} = \sigma \sqrt{\frac{1}{n} + \frac{(X_0 - \bar{X})^2}{\sum_{i=1}^{n} X_i^2 - \left(\sum_{i=1}^{n} X_i\right)^2 / n}}$$

Estimating σ by $S_{Y \cdot x}$ leads to

$$S_{\hat{Y}_0} = S_{Y \cdot x} \sqrt{\frac{1}{n} + \frac{(X_0 - \bar{X})^2}{\sum_{i=1}^{n} X_i^2 - \left(\sum_{i=1}^{n} X_i\right)^2 / n}} \tag{14-10}$$

The standard error of increases with (1) the standard error of estimate $S_{Y \cdot x}$, (2) the reciprocal of the sum of squared deviations, (3) the sample size n, and (4) the difference between the value X_0, and the mean $\bar{X}$. The standard error of estimate and the sum of squared deviations of X are important for the same reasons outlined for S_b. A larger sample size is always better for estimating a mean such as $\hat{Y}_0$.

The implications of the final factor, the difference between X_0 and $\bar{X}$, are illustrated in Figure 14-6. The confidence interval is narrowest at $X_0 = \bar{X}$. The farther X_0 lies from $\bar{X}$, the wider the interval. There is a simple explanation for the peculiar shape of these confidence intervals. Suppose that the bands were parallel to the regression line and not shaped as the interval in Figure 14-6. This would imply that the only error in predicting Y derives from mistakes in estimating a. Once it is recognized that both a and b can be in error, then the shape of these limits is easier to understand. The farther the point is from $\bar{X}$, the greater the impact of an error in estimating β on the estimated value $\hat{Y}_0$. The dangers of extrapolation are clear.

Because of the importance of household trip generation equations in making urban area trip forecasts, the confidence interval about $Y = .807 + .335X$ holds special interest. Consider first the confidence interval about the line at the point $X_0 = \bar{X}$. For the average income household of 15.75 (thousands of dollars), the estimated number of trips per day is $Y_0 = \bar{Y} = 6.083$ [recall the regression line always passes through $(\bar{X}, \bar{Y})$]. To develop the confidence interval, we use the data from Table 13-2 in Equation (14-10):

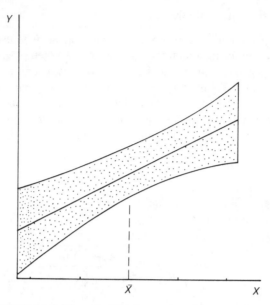

FIGURE 14-6. Characteristic shape of confidence intervals.

$$S_{\hat{Y}_0} = S_{Y \cdot X} \sqrt{\frac{1}{n} + \frac{(X_0 - \bar{X})^2}{\sum\limits_{i=1}^{n} X_i^2 - \left(\sum\limits_{i=1}^{n} X_i\right)^2 / n}}$$

$$= .752 \sqrt{\frac{1}{12} + \frac{0^2}{3237 - (189)^2 / 12}}$$

$$= .752(.288) = .217$$

Now, setting $\alpha = .05$ we use the value of $t_{.025, 10df} = 2.228$ in our confidence interval:

$$\hat{Y}_0 - t_{\alpha/2}(S_{\hat{Y}_0}) \leq \mu_{\hat{Y}_0} \leq \hat{Y}_0 + t_{\alpha/2}(S_{\hat{Y}_0})$$

$$6.083 - 2.228(.217) \leq \mu_{\hat{Y}_0} \leq 6.083 + 2.228(.217)$$

$$5.600 \leq \mu_{\hat{Y}_0} \leq 5.567$$

Table 14-4 summarizes the estimates for $\hat{Y}_0$ at several other values of X within the observed range of the independent variable X. The interval is narrowest at $X = \bar{X}$ and wider for values of X greater than or less than $\bar{X} = 15.75$. This pattern is depicted most clearly in Figure 14-7.

Prediction Interval for Y_{new}

Often we are not interested in making inferences about the mean of the conditional distribution of Y_0 at any given level of $X, X = X_0$. Instead, we wish to make inferences about

TABLE 14-4
Confidence Intervals for $\hat{Y}$

Annual household income $000	Estimated mean trips per day	95% confidence interval		Upper-lower estimate
		Lower estimate	Upper estimate	
10	4.156	3.386	4.925	1.539
12	4.826	4.205	5.448	1.243
14	5.491	4.979	6.014	1.035
15.75	6.083	5.600	6.567	0.967
18	6.838	6.300	7.375	1.075
20	7.507	6.852	8.162	1.310
22	8.179	7.368	8.989	1.621

the value of a single new observation of X, say X_{new}. Suppose we must predict the number of trips per household per day made by households in a traffic zone with a forecasted income of $X = \$20,000$. We cannot use the confidence interval for this purpose since it is based on the expected mean number of trips made by households in several traffic zones with an income of $20,000. It is much more difficult to estimate a *single* value than a mean, so we expect that this *prediction interval* will be wider than the corresponding *confidence interval* at any value of X.

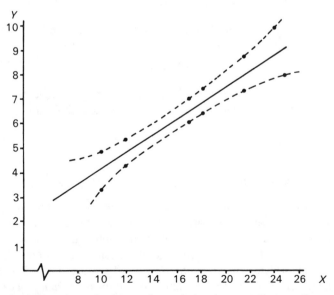

FIGURE 14-7. Confidence intervals for trip generation equation.

The construction of a *prediction interval* is virtually identical to the procedure followed for a confidence interval. The only difference is the standard error of $\hat{Y}_{new}$ is

$$S_{\hat{Y}_{new}} = S_{Y \cdot X} \sqrt{1 + \frac{1}{n} + \frac{(X_{new} - \bar{X})^2}{\sum\limits_{i=1}^{n} X_i^2 - \left(\sum\limits_{i=1}^{n} X_i\right)^2 / n}} \qquad (14\text{-}11)$$

This standard error is almost identical to $S_{\hat{Y}_0}$. In fact, it is larger by an additional $S_{Y \cdot X}$. To distinguish (14-10) from (14-11), the quantity $S_{\hat{Y}_{new}}$ is often called the *standard error of the forecast*. The $100(1 - \alpha)$ percent *prediction or forecast interval* is

$$\hat{Y}_{new} - t_{\alpha/2}(S_{\hat{Y}_{new}}) \le \hat{Y}_{new} \le \hat{Y}_{new} + t_{\alpha/2}(S_{\hat{Y}_{new}}) \qquad (14\text{-}12)$$

Consider now the specification of prediction intervals for the trip generation equation. We begin the specification of the prediction interval for $X_{new} = \bar{X}$ by calculating

$$S_{\hat{Y}_{new}} = S_{Y \cdot X} \sqrt{1 + \frac{1}{n} + \frac{(X_{new} - \bar{X})^2}{\sum\limits_{i=1}^{n} X_i^2 - \left(\sum\limits_{i=1}^{n} X_i\right)^2 / n}}$$

$$= .752 \sqrt{1 + \frac{1}{12} + \frac{(0)^2}{3237 - (189)^2 12}} = .752(1.04) = .782$$

Now, setting $\alpha = .05$ we use the value of $t_{.025,10df} = 2.28$ in our prediction interval:

$$\hat{Y}_{new} - t_{\alpha/2}(S_{\hat{Y}_{new}}) \le \mu_{\hat{Y}_{new}} \le \hat{Y}_{new} + t_{\alpha/2}(S_{\hat{Y}_{new}})$$

$$6.083 - 2.228(.782) \le \mu_{\hat{Y}_{new}} \le 6.083 + 2.28(.782)$$

$$4.341 \le \mu_{\hat{Y}_{new}} \le 7.826$$

Prediction intervals for several values of X are summarized in Table 14-5. Figure 14-8 compares the confidence and prediction intervals for the household trip generation equation. Note the greater width of the prediction interval.

14.3. Violation of the Assumptions of the Linear Regression Model

Specific statistical tests are available for checking whether the assumptions underlying the use of the linear regression model are satisfied by the data utilized in any empirical study. These tests employ the residuals of the sample regression equation $Y_i = a + bX_i + e_i$ as estimates of the error term of the true population regression model $Y_i = \alpha + \beta X_i + \varepsilon_i$. Before presenting these tests in detail, we examine a widely used technique that can provide a quick, but not foolproof, check of these assumptions. Often, a *residual plot* can reveal clear or possible violations of these assumptions. Two forms of these residual plots are com-

TABLE 14-5
Prediction Intervals for $\hat{Y}_{new}$

Annual household income $000	Estimated mean trips per day	95% prediction interval		Upper-lower estimate
		Lower estimate	Upper estimate	
10	4.156	2.310	6.001	3.691
12	4.826	3.037	6.615	3.578
14	5.491	3.742	7.252	3.510
15.75	6.083	4.341	7.826	3.485
18	6.838	5.076	8.599	3.523
20	7.507	5.709	9.305	3.596
22	8.179	6.316	10.041	3.725

monly used. First, an *overall plot* of the residuals can be made as a dot diagram or histogram, as in Figure 14-9. Consider the 12 residuals to the household trip generation problem. To one decimal place these residuals are $-1.2, 0.2, 0.5, 0.2, 0.5, 0.8, 1.2, 0.2, -0.2,$ $-0.8, -0.5,$ and -0.8. In Figure 14-9(a) these observations are depicted as dots along an axis measuring the value of the residual e_i. This procedure can be awkward if there are a large number of residuals, and it is usual to graph the residuals from larger data sets by using a histogram. The histogram format is illustrated in Figure 14-9(b).

Overall plots can also be *standardized* so that comparisons among residual plots are possible. This is done by standardizing the sample residuals by dividing by the standard

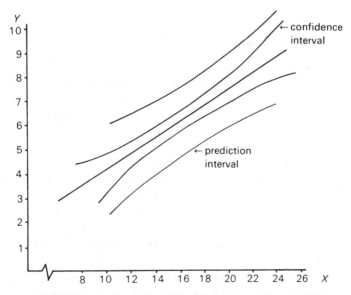

FIGURE 14-8. Prediction interval for trip generation equation.

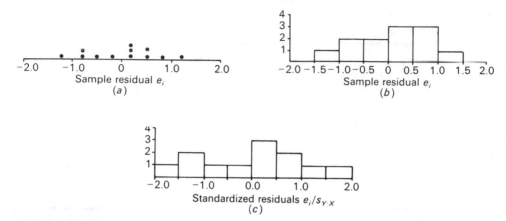

FIGURE 14-9. Residual plots for trip generation equation. (a) dot diagram; (b) histogram; (c) histogram of standardized residuals.

error of the estimate. Now under Assumption 4 of the linear regression model, the residuals from the sample regression line should be normally distributed with mean 0 and variance σ^2. The standardized residuals are thus also normally distributed with zero mean, but have a unit variance and standard deviation. What should we look for in a standardized residual plot? Roughly 95% of the observations should fall in the range $[-2, +2]$, or more exactly $[-1.96, +1.96]$. In Figure 14-9(c), the standardized residuals are plotted for the household trip generation problem. Notice that 100% of observations lie within the range $[-2, +2]$. Given the small sample size of $n = 12$, this is not unexpected. Checking for outliers is also convenient when the residuals are expressed in this form.

Examination of Residual Plots

A more common type of residual plot graphs the sample residuals e_i (or standardized residuals) versus the predicted value of Y, denoted $\hat{Y}_i$. How should such a plot appear? If the first three assumptions of the linear regression model hold, residual plot should be rectangular, as in Figure 14-10(a). Any significant departure from this pattern is usually evidence of some violation in one of these three assumptions. There are several common patterns indicative of specific problems.

    1. A plot in which the width of the residual band increases with Y indicates that the assumption of homoscedasticity has been violated. In Figure 14-10(b) the residuals appear to have a greater variance for large values of Y than for smaller values. Corrective action for this condition must be undertaken, and the least squares estimators a and b do not have the properties specified by the Gauss-Markov theorem [see Draper and Smith (1981)]. Although it is less common, the variance of the residuals may decrease with Y or even increase at both extremes.

    2. Unsatisfactory residual plots can also be obtained when the wrong functional form is used for the systematic component of the relationship. In Figure 14-10(c), for example, we notice that positive residuals occur at moderate values of Y and negative residuals for

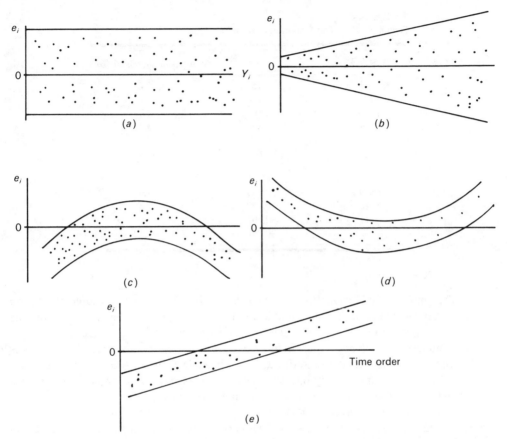

FIGURE 14-10. Residual plots for a regression equation. (a) no violation of assumptions; (b) heteroscedastic; (c) and (d) incorrect functional forms; (e) residuals linearly related to time.

high and low values of Y. This can be corrected by fitting an appropriate nonlinear model. Figure 14-10(d) illustrates another violation of this type. Nonlinear models are briefly introduced in Section 14.4.

3. If the data have been generated by an experiment that took place over time, it is often instructive to plot the residuals in time order. In Figure 14-10(e) residuals appear to increase with time. The appropriate corrective action is to introduce a linear function of time into the regression equation by developing a multiple regression model. A multiple regression equation incorporating a time variable would solve the technical violation illustrated in Figure 14-10(e), but the real task of the researcher is to find some assignable cause for this trend.

The residuals e_i and the fitted values $\hat{Y}_i$ of the household trip generation equation are listed in Table 13-4 and plotted in Figure 13-16(a) and (b). There are only 12 residuals, and it is difficult to find any serious deficiency of the simple linear regression model. Since other variables besides income can affect household trip generation rates, the residuals can hardly be thought of as random.

TABLE 14-6
Residuals from Trip Generation Equation

Traffic zone	e_i	Average household size
1	−1.156	2.2
2	0.174	3.2
3	0.509	3.3
4	0.180	3.4
5	0.503	3.8
6	0.833	3.4
7	1.157	4.0
8	0.162	2.8
9	−0.179	2.9
10	−0.849	2.3
11	−0.502	2.6
12	−0.832	2.5

The final type of residual plot to consider is the plot from the sample regression equation versus other independent variables. For example, a plot of the residuals from the household trip generation equation versus the variable household size reveals the essential nonrandomness of the error component in the simple linear regression model. The residuals and the household sizes for the 12 traffic zones are listed in Table 14-6 and plotted in Figure 14-11. The residuals are strongly linearly related to average household size. Traffic zones with small average household sizes are overpredicted by the regression equation, and those traffic zones with large average household sizes are underpredicted. There is a clear indication that a systematic relationship exists between the error component and household size. It is not unreasonable to expect larger households to make more trips simply because more trips are required to maintain larger households. Or, there may be more than one or

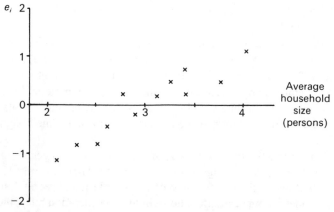

FIGURE 14-11. Residuals and household sizes for 12 traffic zones.

even two working members. In Section 14.5 we see the improvement that can be made in the household trip generation equation by introducing the household size variable.

Testing for Homoscedasticity

When a residual plot gives the impression that the variance increases or decreases in a systematic way, a number of statistical tests can be undertaken to confirm the existence of *heteroscedasticity*. If the data are of the fixed-X variety, a simple, direct test can be undertaken. At each level of X for which data are available, an estimate of σ^2 is made by using

$$s_j^2 = \frac{\sum_{i=1}^{nj}(\hat{Y}_{ij} - \hat{Y}_j)}{n_j - 2} \tag{14-13}$$

where $\hat{Y}_j$ is the predicted value of Y at the level $X = X_j$ and $\hat{Y}_{1j}$, $\hat{Y}_{nj}$, ..., $\hat{Y}_{nij}$, are the n_j residuals from the regression line at $X = X_j$. Note that Equation (14-13) is simply the variance of the residuals at $X = X_j$. It is equivalent to $S_{Y \cdot X}^2$. Similar estimates of σ^2 can be made at each level of X_j for which sufficient replications exist. The constancy of the error variance can then be tested by using an F test on each pair of levels. If the data conform to the random-X model, then a slightly different test is used. First, we divide the observations into two equal parts based on the value of X. The lowest valued observations are put in the first group, and the observations for the highest values of X are put in the second group. We then calculate separate regressions for each data set. The estimates of σ^2 from each half of the data can be compared by using the F test. Many software packages routinely perform a variant of this test. The sample is divided into two equal parts based on the value of X, and an estimate of the variance about the regression is calculated from the *single* regression line generated with all the observations. An F test can then be performed on the two estimates. An even simpler but less powerful alternative exists. The rank correlation between the absolute value of the residuals and the value of the independent variable for each observation is determined. This is tested for significance by using the procedure for rank correlation outlined in Chapter 12.

   If any of these tests indicates the assumption of homoscedasticity cannot be sustained, then appropriate remedial action must be undertaken. A number of different procedures can be followed. These techniques are fully described in Draper and Smith (1981), Pindyck and Rubinfeld (1981), and Neter, Wasserman, and Kutner (1983).

Testing the Normality of the Residuals

The normality of the residuals can be checked by using either of the two goodness-of-fit tests described in Chapter 11: the χ^2 and Kolmogorov-Smirnov tests. The Kolmogorov-Smirnov test is preferred because it is the more powerful. To apply the Kolmogorov-Smirnov test, we simply compare the S-shaped cumulative distribution of the standard normal to the sample cumulative distribution function of the standardized residuals. Although it is possible to determine the sample cumulative distribution function for all residuals, it is common to calculate the cumulative function for the seven integer values of

TABLE 14-7
Kolmogorov-Smirnov Test for Normality of Residuals

Z	Cumulative number of residuals	$S(Z)$ observed cumulative relative frequency	$F(Z)$ normal cumulative probability	Absolute value of difference
−3.0	0	0.0000	0.0013	0.0013
−2.0	0	0.0000	0.0228	0.0228
−1.0	3	0.2500	0.1587	0.0913
0.0	5	0.3750	0.5000	0.1250
1.0	11	0.8333	0.8413	0.0080
2.0	12	1.0000	0.9772	0.0228
3.0	12	1.0000	0.9987	0.0013

Z starting at −3.0. The required calculations for the trip generation equation are summarized in Table 14-7. Of the 12 residuals, three have standardized values below $Z = -1$, and therefore $S(-1) = 3/12 = .2500$. The two cumulative functions are illustrated in Figure 14-12. The test statistic $D = .1250$ is well below the maximum allowable difference of .483 for $n = 7$ at $\alpha = .05$. There is insufficient information to reject the assumption of normality.

Testing for Spatial Autocorrelation

Where the observations used in a regression problem represent contiguous areas of a map, the residuals can be plotted on a map and examined for randomness. Any of the auto-correlation measures described in Chapter 12 can be used for this purpose. The simple two-color test for contiguity, though by no means the most powerful test, is the easiest to apply. Because this test uses a nominal variable, the level of the measurement of the residuals is reduced by dividing them into two nominal categories: positive and negative residuals. The map is developed by coloring the areas corresponding to positive residuals black and the areas corresponding to negative residuals white. Referring once more to the household trip generation equation, we see that this procedure generates the two-color map

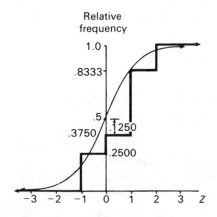

FIGURE 14-12. Two cumulative functions.

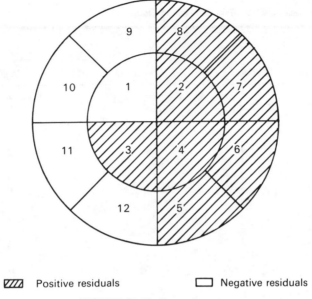

☑ Positive residuals   ☐ Negative residuals

FIGURE 14-13. Two-color map.

of Figure 14-13. Note that all seven positive residuals are contiguous, as are the five negative residuals. Apparently there is strong positive autocorrelation since adjacent traffic zones tend to have identically signed residuals. The join structure is summarized in Table 14-8. There are $L = 20$ total joins, 6 of these BW (black-white), 9 are BB, and 5 are WW.

The required calculations to generate the sampling distributions of the three join types are summarized in Table 14-9. From the visual impression of Figure 14-14, we expect too many BB and WW joins and too few BW joins. Although this is the case, none of the results are statistically significant. This somewhat surprising result can be partially explained by the rather small sample size. Usually, at least 40 observations are required before the assumption of normality of the sampling distributions can be accepted. The loss of information resulting from the reduction in level of measurement of the residuals from interval to nominal is also a problem. More sophisticated tests are described in Chapter 12 and in detail by Cliff and Ord (1973) and Lewis (1977).

14.4. Nonlinear Regression Models

If regression analysis were to be applicable only to purely linear functions, it would be of limited value. Fortunately, this is not the case. Nonlinear models can also be utilized in regression analysis. Nonlinear models can be divided into two types: *intrinsically linear* models and *intrinsically nonlinear* models. A nonlinear model is said to be intrinsically linear if it can be expressed in the standard linear form given by Equation (14-2) by using a suitable transformation of one or both variables in the model. If such a transformation cannot be found, then the model is intrinsically nonlinear. Intrinsically linear models are of

TABLE 14-8
Join Structure for Residuals from Trip Generation Example

Traffic zone	Color	Joins L_i	$L_i(L_i - 1)$	Joins BB	Joins WW	Joins BW
1	W	4	12	0	2	2
2	B	4	12	3	0	1
3	B	4	12	1	0	3
4	B	4	12	4	0	0
5	B	3	6	2	0	1
6	B	3	6	3	0	0
7	B	3	6	3	0	0
8	B	3	6	2	0	1
9	W	3	6	0	2	1
10	W	3	6	0	3	0
11	W	3	6	0	2	1
12	W	3	6	0	1	2
		$\sum_{i=1}^{12} L_i = 40$	$\sum_{i=1}^{12} L_i(L_i - 1) = 96$	18	10	12

$$L = \sum_{i=1}^{12} \frac{L_i}{2} = 20 \qquad K = \frac{1}{2}\sum_{i=1}^{12} L_i(L_i - 1) = 48$$

$$\text{BB} = \frac{18}{2} = 9 \qquad \text{WW} = \frac{10}{2} = 5 \qquad \text{BW} = \frac{12}{2} = 6$$

particular interest because the normal least squares estimating procedures can be used to determine the two parameters in the equation. Linearity can be produced by transforming the independent variable X, the dependent variable Y, or both X and Y. The variety of possibilities is illustrated by using models of each type.

Models Obtained by Transforming the Independent Variable

Consider the following three potential functional forms describing the regression relationship:

$$Y_i = \alpha + \frac{\beta}{X_i} + \varepsilon_i \tag{14-14}$$

$$Y_i = \alpha + \beta X_i^r + \varepsilon_i \tag{14-15}$$

$$Y_i = \alpha + \beta \log X_i + \varepsilon_i \tag{14-16}$$

Each of these functions can be made linear by an relatively simple transformation. In Equation (14-14) if we replace $1/X_i$ with Z_i, the equation becomes $Y_i = \alpha + \beta Z_i + \varepsilon_i$. This is the reciprocal transformation. For (14-15), the appropriate substitution is $Z_i = X_i^r$. This leads to a family of regression models depending on the value of r. If $r = 1$, Equation (14-15) reduces to the simple linear regression model; for $r = 1/2$, it specifies a square root

TABLE 14-9
Tests of Spatial Autocorrelation for Trip Generation Residuals

$$p = \frac{7}{12} = .5883 \qquad q = \frac{5}{12} = .4167 \qquad L = 20 \qquad K = 48$$

$\mu(BB) = p^2 L = (.5833)^2 (20) = 6.805$

$\sigma(BB) = \sqrt{p^2 L + p^3 K + p^4(L + K)}$

$$= \sqrt{\begin{array}{l} 6.805 + (.5833)^3(48) \\ - (.5833)^4(20 + 48) \end{array}}$$

$= 2.908$

$Z(BB) = \dfrac{9 - 6.805}{2.908}$

$= .755$

$\mu(BW) = 2pqL = 2(.5833)(.4167)(20) = .6805$

$\sigma(BW) = \sqrt{2pqL + pqK - 4p^2q^2(L + K)}$

$$= \sqrt{\begin{array}{l} 9.722 + (.5833)(.4167)(48) \\ - 4(.5833)^2(.4167)^2(20 + 48) \end{array}}$$

$= 2.306$

$Z(BW) = \dfrac{6 - 9.722}{2.306}$

$= -1.614$

$\mu(WW) = q^2 L = (.4167)^2 (20) = 3.473$

$\sigma(WW) = \sqrt{q^2 L + q^3 K + q^4(L + K)}$

$$= \sqrt{\begin{array}{l} 3.473 + (.4167)^3 (48) \\ - (.4167)^4 (20 + 48) \end{array}}$$

$= 2.213$

$Z(WW) = \dfrac{5 - 3.473}{2.213}$

$= .690$

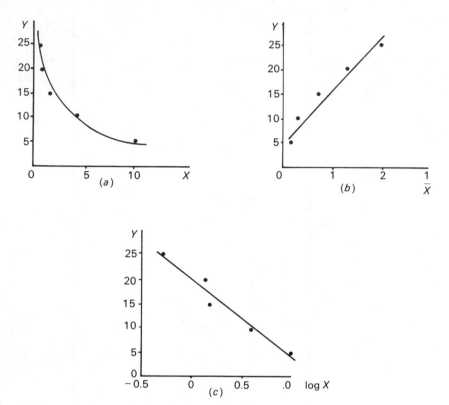

FIGURE 14-14. (a) Nonlinear function; (b) reciprocal transformation of (a); (c) logarithmic transformation of (a).

transformation; and for positive integer powers of r, it specifies a series of power functions. Equation (14-16) specifies a logarithmic transformation of the independent variable. These three equations are all intrinsically linear since proper transformation produces a linear equation of the form (14-2). These models are termed intrinsically linear since they are linear in the regression model parameters α and β.

The effects of these transformations are best illustrated by a simple numerical example. Suppose the entire data set consists of the following five observations:

X	0.50	0.75	1.50	4.0	10.0
Y	25	20	15	10	5

When they are graphed in Figure 14-14(a), the data clearly suggest that the relationship between the two variables is nonlinear. If the transformation $Z = 1/X$ is applied to the data, the input to the regression model is

Z	2.0	1.33	0.67	0.25	0.10
Y	25	20	15	10	5

This transformation yields the scatter diagram of Figure 14-14(b). There appears to be reasonable cause to fit a linear function to these data. The intercept and regression coefficient obtained from the regression equation $Y = \alpha + \beta Z$ can be substituted into the reciprocal equation $Y = \alpha + \beta (1/X)$.

Although the reciprocal transformation seems to have solved the problem, caution is in order. The goal is not necessarily to find a transformation of the variables that creates a scattergram suggesting a linear relation. Many different transformations can generate virtually identically shaped functional relations. For example, if the same observations are transformed by substitution $Z = \log X$ following data are obtained:

Z	−0.30	−0.12	0.18	0.6	1.0
Y	25	20	15	10	5

These data are graphed in Figure 14-14(c). Clearly, the logarithmic transformation might also be used to produce a linear relation. In many instances the data might be subjected to several different transformations with near identical results. The choice among these transformations is best made on the basis of a priori knowledge of the nature of relationship linking the two variables.

Models Obtained by Transforming the Dependent Variable

It is also possible to transform the dependent variable. As an example, consider the function

$$Y_i = \frac{1}{1 + e^{\alpha + \beta X_i + \varepsilon_i}} \tag{14-17}$$

Taking the reciprocals of both sides and then subtracting 1 yields

$$\frac{1}{Y_i} - 1 = e^{\alpha + \beta X_i + \varepsilon_i} \tag{14-18}$$

Taking the natural logarithm, denoted ln, of each side produces

$$\ln\left(\frac{1}{Y_i} - 1\right) = \alpha + \beta X_i + \varepsilon_i \tag{14-19}$$

which is linear in the two parameters α and β. The effect of this transformation is illustrated by using another set of hypothetical data:

X	Y	ln(1/Y − 1)
3	0.1	2.2
6	0.2	1.4
7	0.5	0.0
8	0.5	−1.4
10	0.9	−2.2

In Figure 14-15 we see that the logistic S-shaped function suggested in the scatter-gram of the raw (X, Y) data is changed to a near-linear relation by this transformation. Logistic functions are typically found in the description of the temporal pace of the diffusion of an innovation. Variable Y is the percentage population adopting some in-novation, and X is the time or year of adoption. The innovation is slowly adopted in the initial stages of the diffusion process. Then, large numbers of individuals adopt the innovation. Finally, the process ends as the few remaining holdouts adopt the innova-tion.

Models Obtained by the Transformation of Both X and Y

In many situations in geography, a *multiplicative* model such as a simple power function proves to be an adequate representation of the relationship between two variables. Consider the regression model

$$Y_i = \alpha X_i^{\beta} \, \varepsilon_i \qquad\qquad (14\text{-}20)$$

where α and β are unknown parameters and ε_i is the usual population error. By taking the logarithms of both sides of (14-20), the model is converted to:

$$\log Y_i = \log \alpha + \beta \log X_i + \log \varepsilon_i \qquad\qquad (14\text{-}21)$$

This can be estimated by normal least squares procedures. There is one important difference between the multiplicative model (14-20) and (14-2). For the requirements for hypothesis testing and confidence interval construction to be met, the log of the error term must be normally distributed. An alternate specification of $Y_i = \alpha X_i^{\beta} + \varepsilon_i$ has an additive error term. This model is *intrinsically nonlinear* since there are no substitutions for X and Y that can reduce it to the form of (14-2).

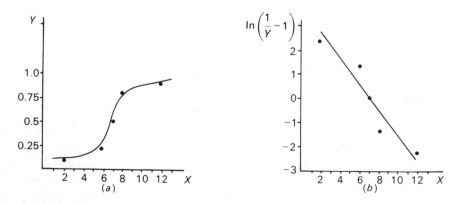

FIGURE 14-15. (a) Logistic function; (b) linear transformation of (a).

Nonlinear Regression Models in Geography

Many regression models used by geographers require some sort of variable transformation. Consider the following example drawn from river channel morphology. All existing evidence indicates that the greater the quantity of water that moves through a river channel, the larger the cross section of the channel. For example, many different studies have demonstrated that a multiplicative model can be used to relate water surface width b of most rivers to the increase in the mean annual discharge Q_m in a downstream direction:

$$b = kQ_m^{\beta} \tag{14-22}$$

Usually the coefficients k and β are different for each river, but several studies have utilized field data collected from a large number of different rivers in roughly the same geographic area. Research has demonstrated that for most rivers an equation of the form $b = kQ^{0.5}$ could be derived, with each river having a different value of the coefficient k. In a study using data aggregated for several different streams, the relation $b = 2.38Q_{2yr}^{0.527}$ was shown to provide an excellent statistical fit. In this case the independent variable was not mean annual discharge but the discharge at the river reach for a flood with a 2-yr recurrence interval. Differences in the values of the parameters for these hydraulic relations depend on the exact specification of the dependent and independent variables as well as the characteristics of the sediment load of the rivers involved. In turn, characteristics of the sediment load depend on the river bed material. Coarse-load gravel-fed streams tend to have channels with a high width/depth ratio, whereas fine sediment beds tend to produce narrow and deep cross sections.

Many different hydraulic relations tend to be described best by using power functions requiring a log-log transform of the variables in the equation. For example, research into stream meandering has shown that many meander dimensions are related to river channel discharge. Meander wavelength appears to vary with discharge according to a simple multiplicative law similar to (14-25). All such relations appear as straight lines when they are graphed on log-log paper.

14.5. Introduction to Multiple Regression Analysis

Seldom in the social or physical sciences is it possible to satisfactorily explain the variations in a dependent variable by using a single independent variable. Fortunately, it is possible to extend regression analysis to situations in which several independent variables are used to account for the variability of a single response variable. This is termed *multiple regression analysis*. From the point of view of prediction, it is an obvious advantage to be able to use the information in two or more variables in an explanatory equation. We should be able to provide better predictions by using multiple regression analysis. How can appropriate variables be found? At times, a theory linking the two variables might suggest the proper specification of a multiple linear equation. In other instances previous empirical research might suggest logical variables to be included. For example, household trip generation equations generally include other variables besides household income. The number of cars available to the household and the size of the household are two frequently used predictor variables for household trip generation rates. Larger households with two or more cars tend to make more trips than smaller households with one or no cars available.

There are two ways of identifying or verifying the usefulness of potential variables. First, any residual plot that indicates a linear relation between a new variable and the residuals of a simple regression equation usually points to the need for a new variable to be included in the model. Figure 14-11 clearly illustrates the need to introduce household size as an explanatory variable in the trip generation equation. Second, it is possible to calculate the simple correlation coefficient between the residuals and any new variable. Whenever a significant correlation exists, a multiple regression equation is usually necessary for proper specification.

Although a complete discussion of multiple linear regression is beyond the scope of this text, a few of the concepts are introduced in the context of household trip generation. Relabel the independent variable of average household income as X_1, and let X_2 be the average household size in the traffic zone. For these variables, it is now possible to fit the following multiple linear regression equation:

$$Y_i = a + b_1 X_{1i} + b_2 X_{2i} + e_i \qquad (14\text{-}23)$$

The subscript i refers to the observation number or index of the traffic zone. Least squares can still be applied to the estimation of the three parameters a, b_1, and b_2. Consider Figure 14-16. The dependent variable Y (household trip generation rate) is illustrated as the vertical axis in the three-dimensional space formed by including both household income X_1 and household size X_2, as predictors. Each observation can be located as a single point in this

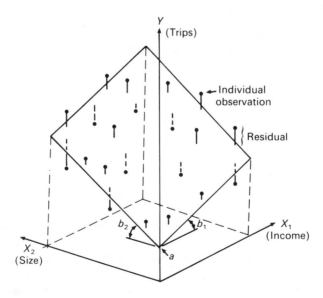

FIGURE 14-16. Least squares estimation of a multiple regression equation with two independent variables.

three-dimensional space. The regression equation $Y = a + b_1X_1 + b_2X_2$ defines a plane in this space. The exact orientation of this plane is determined by the values of the three parameters a, b_1, and b_2. Parameter a is the intercept on the Y axis, where $X_1 = X_2 = 0$. The regression coefficient b_1 defines the slope of the plane with respect to the X_1 axis, and the regression coefficient b_2 controls the orientation with respect to the X_2 axis. The values of these three parameters are chosen so that the sum of squared deviations $\Sigma_{i=1}^{n}(Y_i - a - b_1X_{1i} - b_2X_{2i})^2$ is minimized. These deviations represent the vertical distances from the observations to the regression plane. As usual, observations above the plane appear as positive residuals and those below the plane as negative residuals.

The least squares regression equation for the household trip generation model

$$Y = -3.01 + 0.350X_1 + 1.18X_2 \qquad (14\text{-}24)$$

indicates that both household size and household income are positive influences on trip generation rates. The intercept of -3.01 is clearly out of the range of the input data and can be ignored. It is meaningless to suggest that households with no members and no income make -3.01 trips per day! How has the multiple regression equation improved the prediction of trip generation rates? First, note in Table 14-10 that the residuals are generally smaller once the household size variable has been included. Note particularly the significant

TABLE 14-10
Goodness-of-Fit Measures for Multiple Regression Trip Generation Example

Traffic zone	Y_i	$\hat{Y}_i$	$(Y_i - \hat{Y}_i)$	$(Y_i - \hat{Y}_i)^2$	$(Y_i - \bar{Y}_i)^2$
1	3	3.09	-0.09	0.0081	9.505
2	5	4.97	0.03	0.0009	1.173
3	5	4.73	0.27	0.0729	1.173
4	4	4.15	-0.15	0.0225	4.339
5	6	6.37	-0.37	0.1369	0.007
6	7	6.60	0.40	0.1600	0.841
7	9	9.06	-0.06	0.0036	8.509
8	7	6.60	0.40	0.1600	0.841
9	8	8.12	-0.12	0.0144	3.675
10	8	8.11	-0.11	0.0121	3.675
11	6	6.01	-0.01	0.0001	0.007
12	5	5.19	-0.19	0.0361	1.173
			0.0	0.6276	34.918

$$S_{Y \cdot X_1 X_2} = \sqrt{\frac{.6276}{12 - 3}} = .264$$

$$R^2 = \frac{34.918 - .6276}{34.918}$$

$$= .982$$

decline in the sum of the squared residuals ESS. It has fallen from 5.667 to 0.6285. Second, we notice a similar improvement in the standard error of estimate.

The superiority of the new equation is best illustrated with the use of the coefficient of determination. No matter how many variables are added to the regression equation, the total variation in Y, or TSS, remains constant at $\Sigma_{i=1}^{n}(Y_i - \bar{Y})^2$. As we have just witnessed, the addition of a new variable reduces the ESS and increases the RSS. At worst, the introduction of a new variable can offer no change in ESS and RSS. Except in the most unlikely of cases, the new variable will cause some reduction in ESS. Define the *coefficient of multiple determination* as

$$R^2 = \frac{\text{TSS} - \text{ESS}}{\text{TSS}} = \frac{\text{RSS}}{\text{RSS}}$$

We have used uppercase for this statistic to distinguish it from the simple coefficient of determination r^2. The formula for R^2 is exactly the same as for r^2. The only difference is that the ESS in the multiple regression equation is based on the residuals from the multiple linear regression equation. For this case, the reduction in ESS from 567 to 0.6285 leads to an R^2 value of .982. The obvious statistical questions are:

1. Has the new variable improved the prediction of household trip generation rates?
2. Is the new equation superior to the simple linear regression?
3. Does this new equation satisfy the assumptions of regression analysis?

Multiple regression analysis is probably the single most widely used statistical technique in the social sciences. The generalization from simple to multiple regression introduces a number of problems beyond the scope of this text. An especially readable, yet technically excellent, presentation is contained in Neter, Wasserman, and Kutner (1983).

14.6. Summary

In this chapter we have extended the role of regression analysis beyond the simple descriptive function introduced in Chapter 13. First, the descriptive role of regression analysis is identified as but one function undertaken in a more comprehensive five-step model building procedure. Testing the statistical significance of the equation using the derived model for prediction purposes requires four specific assumptions. With these assumptions it is possible to test the hypothesis that the slope of the regression line equals zero. If this hypothesis can be rejected at a given level of significance, then the model is necessarily useful for purposes of prediction. For this test to be valid, certain other assumptions must be met. Statistical tests of assumptions—normality, homoscedasticity, randomness of the error term—have also been discussed. We have briefly introduced two other modifications of regression analysis. First, it is possible to use conventional least squares regression analysis to fit some nonlinear relationships by using variable transformations. Second, an introduction to multiple regression analysis suggests the wide applicability of statistical technique.

REFERENCES

A. D. Cliff and J. K. Ord, *Spatial Autocorrelation* (London: Pion, 1973).

N. Draper and H. Smith, *Applied Regression Analysis*, 2d ed. (New York: Wiley, 1981).

P. Lewis, *Maps and Statistics* (London: Methuen, 1977).

J. Neter, W. Wasserman, and M. Kutner, *Applied Linear Regression Models* (Homewood, Ill: Irwin, 1983).

R. Pindyck and D. L. Rubinfeld, *Econometric Models and Economic Forecasts*, 2d ed. (New York: McGraw-Hill, 1981).

FURTHER READING

Examples of the use of regression analysis in geographical research can be found in several sources. Recent issues of many journals in the field contain summaries of research findings, many of which are based on the use of regression analysis. Introductory textbooks also describe several examples of the use of regression analysis. Useful references are listed in Chapter 2. The more advanced textbooks cited in the References section of this chapter also contain relevant examples. Finally, the problems at the end of this chapter, as well as the examples contained in Chapters 12 and 13, may be consulted. Note that most of the applications in the technical journals utilize quite advanced forms of regression analysis. Many of these techniques have been only mentioned in passing in this introductory text.

For the specific topics of this chapter, useful references include Cliff and Ord (1973) for spatial autocorrelation and Leopold, Wolman, and Miller (1984) for hydraulic relations of streams.

L. Leopold, G. Wolman, and J. Miller, *Fluvial Processes in Geomorphology* (San Francisco: Freeman, 1984).

PROBLEMS

1. Explain the meaning of the following terms:
   a. Measurement error
   b. Stochastic error
   c. Homoscedasticity assumption
   d. Point estimates
   e. Residual plot
   f. Spatial autocorrelation

2. Briefly explain the assumptions of the linear regression model and the implications of the Gauss-Markov theorem.

3. Differentiate between a confidence and prediction interval in the context of a regression equation.

4. Differentiate between a nonlinear regression model and an intrinsically linear model.

5. Consider the sound data of Problem 5 in Chapter 13.
   a. What is the PROB-VALUE that the true regression coefficient is zero? less than zero?
   b. What is the PROB-VALUE that the regression intercept is zero?
   c. Construct and evaluate the ANOVA table for this regression equation.
   d. Construct a 95% confidence interval for the mean decibel level at a distance of 200 ft.
   e. Construct a 99% prediction interval for the decibel level at a distance of 100 ft.

6. Using the crime rate data of Problem 6 in Chapter 13
   a. What is the PROB-VALUE that true regression coefficient is zero? less than zero?
   b. What is the PROB-VALUE that the regression intercept is zero?
   c. Construct and evaluate the ANOVA table for this regression equation.
   d. Construct a 95% confidence interval for the mean crime rate for a city of 400,000 persons.
   e. Construct a 99% prediction interval for the crime rate of a city with a population of 1,000,000 persons.

7. Test the residuals from the multiple regression model of household trip generation rates (Table 14-14) for
   a. Autocorrelation, using the contiguity test for two-color maps described in Chapter 12.
   b. Normality using the χ^2 test described in Chapter 12.
   c. Normality using the runs test described in Chapter 12.

8. Using the data from the restaurant profitability example of Problem 7 in Chapter 13
   a. Calculate the PROB-VALUE that the regression coefficient for number of households is a significant determinant of restaurant profitability for mall locations.
   b. Calculate the PROB-VALUE that the regression coefficient for number of households is a significant determinant of restaurant profitability for highway locations.
   c. Generate 95% confidence intervals on each of these coefficients.
   d. Based on the results of part (c), do you think that the impact of the number of households is different by location of restaurant? Why or why not?
   e. Generate ANOVA tables for the three regression equations estimated in Problem 13-7.
   f. Generate residual plots for the equations. Are any assumptions clearly violated?
   g. One way of testing the importance of location on restaurant profitability is to calculate a correlation coefficient between the residuals of the equation from part a of Problem 13-7 with location expressed as a nominal variable. What correlation coefficient can you use? Is it significant?

9. Is there a significant time trend of revenue-passenger miles in the data of Problem 13-8?

10. Using the data for residential densities of Problem 13-9
    a. Examine the residuals from the first equation. Does there appear to be any pattern to them?
    b. Compare these residuals to those of the second equation. Does it appear that the transformation has improved the situation?
    c. Place the first six observations in Group A and the last six in Group B. Test the equation for heteroscedasticity using the procedure described in Section 14-3.

15

Time Series Analysis

Chapter 3 presented fundamental concepts of time series, and described simple smoothing of observed series. The approach there was largely heuristic, in the sense that we presented rules of thumb having intuitive rather than theoretical appeal. In this chapter, we extend the treatment of time series in two ways. First, we will take a more formal statistical approach, one based on modeling the time-varying behavior of an observed series. At the risk of stating the obvious, we emphasize that this modeling is empirical in nature, meaning that the models are developed from observations rather than theory. We will not be assuming a priori knowledge of the mechanisms giving rise to the observations, but instead attempt to discover statistical regularities consistent with whatever physical or social processes happen to be operating. Obviously, where theory is available, it should be used to restrict candidate models to those admitted by the theory. We also emphasize that despite the central role played by the observations, researchers are by no means free from having to make assumptions about mechanisms underlying the observations. As a matter of fact, it will sometimes be necessary to make strong assumptions.

This chapter will also return to the topic of smoothing, mainly with the idea of placing it in the larger context of time series filtering. As a first step, we will cover the simple running means described earlier in Chapter 3. As will be seen, they are less than optimal in several regards. We will therefore discuss methods for designing filters better suited to smoothing, as well as filter design for other time series tasks.

To keep the chapter short, we will not provide a detailed treatment of these topics, both of which are often taught as one-semester courses. The sections that follow should therefore be considered an introductory discussion, aimed primarily at presenting the rationale behind and potential use of some commonly used techniques. Additionally, we want to provide enough coverage that should time series analysis be useful in a student's own research, he/she will have at least a few tools at hand.

15.1. Time Series Processes

In Chapter 3, we drew a distinction between a time series *process* and an observed time series, called a *realization* of the process. Loosely speaking, we think of the process as giving rise to the observed series, in much the way we think of a random variable giving rise to a sample. Imagine, for example, a closed lake or reservoir subject to variable inflow gains and evaporative losses. In some years there is a net gain to the lake; whereas in other years evaporation exceeds inflow, and lake level decreases. To simplify matters, we will collect gains and losses in a single term a_t, representing the addition for year t. Obviously, a_t will be positive or negative depending on whether or not gains are larger than losses for year t. Let us imagine that the climate is such that on average the gains equal losses. To be consistent with the climatic assumption, the average of a taken over all time must be zero. Let us further assume that the increments a_t are statistically independent of one another, and that they come from the same probability distribution. (Random variables with these properties are said to be independently and identically distributed, or i.i.d.). Given some initial value for lake level, we can write the level for any year t as

$$Y_t = Y_{t-1} + a_t$$

The sequence above is called a *random walk* process, with a_t representing random steps, or shocks, acting to move Y up or down. This is a very simple example of a time series process, but it nevertheless illustrates a number of important concepts. First, we note that because the additions a_t are random, every sequence of Y will contain random variation. However, successive values of Y will not be uncorrelated, because of carryover from the previous year. Second, not only is there random variation from year to year, but the value in a particular year is a also a random variable. That is, because a_t is random, we must be willing to admit many values are possible for a given year t. This is analogous to saying that many values are possible for the ith value in a sample, and that X_i is a random variable. The difference is that with regard to a sample, we take pains to be sure that the sample values are uncorrelated, whereas in time series analysis we are expressly interested in whatever temporal correlation might exist.

In time series analysis, Y is called a *stochastic process*. It represents a family of random variables indexed by time. Each member of the family, a sequence of Y, is a separate realization of the process. When we observe a time series, we obtain one realization of the underlying stochastic process.

DEFINITION: STOCHASTIC PROCESS

A stochastic process Y_t is a family of random variables indexed by time. The process generates values of Y serially in time, with many sequences possible for a given interval of t. An observed time series (or realization) is a particular sequence of Y, corresponding to one outcome of the stochastic process.

15.2. Properties of Stochastic Processes

Stationarity

Notice that a stochastic process generates multiple values of Y for any given t. It is quite natural, therefore, to ask questions about the probability distribution of Y at various t. For example, we might wonder if the mean and variance of Y are constant for all t. In the lake example, the "climate" (a_t) is clearly unchanging, but we might wonder if there is any upward or downward trend in the average lake level, or if there is a tendency for lake levels to become more variable as time advances. We define the mean and variance just as before, except that we must allow for the possibility of change over time. With this small additional complication, we again rely on expected values for the definitions:

$$\mu_t = E(Y_t)$$

$$\sigma_t^2 = E(Y_t - \mu_t)^2$$

To evaluate these functions, we need information about the probability distribution of Y for every t. Stochastic processes whose probability distributions are the same for all time are called *strictly stationary* processes. In this text we will work with a less stringent concept, call *weak stationarity*. A process is weakly stationary if its moments (e.g., mean, variance, etc.) are the same for all time. The distinction between strict and weak stationarity is subtle and of no real interest here; thus we will use the term stationarity to mean weak stationarity.

DEFINITION: STATIONARITY
A stochastic process is stationary if its statistical moments are invariant over time.

Note that varying degrees, or order, of stationarity are possible. For example, a process might be stationary in the mean, but not in variance. As a matter of fact, the random walk described above is one such process. Although the mean is constant for all time, the variance is proportional to the length of the series. Going back to the lake level example, we would have unchanging *mean* lake levels, but the variability of levels would increase through time. It is thus first-order stationary, but not second-order stationary. It should be mentioned that stationarity at a given order requires stationarity at all lower orders. Unless we state otherwise, we will always assume second-order stationarity.

Autocorrelation

Another property of interest is temporal autocorrelation in Y. As was the case for spatial autocorrelation, we are once again interested the relationship between nearby values of Y. Here, of course, the observations are arranged in time rather than space; thus "nearby" means "near the same time." Considering adjacent observations first, we are led to correlate observations at time t with observations at time $t + 1$. This is the *lag-one* autocorrelation defined by

$$\rho(t, t + 1) = \frac{E(Y_t - \mu_t)(Y_{t+1} - \mu_{t+1})}{\sigma_t \sigma_{t+1}} \tag{15-1}$$

This is virtually identical to Equation (5-40), where we defined the bivariate correlation coefficient as the covariance divided by the product of the standard deviations. Conceptually, it is as if we are correlating one column of Y with another indentical column whose entries have been shifted upward one unit of time. The coefficient once again ranges from -1 to 1 and measures linear dependency. A value near $+1$ indicates strong positive association, whereas a large negative value means Y_t tends to oscillate, with values above average immediately followed by values below average. An autocorrelation near zero indicates no linear relationship between adjacent values.

Higher order autocorrelations are similarly defined. In general, we will write $\rho(t, t + k)$ for the lag-k autocorrelation. It will also be useful to have a symbol for autocovariance, $\gamma(t, t + k)$. These are formally defined by

$$\gamma(t, t + k) = E(Y_t - \mu_t)(Y_{t+k} - \mu_{t+k})$$

$$\rho(t, t + k) = \frac{E(Y_t - \mu_t)(Y_{t+k} - \mu_{t+k})}{\sigma_t \sigma_{t+k}} = \frac{\gamma(t, t + k)}{\sigma_t \sigma_{t+k}}$$

For a stationary process, μ and σ are constant, and these definitions simplify to

$$\gamma_k = E(Y_t - \mu)(Y_{t+k} - \mu) \tag{15-2}$$

$$\rho_k = \frac{E(Y_t - \mu)(Y_{t+k} - \mu)}{\gamma_k} = \frac{\gamma_k}{\sigma^2} \tag{15-3}$$

As an example of how these might be used, consider a familiar problem of finding the variance of the sample mean. For independent Y, we have seen that $\sigma_{\bar{Y}}^2 = \sigma^2/n$. In the case of a time series containing autocorrelation, the Y values are clearly not independent of one another; thus, the standard formula σ^2/n requires modification. In particular, we will need an additional term reflecting the aggregate effect of covariances at all lags. This additional term can be positive or negative, depending on the value of the individual covariances. The appropriate formula is

$$\sigma_{\bar{Y}}^2 = \frac{\sigma^2}{n} + \frac{2}{n} \sum_{k=1}^{n-1} \left(1 - \frac{k}{n}\right) \gamma_k \tag{15-4}$$

Notice that if the autocovariances are positive, the sum will be positive, indicating that σ^2/n underestimates the standard error of $\bar{Y}$. In order to reliably estimate sampling error in $\bar{Y}$, we need a larger number of observations than would be needed in the absence of autocorrelation. Given some thought, this is not surprising. We expect a process with positive autocorrelation to have periods of unusually high values followed by periods of unusually low values. (After all, nearby values are positively correlated). A sample occurring within one of these episodes will reveal within-episode variability, but will not reflect the transition from one episode to another. Looking at this another way, the presence of autocorrelation means that we do not have n independent pieces of information. It is as if the effective sample size is smaller, and more observations are necessary to obtain an estimate of equal reliability.

On the other hand, if the autocovariances are negative, the sum in Equation (15-4) will be negative, indicating that we expect more variance in a short sample than would occur if the observations were independent. In either case, positive or negative γ, the implication is clear: when computing the standard error of $\overline{Y}$, we need to account for autocorrelation. The standard formula, σ^2/n, will not work in the presence of strong autocorrelation, whether negative or positive.

Partial Autocorrelation

The autocorrelation function ρ_k indexes the association between Y values separated by k intervals of time. For example, ρ_3 measures the correlation at lag 3. It seems obvious that the correlation at lags 1 and 2 must influence the correlation at lag 3, and ρ_3 therefore includes the effects of intervening values. The *partial autocorrelation* function is useful for removing the influence of intermediate lags. It provides the correlation between values of Y at lag k, controlling for values at lag $k - 1$, $k - 2$, and so forth. The partial autocorrelation function for lag k is denoted ϕ_{kk} and can be understood by imagining a somewhat involved multiple regression procedure.

The procedure calls for values of Y_{t+k} to be regressed on preceding values of Y. For example, if we are interested in ϕ_{33}, we would regress Y_{t+3}, on Y_{t+2} and Y_{t+1}. The residuals from this prediction equation correspond to variance in Y_{t+3}, which cannot be explained by variance in the preceding two periods. We would also regress Y_t on Y_{t+2} and Y_{t+1}. Errors of this prediction represent variance in Y_t that cannot be explained by variance in the following two periods. The correlation of the two sets of residuals gives ϕ_{33} and represents correlation remaining at lag 3 after correlations at shorter lags have been removed.

The regression procedure above is useful for understanding how the partial correlation function comes about, but is not likely to be used in actual practice, as much more convenient expressions exist. For example, an expression for ϕ_{22} is

$$\phi_{22} = \frac{\rho_2 - \rho_1^2}{1 - \rho_1^2} = \frac{\rho_2 - \phi_{11}\rho_1}{1 - \phi_{11}\rho_1}$$

By writing $\rho_1 = \phi_{11}$, we see that the partial correlation can be computed from the simple correlations up to order 2 and partial correlations of lower order. In other words, given the first-order partial coefficient, one can find the second-order partial coefficient without going through a regression procedure. For higher order partial coefficients ϕ_{33}, ϕ_{44}, etc., similar recursive equations can be used (see Cryer, 1986 p. 109).

15.3. Types of Stochastic Processes

Having defined a stochastic process, we turn to several specific classes of processes that form the basis of most time series modeling. These processes have the advantage of being easy to understand, yet provide for a very wide range of behavior, especially when combined. In order to simplify the mathematics, we assume that the mean has been subtracted, so that μ_Y is zero.

Autoregressive (AR) Processes

Autoregressive processes are those for which the current value Y_t can be found from past values plus a random shock a_t. The simplest is a first-order process, where only the value immediately preceding Y_t contributes:

$$Y_t = \phi\, Y_{t+1} + a_t$$

This is similar to the random walk model described earlier, except for the parameter ϕ. This coefficient determines how strongly the past (Y_{t-1}) influences the present (Y_t). Clearly, if ϕ is zero, Y is governed completely by the random component a_t. On the other hand, large values of ϕ imply that previous values carry over, and are "remembered" by the system in subsequent years. The amount of memory is variable, but stationarity requires that $-1 < \phi < 1$. For ϕ outside this range, it can be shown that the variance of Y is a function of t and therefore does not meet the stationarity requirement. (The random walk process has ϕ exactly equal to 1; thus it is an example of a nonstationary process.) A first-order auto-regressive process is abbreviated AR(1).

Figure 15-1 shows time series from two AR(1) processes. They are based on the same sequence of a_t, but have different values for ϕ and μ. Both curves vary about their mean with no overall long-term trend, but the pattern of variance is very different. Notice how the large positive value for ϕ gives excursions from the mean that persist for considerable periods. It is easy to imagine that if we were to observe only a part of the record, say Y_{50} to Y_{75}, we might conclude that a deterministic trend is present. On the other hand, the process with negative ϕ shows a pronounced tendency to oscillate upward and downward from one period to the next.

The AR(1) process extends easily to higher orders. The general equation for an AR(p) process is

$$Y_t = \phi_1 Y_{t-1} + \phi_2 Y_{t-2} + \phi_3 Y_{t-3} + \cdots + \phi_p Y_{t-p} + a_t \tag{15-5}$$

Though we are free to choose any order p, model orders higher than 3 are not frequently encountered. The stationarity requirements for higher order models are more complicated, and we will not state them here (see Wei, 1990, p. 46). Autoregressive models are a natural choice when a system contains components able to store quantities like energy or mass or information from one observing period to the next. The issues of how to choose the order and estimate coefficients will be covered later. For now we simply note that AR models treat the present as a linear combination of the past, subject to random shocks. As might be expected, the size of the random shocks has a direct influence on the ability to predict Y from its past. If the random component is relatively small, Y will be highly predictable.

Moving Average (MA) Processes

A moving average (MA) model assumes Y_t is a linear combination of past shocks plus a new a_t. The first-order model is just

$$Y_t = a_t - \theta a_{t-1}$$

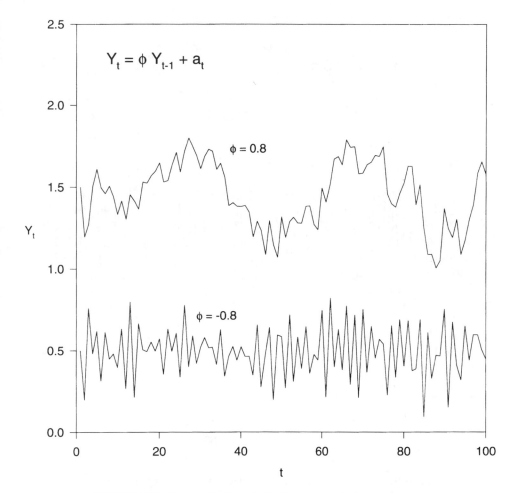

FIGURE 15-1. Time series from first-order autoregressive processes.

The name "moving average" seems ill-chosen, as it does not seem that we are averaging anything. To appreciate the moving average aspect, we must view the a_t as data and notice that values of Y are weighted sums of nearby a. For the first-order moving average model, the weights in the running mean are 1 and $-\theta$. An example of an MA(1) process is shown in Figure 15-2 using θ of .5. The model produces a very noisy record, with large jumps possible in a single time step. We saw previously that the need for stationarity placed a restriction on ϕ in the AR(1) model. For MA models, the concern is to ensure *invertibility*. Looking ahead to the problem of identifying a model based on an observed series, we will want to choose models that can be uniquely identified from their autocorrelation function. Such models are called invertible, because their autocorrelation function can be "inverted" to find the underlying process. It so happens that if θ is outside the range $-1 < \theta < 1$, different processes (θ values) have the same autocorrelation structure, and hence are not invertible. Knowing the autocorrelation would not allow one to identify the model. The

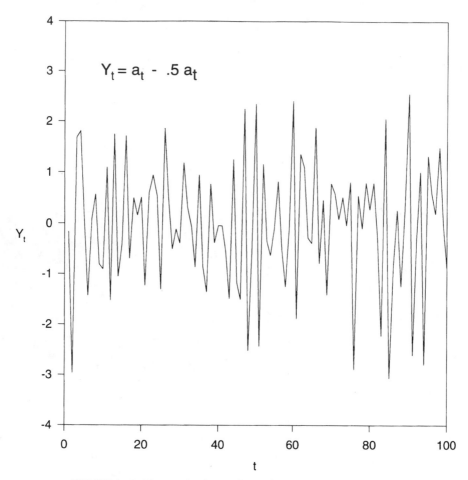

FIGURE 15-2. Time series from a first-order moving average process.

need for invertibility therefore restricts the admissible values of θ, just as stationarity did for ϕ.

As is true for AR models, higher order MA models are available. The order is symbolized by q, and the general equation is

$$Y_t = a_t - \theta_1 a_{t-1} - \theta_2 a_{t-2} - \cdots - \theta_q a_{t-q} \tag{15-6}$$

In many applications, it will be difficult to interpret a moving average model in terms of physical or social processes. Some help is provided by thinking of a_t as a prediction error, the random shock at time t for which we have no predictive ability. With this view, Equation (15-6) can be seen as

Observed = Error + Predicted

Notice that the predicted values are a linear combination of past values of a, each of which is itself a prediction error. In other words, moving average models make predictions for the future based on past errors of prediction. It is a bit like using an instrument known to give false readings. One could use past errors to guess about true values for the present. Obviously, this interpretation is somewhat tortured, and not easily related to commonly encountered processes. Given this seeming dearth of obvious applications, one might wonder why MA models are ever used. The answer lies in their ability to very succinctly encapsulate complicated temporal structure. In time series modeling, as in other modeling, one generally looks for models with as few parameters as possible. Faced with a choice between a low-order MA model or a high-order AR model with many coefficients, the MA model might be preferred despite the difficulty of interpretation.

Autoregressive-Moving Average (ARMA) Processes

To give still more flexibility, we can mix autoregressive and moving average terms in a single model. These are called autoregressive-moving average models, abbreviated ARMA(p, q) where p and q are the respective orders for the AR and MA parts. The general equation is

$$Y_t = \phi_1 Y_{t-1} + \phi_2 Y_{t-2} + \phi_3 Y_{t-3} + \cdots + \phi_p Y_{t-p}$$
$$+ a_t - \theta_1 a_{t-1} - \theta_2 a_{t-2} - \cdots - \theta_q a_{t-q} \tag{15-7}$$

For mixed models, it is necessary to ensure both stationarity and invertibility. As expected, this places restrictions on values for the parameters appearing in Equation (15-7).

15.4. Removing Trends—Transformations to Stationarity

Suppose one is working with a process known to be nonstationary. For example, perhaps siltation is causing reservoir levels to slowly increase, or inflation is affecting bank balances. These cause drift in the mean value, violating stationarity. Alternatively, perhaps the time series is subject to seasonal variation, so that the mean changes from one season to the next. Situations like this are obviously extremely common and call for special treatment before modeling. Fortunately, a number of techniques are available for transforming a nonstationary series into one that is suitable for modeling. Two basic methods are used, called the *deterministic* approach and the *stochastic* approach.

The deterministic approach is appropriate when one believes that the disturbance from stationarity applies for all time without change. For example, we might believe that the seasonal temperature cycle operates throughout the record with unchanging amplitude. Or perhaps we believe that the siltation rate is constant. In these cases one is justified in fitting a trend equation to the data and using the equation to remove the trend. For example, one might fit a linear or quadratic function to a time series. By subtracting the fitted values from each observation, one obtains a "detrended" time series, presumably now stationary. Notice that the trend equation provides the mean for each observation time; thus, we have estimated the so-called *mean value function* for the time series. Evaluating the function for a particular t gives the mean for that t.

In the case of a seasonal trend, we must believe that the mean for each season is the same for all years. Working with monthly data, we would calculate the mean January value for all years, the mean February value, etc. By doing so, we are effectively fitting another mean value function, this time one that varies seasonally. Once again the mean value function is subtracted from the original series, giving a detrended series. There are other ways of fitting a mean value function, but all are similar to the above in that they yield μ as a function of time. Moreover, one need not fit the mean value function before constructing the model. For example, we can include a constant δ in a general ARMA(p, q) model:

$$Y_t = \delta t + \phi_1 Y_{t-1} + \phi_2 Y_{t-2} + \phi_3 Y_{t-3} + \cdots + \phi_p Y_{t-p}$$
$$+ a_t - \theta_1 a_{t-1} - \theta_2 a_{t-2} - \cdots - \theta_q a_{t-q} \qquad (15\text{-}8)$$

At each time period δ increments Y; thus, inclusion of the constant amounts to adding a linear trend.

Still other approaches can be used in special circumstances. Suppose, for example, than a measuring device is known to have changed at some point during the observing period. Perhaps the instrument was moved to a new location, or perhaps a different type of instrument was installed. Such changes might well lead to a discontinuity in the mean and provide justification for calculating the mean separately for the two recording periods. By expressing observations as deviations from their respective subperiod mean, the discontinuity is removed.

It is important to stress that one must have strong reasons for using a deterministic mean value function. It is not sufficient to look at the time series and conclude that the mean is changing and that detrending is required. As evidence for this, recall Figure 15-1, which showed several apparent "trends," all of them fictitious.

The alternative to a deterministic approach is a stochastic transformation, in which the source of nonstationarity is *not* assumed to be uniform over all time. It might be, for example, that the mean value changes slowly over time, with the changes determined by a random walk. Because changes in the mean are unpredictable, there is no deterministic function of time that can remove them. The solution is to subtract each observation from its neighbor, a process called differencing, denoted by the difference operator ∇. Instead of using the observations Y, one builds a time series for the differenced series. The new (differenced) series is just

$$Z_t = \nabla Y_t = Y_t - Y_{t-1}$$

To see how this removes nonstationarity in the mean, imagine that μ has increased by a random amount b_t between times $t - 1$ and time t. Differencing the Ys leaves behind the random increment, which is accommodated in a time series model for Z. In some cases the time series will not be stationary after differencing, and it may be useful to take differences of the differenced series. Still higher differencing can be done as well, but in practical applications first or second differences will nearly always suffice. Models that are stationary after differencing are called *integrated* autoregressive-moving average models, or ARIMA models. They are denoted ARIMA(p, d, q), with d indicating the order of differencing.

In the ARIMA models, adjacent values of Y are differenced to produce a stationary series. In modeling nonstationary seasonal processes, one can use seasonal differences to produce a stationary series. Using monthly data, for example, one can form the differences $Y_t - Y_{t-12}$. This is appropriate when the seasonal component is subject to random variation, as would be the case when the seasonal variation is governed by a random walk process. Once again, we are not restricted to first differencing, but can use second or even higher-order differencing. Seasonal series sometimes present another challenge, in that the value for a month might be related to the immediately preceding months and to the same month in preceding *years*. This leads to models that employ autoregressive and moving average terms from corresponding seasons in previous years. Such models are called *multiplicative seasonal* ARIMA models and are widely used in modeling economic time series. The order of the seasonal submodels can be different from the nonseasonal submodels; thus the number of possibilities is nearly endless. The references provide details.

15.5. Model Identification

It is clear from the foregoing that a very broad range of time series models exist and that there will be few occasions when one has sufficient knowledge to choose an appropriate model on theoretical grounds. In time series analysis, one is therefore almost always faced with the problem of identifying a model from the data. By this we mean that one must somehow decide on the order of the moving average and autoregressive components. (Note that choosing an order of zero amounts to excluding one or the other from the model.) Additionally, one must decide whether or not to detrend, and, if so, how to do so (deterministic vs. stochastic, etc.). In this section we offer some general guidelines, along with the suggestion that one should always experiment with a number of candidate models.

The most powerful aids to model identification are the autocorrelation and partial autocorrelation functions. A quick look will often narrow the possibilities considerably. Consider, for example, Figure 15-3, which shows the autocorrelation function calculated from one of the AR(1) models plotted in Figure 15-1. We see the autocorrelation values alternating in sign from lag to lag and decreasing geometrically. This is diagnostic of all AR(1) processes having a negative coefficient. If we had only the graph to go on, we would certainly look closely at AR(1) models as a candidate for the time series. As implied by the example, AR and MA models have their own characteristic pattern of autocorrelation and partial autocorrelation. The following relations are especially useful.

- AR(1) models
  - ρ_k The autocorrelation function is simply $\rho_k = \phi^k$. As a result, the autocorrelation decreases by a factor of ϕ for every increase in lag. Note that negative ϕ causes ρ_k to alternate in sign from lag to lag, with odd lags having negative autocorrelation.
  - ϕ_{kk} The partial autocorrelation function is zero for all lags greater than 1. That is, $\phi_{kk} = 0$ for $k = 2,3, \ldots$. This is expected, given that Y_t depends only on Y_{t-1}. Any correlation at higher lags arises from the first-order correlation, which is absent from ϕ_{22}, ϕ_{33}, etc

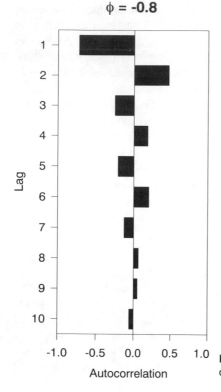

$\phi = -0.8$

Lag (vertical axis, values 1 through 10)

Autocorrelation (horizontal axis: -1.0, -0.5, 0.0, 0.5, 1.0)

FIGURE 15-3. Autocorrelation functions for the AR(1) process $Y_t = -0.8Y_{t-1} + a_t$

- AR(p) models
  - ρ_k The autocorrelation function can be written $\rho_k = \phi_1\rho_{k-1} + \phi_2\rho_{k-2} + \cdots + \phi_p$. As for AR(1) models, ρ tails off geometrically.
  - ϕ_{kk} The partial autocorrelation function is zero for all lags greater than p.
- MA(1) models
  - ρ_k The autocorrelation function is zero for $k > 1$. Only Y_{t-1} is correlated with Y_t, the value of ρ_1 being $-\theta/(1 + \theta^2)$.
  - ϕ_{kk} The partial autocorrelation function decreases slowly with increasing lag.
- MA(q) models
  - ρ_k The autocorrelation function is zero for all lags greater than q. For lower lags, no general pattern exists, as ρ_k is a complicated function of the θ values in the model. (ϕ_k is a rational polynomial in θ.)
  - ϕ_{kk} The partial autocorrelation decreases slowly with increasing lag.
- ARMA(q, p) models
  - ρ_k After the first $q - p$ lags, the autocorrelation function decreases in a complicated fashion, with a mixture of geometric decrease superimposed on a sinusoid of decreasing amplitude.
  - ϕ_{kk} After the first $q - p$ lags, the partial autocorrelation shows the same sort of complex behavior as ρ_k.

In using the autocorrelation and partial autocorrelation functions, we will of course have to work with sample estimates of these functions. Because of sampling error, there will be departures from the patterns described above. For example, the sample partial correlations ϕ_{22}, ϕ_{33} ... in an AR(1) model will be near zero, but not exactly zero. To decide whether or not a function has decayed to zero, it will be necessary to consider the statistical significance of the sample values. Fortunately, the sampling distributions of ρ_k and ϕ_{kk} are known for large n. As a diagnostic aid, computer programs routinely provide confidence intervals along with estimate of ρ_k and ϕ_{kk}.

It should be pointed out that in arriving at the standard errors, it is necessary to assume that the a_t are independently and identically distributed. This will be the case for any stationary ARMA model, provided that model has been correctly specified. After all, if the a_t are not i.i.d., some temporal structure remains in the residuals, which implies that we do not have the correct model form. This leads to another tool in model identification, residual analysis. Having fit a candidate model, one should examine the model residuals, looking for autocorrelation, changes in variance with t, and other departures from randomness. Many of the issues raised in the regression chapter are applicable here and can be used without modification if time t is substituted for the independent variable X.

The residual mean square is commonly used in comparing models, the idea being that a successful model will reproduce the observed series and thereby give a_t values that are on average small. Thus, everything else being equal, one would choose a model giving a small mean square error over one with larger error. It must be remembered, however, that as the number of parameters increases, explained variance cannot decrease. Because of this, adding terms to a model will improve the fit as measured by residual mean square. To avoid overfitting, it is necessary to "correct" residual error for the number of parameters present. For example, rather than minimize the raw mean square error, a number of authors prefer Akaike's index

$$\text{AIC} = \hat{\sigma}_a^2 \left(1 + \frac{p}{n} \right)$$

where p is the number of parameters, n is the number of observations, and $\hat{\sigma}_a^2$ is an estimate of the residual variance. Most computer programs report AIC or other similar indices, allowing the user to compare models with differing numbers of parameters.

15.6. Model Fitting

Assuming one has a candidate model, how does one go about assigning parameters? This is a question of model fitting, in which we must choose parameter estimates that are in some sense optimal. We faced this problem in regression analysis, and relied on the principle of least squares. In time series several variants of least squares are used. To understand them, it is helpful to imagine that we wish to fit an AR(1) model. Let us suppose that the mean Y is zero, or has been subtracted from the data before analysis. Given a series of length n, we can predict Y for $t = 2, 3, \ldots n$. The least squares approach says we find $\hat{\phi}$, which minimizes

$$\text{RSS} = \sum_{t=2}^{n} (Y_t - \hat{\phi} Y_{t-1})^2$$

The solution is

$$\hat{\phi}_{LS} = \frac{\sum_{t=2}^{n} Y_{t-1} Y_t}{\sum_{t=2}^{n} Y_{t-1}^2}$$

The least squares estimate $\hat{\phi}_{LS}$ is unbiased, but is not necessarily between -1 and 1. Because of this, we cannot be sure that the resulting model will be stationary. More commonly used is the Yule-Walker estimate

$$\hat{\phi}_{YW} = \frac{\sum_{t=2}^{n} Y_{t+1} Y_t}{\sum_{t=1}^{n} Y_t^2}$$

Notice that the only difference is in the denominator, where the sum runs from 1 to n. This gives a model that is always stationary, but at the expense of an estimate for ϕ that is biased low, towards zero. Another approach is the so-called maximum entropy algorithm, which gives unbiased estimates ϕ_{ME} such that $-1 \le \phi_{ME} \le 1$. The method is based on predicting both forward and backward in time, rather than just forward as in the above methods. In other words, we compute an additional set of residuals, this time using Y_{t+1} to predict Y_t. The ME method in effect combines the forward and backward least squares estimates. For a first-order model, the formula for the estimate is

$$\hat{\phi}_{ME} = \frac{\sum_{t=2}^{n} Y_{t-1} Y_t}{Y_1^2/2 + \sum_{t=2}^{n-1} Y_{t-1}^2 + Y_n^2/2}$$

Notice that again only the denominator has changed from the least squares estimate. For a long time series, the differences will hardly affect the estimate at all; thus the methods are nearly equivalent for large samples. In addition to the least squares methods, many computer programs use something called the maximum likelihood method. For the AR(1) model, we find $\hat{\phi}$, which minimizes

$$l(\hat{\phi}_{ML}) = n \ln(RSS) - \ln(1 - \hat{\phi}_{ML})$$

Though not based on least squares directly, once again the residual sum of squares figures prominently. Not surprisingly, for reasonably large n, the maximum likelihood estimate will be very close to those mentioned earlier.

Fitting models with moving average terms is more complicated on two grounds. First, the moving average parameters θ_κ enter nonlinearly into the residual sum of squares, so that MA models are intrinsically nonlinear. Iterative solution methods must be used, in which parameter values are repeatedly adjusted in a trial and error search procedure. A second complication is that values of a are needed for times before the observation period. To see this, recall that Y_1 is based partly on the prediction error a_0. There is no prediction error for time t_0, as we do not even have Y_0. One option is to simply take these to be zero, which is their expected (average) value. Another option is to generate forecasts backwards in time for the needed a_t. The method used to start the sequence of a is of little importance for long records, but can make a significant difference when working with small samples.

EXAMPLE 15.1. Figure 15-4 shows a long sequence of maximum annual streamflows. Clearly visible are periods of consistently above and below average floods. The data graph certainly gives the impression of memory in the system, a fact that is confirmed by the autocorrelation plot (Figure 15-5). We will use the data to construct a time series model, hoping to learn more about temporal structure, especially how much predictability is present.

The data standard deviation is 431 cubic meters per second, implying that if we were to use the mean as a forecast for every year, we would obtain a root-mean-square forecast error of 431 m³/s. This corresponds to a mean square error of 185,761 (431^2) and provides a standard against which we will measure model performance. Any model that can't improve upon the mean is of little use in forecasting.

As a first step, we try an AR(1) model. The autocorrelation plot (Figure 15-5) does not suggest this will be completely successful, as ρ_k does not decay consistently towards zero with increasing lag. Nevertheless, for the sake of exposition, we persist. We obtain a model $Y_t = 0.919 \, Y_{t-1} + a_t$ (Table 15.1). The residual mean square is 74,890, about 40% of the baseline figure. In other words, the AR(1) model explains about 60% of the variance

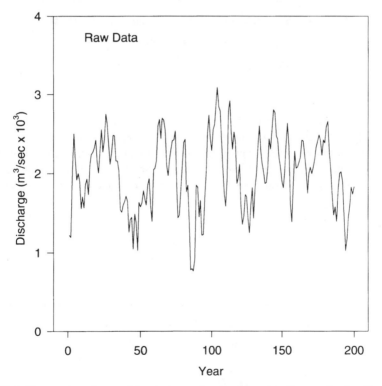

FIGURE 15-4. Time series of annual floods (or peak annual flows) in thousands of cubic meters per second.

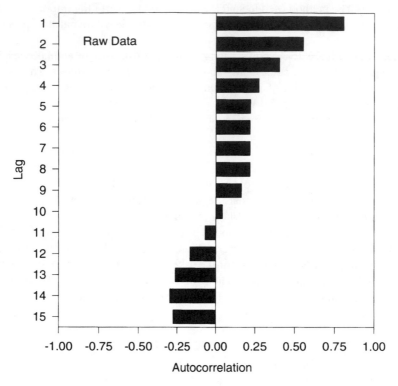

FIGURE 15-5. Autocorrelation of streamflow data.

in the original time series ($R^2 = 0.59$). The coefficient estimate $\hat{\phi}$ is about 35 times larger than its standard error, so is highly significant. We have every confidence that ϕ is not zero, under the assumption that an AR(1) is appropriate. However, when the residuals from the model are examined (Figure 15-6), we see large partial autocorrelations at lags 2 and 4, which are significantly different from zero. We cannot conclude that the residuals are truly random.

Trying an MA(1) model again yields a significant parameter estimate, but the predic-

TABLE 15.1
Statistics for Various Models Fit to Streamflow Data

Model	Parameter	Estimate	Standard error	Model MSE	Model R^2
AR(1)	($\hat{\phi}$)	.919	.027	74,890	.59
MA(1)	($\hat{\theta}$)	−.825	.042	89,220	.52
ARMA(1,1)	($\hat{\phi}$)	.851	.039	67,520	.63
	($\hat{\theta}$)	−.417	.069		

tive power of the model is somewhat less than the AR(1) model. The residual MSE has increased, while R^2 falls slightly, to .52. The residual partial autocorrelations for lags 1 and 2 are both significant, implying that this model is not correct (Figure 15-7).

Finally, we turn to an ARMA(1,1) model. Both coefficients agree in sign with the previous models, but are smaller in magnitude. In other words, like the others, this model finds a positive autoregressive and a negative moving average term. None of the residual partial autocorrelations are significantly different from zero (Figure 15-8). With this result, we would be hard-pressed to consider more complicated models.

The residual MSE for the ARMA model is smaller than for either of the other two models, and R^2 increases to .63. This figure sounds impressive and certainly represents real skill in forecasting floods. Comparing this to the AR(1) results leads to some interesting points. First, we see that although the AR(1) model is certainly wrong, replacing it with a (presumably) correct model does not lead to a dramatic improvement in forecast skill, as explained variance is only 4% larger. Looking at the time series of forecasts (Figure 15-9), we see large forecasts errors in many years.

Though not shown on the graph, the gap between the data and ARMA model is much larger than the difference between models. If forecast error alone is important, the models are nearly equivalent. On the other hand, if we are interested in temporal structure, the models are quite different, but MSE does not bring out the difference. In other words, on the basis of MSE alone, we have little reason to conclude there is a moving average

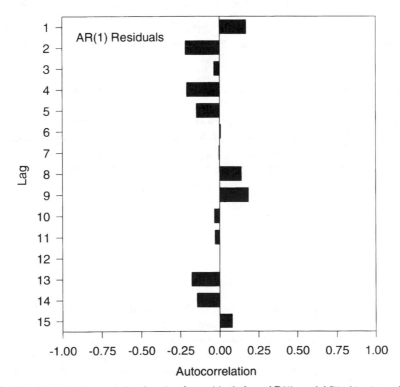

FIGURE 15-6. Partial autocorrelation function for residuals from AR(1) model fitted to streamflow data.

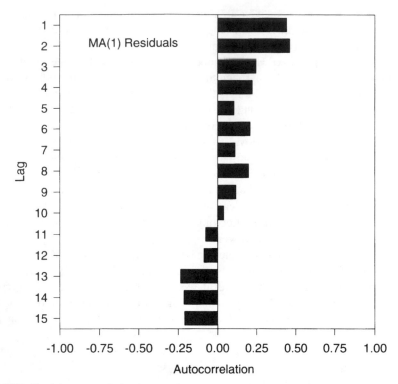

FIGURE 15-7. Partial autocorrelation function for residuals from MA(1) model fitted to streamflow data.

component in the data. To detect the moving averaging component, we must examine the residuals.

A final point to make about prediction error is that we are computing "errors" using data from which the model was constructed. We must wonder how well the model will do when asked to forecast new values in the future. After all, our error statistics measure failure on the data used for fitting (calibrating) the model. We would have much more confidence in the performance measures if they were based on forecasts of *independent* data. This raises the issue of cross-validation, a topic covered in Chapter 17. Here we will simply point out that MSE, R^2, and other error measures based on the calibration data are likely to understate true forecast error.

15.7. Times Series Models, Running Means, and Filters

Though seemingly unrelated, there is a parallel between the time series models described above and the running means we used for smoothing in Chapter 3. In smoothing a time series, we apply an operation that is designed to remove the random component, leaving behind the systematic component. In time series modeling, on the other hand, we do just the reverse. From an observed time series, we build a model designed to capture whatever systematic variation exists in the data. Given a sequence of data Y_t containing systematic and

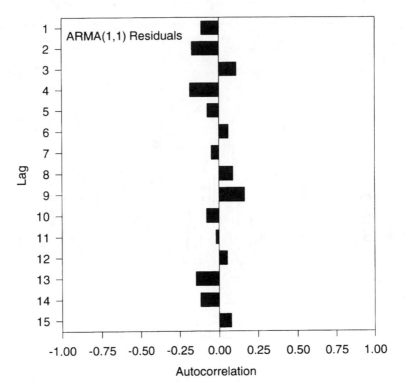

FIGURE 15-8. Partial autocorrelation function for residuals from ARMA(1, 1) model fitted to streamflow data.

random variation, a time series model produces a new sequence a_t containing nothing but random variation. In addition to this conceptual similarity, there is a computational similarity as well. For example, recall the weighted moving average described in Chapter 3:

$$\hat{Y}_t = \frac{1}{8} Y_{t-2} + \frac{2}{8} Y_{t-1} + \frac{2}{8} Y_t + \frac{2}{8} Y_{t+1} + \frac{1}{8} Y_{t+2} \qquad (15\text{-}9)$$

where $\hat{Y}_t$ is the smoothed value at time t. Noticed how observed values near t are multiplied by a coefficient and summed to give $\hat{Y}$, which is the systematic component Y. Now consider an AR(2) model

$$Y_t = \phi_1 Y_{t-1} + \phi_2 Y_{t-2} + a_t = \hat{Y}_t + a_t$$

where $\hat{Y}_t$ and a_t are the predicted and residual values at t. Considering just the predicted portion, which we hope is the systematic part of Y, we can write

$$\hat{Y}_t = \phi_1 Y_{t-1} + \phi_2 Y_{t-2}$$
$$= \phi_2 Y_{t-2} + \phi_1 Y_{t-1} + 0Y_t + 0Y_{t+1} + 0Y_{t+2} \qquad (15\text{-}10)$$

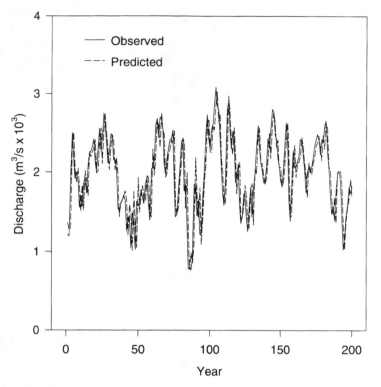

FIGURE 15-9. Observed and predicted streamflows. Predictions are based on an ARMA(1, 1) model fitted to observed series.

Once again we are applying weights to values near t in hopes of uncovering the systematic component. In other words, the time series model can be seen as a type of running weighted mean. Of course, this particular running mean considers only values before t in obtaining $\hat{Y}_t$. We will return to the relationship between time series models and running means later. For the moment, we will focus on the smoothing operation. Equations (15-9) and (15-10) are known as *linear combinations* of Y. The observed values are combined with one another after multiplying by a coefficient. To write a more general expression for this operation, let c_k stand for a set of coefficients. The subscript k identifies each element in the array, just as i identified each sample element x_i. For our example (Equation 15-9) the coefficient array is $c = [1/8, 2/8, 2/8, 2/8, 1/8]$. Instead of going from 1 to 5, we will let the subscripts k run from −2 to 2. With this notation we can write the weighted mean as

$$\hat{Y}_t = \frac{1}{8}Y_{t-2} + \frac{2}{8}Y_{t-1} + \frac{2}{8}\,Y_t + \frac{2}{8}\,Y_{t+1} + \frac{1}{8}Y_{t+2}$$

$$= c_{-2}Y_{t-2} + c_{-1}Y_{t-1} + c_0Y_{t-0} + c_1Y_{t+1} + c_2Y_{t+2}$$

$$= \sum_{k=-2}^{2} c_k Y_{t-k}$$

To smooth the observed series, we move the coefficient array c through the data, forming a weighted sum above for every t. This operation is known as filtering the data, and the array c is called a filter. Technically speaking, the filter is *convolved* with the data. Notice that the AR(2) model is also a filter, one whose coefficients are ϕ_2, ϕ_1, 0, 0, 0. Many other operations can also be seen as filters, though we often do not think of them that way. For example, when we interpolate linearly between two points we are using a filter with coefficients [1/2, 1/2]. Other examples of filtering include slope estimation, numerical integration, cartographic line generalization, and edge detection, to name just a few.

DEFINITION: FILTERING
A filter is an array of constants applied to the values of a time series. Each element of the filter is paired with and multiplied by an element of a time series. The products are summed to give a filtered value.

We do not propose to discuss all types of filters here, but will consider only filters that can be expressed as a linear combination of the original data. Moreover, for reasons explained later, we will restrict our attention to filters that are symmetric about the central value, like the running mean above. This means that the coefficients will appear as pairs surrounding the central weight. If c_m is the element farthest away from the center, the array is

$$c = [c_{-m}, c_{m+1}, \ldots, c_{-1}, c_0, c_1, \ldots, c_m]$$

The filter contains a central value c_0, and m weights to either side. The length is therefore $2m + 1$. Because of symmetry, $c_1 = c_{-1}$, $c_2 = c_{-2}$ and so forth. This allows the filtering operation to be simplified:

$$Y_t = \sum_{k=-m}^{m} c_k Y_{t-k}$$

$$= c_0 Y_t + \sum_{k=1}^{m} c_k (Y_{t-k} + Y_{t+k}) \tag{15-11}$$

Because of symmetry in the filter, Y_{t-k} and Y_{t+k} are multiplied by the same value, c_k. This means they can be added together before multiplying, saving some work. If we were working by hand, or if computational efficiency were an issue, we would use the lower expression in Equation (15-11). The upper expression applies to both symmetric and asymmetric filters.

It is obvious that the way a filter affects a time series depends on the coefficients that comprise the filter. But how can we know what a particular filter will do? For example, both the running mean and the AR(2) filters are supposed to isolate systematic variation in Y. Why would we prefer one over the other? How does a running mean with uniform weights compare to one like Equation (15-9)? To answer these questions, we need some way of describing a filter in terms of the effects it produces. This requires that we adopt a different perspective, outlined in the next section.

15.8. The Frequency Approach

As mentioned in Chapter 3, a time series is smoothed with the goal of uncovering slowly varying trends. In smoothing the series, we are making the implicit assumption that the rapidly varying components are not of interest and need to be removed. In other words, we view variations that last only a short period as *noise*, something that confounds our ability to detect the important, long-period changes. The running mean succeeds if it removes short-period variation while leaving long-period variation intact. This breakdown of variance into different lengths of persistence lies at the heart of the frequency approach to time series. It requires that we conceive of the time series as containing variance at different periods. To pick an obvious example, consider a time series of air temperature measured every hour. Most of the variance in temperature will be a result of the diurnal (daily) and annual cycles. To a good approximation, the temperature record could be reproduced by adding together two periodic functions: one representing the daily rise and fall of temperature, the other representing annual variations.

Alternatively, suppose we have a time series of monthly rainfall. We would not see any diurnal variation, but the annual cycle would be represented, as would variation at still longer periods. If we were interested in searching for solar cycles, we might want to remove the annual cycle so that any 11-yr cycle could emerge. Figure 15-10 shows something similar for a hypothetical time series with variance at just two periods, 1 yr and 10 yrs in duration. The 10-yr oscillation, here labeled signal, repeats itself every decade. We would therefore describe it it as a *periodic* or *sinusoidal* function, with a fundamental period of 10 yr. Equivalently, we would say its frequency is 1 cycle/decade, or .1 cycles per year. Note that the period of a function is measured in units of time (days, years, etc.), whereas frequency has dimensions of cycles per unit time (cycles/day, cycles/yr, etc.). These are reciprocals of each other; thus, low-frequency oscillations correspond to long-period oscillations, and vice versa.

Notice also that the amount of variation is different for signal and noise in Figure 15-10. The signal curve ranges from 5 to 25, a difference of 20. The variation in the noise curve is only half as large, running from −5 to 5, a difference of 10. The variation of a sinusoid is summarized by its amplitude, formally defined as the difference between the maximum and minimum divided by two. Our hypothetical signal and noise curves have amplitudes of 10 and 5, respectively.

DEFINITION: PERIOD, FREQUENCY, AND AMPLITUDE

The period of a sinusoid is the length of time required for the function to complete one cycle. The frequency of a sinusoid is the number of cycles completed per unit time. Frequency and period are reciprocals: Frequency = 1/Period. The amplitude of a sinusoid is one-half the difference between maximum and minimum.

Suppose we filtered the upper curve of Figure 15-10 with a 5-yr running mean. We would expect the annual cycle to be removed, leaving behind the 10-yr cycle (bottom curve). A 20-yr running mean would, of course, produce a different result. We see that different filters remove and pass different frequencies, and that knowing which frequencies are removed is helpful in knowing which filter to choose. Here, then, is the rationale for taking a frequency approach to filtering:

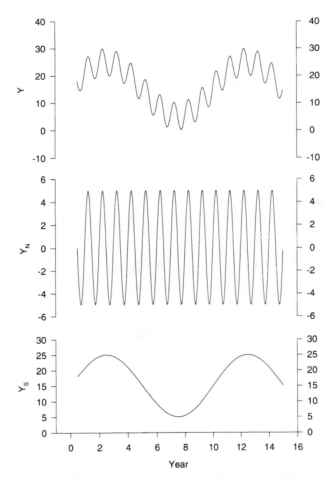

FIGURE 15-10. A hypothetical series Y, composed of rapidly varying noise Y_N superimposed on a slowly varying signal Y_S.

- Filters selectively modify the frequencies present in a time series—only by understanding what a filter accomplishes from a frequency standpoint can we understand how the observed series is modified.

In adopting the frequency approach, we describe the action of a filter in terms of its effects on periodic functions. It must be emphasized that we need *not* believe that the time series itself is periodic, or that it arises from the action of processes that are periodic. The only requirement is that we desire to separate the time series into components with varying time scale. In most cases the high-frequency components are considered noise, so that we seek smoothing filters that allow low-frequency variation to pass while removing high frequencies. There is, however, nothing about the filtering process that demands this posture. If appropriate, we can think of noise as appearing at low frequencies and can design filters that pass only high frequencies. Another alternative is a band-pass filter, which passes frequencies only within a specified range.

The Amplitude Response Function

As we have observed, a given filter will remove some frequencies and leave others untouched. When a frequency is removed, its amplitude is reduced to zero. If it is unaltered, the amplitude remains at 100% of the original amplitude. A filter's pattern of removal, reduction, etc. is called the *amplitude response* of the filter. The response function gives the proportion of amplitude that remains after the filter is applied. Thus, if the response function is zero for some frequency, the filter will completely remove variation at that frequency. If the response is unity, variation is unchanged. If the response is .25, it means the amplitude would be reduced to only 25% of the original, a reduction of 75%.

DEFINITION: AMPLITUDE RESPONSE FUNCTION

The amplitude response for a filter provides information about which frequencies are passed by a filter and which frequencies are removed. If the value is unity for some frequency, the filter does not affect that frequency, and the amplitude at that frequency is unchanged by the filter. A response value of zero means that the amplitude is reduced to zero.

Before showing how the response function is calculated, we will describe the response for the 5-point weighted mean used before (Equation 15-9). The amplitude response for that filter is plotted in Figure 15-11.

Following common practice, frequency is plotted as increasing from left to right. The left-hand part of the diagram therefore refers to long-period, slowly varying components of variation, whereas the right-hand side refers to high-frequency variations. The frequency axis is plotted in units of cycles per observation, rather than cycles/year or cycles/day or some other time-based unit. This is because we are free to use the filter on yearly data, on daily data, or on data collected at any other interval. By plotting frequency as cycles/observation, we make the graph useful for all data intervals. Suppose we happen to have yearly data. In that case, a frequency value of .1 from the graph represents .1 cycles/yr, or a period of 10 yr. If we have daily data, .1 means .1 cycles/day, or a period of 10 days.

The highest frequency shown is .5 cycles/observation. Why is that? Does it mean the filter only affects lower frequencies? Or does it mean higher frequencies are completely removed? In point of fact, neither of these is true. The frequency range is restricted because only frequencies below .5 can be captured by sampling a process at discrete moments in time. There is a very important theorem, called the *sampling theorem*, which states that we must have at least two observations for every cycle present in the process. In other words, the shortest cycle we can detect has a period that is twice the sampling interval. A period of twice the sampling interval corresponds to a frequency of 1/2 cycle per observation.

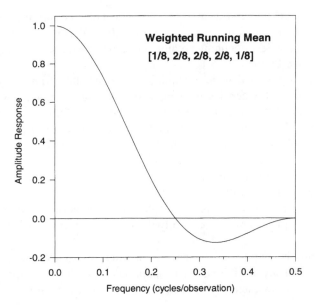

FIGURE 15-11. Amplitude response function for the five-point weighted mean [1/8, 2/8, 2/8, 2/8, 1/8].

DEFINITION: SAMPLING THEOREM FOR TEMPORAL DATA

Suppose a periodic function is sampled (measured) at uniform points in time. In order for the sample points to reproduce the function, at least two points are required for every cycle. The sampling frequency must be at least twice the frequency of the function.

The highest frequency detectable is known as the *Nyquist frequency* and is always .5 cycles/observation. It's interesting to think about what happens to frequencies above the Nyquist frequency. In other words, what happens if we violate the sampling theorem and do not sample often enough? Do we simply miss those undetectable frequencies? The answer, unfortunately, is no. Instead, oscillations at higher frequencies will appear in the sample as low-frequency oscillations. What we see as low-frequency variation will therefore be a combination of true variation and contamination from higher frequencies. This problem, called *aliasing*, is very important. It says that when we do not sample frequently enough, we don't just miss features, but rather we produce a sample series *distorted* by what is undetectable. It must be emphasized that there is no way to remove these distortions once they occur—no amount of processing can undo the damage done by aliasing.

Returning to Figure 15-11, we see the curve mostly decreases with increasing frequency, indicating that the filter is a smoothing filter; that is, like we hoped, the filter preferentially removes high frequencies, but passes low frequencies.

Notice that the curve is unity only at frequency zero, so this is the only frequency that is not reduced in amplitude. A frequency of zero corresponds to a period of infinity; thus, the figure is saying that only infinite-period sinusoids are unchanged. But what is an infinite-period sinusoid? It would seem to be a curve that changes infinitely slowly, so that it does not change at all and is therefore a horizontal line. This is, in fact, the case. If we use the filter on a time series containing nothing but constant values, we will get the original time series back. (See Problem 15.5.)

For frequencies greater than zero, amplitude response falls from unity toward zero. The response is such that frequencies are damped out by various amounts, with the amount of damping increasing with increasing frequency. Finally, at a frequency of .25, the curve reaches zero. This is the first frequency completely removed by the filter—lower frequencies are only partly attenuated. Notice that amplitudes are negative at frequencies above .25. For example, the amplitude is about −.1 at frequency .35. This means that oscillations are reduced to 10% of their original amplitude but are reversed! Where there is a peak in the original series, there will be a valley in the filtered series. For a smoothing filter, this is not at all what we are after. We would prefer a filter that does not pass *any* high-frequency information. If that isn't possible, we certainly do not want our filter to turn noise maxima into minima.

By examining the amplitude response of our filter, we see exactly what it is able to accomplish, and what are its failings. The amplitude response is surprisingly easy to compute. Valid for any frequency f between 0 and 1/2, the equation is

$$R(f) = \sum_{k=-m}^{m} c_k \cos(2\pi k f), \qquad 0 \le f \le 1/2 \tag{15-12}$$

When using this equation, it is very important that frequency f be expressed as cycles/observation. This will sometimes require conversion from a more natural unit of frequency. For example, perhaps we are interested in cyles/year, but have data measured every month. To obtain the response at some frequency in cycles/year, we would divide f by 12 before applying Equation (15-12).

Amplitude Response of Running Means

Equation (15-12) allows us to compute the amplitude response for any filter. Without question the most common type of filter is a running mean of length L, where L successive observations are averaged together (all $c_k = 1/L$). Given their widespread use, it is important to understand their behavior.

Figure 15-12 shows the amplitude response for 3-point through 11-point running means. All are low-pass smoothing filters, removing high frequencies much more than low frequencies. First, notice where each curve passes through zero. The 11-point mean reaches zero soonest, at $f = 1/11$, then the 9-point mean at $f = 1/9$, and so forth. The 3-point mean is zero only at $f = 1/3$. The implication is that a filter of length L *completely* removes variation at a frequency of $1/L$, equivalent to a period of L. This means, for example, that an 11-yr average completely suppresses any 11-yr cycle in the time series. In deciding on how long a filter to use, we would probably match our definition of "noise" to the frequencies removed. If an application called for "noise" to be anything at frequencies above .2 cyles/observation, we would want one of the longer filters shown. On the other hand, if a frequency of .2 were considered to be signal, we would have to choose the 3-point filter.

The graph shows that running means remove variation at a period of *exactly* L, but

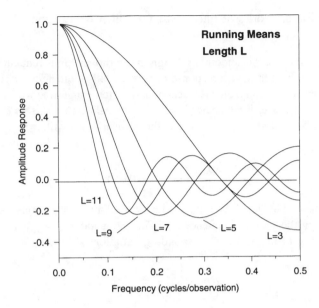

FIGURE 15-12. Response function for running means of various lengths.

do not remove all variation at periods shorter than L. Frequencies higher than $1/L$ are damped, but not removed. Sometimes the damping is such that maxima and minima are reversed in the filtered series (negative R). All of the running means show similar behavior, an amplitude response with wiggles for f above $1/L$. Depending on how large the noise is relative to the signal, the noise could dominate the filtered series. This is the primary objection to running means. Though simple to construct and use, they are far from optimum in regards to their amplitude response. In other words, they do not accomplish very well what we are after, namely the removal of high frequencies.

The failure of running means suggests that we ought to turn our thinking around. Instead of looking at how close a filter comes to our goal, we ought to design a filter from the outset to meet a particular goal. Rather than compute the response function from the filter, we ought to begin with the response function and choose the filter coefficients accordingly.

15.9. Filter Design

In designing a filter, we are going to begin with a specification in terms of frequency. That is, we will specify what the filter is supposed to accomplish from the frequency standpoint. Having done so, we will then select filter weights that come as close as possible to our ideal. It will be seen that compromise is necessary in the design process, and thus the filter obtained will not be perfect. Though we will not produce a perfect filter, we will have control over its failings and can make the compromises in a way that minimizes impact on the findings. Naturally, the compromises made will depend on the time series at hand: its length, the pattern of variation it contains, and what we hope to learn.

Several basic categories of filters exist for which we can state design goals quite easily:

1. Low-pass: A classic smoothing design that removes high-frequency variation. An ideal low-pass filter would have a response of zero for all frequencies *above* some cut-off frequency f_{cut}. The cut-off frequency is a design parameter, something we choose depending on our definition of noise. Frequencies below f_{cut} are called the *passband*, whereas those above f_{cut} are the *stopband*. Mathematically, the ideal low-pass filter is

$$R(f) = \begin{cases} 1 \text{ if } f \le f_{cut} \\ 0 \text{ otherwise} \end{cases} \qquad (15\text{-}13)$$

2. High-pass: The ideal high-pass filter would have a response of zero for all frequencies *below* the cut-off frequency f_{cut}. The passband is $f \ge f_{cut}$, ideally having an amplitude response of unity. The corresponding equation is

$$R(f) = \begin{cases} 1 \text{ if } f \ge f_{cut} \\ 0 \text{ otherwise} \end{cases} \qquad (15\text{-}14)$$

3. Band-pass: We are interested only in frequencies within a band running from f_{low} to f_{high}. The ideal band-pass filter would have a response of zero for all frequencies below f_{low} *and* above f_{high}. Its response would be unity between these values:

$$R(f) = \begin{cases} 1 \text{ if } f_{low} \le f \le f_{high} \\ 0 \text{ otherwise} \end{cases}$$

Graphs of the three filter types are shown in Figure 15-13. As much as possible, we must choose filter coefficients that will produce those ideal response curves. Fortunately, we do not need a separate design strategy for each class of filter. Given a low-pass filter, one can easily construct a high-pass filter corresponding to its inverse. Additionally, it will be seen that band-pass filtering can be performed using a low-pass and high-pass filter in tandem. The basic problem, therefore, is to come up with a good low-pass filter. At a minimum, we want to observe the following guidelines:

- The coefficients should sum to unity. This will ensure that the response will be unity at $f = 0$ (the mean of Y will not be changed). A filter whose coefficients sum to unity is said to be *normalized*.
- The coefficients should be symmetric, such that $c_k = c_{-k}$. This will ensure that the filter does not displace maxima and minima in the original series. Waveform amplitudes will be modified, but their position in time will be preserved. Filters with this property are called *zero phase* filters—they do not produce any phase shifts in Y.
- The filter should be as long as is practical. Increasing L will improve the design, but will also lead to greater data loss at the beginning and end of the series. For a short time series, where one can afford to sacrifice only a few points, it may be necessary to compromise on the design.

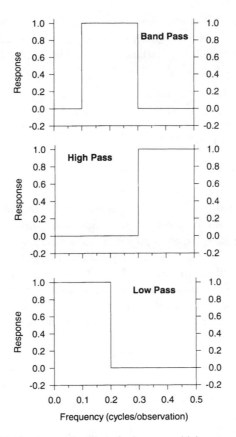

FIGURE 15-13. Ideal response functions for low-pass, high-pass, and band-pass filters.

A number of methods are available for constructing filters consistent with the above. We will describe one technique, the Lanczos method, which is both flexible and uncomplicated. Although its derivation requires advanced mathematics, the procedure used to construct a Lanczos filter is straightforward. For a low-pass design, the steps are the following:

1. Choose the cut-off frequency f_{cut}. This will mark the center of the transition between frequencies passed $(f \leq f_{cut})$ and frequencies removed $(f > f_{cut})$.
2. Choose the length of filter $L = 2m + 1$. There will be m pairs of coefficients, plus one central coefficient c_0. Filtered values will be unavailable for the beginning and ending m points.
3. Find the preliminary filter coefficients b_k from

$$b_k = \frac{\sin(2\pi f_{cut}k)}{\pi k} \cdot \frac{\sin[\pi k/(m + 1)]}{\pi k/(m + 1)} \tag{15-15}$$

for $k = 1, 2, \ldots, m$. Use $b_0 = 2 f_{cut}$ for the central coefficient.

4. The preliminary filter coefficients will not sum to unity. To normalize the filter, we need to divide each preliminary coefficient by the filter sum. We must to sum all the weights, remembering that the filter contains b_{-2}, b_{-1}, etc. Taking advantage of symmetry, the sum is

$$B = \sum_{k=-m}^{m} b_k = b_0 + 2 \sum_{k=1}^{m} b_k$$

5. Compute the final coefficients

$$c_k = b_k/B \tag{15-16}$$

Lanczos filters have two primary advantages over running means. First, they let us adjust filter length independently of cut-off frequency. With running means, the only way to increase f_{cut} is to decrease L, giving a poorer amplitude response. Here we choose L on the basis of how much data we can afford to lose, without regard to f_{cut}. As L is increased, our filter comes closer and closer to the ideal. This effect is apparent in Figure 15-14, which shows three Lanczos filters all using a cut-off frequency of .25. Notice that for all three filters, the amplitude response is about .5 at $f = .25$. In the Lanczos design, the cut-off frequency becomes the midpoint of the transition from the passband ($R(f) = 1$) to the stopband ($R(f) = 0$). Increasing L causes the transition band to become narrower, so the actual response comes closer to the ideal response.

Another very important advantage of Lanczos filters is the nearly flat response in both the pass- and stopbands. Ripples at high frequencies, so prominent with running means, are nearly absent. This improvement is not happenstance, but is explicitly part of the Lanczos design. As a matter of fact, the second term in Equation (15-15) appears solely to smooth out unwanted ripples in both the stopband and passband. Here also an increase in L improves the design, as the response function flattens with increasing filter length.

We see that the design strategy is fundamentally one of trading off loss of data against improvements in filter performance. Good practice calls for experimenting with various values of L and plotting the response function for each filter produced. The primary disadvantage to Lanczos filters is the rather slow transition from passband to stopband. This can be improved by increasing L, of course, but for a given L Lanczos filters do not provide a very steep transition. If this is important in a particular application, some other design might be preferred. A number of other strategies exist, optimized in other ways. Many are described in the Hamming (1982) text.

It was suggested above that a low-pass design could serve in high-pass and band-pass filtering. Suppose we are high-pass filtering a time series. We want a new series containing frequency components above f_{cut}, with low-frequency components removed. Looking at the equations for the ideal response functions, 15-13 and 15-14, we see that they add to unity at every frequency. Given the low-pass response $R(f)$, we could subtract from unity and obtain the high-pass response as $1 - R(f)$. In other words, the high-pass response is the "inverse" of the corresponding low-pass filter. There are two ways to arrive at the inversion.

One way is to invert the filter itself. The process is simple and calls for negating all

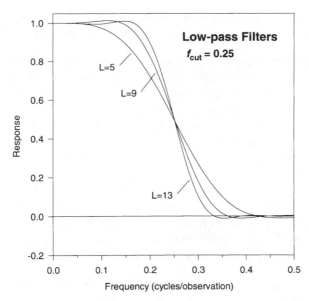

FIGURE 15-14. Response functions for three Lanczos filters of various lengths.

but the central coefficient. To be specific, given a low-pass filter c_k, we obtain a high-pass filter d_k from

$$d_k = \begin{cases} 1 - c_0 & \text{for } k = 0 \\ -c_k & \text{for } k > 0 \end{cases}$$

If the response function for c is $R(f)$, the response function for the new filter d will be $1 - R(f)$, as it should be.

As an alternative, we can accomplish the same thing by inverting the time series. We design a low-pass filter and apply it to the data. We then subtract each filtered value from the corresponding original observation. This will produce a new series containing only high frequencies. The rationale is simple: if low frequencies are removed from the original series, one will be left with only high frequencies.

This same approach can be used to produce a band-pass filtered series. We filter the series with a low-pass filter *and* a high-pass filter. In both cases the original series are filtered. We subtract the two filtered series from the original, leaving behind band-pass filtered data. Writing this out in steps, the procedure is:

1. Filter the observed data with a low-pass filter using a cut-off frequency of f_{low}.
2. Filter the observed data with a high-pass filter using a cut-off frequency of f_{high}.
3. Subtract the series obtained in Steps 1 and 2 from the original data. This will produce a filtered series containing frequencies between f_{low} and f_{high}.

Of course, the low- and high-pass filters should have the same length, so that the loss of data is the same in both steps and the filters are of comparable quality.

EXAMPLE 15.2. Let us suppose we want to filter the streamflow data shown in Figure 15-4. Recall that we used these data previously to construct several time series models. Because we can think of a time series model as a filter, it seems we have already filtered the streamflow data! The results using an ARMA(1,1) model were shown in Figure 15-9, labeled "Predicted." These are observed values, filtered by the model coefficients. Notice how many of the short-term excursions are absent in the forecast series. This shows that the model is acting something like a low-pass filter, removing some of the fine-scale structure in the observations. Because it is one-sided, looking backward in time, the model obviously fails to meet our recommendation for symmetry in filter design. Lack of symmetry suggests that we should expect the filter to displace maxima and minima from their original locations. Looking carefully at the filtered series, we do see these phase shifts. The major excursions in the predicted values lag those in the original data. A one-sided filter like this is necessary for forecasting, but it is far from ideal for smoothing.

EXAMPLE 15.3. The time series model filtered the series, but did not give us control over the amplitude response. We will therefore use the design procedure outlined above, constructing both a low-pass and high-pass filter for the streamflow data.

1. The time series shows a slowly varying component of variance having a period of about 25 years. We want a cut-off frequency that separates this from what appears to be largely random variation. Being conservative, we will choose f_{cut} = .125 cycles/yr, corresponding to a period of 8 yrs.
2. We decide to sacrifice 10 observations at both ends, totaling about 10% of the sample values. This gives a filter length of $L = 2m + 1 = 21$.
3. We use Equation (15-15) to get the preliminary coefficients b_k (see Table 15.2).
4. Summing the preliminary filter gives the normalization value $B = .9915465$.
5. We dividing each b_k by B to obtain the final normalized low-pass filter c_k, $k = 0$, 1, . . . , m.

Though we will not use it in this example, Table 15.2 also shows a high-pass filter constructed from the low-pass coefficients. Notice that for all but d_0, we have simply negated the low-pass coefficient. The response functions for both c and d are shown in Figure 15-15. Visual inspection confirms our design goal.

Applying the low-pass filter to the streamflow data, we get the series shown in Figure 15-16. High-frequency components have been effectively removed, leaving behind the slowly varying components. Notice that the filter has modified amplitudes in the original series, but has not introduced phase shifts. Peaks and valleys appear in the filtered series at the same times as in the raw data. In other words, low-flow and high-flow episodes maintain their position in the filtered series.

15.10. Summary

This chapter considered two related forms of time series analysis, time series modeling and time series filtering. Time series models attempt to capture the dynamic behavior of a series

TABLE 15.2
Coefficients of 21-Point Lanczos Filter

k	Preliminary b_k	Low-pass c_k	High-pass d_k
0	.2500000	.2521314	.7478686
1	.2220317	.2239247	−.2239247
2	.1506405	.1519248	−.1519248
3	.0661780	.0667422	−.0667422
4	.0000000	.0000000	.0000000
5	−.0312029	−.0314689	.0314689
6	−.0306441	−.0309054	.0309054
7	−.0146301	−.0147548	.0147548
8	.0000000	.0000000	.0000000
9	.0052602	.0053050	−.0053050
10	.0031400	.0031668	−.0031668
Sum:	.9915465	1.00000000	.0000000

using a small number of parameters. The models are developed from the observed series and typically rely on some variant of least squares to estimate parameters. Two broad classes of models were considered, autoregressive and moving-average models. Combining the two, possibly with differencing, provides a very general model framework, able to accommodate a wide range of observed series. These so-called ARIMA models are commonly used in human and physical geography, serving as forecast devices and as aids in understanding the structure of a time series.

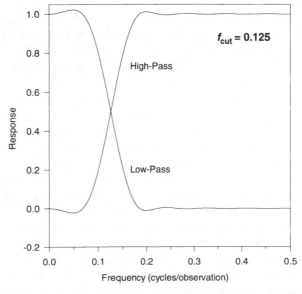

FIGURE 15-15. Amplitude response functions for the filters of Table 15.2.

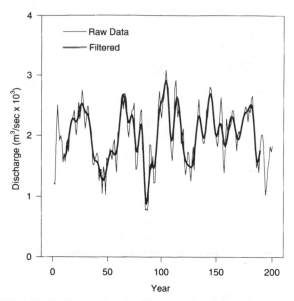

FIGURE 15-16. Streamflow data filtered with low-pass filter of Table 15.2.

Filtering is a process in which a set of coefficients is convolved with a time series, giving a new, filtered series. A special case of filtering is smoothing, where the goal is to remove noise associated with short-period departures from long-term trends. Many other operations, including time series forecasting, can also be written as a filtering operation. To understand how a given filter modifies a time series, we adopted the frequency approach. A key concept is the amplitude response of a filter, which shows how amplitudes at various frequencies are modified by the filter. For example, by examining the amplitude response of simple running means, we saw that they are deficient in several regards.

The preferred approach in filtering is to begin with a design in the frequency domain—a specification of what the filter is supposed to accomplish from the frequency standpoint. Classic designs include low-pass, high-pass, and band-pass filters. Low-pass filters are most common, as they conform to the usual assignment of noise to high frequencies. A number of design strategies exist, one of which we covered, the so-called Lanczos method. It has the advantages of being easy to implement while still giving acceptable results. High-pass filtering is perhaps less common, but easily accomplished. One can either invert a low-pass filter, or subtract a low-pass filtered series from the observed series. For band-pass filtering, two convolutions are performed, one using a low-pass filter, the other using a high-pass filter. By subtracting both filtered series from the original, one obtains a series that is effectively band-pass filtered.

This chapter has considered observations arranged serially in time. It is worth mentioning that many of the issues discussed apply equally well to spatial data. Rather than a time series, we might have data collected along a transect. Instead of using a times series model to study relations between future and past values, we would focus on relations between neighboring values. Likewise, rather than filter to selectively remove features according to their duration (period), we would filter in order to remove features according to their spatial scale, that is, their size.

Of course, true spatial data are typically two-dimensional, not one-dimensional like time series. It so happens that both the models and filters described above can be generalized to two dimensions, thus can be used in common geographic contexts. In constructing a spatial time series, one might express observed values as a linear combination of surrounding values. The coefficients comprising the model encapsulate the spatial structure of the data, in the same way that temporal structure is encapsulated by the coefficients of a one-dimensional time series model. Similarly, when filtering geographic data, one will use a two-dimensional array of filter coefficients. The array is moved throughout the *area* of interest, much like one-dimensional filters are moved from end to end across the time series. Not surprisingly, two-dimensional models and filters are mathematically much more complicated than their one-dimensional analogs. Interested readers are referred to the references for details.

FURTHER READING

R. J. Bennett, *Spatial Time Series* (London: Pion Limited, 1979).

J. D. Cryer, *Time Series Analysis* (Boston: Duxbury Press, 1986).

D. Graupe, *Time Series Analysis, Identification and Adaptive Filtering* (Malabar, Fla.: Robert E. Krieger Publishing, 1984).

R. W. Hamming, *Digital Filtering* (Englewood Cliffs, N.J.: Prentice-Hall, 1983).

W. S. Wei, *Time Series Analysis: Univariate and Multivariate Methods* (Redwood City, Calif.: Addison-Wesley Publishing, 1990).

PROBLEMS

1. Explain the meaning of the following terms:
   a. Random walk
   b. Stationarity
   c. Partial autocorrelation
   d. Moving-average model
   e. Symmetric filter
   f. Low-pass filter
   g. Stochastic process
   h. Autocorrelation
   i. Autoregressive model
   j. Smoothing
   k. Amplitude response function

2. Why are moving average models sometimes preferred over autoregressive models?

3. Explain why the familiar formula for the standard error of $\bar{X}$ ($\sigma_{\bar{X}} = \sigma/\sqrt{n}$) is inappropriate in the presence of strong autocorrelation.

4. In transforming for stationarity, what is the difference between the deterministic and stochastic approaches?

5. Suppose a stochastic process is first-order autoregressive with a coefficient of $-.7$. How would its autocorrelation function appear? How would its partial autocorrelation appear?

6. In selecting a time series model, does one always choose one with minimum residual error? Why or why not?

7. Given the following time series of 100 values,

−.7784	.7500	−.3678	−.5595	−.8068	.6306	−.2161	−.5251
.0090	.3766	−.3266	−.2225	.5015	.1136	−.3976	.0820
.2030	.2598	−.1462	.3907	−.0836	.4817	−.0666	.4311
−.00005	.4766	−.4059	−.2351	−.1060	.1564	−.6055	−.1502
−.4424	.1580	.0465	−.6030	−.2929	.0327	.0628	−1.2191
−.4335	−.4555	.1592	−1.2033	−.1421	−.4881	.6016	−1.2358
−.0693	−.3998	.6816	−1.0855	−.2658	−.3522	.4368	−.1493
.1064	−.2848	.6976	−.1703	.2378	−.3518	.6744	−.5485
.2665	−.7421	.5483	−.3396	.3229	−.5831	.1049	−.7798
.4224	−.5528	−.0297	−.7729	.1245	−.9545	.1793	−.3034
.0054	−.5092	.2993	.0193	.2788	−.6221	.4148	.5104
.5560	−.9757	.4210	.7366	.2686	−.3631	.5365	.4295
.4316	−.4213	−.1576	.2863				

a. Compute autorcorrelations and partial autorcorrelations out to lag 15.
b. Plot the sample autocorrelation and partial autocorrelation functions.
c. What type and order of time series models are suggested by the results above?

8. Record maximum temperatures for the last 30 days. Compute autocorrelations and partial autorcorrelations for lags one and two. Explain the differences.

9. Using your temperature data from the previous question,
a. Construct a first-order AR model.
b. Construct a first-order MA model.
c. How do the two models compare in terms of MSE?
d. Is there reason to believe that an ARMA model is appropriate? Are higher order models called for?

10. What are the consequences of violating the sampling theorem?

11. Graph the amplitude response function for a 5-point running mean with coefficients [.1, .2, .4, .2, .1].

12. Why are symmetric filters so common?

13. Construct a 21-point Lanczos filter with a cutoff frequency of .2.

14. Graph the amplitude response function of the above. What is the response at $f = .1$?

16

Exploratory Data Analysis

The various measures and graphical devices described in Chapter 2 represent what might be called the conventional, or traditional, core of elementary descriptive statistics. Virtually every textbook developed for introductory statistics courses includes nearly all of this material. The standard computer packages for statistical analysis include routines for calculating these measures and graphical summaries. Over the past 15 to 20 years, a wealth of novel, and some say ingenious, methods of analyzing data have been developed to supplement this traditional approach. These approaches are now usually termed *exploratory data analysis* (EDA), a name taken from the influential book of the same name written by John Tukey in 1977. The motivation underlying the development of these techniques was the establishment of a set of new devices that could be used to uncover important features of a set of numbers we wish to understand—features that may go unnoticed if traditional techniques are employed.

Many introductory textbooks give the false impression that there is a *single, correct* way to analyze data. In EDA, on the other hand, there is an explicit recognition that there may be several different ways to confront numerical data, each of which has appropriate strengths and weaknesses. EDA is more of a practical approach to data analysis. Rather than being limited to a set of conventional techniques derived from a particular philosophy of statistics, the *data* themselves are supposed to *guide* students in the choice of techniques appropriate for the problem at hand. While the goal remains that of trying to understand the data, the data are supposed to lead the statistician, not vice versa.

Because their origin is still relatively recent, EDA techniques are not as widely available in standard computer packages for statistical analysis. In fact, some would argue that many of the advantages to be gained from confronting a set of data with these techniques would be lost if they were forced into the mechanical mold of computer-programming algorithms. Because these techniques remain *exploratory* in nature, there are fewer rules of thumb available to the student. Analysis of data can be thought of as part art *and* part science. As experience with these techniques is gained, more "rules" may emerge. In a sense though, the development of "rules" of analysis is almost against the spirit of exploratory data analysis: They are like handcuffs limiting our reach and inhibiting our freedom to understand the important qualities of the data at hand.

In this chapter, we will explore five elementary techniques of exploratory data analysis, drawing comparisons to the equivalent conventional technique where appropriate. The first three techniques, the *stem-and-leaf display*, the *letter value display*, and *boxplots* have clear parallels to the preliminary analyses undertaken in descriptive statistics. Resistant lines represent an alternative to finding a regression line linking two variables X and Y. Finally, *smoothing techniques* used to search for more complex relationships between two variables X and Y are illustrated for regularly spaced data series. They represent alternatives to the moving average and related techniques discussed in the introduction to time series in Chapter 3.

16.1. The Stem-and-Leaf Display

Consider a collection of data values of some variable, or what an exploratory data analyst prefers to call a *batch*. It is not usual to call our data a *sample*, since we are seldom interested in inferential techniques when we apply EDA. The first step in conventional descriptive statistics is to reduce a large number of data values into a convenient framework, usually by constructing a histogram and a frequency tableau. The first step in EDA has the same purpose but uses different techniques. These graphical techniques are known as *displays*. The first display to be constructed is usually the stem-and-leaf display. The purpose of a stem-and-leaf display is to illustrate the following:

1. The range of values within the data,
2. Where concentrations of values occur,
3. How many concentrations of data values there are within the batch,
4. Whether there are gaps in the values,
5. Whether there is symmetry in the batch,
6. Whether there are extreme values that differ markedly from the remaining values in the batch, and
7. Any other data peculiarities.

This is very similar to the rationale for constructing a frequency table and histogram. The unique feature of the stem-and-leaf display is that the actual digits that make up the data values in the batch are used to construct the display. This means that it is *possible to reconstruct the original data* from the stem-and-leaf display, a construction not possible with the histogram. Also, the actual digits of the data batch are used to guide the sorting and display process.

Stems and Leaves

The first step in the construction of a stem-and-leaf display is to divide the digits of the data values in the batch into three classes: sorting digits, display digits, and digits that can be ignored. Suppose that the batch consists of a set of elevations between 12,000 and 20,000 ft above sea level. Let us examine several ways of splitting a typical observation such as 15,328. One way would be

Sorting digits:	Display digits:	Digits to be ignored:
15	3	28

In this case there are two sorting digits, the hundreds digit is the display digit, and the last two digits are ignored. There is some choice in the number of digits to be assigned and the number to be ignored. It is possible to use all the digits and ignore none. A perfectly legitimate split would be

Sorting digits:	Display digits:	Digits to be ignored:
1532	8	none

Just as we must make a choice for the length of the interval in constructing a histogram to best illustrate properties of some variable, so, too, must the correct choice of display digits be made in a stem-and-leaf construction. When the digits are divided 15-3-28, we are saying that the elevation in hundreds of feet is important, but tens of feet and feet are unimportant. If we divide the digits 1532-8-0, we are indicating that the unit digit is important in the display. This might be desirable if all of the observations are between 15,320 and 15,330. The effects of digit classification on the appearance of stem-and-leaf displays are illustrated in the examples of this section.

The leading, or sorting, digits are called the stem of the observation, and the display digit is referred to as the *leaf*. Only four steps are required to construct a stem-and-leaf display:

1. Sort the data from lowest to highest value.
2. Examine the data and select the pair of digits to divide the data values into stem, leaf, and ignored portion.
3. In a column, list *all possible* sets of leading digits, from lowest to highest. Note that we must include sets of leading digits that might have occurred, but don't happen to be in this particular batch. The first stem is always the lowest of the batch, and the last stem is always the highest stem of the values in the batch. We don't need to go beyond these bounds.
4. For each observation, or data value, write the display digit on the stem line corresponding to its sorting digits or stem. When all data values in the batch are put into the display, there should be one leaf for each observation.

Although this may seem quite complicated, a quick examination of the overall appearance of several stem-and-leaf displays from several examples will clarify the use of this algorithm.

EXAMPLE 16-1. As a first example, let us portray the DO batch of Table 2-1 as a stem-and-leaf display, We begin by sorting the data values from lowest to highest as in Table 16-1. The data are divided at the decimal point leaving only three stems, 4, 5, and 6. The tenths' digit is the display digit and the leaves of the table, and no digits are ignored. When all the values in the batch contain two digits, this is the only logical construction. The stem-and-leaf display for this format is illustrated in Table 16-2. There are nine leaves for stem 4, corresponding to the first nine values of Table 16-1: 4.2, 4.3, 4.4, 4.5, 4.7, 4.7, 4.8, 4.8, and 4.9. Note that we can generate these nine values from the first row of the stem-and-leaf display itself! There are 30 leaves for stem 5 and 11 leaves for stem 6. This is not a good example of a stem-and-leaf display, as there are too few stems for the 50 leaves. Nevertheless, we can still see that this batch of values is symmetric, ranges from 4.2

TABLE 16-1
Dissolved Oxygen Values of Table 2-1, Sorted by Value, mg/L

Rank	Value	Rank	Value
1	4.2	26	5.6
2	4.3	27	5.6
3	4.4	28	5.7
4	4.5	29	5.7
5	4.7	30	5.7
6	4.7	31	5.7
7	4.8	32	5.8
8	4.8	33	5.8
9	4.9	34	5.8
10	5.1	35	5.8
11	5.1	36	5.9
12	5.1	37	5.9
13	5.2	38	5.9
14	5.3	39	5.9
15	5.3	40	6
16	5.4	41	6.1
17	5.4	42	6.2
18	5.4	43	6.3
19	5.4	44	6.4
20	5.5	45	6.4
21	5.5	46	6.4
22	5.6	47	6.6
23	5.6	48	6.7
24	5.6	49	6.8
25	5.6	50	6.9

to 6.9, and is concentrated at values between 5.0 and 5.9. A number of ways of improving the appearance of this display are illustrated in subsequent variations of this basic stem-and-leaf display.

EXAMPLE 16-2. Table 16-3 contains a list of 40 houses, their addresses, and their dates of construction. These houses are all contained within a three-square-block area of a city and are the subjects of an intensive study of housing maintenance. The stem-and-leaf display for this batch of construction dates is shown in Table 16-4. In this case, there are

TABLE 16-2
Stem-and-Leaf Display for the Data of Table 2-1

Stems	Leaves
4	234577889
5	111233444455666666777788889999
6	01234446789

three digits in the stem, and the last digit is the leaf. What does this stem-and-leaf display reveal? First, we note two data extremes—one, a house constructed four decades earlier than the second oldest house in the batch, and the other, built almost two decades after the second newest house in the neighborhood. Second, examine the dates of the houses built in the 1920s. All were built in 1924. This peculiarity would not be picked up in a conventional histogram. But, by using the actual digits of the data values in the display, this

TABLE 16-3
Addresses and Construction Dates of a Batch of Houses

Observation	Address	Construction date
1	67 Pine Street	1924
2	73	1924
3	77	1924
4	82	1924
5	85	1924
6	91	1924
7	92	1936
8	96	1946
9	1100 Fifth St	1945
10	1102	1934
11	1104	1934
12	1105	1936
13	1106	1868
14	1113	1980
15	1128	1911
16	1131	1955
17	1143	1933
18	1156	1956
19	1178	1958
20	1191	1960
21	1198	1961
22	221 Magnolia Boulevard	1914
23	228	1962
24	264	1962
25	279	1935
26	282	1932
27	291	1951
28	297	1952
29	300	1953
30	1433 Reading Road	1908
31	1434	1947
32	1440	1947
33	1443	1947
34	1451	1932
35	1453	1946
36	1457	1946
37	1479	1910
38	1482	1938
39	1486	1937
40	1498	1931

TABLE 16-4
Stem-and-Leaf Display for Batch of Table 16-3

Stems	Leaves
186	8
187	
188	
189	
190	8
191	14
192	444444
193	12234456678
194	5666777
195	123568
196	0122
197	
198	0

Note: Unit = 1. And 186 8 represents 1868.

potentially important characteristic of the batch becomes visible. We might also note that the all construction in the 1940s was in the postwar period.

Another useful addition to a stem-and-leaf display is to indicate the units of the data values, particularly where they may not be obvious. Table 16-4 is labeled Unit = 1, and an example relating the stem-and-leaf display to the actual data given as 186 8 represents 1868. When the data include a decimal point, this unit specification becomes mandatory. The note for the display of Example 16-1 illustrated in Table 16-1 would be Unit = 0.1 where 4 2 represents 4.2.

Multiple Lines per Stem

Sometimes, the stem-and-leaf display for a batch can be ineffective if it is too stretched out or too compressed. The stem-and-leaf-display of the DO batch illustrated in Table 16-2 represents a poor display. There are only three stems, and it is very difficult to say whether or not the batch is symmetric around some central value or not. There are two ways of constructing a display to overcome this problem. First, we can use two lines per stem, doubling the number of lines in the display. Normally, this also helps to alleviate the appearance of overcrowding on the display. Each stem is converted to two lines by associating leaf values 0, 1, 2, 3, and 4 with the first line of the stem and the values 5, 6, 7, 8, and 9 with the second line of the stem. The first line is usually labeled with an asterisk and the second line with a dot. Table 16-5 illustrates a revised version of the stem-and-leaf display of the DO batch. Expressed in this way, the symmetry of the batch of dissolved oxygen values is much more apparent than in the display shown in Table 16-2. Of course, histograms with too few intervals have the same difficulty in adequately supplying a useful graphical summary of a variable.

An alternative to the two lines per stem format is the five lines per stem display

TABLE 16-5
A Two-line Stem-and-Leaf Display for the Lake DO Batch

Stems	Leaves
4*	234
4·	577889
5*	1112334444
5·	55666666777788889999
6*	0123444
6·	6789

Note: The first line is labeled with an asterisk and the second line is labeled with a dot.

illustrated in Table 16-6. In this construction, the five lines for each stem are denoted by asterisks for leaf values 0 and 1, by a T for leaves 2 and 3 (i.e., T̲wo and T̲hree), an F for values 4 and 5 (i.e., F̲our and F̲ive, and S for 6 and 7 (i.e., S̲ix and S̲even), and a dot for leaves 8 and 9. The symmetry of the DO batch is again prominent, as is the concentration of values between 5.4 and 5.9. In this example, there appears to have been a definite advantage to moving from a display of one line per stem to two lines per stem (that is, summarizing the batch with Table 16-5 instead of 16-2), but limited gains are achieved by extending the display to five lines per stem as in Table 16-6. The two-line display is the simplest display to illustrate (1) the symmetry of the batch, (2) the concentration of values between 5.5 and 5.9, and (3) the number of repeated values in this range. Another basic principle of EDA is to use the *simplest* version of any display that adequately illustrates the salient features of a batch.

TABLE 16-6
A Five Line Stem-and-Leaf Display of the DO Data

Stems	Leaves
4T	23
4F	45
4S	77
4·	889
5*	111
5T	233
5F	444455
5S	6666667777
5·	88889999
6*	01
6T	23
6F	444
6S	67
6*	89

Listing Stray Values

Data values that are far removed from the remainder of the batch are known as *strays*. That a batch contains such values seems a sufficiently important feature to warrant a special designation in a stem-and-leaf display. The stem-and-leaf display of the house construction dates summarized in Table 16-4 contains two such data values, 1868 and 1980. To include them in the stem-and-leaf display requires four additional stems for 187, 188, 189, and 197, but there are no data values (and therefore leaves) for these stems. Once it is determined that there are strays in a batch, it is useful to designate them with the labels HI and LO depending on the end of the display to which they belong. To emphasize the gap between these values and the rest of the batch, one blank line is left between the list and the rest of the display. Table 16-7 is a re-expression of Table 16-4 using this convention.

Positive and Negative Values

If the data batch contains both and negative values, then the stems near zero require special designation. Numbers slightly less than zero are attached to stem −0, and numbers slightly greater than zero are give stem +0. For example, the display

Unit = 0.1	
−1.2 Represents −1.2	
−1	2
−0	78
+0	456
1	3

TABLE 16-7
A Re-expression of Table 16-4, Using the Convention for Strays

Stems	Leaves
LO	1868
190	8
191	014
192	444444
193	12234456678
194	5666777
195	123568
196	0122
HI	1980

Note: Unit = 1, and 190 8 represents 1908.

contains the values −1.2, −0.7, −0.8, 0.4, 0.5, 0.6, and 1.3. A value of exactly 0.0 can be put on either 0 stem. If there are several different 0.0 values, then it is best to divide them as equally as possible between the two zero stems. This preserves the appearance of the display.

16.2. Letter Value Displays

After we have taken an initial look at our batch, the next step is to develop summary measures for the central value and spread of the numbers. This idea is not new; it is usual to calculate several measures of central tendency and variability in conventional statistics. In EDA, we also want to describe more precisely the shapes and patterns expressed in our stem-and-leaf displays, but we generally pay special attention to extraordinary values. In practice, researchers sometimes purge their data sets of extreme values—whether or not there is a reason to believe that the value is a flawed observation. These extreme values often tell us more about the process that generates the data than observations in the center of the batch. For example, if we were to find that the dissolved oxygen value in a particular lake was very low, and knew it was the site of a particular industrial plant, we might be led to check the effluent sources from the plant to determine the particular chemical process absorbing lake oxygen. If we were to eliminate this observation, our study might miss this important feature of dissolved oxygen variability. *Letter value displays* are summaries of letter values used to describe the center and spread of a data batch.

Medians, Hinges, and Other Summary Values

To determine the letter values, begin by ordering the data batch from lowest value to highest. Note that this ordered list could easily be generated from a stem-and-leaf display, which can be used to reconstruct all data values in a batch. Once the batch is ordered, a set of suitably selected data values can be used to summarize many important features of the batch. *Letter values* are one simple set of these data values.

Ordering observations in a data set is usually done sequentially from highest to lowest. Instead, let us consider how far an observation lies from either the low or high end of the data set.

DEFINITION: DEPTH
The depth of each data value in a batch is the value's position in an ordered list of values starting at the *nearer* end of a batch.

Each extreme value is the first value in such an enumeration and thus has a depth of 1; the second highest and second lowest have a depth of 2; and so on. In general then, there are two data values with depth i, unless depth i is the middle of the values. Consider the five observations 2, 3, 5, 7, and 8. Our ordered list with depths would be

Order from top	Order from bottom	Depth	Data value
1	5	1	2
2	4	2	3
3	3	3	5
4	2	2	7
5	1	1	8

If the number of observations n is odd, there is a "deepest" data value, one which is equally far from either end of the ordered batch As in conventional descriptive statistics, this is the median.

DEFINITION: MEDIAN

The median value of a batch is the data value with the highest depth of the values in a batch.

The depth of the median is simply $(n + 1)/2$. In the simple five number batch shown above, the median has a depth of $(5 + 1)/2 = 3$. It is common to express this as

$$d(M) = (n + 1)/2 \qquad (16\text{-}1)$$

If n is even, then there are two middle values in the batch, i.e., two values with the highest depth. If we use this formula, then the depth of the median will have a fractional part of $\frac{1}{2}$, a depth that points midway between two data values. For example, the batch of six values 47, 103, 142, 177, 192, and 221

Depth	Data value
1	47
2	103
3	142
3	177
2	192
1	221

yields $d(M) = (6 + 1)/2 = 3\frac{1}{2}$, meaning that the median is the average of the two middle data values with a depth of 3: $M = (142 + 177)/2 = 159\frac{1}{2}$. For simplicity, we label the median with letter M.

Knowing that the median divides an ordered batch into halves, we might wish to continue this process and determine the middle of these two halves.

DEFINITION: HINGES, QUARTERS, OR FOURTHS

The hinges, quarters, or fourths of a data batch are the summary values in the middle of each half of the data batch.

They are roughly one-quarter of the way in from each end of the data set. They can be found in much the same way as the median. *Beginning with the integer part of the median*, we simply add 1 and divide the result by 2:

$$d(H) = (d^*(M) + 1)/2 \qquad (16\text{-}2)$$

where $d^*(M)$ is used to distinguish the integer component of the depth of the median from the computed depth $d(M)$. Hinges are quite similar to quartiles, but not exactly the same, since the hinges are calculated directly from depth of the median, not in relation to the entire data set.

We can continue this process and determine the eighths using the integer component of the depth of the hinges:

$$d(E) = (d^*(H) + 1)/2 \qquad (16\text{-}3)$$

At this point we have computed the depths of five summary values for a batch: one value for M, two for H, and two for E. Though we could continue, there is little to gain from doing so. The advantage of simplicity is soon lost. For example, for a batch with 21 data values, the median M has a depth of 11, the hinges H a depth of 6, the eighths E a depth of 3.5, the letter D a depth of 2, and the *extremes* with a depth of 1.

Displaying the Letter Values

The summary values that we have been calculating using the depths of the values within a batch are the start of the sequence of *letter values*, which can be used to provide a numerical summary of a data batch. The simplest display is the *five-number summary*, which has the following form:

	M	
H		H
Lo		Hi

which places the median in the center of the first row, the hinges offset from the median on the second row and the extremes on the third row. From this summary, we note (1) the *range* of the values in the batch (HI − LO), (2) the center of the data values as the median, and (3) 50% of all the values in the batch are between the hinges. Also, there is a quick check for symmetry. If the hinges are not located at an equal distance from the median, there is asymmetry in the batch.

EXAMPLE 16-3. The monthly rainfall, in millimeters, during March in the last 20 years at a climatic station are as follows:

0.0	1.8	0.1	19.6	0.2	6.1	5.1	7.9	30.5	9.7
11.7	24.1	37.8	6.3	3.0	22.8	14.2	28.7	0.5	12.2

Table 16-8 illustrates the calculation of the letter values for this batch. From this table, we can extract the following five-letter summary:

$$
\begin{array}{c c}
 & 8.8 \\
2.4 & 21.2 \\
0 & 37.8
\end{array}
$$

March rainfall at this station varies between 0.0 and 37.8 mm and is between 2.4 and 21.2 mm 50% of the time. The batch is not symmetric, being positively skewed. This is apparent since the lower hinge is only 6.4 mm below the median, but the upper hinge is 12.4 mm above the median.

It is also possible to embellish the letter value display to provide more information. First, we note that, except for the median, the letter values are always in pairs. A value midway between two letter values is a *midsummary*. The *midrange* or *midextreme* is the numerical average of the extremes, the *midhinge* lies midway between the hinges, the *mideighth* lies midway between the eighths, and so on. Also, we can calculate the spread as the difference between the letter values in a pair. The *range* is the difference in the extremes, the *H-spread* is the difference between the hinges, and the *E-spread* is the

TABLE 16-8
Locating and Calculating the Letter Values for Example 16-3

Depth of letter values	Depth	Data value	Letter values
	1	0	Lo = 0.0
d(D) = (3 + 1)/2 = 2	2	0.1	D = 0.1
d(E) = (5 + 1)/2 = 3	3	0.2	E = 0.2
	4	0.5	
	5	1.8	
d(H) = (10 + 1)/2 = 5.5			H = 2.4
	6	3	
	7	5.1	
	8	6.1	
	9	6.3	
	10	7.9	
d(M) = (20+1)/2 = 10.5			M = 8.8
	10	9.7	
	9	11.7	
	8	12.2	
	7	14.2	
	6	19.6	
d(H)			H = 21.2
	5	22.8	
	4	24.1	
d(E)	3	28.7	E = 28.7
d(D)	2	30.5	D = 30.5
	1	37.8	Hi = 37.8

difference between the eighths. How can we use these midsummaries and spread values? By comparing the midsummaries to the median, we can look for either positive or negative skewness. If the midsummaries continually increase, then the batch is positively skewed; if they continually decrease, then the batch is negatively skewed. The degree of skewness is indicated by the point at which the midsummaries change and the rate at which they do so.

The variability in a batch is indicated by the spread values and are interpreted in the context of the particular letter value. For example, we have already noted that the values of the two hinges enclose the middle 50% of the values of a batch. The E-spread gives the range of the middle three-quarters of the data, the D-spread encloses the middle seven-eighths of the data, and so on. All of these spreads are responsive to changes in the variability of the data in a batch. A more complete letter value display for the March rainfall data of Example 16-3 has the following form:

Letter value	Depth	Lower		Upper	Mid	Spread
M	10.5		8.8		8.8	
H	5.5	2.4		21.2	11.8	18.8
E	3.5	0.2		28.7	14.5	28.5
D	2	0.1		30.5	15.3	30.4
1		0.0	$n = 20$	37.8	18.9	37.8

The gradually increasing value of the midsummaries indicates a skewness towards the high side, not surprising since the batch has a natural minimum of 0. Seventy-five percent of the values are between 0.2 and 28.7 mm, and it is easily seen that all values are less than 37.8 mm. Compared with the frequency table, the letter value display seems slightly more useful in extracting simple descriptive information about a variable.

16.3. Boxplots

While letter value displays provide a neat summary of a data batch, they have one serious drawback. They are not graphical. It is thus very difficult to compare the values in several similar batches without a painstaking examination of a series of letter-value displays. What a useful graphical display should be able to illustrate is (1) where the middle or center of the batch is, (2) how spread out the middle of the batch is, (3) how the tails of the data relate to the middle, and (4) whether there are strays. *Boxplots* serve these purposes.

The Skeletal Boxplot

The simplest boxplot, often termed a *skeletal boxplot* or *box-and-whiskers plot*, is derived from the simplest letter value display, the *five-number summary*, and uses the median, hinges, and extremes. First, it is useful to draw a vertical scale for the batch clearly indicating the units being measured. Second, draw a long, thin box that stretches from the lower to the upper hinge, with an internal bar across the box placed at the median. From each end of the box, we extend a line (the "whisker") extending to the extreme value in each direction. One quick glance at such a plot allows us to form impressions of (1) the overall level, (2) the spread, and (3) the symmetry of the data. For instance, for the box-and-whiskers plot for the house construction dates of Example 16-2 shown in Figure 16-1, we see that median date of house construction in this neighborhood was roughly 1935, and 50% of all the houses were built between 1925 and around 1950. It is also clear that house construction has taken place over a period exceeding a century, from roughly 1870 to 1980.

Outliers

One of the defects in the box-and-whiskers plot is that it fails to adequately treat outliers, values that are either so high or so low that they appear to be isolated from the rest of the data batch. A simple modification of the box-and-whiskers plot is needed to direct our attention to these outlying values. When we detect these outliers, we may want to examine them very closely. They may be indicative of data collection errors or they may give us valuable insights into the nature of the batch. The question now facing us is "how do we unambiguously identify outliers?" One simple rule that we might use to define outliers is

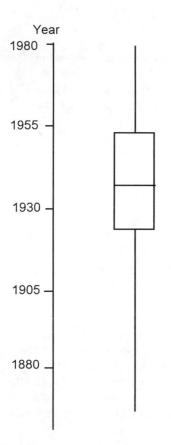

FIGURE 16-1. Box and whiskers plot of construction dates.

to develop an expression based on the location of the hinges and their corresponding H-spread.

DEFINITION: INNER FENCES AND OUTER FENCES

Inner fences are defined as

$$\text{lower hinge} - (1.5 \times H\text{-spread}) \text{ and}$$
$$\text{upper hinge} + (1.5 \times H\text{-spread})$$

and outer fences are defined as

$$\text{lower hinge} - (3 \times H\text{-spread}) \text{ and}$$
$$\text{upper hinge} + (3 \times H\text{-spread})$$

Next, we define any data value beyond either inner fence as *outside,* and data value beyond either outer fence as *far outside.* The outermost actual data value that is still not beyond the corresponding inner fence is known as an *adjacent value.* Any data value outside the adjacent value can be defined as an outlier.

For the house construction date batch, we note that the lower hinge is 1924 and the upper hinge is 1951.5. Therefore, the H-spread is 1951.5 − 1924 = 27.5. The inner fences are

$$1924 - (1.5 \times 27.5) = 1882.75$$
$$1951.5 + (1.5 \times 27.5) = 1992.75$$

and the outer fences are at

$$1924 - (3 \times 27.5) = 1841.5$$
$$1951.5 + (3 \times 27.5) = 2034$$

The corresponding adjacent values, just inside the inner fences, are 1908 and 1980. On this basis, we note that we have only one outlier—the house at 1106 Fifth St., built in 1868.

Boxplot Construction

A boxplot is a modified version of a box-and-whiskers plot. The inner box is constructed in an identical matter; solid lines are used to mark off a box with a length from lower to upper hinge and a solid line across the box to locate the median. The "whiskers" are drawn as dashed lines and, unlike the box-and-whiskers plot, are drawn out from each hinge only to the corresponding *adjacent value.* We do not choose the fences because they may not be data points, whereas we know that the adjacent values will always be a data value within the batch. Each outlier is shown individually, and labeled appropriately; far outside values should be particularly prominent. One possibility is to display these values in upper case or using a different color.

Figure 16-2 illustrates a boxplot of the construction date batch. The distinguishing feature of the boxplot is the attention that is given to the outlier, the house at 1106 Fifth St. We might be tempted to examine Fifth St. more closely. If we did, we would note that

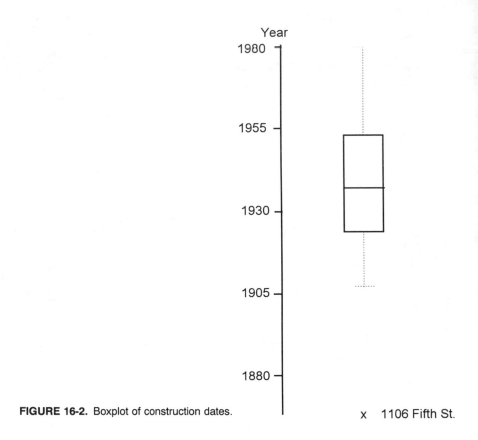

FIGURE 16-2. Boxplot of construction dates.

x   1106 Fifth St.

it contains houses built in 1868 and 1911, but also two homes built in the early 1960s and one in 1980. Moreover, it appears there were homes built in three other decades. By contrast, virtually all of Pine St. was built in 1924.

Comparing Boxplots

In order to compare several different batches, we can simply place the boxplots side by side. For example, we might compare the construction histories of several neighborhoods by developing individual boxplots and placing them on an identical scale as in Figure 16-3. Several comparative statements can be drawn from this figure. First, we note that the neighborhood labeled Fernwood is the most recently developed of the four sample neighborhoods, except for a single house built well before a more or less continuous period of house construction. The neighborhood labeled Rosedale is noteworthy because it was apparently developed over a very short period and has the appearance of being a post WW II suburban bedroom community. The recent outlier led the researcher to uncover a new home, rebuilt after a fire. By contrast, Stowe and Van Pelt have longer development histories, with no apparent outliers. The development history of Van Pelt is so long one might conjecture that it is actually a neighborhood with two separate periods of house

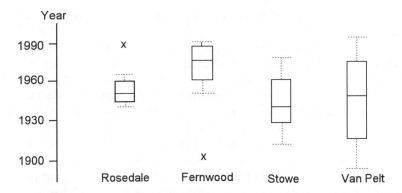

FIGURE 16-3. Boxplot of construction dates for several neighborhoods.

construction activity. Some of the older houses may have been demolished and replaced by newer houses leading to a neighborhood with mixed housing. This would have to be checked through a more detailed analysis of the stem-and-leaf display. In sum, comparison of boxplots by placing several batches on an identical scale appears to be a useful technique for illustrating the differences and similarities among sets of data.

16.4. Resistant Lines

In Chapter 14, we focused our attention on techniques useful for relating a dependent variable, Y, to an independent variable X. The most commonly used technique is based on the least squares criterion, a criterion that is known to minimize the sum of squared deviations of the data points to the regression line. From the Gauss-Markov theorem, we also know that these estimators are important in developing the inferential aspects of regression analysis and testing its goodness-of-fit.

Is there a simpler approach? Well, let us begin by recalling that any two points can be used to determine the two parameters a and b of the line $Y = a + bX$. In Appendix 13A, we showed that we can easily use any two points with coordinates $P_1(X_1, Y_1)$ and $P_2(X_2, Y_2)$ to calculate unique values of these two parameters using the equations:

$$b = (Y_2 - Y_1)/(X_2 - X_1) \tag{16-3}$$

and

$$a = Y_1 - bX_1 \text{ or } a = Y_2 - bX_2 \tag{16-4}$$

In regression analysis, we used all n points to determine a and b. One of the drawbacks to this, is that outliers in the data may unduly influence the location of the line. (See Section 13.5.) In exploratory data analysis, we try to prevent outliers from unduly influencing the location of this line. This can be done by using a method to select the two points P_1, P_2 that prevents us from choosing a "poor" pair of points, i.e., a selection that would lead to an unrepresentative summary line. We do this by selecting a representative point near each end of the line.

First, divide the x-values into three regions—points with low x-values on the left, points with middle x-values, and points with high x-values on the right. This is most easily done by placing one-third of the points in each category. If the number of points is divisible by 3, then this is easy. If there is one extra point, place it in the middle; if there are two extra points, place one in each of the two outer portions of the partition. Whenever several data points have the same x-value, they must be placed in the same portion of the partition. These ties may make it difficult to come to an equal allocation of points in each partition, but *judgment* can be used to make the best allocation of points.

Within each portion or third of the points, we determine the *median* point. This is done independently for the x-values and y-values of the points in each partition, using the same method described in Section 4.2 for spatial data. Denote these three *summary points* for the left, middle, and right portions of the data as (x_L, y_L), (x_M, y_M), and (x_R, y_R), respectively. Note that by using the medians within these three regions we make our point selection resistant to the existence of *outliers* or *stray* values.

EXAMPLE 16-4. Consider again the data for household trip generation rates of Example 13-4.

Traffic zone	Average household income $000	Trips per household per day
1	9	4
2	10	3
3	11	5
4	12	5
5	14	6
6	15	5
7	16	7
8	17	6
9	18	7
10	21	9
11	22	8
12	24	8

Since there are 12 data points, we place four points in each of three portions of the plot. Figure 16-4 illustrates the three summary points for the trip generation data. For the left portion of the data, the summary point is the median of the four points with coordinates (9, 4), (10, 3), (11, 5), and (12,5). We analyze each direction independently, so our estimate of (x_L, y_L) is (10.5, 4.5). Similarly, (x_R, y_R) can be computed as (21.5, 8) and (x_M, y_M) as (15.5, 6).

Once we have found the summary points, we can easily calculate the slope of the line that passes through the *left* and *right* summary points of the data batch. For this data, $b = (8 - 4.5)/(21.5 - 10.5) = .3182$. Now that we have determined the slope of the line, we need

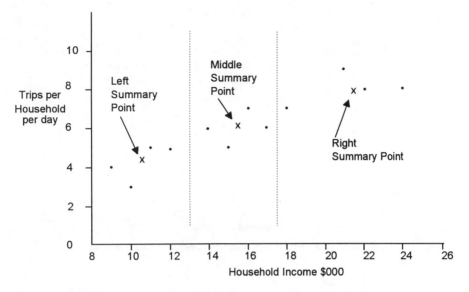

FIGURE 16-4. Dividing a plot into three parts and finding summary points.

to determine the value of the intercept a to determine the location of the *resistant line* that summarizes this data batch.

DEFINITION: RESISTANT LINE

A resistant line summarizes an X-Y data batch and is resistant to the existence of strays or outliers in both the X and Y directions.

There are a number of different ways we might determine the value for a. First, we might adjust the line to pass through the middle summary point (x_M, y_M) by computing $a_M = y_M - bx_M$. But, rather than allowing the middle summary point itself to determine the location of the intercept and hence the line, it may be better to use *all three* summary points. To do this, simply take the three separate estimates of a from each of the three summary points and average them:

$$a_L = y_L - bx_L \quad a_M = y_M - bx_M \quad a_R = y_R - bx_R$$

and therefore

$$a = 1/3(a_L + a_M + a_R) = 1/3[(y_L + y_M + y_R) - b(x_L + x_M + x_R)] \qquad (16\text{-}5)$$

For the trip generation data batch,

$$a = 1/3(1.1591 + 1.0682 + 1.1591) = (1/3[(4.5 + 6 + 8) - .3182(10.5 + 15.5 + 21.5)]$$

$$a = 1.1288$$

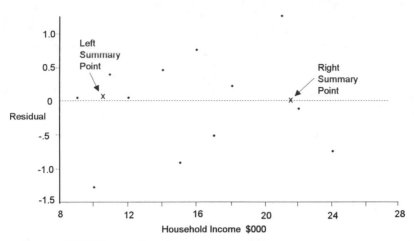

FIGURE 16-5. Plot of residuals against income, first iteration.

and the initial fitted line is $Y = 1.1288 + .3182\,X$. Just as in conventional regression analysis, we can now compute the residuals of the data values from the batch. These residuals are listed in the first column of Table 16-9 and are illustrated in Figure 16-5. Some of these residuals are positive, and some are negative. If our line is representative of the data, we would expect these residuals *to have a slope of zero*. To test this, we find the median values in the three parts of the data (10.5, .0303), (15.5, −.0606), and (21.5, .0076). The values used to determine these median values are shown as bold text in Table 16-9. The slope of the residuals is obtained from the right and left points using Equation (16-3):

$$b' = (.0076 - .0303)/(21.5 - 10.5) = -.0021$$

Since this is very close to zero, we could stop here and use the computed line as our estimate of the resistant line. Or, we could try to improve the fit by making a new estimate of the slope, correcting for this slight error:

$$b_2 = .3182 - .0021 = .3161$$

We use this new estimate of the slope to calculate the residuals listed in the column labeled "2nd residual" in Table 16-9. The slope of these residuals is again tested and found to be

$$b' = (.0489 - .0520)/(21.5 - 10.5) = -.0003$$

Using a new estimate of $b_3 = .3158$ leads to a third set of residuals. When the slope of these residuals is calculated, it turns out to be zero, so we can assume that we have the

TABLE 16-9
Calculating the Resistant Line for the Household Trip Generation Data

	Trips per household/day	1st residual	2nd residual	3rd residual	Final residual
	Calculating the resistant line for the household	0.01	0.03	0.03	0
	Trip generation data	-1.31	-1.29	-1.29	-1.32
$X_L = 10.5$	5.5	0.03	0.05	0.06	0.03
11	5	0.37	0.39	0.4	0.39
12	**5**	**0.05**	**0.08**	**0.08**	**0.05**
14	**6**	**0.42**	**0.45**	**0.45**	**0.42**
15	5	-0.90	-0.87	-0.87	-0.89
$X_M = 15.5$	6	-0.06	-0.03	-0.02	-0.05
16	7	0.78	0.81	0.82	0.79
17	**6**	**-0.54**	**-0.50**	**-0.5**	**-0.53**
18	**7**	**0.14**	**0.18**	**0.19**	**0.16**
21	9	1.19	1.23	1.24	1.21
$X_R = 21.5$	8	0.01	0.05	0.06	0.05
22	**8**	**-0.13**	**-0.08**	**-0.08**	**-0.11**
24	8	-0.77	-0.72	-0.71	-0.74

best slope estimate. If the slope of the residuals changes from positive to negative (or vice versa) we can assume that the true slope lies between the two estimates of b used to derive the residuals. Let us label the two slope estimates b_1 and b_2, and the slopes of the residuals based on these values b_1' and b_2'. An efficient formula to derive a new estimate of the slope is

$$b_{new} = b_2 - b_2'[(b_2 - b_1)/(b_2' - b_1')] \qquad (16\text{-}6)$$

In this case we do not need to use it as we have already obtained a satisfactory estimate of the slope. We can continue the process until the slope of the residuals is either zero or small enough to ignore. Using a spreadsheet program on a PC, we can easily compute as many iterations as is necessary, though calculations by hand are tedious.

We can now use the summary points of the third iteration to determine the final value of our intercept a. We simply take our average miss of the three summary points and add it to the first estimate of the intercept. Our final fit therefore becomes

$$y = (1.1288 + .315x) + 1/3(.0554 + .0554 - .0235) = 1.1579 + .3158x$$

We can use this line to compute the final residuals shown in the last column of Table 16-9. It is interpretable in the same way as our normal regression lines. We would expect a household with an income of $20,000 to average $1.1579 + .3158 (20) = 7.4739$ trips per day. Every $10,000 of increased income leads to just over 3 (10 times $.3158 = 3.158$) more trips by the household in a average day.

Nonlinear Relationships

While straight-line summaries of relationships are certainly simple, they may not be the best ways of expressing the relationship between two variables X and Y. One way of determining whether or not a nonlinear representation might be better for a data batch is to examine the pattern in the residuals. Just as in regression analysis, any sign of a bend in the pattern of the residuals may indicate a nonlinear relationship in the original data. Of course to do so we have to calculate a resistant line in the first place!

There is an even simpler way of looking at the data values in a batch to see if they are likely to be linear before we go to the trouble of fitting a resistant line. We begin by computing our three summary points. We can approximate the slope of the left portion of the batch using the left and middle summary points, and the slope of the right portion of the batch using the middle and right summary points.

DEFINITION: HALF-SLOPES

The left and right half-slopes of a data batch are calculated as

$$b_L = \frac{y_M - y_L}{x_M - x_L} \quad \text{and} \quad b_R = \frac{y_R - y_M}{x_R - x_M}$$

and the half-slope ratio is defined as b_L / b_R .

If the half slopes are equal, then the relationship is straight and the half-slope ratio is 1. Any departure from a value of 1 for the half-slope ratio indicates nonlinearity. When it is not close to 1, say it has a value of 3, we might wish to examine other ways of introducing linearity, perhaps by fitting two resistant lines to two different parts of the batch. Just where the break occurs can be estimated by examining the basic X-Y plot. Transformation of one of the variables is another alternative. Taking the log or the power of one or the other of our variables can often transform a seemingly nonlinear relationship into a linear one. For example, if the half-slope ratio is very low, the scatter has a J-shape, and we might wish to transform variable X by taking X^2.

16.5. Smoothing

In Chapter 3, we examined various ways of smoothing time series observations in order to remove linear, seasonal, or even irregular trends. In EDA, *data smoothers* use simple medians and averages to summarize series of observations in a batch; however, the fit that these smoothers produce need not necessarily follow any specific formula. The goal is to capture the overall behavior of the series, where it rises, where it falls, and whether there are any regularities or cycles in the data. Fluctuations between individual data points in the sequence should appear as residuals from this smoothed pattern. In the vernacular of EDA, we distinguish the *rough* from the *smooth*, realizing that our interest should lie in both the fit and the residuals. In order to extract these smooth patterns from our data, we pay special interest to strays and outliers. In essence, we try to minimize their effects on the underlying

TABLE 16-10
Coal Production (Millions of Metric Tonnes) in a Region over a 30-Yr Span

Year	Tonnes	Year	Tonnes
1	59	16	60
2	51	17	60
3	51	18	70
4	57	19	70
5	57	20	56
6	46	21	69
7	54	22	70
8	70	23	95
9	60	24	53
10	50	25	66
11	59	26	70
12	48	27	56
13	50	28	54
14	50	29	70
15	60	30	60

pattern. As we have seen in several other instances, *medians* rather than *means* tend to be less subject to substantial variations due to a single outlier. Again, they will play an important role in the smoothing procedures discussed below.

EXAMPLE 16-5. The production of coal in millions of metric tonnes over a 30-yr period is shown in Table 16-10 and illustrated in Figure 16-6. At first glance, the batch does not have any obvious cycles, appearing more random than anything else. Production begins and ends at about 60 million metric tonnes and ranges from a low of 46 million metric tonnes to almost 100 million metric tonnes. However, as we shall see, the use of EDA smoothers will reveal a simpler pattern that is much easier to describe.

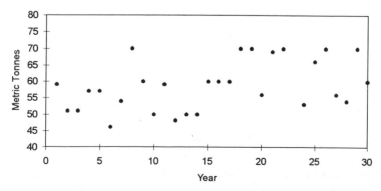

FIGURE 16-6. Scattergram of coal production over 30-yr period.

Elementary Smoothers

Whenever we examine a smooth sequence of values, we are struck by how each point in the sequence is much like its neighbors. Changes in the data values are not sudden. One simple procedure for smoothing is to achieve this property by replacing each value in a series with the median of the three y-values: itself, its preceding value, and its successor. Any value in the sequence that is out of step will be replaced by one of the other two values, whichever is closer to it. We could also replace any y-value with the median of its nearest five neighbors, seven neighbors, etc. These types of smoothers are usually termed *running medians* because we "run" along the data sequence as we generate the medians for that particular value in the sequence.

For the initial values in a sequence, we can run into problems. The first value in a sequence has no predecessor, and we obviously cannot determine the median of three. Similarly, we cannot determine the median of five for the second value in a sequence because it does not have two preceding values. For medians of three and five, it is usual to accept the two end values in the sequence. For a median of five, we can replace the second and next-to-last with medians of three, and so on for longer sequences.

Running medians of three and five for the coal production data of Example 16-3 are illustrated in Figures 16-7 and 16-8, respectively. In comparison to the raw data of Figure 16-6, we can see that the application of a running median of three has smoothed some of the abrupt changes in the original series. The medians of five are slightly smoother, but less like the original data batch. Fortunately, we can do better. The answer lies in using medians of even length, which use the numerical average of the two central values. For a median of length four, we drop the two extreme values, the lowest and highest y values, and average the two middle y values, whether or not they are consecutive. When we use even length medians, we must also average the X (or time) variable. These centered values will not be at one of our original X values, but midway between consecutive values. We can recenter them onto our original data values by taking the average (or median) of these smoothed values.

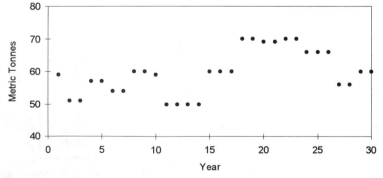

FIGURE 16-7. Running medians of three data points for coal production over a 30-yr period.

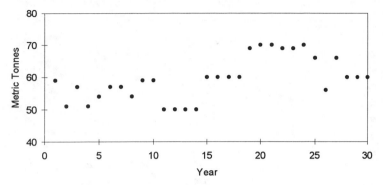

FIGURE 16-8. Running medians of five data points for coal production over a 30-yr period.

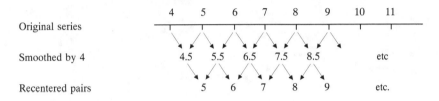

By convention, smoothing operators are summarized by one-digit values corresponding to their length. Therefore, the recentered pairs shown above can be abbreviated to a 42 smoothing. The sequence of operations leading to a 42 smoothing of the coal production data is illustrated in Figure 16-9. Note that the sequence appears smoother, and the underlying pattern is emphasized. While it is useful to smooth sequences using simple running averages, it is also possible to smooth a sequence by replacing each value with a running *weighted* average. For example, we can smooth three consecutive values x_{t-1}, x_t, and x_{t+1} by replacing x_t with, say $\frac{1}{4}x_{t-1} + \frac{1}{2}x_{t-1} + \frac{1}{4}x_{t+1}$. Of course any three weights that sum to 1 could be used in this type of smoothing, but this simple form known as *hanning* has the added advantage of simplicity. Hanning can be used at any time in a smoothing procedure, but is especially useful when outliers have been replaced by more moderate values. Longer span medians are more resistant to outliers than are shorter ones.

Compound Smoothers

Applying one smoother to the results of another smoother is known as resmoothing. We have already seen that a 42 smoother can generate a smooth representation from a data sequence. We can also combine elementary smoothers by repeatedly applying them to the data sequence until there is no change in the resulting sequence. The smoothers 3R and 5R refer to the repeated application of the running median of length 3 and 5, respectively.

Running medians have one drawback. They can sometimes smooth a sequence of data values *too* much, often removing interesting patterns. When we dissect a series into two parts using a smoother

$$\text{data} = \text{smooth} + \text{rough}$$

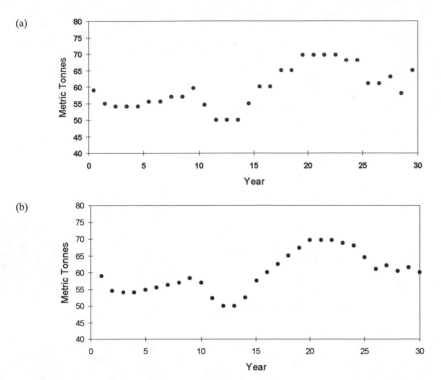

FIGURE 16-9. Coal production (a) smoothed by running medians of four data points and (b) followed by running medians of two data points.

and continually work on the smooth, we lose sight of additional information that may be in the "rough" or the residuals. We can perform a complementary smoothing of the rough and add the result to the original smoothing.

Table 16-11 shows the calculations for the running medians of 5, followed by a calculation of the rough, and then this rough smoothed by 5. When added to the original smoothed sequence, it is common to call such a smoother 5,twice to indicate that the both the original sequence and the rough have been smoothed. The resultant series is illustrated in Figure 16-10. Normally, the same elementary smoother is applied to both sequences, the original data and the rough, but this need not be so.

DEFINITION: COMPOUND SMOOTHER

A compound smoother is a smoother generated by the application of several elementary smoothers to the original sequence and the rough of the original sequence.

Certain combinations of smoothers seem to perform remarkably well; others, only in specific instances. Two combinations that seem to be quite robust in smoothing irregular sequences are the 4253H and 4253H,twice compound smoothers. In this case, we smooth by 4, recenter by 2, smooth by 5 and then 3, and then utilize hanning. If we perform this compound smoother on both the original sequence and the rough, we arrive at the 4253H,

TABLE 16-11
Coal Production Smoothed by 5, Reroughed by 5

Year	Tonnes	Smoothed by 5	Rough	Rough smoothed by 5	5,twice
1	59	59	0	0	59
2	51	51	0	0	51
3	51	57	−6	0	57
4	57	51	6	0	51
5	57	54	3	−3	51
6	46	57	−11	3	60
7	54	57	−3	1	58
8	70	54	16	−3	51
9	60	59	1	1	60
10	50	59	−9	1	60
11	59	50	9	0	50
12	48	50	−2	0	50
13	50	50	0	0	50
14	50	50	0	0	50
15	60	60	0	0	60
16	60	60	0	0	60
17	60	60	0	0	60
18	70	60	10	0	60
19	70	69	1	0	69
20	56	70	−14	1	71
21	69	70	−1	1	71
22	70	69	1	−1	68
23	95	69	26	0	69
24	53	70	−17	1	71
25	66	66	0	0	66
26	70	56	14	−6	50
27	56	66	−10	0	66
28	54	60	−6	0	60
29	70	60	10	0	60
30	60	60	0	0	60

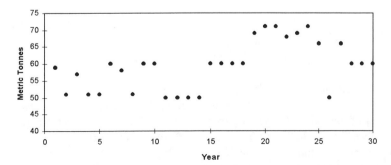

FIGURE 16-10. Coal production smoothed by 5,twice.

TABLE 16-12
Smoothing Coal Production by 4253H

Year	Tonnes	Smoothed by 4 recentered by 2	Smoothed by 5	Smoothed by 3 (ends adjusted)*	Hanning H	Rough
1	59	59.00	59.00	54.50	54.5000	4.5
2	51	54.50	54.50	54.50	54.5000	−3.5
3	51	54.00	54.50	54.50	54.5000	−3.5
4	57	54.00	54.50	54.50	54.5625	2.44
5	57	54.75	54.75	54.75	54.8750	2.13
6	46	55.50	55.50	55.50	55.5000	−9,5
7	54	56.25	56.25	56.25	56.2500	−2.25
8	70	57.00	57.00	57.00	56.8125	13.19
9	60	58.25	57.00	57.00	57.0000	3
10	50	57.00	57.00	57.00	55.8125	−5.81
11	59	52.25	52.25	52.25	53.4375	5.56
12	48	50.00	52.25	52.25	52.2500	−4.25
13	50	50.00	52.25	52.25	52.3125	−2.31
14	50	52.50	52.50	52.50	53.6875	−3.69
15	60	57.50	57.50	57.50	56.8750	3.13
16	60	60.00	60.00	60.00	60.0000	0
17	60	62.50	62.50	62.50	62.5000	−2.5
18	70	65.00	65.00	65.00	64.9375	5.06
19	70	67.25	67.25	67.25	67.2500	2.75
20	56	69.50	69.50	69.50	68.9375	−12.94
21	69	69.50	69.50	69.50	69.5000	−0.5
22	70	69.50	69.50	69.50	69.3125	0.69
23	95	68.75	68.75	68.75	68.7500	26.25
24	53	68.00	68.00	68.00	67.3125	−14.31
25	66	64.50	64.50	64.50	64.7500	1.25
26	70	61.00	62.00	62.00	62.5000	7.5
27	56	62.00	61.50	61.50	61.5000	−5.5
28	54	60.50	61.00	61.00	61.0000	−7
29	70	61.50	60.50	60.50	60.5000	9.5
30	60	60.00	60.00	60.00	60.0000	0

*The endpoints have been smoothed using a procedure outlined in the following section.

twice smoother. The calculations for this smoother are given in Tables 16-12 and 16-13. To see the effects of each step of the smoother, the calculations are shown in sequence. Each column represents the results of applying the smoother named at the top of the column to the preceding column.

The last column of Table 16-12 is the rough or the residual from the first pass of 4235H. Consider the value for year 3 in the column for figures smoothed by 5, 54.50. It is the median of the values for years 1 to 5 in the previous column. The median is 54.50 in the sequence 59.00, 54.50, 54.00, 54.00, and 54.75. The value of 54.5625 in the H column for year 4 is calculated as $\frac{1}{4}(54.50) + \frac{1}{2}(54.50) + \frac{1}{2}(54.75) = 54.5625$. In Table 16-13, the value for year 4 in the final results column is the sum of the H column values in Tables 16-12 and 16-13, $54.5625 - .375 = 54.1875$.

TABLE 16-13
Reroughing of Coal Data of Table 16-12 using 4253H,twice and Final Smoothed Results

Rough*	Smoothed by 4 recentered by 2	Smoothed by 5	Smoothed by 3 (ends adjusted)**	Hanning H	Final smooth
4.5000	4.5	4.5	0.7	0.7	55.2
−3.5000	−0.02	−0.02	−0.02	0.07	54.57
−3.5000	−0.61	−0.38	−0.38	−0.29	54.21
2.4375	−0.69	−0.38	−0.38	−0.38	54.19
2.1250	−0.38	−0.38	−0.38	−0.3	54.58
−9.5000	−0.06	−0.06	−0.06	−0.09	55.41
−2.2500	0.16	0.16	0.16	0.16	56.4
13.1875	0.38	0.38	0.38	0.32	57.13
3.0000	2.33	0.38	0.38	0.38	57.38
−5.8125	1.83	0.38	0.38	−0.21	55.61
5.5625	−1.95	−1.95	−1.95	−1.4	52.04
−4.2500	−3.14	−2.08	−2.08	−2.05	50.2
−2.3125	−3	−2.08	−2.08	−2.08	50.23
−3.6875	−2.08	−2.08	−2.08	−1.86	51.83
3.1250	−1.2	−1.2	−1.2	−1.08	55.79
0.0000	0.16	0.16	0.16	−0.07	59.93
−2.5000	1.47	0.63	0.63	0.51	63.01
5.0625	0.75	0.63	0.63	0.63	65.56
2.7500	0.63	0.63	0.63	0.62	67.87
−12.9375	0.61	0.61	0.61	0.59	69.53
−0.5000	0.09	0.53	0.53	0.55	70.05
0.6875	0.09	0.53	0.53	0.53	69.84
26.2500	0.53	0.53	0.53	0.53	69.28
−14.3125	2.67	0.53	0.53	0.53	67.84
1.2500	1.13	0.53	0.53	0.26	65.01
7.5000	−2.13	−0.56	−0.56	−0.29	62.21
−5.5000	−0.56	−0.56	−0.56	−0.56	60.83
−7.1250	−0.88	−0.56	−0.56	−0.55	60.94
9.5000	1	0	0	−0.14	60.35
0.0000	0	0	0	0	60

*The first column of this table is identical to the last column of Table 16-12. The last column adds the *H* column to the *H* column of Table 16-12.

**The endpoints have been smoothed using a procedure outlined in the following section.

Figure 16-11 compares the two smoothings 4253H and 4253H,twice on the coal production data. The two smoothed series are remarkably similar, though the addition of the smoothed rough in the 4253H,twice series increases the depth of the troughs in the series, especially the trough in the year 10 to year 15 section.

Smoothing the Endpoints

To this point, we have virtually ignored the endpoints of our series. It is impossible to directly include these points in our smoothing operations since they are not surrounded by

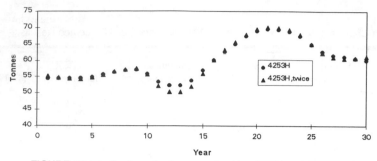

FIGURE 16-11. Coal production smoothed by 4253H and 4253H,twice.

sufficient data points. There are certain obvious substitutions possible. For example, for the second and next to last points in a data series, Y_2 and Y_{n-1}, we can substitute a median of span 3 where a median of span 5 is called for. But what about data points Y_1 and Y_n? Although we can continue to copy their values as the smoothing proceeds, they may become increasingly dissimilar to the smoothed sequence.

The simplest approach is to extrapolate the values for the endpoints from the parts of the series where we have smoothed values. The general approach is shown in Figure 16-12. Let us label our actual data values Y_1, Y_2, . . ., Y_n and our smoothed values Z_1, Z_1, . . ., Z_n. For Y_0, we linearly extrapolate the values of Z_2 and Z_3 "backwards"; and for Y_{n+1}, we linearly extrapolate the values for Z_{n-2} and Z_{n-1} forward. Let us suppose our data are equally spaced with an interval of Δy. At the low end of the series, the line has the slope $(Z_3 - Z_2)/\Delta y$. We are extrapolating a value for Y_0 that is two intervals of Δy beyond Z_2, so the estimated value is

$$\hat{Y}_0 = Z_2 - 2\Delta y(Z_3 - Z_2)/\Delta y$$

$$= 3Z_2 - 2Z_3$$

where the Ys are the smoothed values. An equivalent procedure for the final point yields

$$\hat{Y}_{n+1} = 3Z_{n-1} - 2Z_{n-2}$$

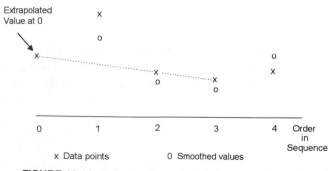

FIGURE 16-12. Adjusting the endpoints in a smoothed series.

Once we have determined values for an extra point beyond both ends of the existing series, we can simply determine the values for Y_1 and Y_n using the median of the extrapolated point, the observed endpoint, and the smoothed point next to the end:

$$Z_1 = \text{Median}(\hat{Y}_0, Y_1, Z_2)$$

$$Z_n = \text{Median}(\hat{Y}_{n+1}, Y_n, Z_{n-1})$$

This adjustment is usually only made once in a compound smoother, usually towards the end of the procedure.

As an example of this adjustment, consider the adjustment for year 30 in the fifth column (smoothed by 3) of Table 16-12. To extrapolate a value for Y_{31}, we begin with the smoothed values for $Z_{29} = 60.50$ and $Z_{28} = 61.00$: $Y_{31} = 3\ (60.5) - 2\ (61.00) = 59.5$. To determine the endpoint, we then choose the median among 59.5, 60.00, and 60.5. The resulting endpoint is therefore 60.00.

16.6. Summary

This chapter has presented a number of techniques used in exploratory data analysis. These techniques differ from their counterparts in conventional statistical analysis in a number of ways. Many were developed so that an analyst with a pencil and a piece of paper could examine a give batch of data. Unlike conventional descriptive statistics, the analysis often requires judgment on the part of the analyst. One important feature of a data set is an outlier or stray. In EDA, the treatment of extraordinary values is generally different than in conventional techniques. First, if we can identify them as strays, they may tell us something about the process giving rise to the data itself. Why did this value arise? Second, an attempt is made in EDA to derive techniques of analysis that resist the influence of these outliers on the summary value or function. For example, the location of resistant lines is generally less affected by a single outlier than is a regression line. Finally, EDA is very pragmatic. Seldom are we interested in the degree of fit between a data set and some theoretically derived distribution such as the normal. There is no attempt to come up with a single correct answer. The data itself and the judgment of the analyst also play critical roles in EDA. Because of this, the impact of the computer on EDA is less significant. While certain statistical packages are capable of performing simple EDA displays, the algorithms used to generate them ignore the source of the data and the judgment of the analyst. Where inference is not an overriding goal, the techniques of EDA often prove to be a useful alternative to conventional descriptive statistics in understanding a data set.

REFERENCE

John W. Tukey, *Exploratory Data Analysis* (Reading, Mass.: Addison-Wesley, 1977).

FURTHER READING

The original volume by Tukey is an excellent source for all of the techniques described in this chapter. Students reading 10 to 15 pages of this text cannot help but notice the dramatically different approach characteristic of EDA. A more conventional textbook presentation of this material is available in Vellemen and Hoaglin (1981).

P. F. Vellemen and D. C. Hoaglin, *Applications, Basics and Computing of Exploratory Data Analysis* (Boston: Duxbury, 1981).

PROBLEMS

1. Explain the meaning of the following terms:

   a. Exploratory data analysis
   b. Batch
   c. Stem-and-leaf display
   f. Strays
   e. Letter value display
   f. Depth
   g. Hinges, quarters
   h. Reroughing

   i. Resistant lines
   j. Half slopes
   k. Midsummaries, midextremes
   l. Five-number summary
   m. Skeletal boxplot
   n. Inner and outer fences
   o. Smooth and rough
   p. Elementary and compound smoothers

2. Why are endpoint adjustments made in smoothing procedures?

3. What are the principal differences between the smoothing operators defined in Chapter 3 for conventional time series analysis and those of EDA?

4. The presence of arsenic, chromium, lead, or silver in concentrations greater than .05 mg/L is usually sufficient to cause rejection of a water supply for use as drinking water. Consider the following sets of measurements, in milligrams per liter, taken in the last year at some location:

Month	Arsenic	Chromium	Lead	Silver
Jan	.08	.07	.02	.04
Feb	.07	.07	.01	.01
Mar	.09	.08	.02	.01
Apri	.05	.05	.05	.05
May	.07	.06	.06	.05
Jun	.02	.01	.14	.18
Jul	.02	.03	.13	.13
Aug	.02	.02	.13	.14
Sep	.01	.02	.10	.11
Oct	.02	.01	.02	.01
Nov	.01	.02	.01	.01
Dec	.06	.06	.02	.02

    a. Examine the data to see whether there are any patterns in these concentrations by element and time period.

    b. Design a stem-and-leaf display that emphasizes the patterns you have uncovered in Part a. Can you think of a way of portraying all the data in one display?

    c. *Experiment* with alternative formats for the display.

    d. Summarize the data using boxplots and compare the concentration levels by element.

5. The monthly rainfall in millimeters during March in Chicken Finger, Arkansas for the last 40 years is as follows:

0.1	2.2	0.2	20.9	0.7	6.2	5.3	8.0	31.2	10.7
12.7	24.6	38.1	6.1	3.2	23,2	14.8	29.1	1.7	12.4
12.4	12.6	128.0	69.6	17.6	14.1	3.8	5.8	0.3	8.6
5.3	6.6	42.9	13.5	24.5	2.7	6.7	14.2	29.3	13.1

    a. Compare two stem-and-leaf displays for this data, one using the tenths' digit as the display digit and one in which the tenths' digit is ignored. Which is better?

    b. Improve each display using the convention for displays.

    c. Generate a boxplot of the data and develop a five-number summary

6. Consider the burglary data of Problem 5 in Chapter 2.

    a. Generate a boxplot of this data

    b. Generate a five-number summary of this data.

    c. What can you say about the variability of burglaries on the basis of this analysis? Is it about the same as if you were to use simple descriptive statistics?

7. Repeat Problem 6 for the corn yield data of Problem 6 in Chapter 2.

8. Consider the time series data of Problem 8 in Chapter 3. Use a number of elementary and compound smoothers to generate smoothed sequences of this data. Compare your smoothers to the smoothed sequences obtained using conventional time series methods.

9. Consider the sound decibel data from Problem 5 in Chapter 13. Develop a resistant line for this data, and compare it to the regression equation.

Spreadsheet Exercises

10. Enter the data for coal production given in Table 16-10 into the first column of a spreadsheet. In successive columns utilize formulas to generate the smooth sequences for the following:

    a. Running medians of length 3, 5, and 7

    b. 3R, 5R, and 7R

    c. 4253H and 4253H, twice

Advanced Spreadsheet Exercises

11. Generate a series of graphs that compares the results of the smoothing using the various smoothers generated in Problem 10 above. Superimpose a few on the original data to examine the impact of the smoother as it proceeds. For example, portray the original data, 4253H and 4253H,twice on a single graph. These graphs are best compared on a full-landscape page presentation.

17

Computer-Intensive Methods

Preceding chapters have introduced a body of techniques that we will call *traditional*, meaning that they have enjoyed widespread use for many years. Though their utility has been proved time and again, several general shortcomings are apparent. First, many of the most popular methods are parametric and require strong assumptions about the probability distribution underlying a random variable. For example, to set up a confidence interval for the correlation coefficient ρ, it is necessary to assume that the random variables X and Y follow a bivariate normal distribution. As is usually the case, this assumption is made for mathematical convenience, not because bivariate normality is especially common. By assuming bivariate normality, one can derive the sampling distribution of r and thereby assign probabilities. However, when the underlying variables are not bivariate normal, the sampling distribution of r can be very different, with the result that the standard confidence intervals are inappropriate. A second shortcoming, shared by both parametric and nonparametric methods, is the relatively small number of statistical questions they consider. Though the list of techniques covered thus far might seem lengthy, the issues they address are in fact limited to a rather narrow set of problems, and obvious questions can be posed for which no method is readily available. For example, it is easy to set up confidence intervals for the difference of two means, $\mu_1 - \mu_2$, but not for the product of two means, $\mu_1 \mu_2$. A final shortcoming is that traditional methods tend to require that one follow a prescribed methodology and do not admit much in the way of improvisation. For example, we can test parameters in a regression model, providing they have been estimated using the principle of least squares. If one prefers another fitting procedure, say, minimum absolute error or perhaps minimum maximum error, the classical hypothesis-testing method is unavailable.

This chapter considers several methods that attempt to overcome the deficiencies cited above. Though varied in purpose, the techniques covered are alike in their reliance on considerably more computing power than traditional methods, hence the term *computer intensive*. Generally speaking, computer-intensive methods replace knowledge (or assumptions) about random variables with sheer volume of calculation. We emphasize, however, that although the amount of calculation may be enormous compared to traditional methods, the burden is not excessive by present computing standards. All of the methods we describe can easily be performed on personal computers, given datasets of moderate size (hundreds to thousands of cases).

Not surprisingly, computer-intensive methods are comparatively new, a fact that complicates their use in at least two important ways. First, they are not routinely available in statistical software packages. We are therefore in the ironic position of finding little software available for methods that are only feasible using a computer! Fortunately, because they are surprisingly simple in principle, only a little programming is needed to implement most methods. Another problem deserving mention is that less is known about these techniques, both from a theoretical standpoint and in applied terms. Not only do some statisticians view these methods as unproven, but also there is not yet the long history of use that typifies traditional methods. Still, because they can be a compelling alternative when traditional methods are clearly inadequate, computer-intensive methods are growing rapidly in both use and acceptance.

17.1. Estimating Probability Density—Kernel Methods

This section considers the problem of obtaining probability density estimates from a sample. We assume that a random sample $(x_1, x_2, \ldots, x_n)$ has been obtained from an unknown continuous probability distribution $f(x)$. The goal is to construct an estimate of this probability distribution, which we denote $\hat{f}(x)$. There are numerous uses for such estimates. For example, we may be interested in knowing something about the shape of $f(x)$—e.g., is it reasonably close to normal, or perhaps bimodal? How different is it from another distribution? Such uses might be satisfied by just a graph of $\hat{f}(x)$. Alternatively, $\hat{f}(x)$ might be used to estimate probabilities. For example, What proportion of the population is below the poverty level, or what are the chances of more than 10 cm of precipitation falling in a month? Such uses call for numerical operations on $\hat{f}(x)$. The operations might be as simple as finding the area between two values of X, or could be very complicated. Regardless of whether the use is graphical or numerical, what is needed is a way of estimating f for any value of X.

Consider, for example, Figure 17-1, which shows tropical cyclone activity in the Atlantic area for the period from 1950 to 1990. A value near 100% is a "normal" year, whereas lower and higher values correspond to years with abnormally light or heavy cyclone activity. With these data in hand, we might want to estimate the probability of a year with less than 50% of normal activity, or more than 200% of normal, or some other probability.

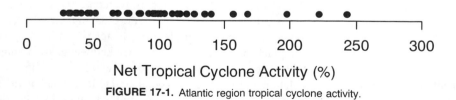

FIGURE 17-1. Atlantic region tropical cyclone activity.

Traditional Approaches

There are, of course, traditional methods that might employed. For example, one could proceed parametrically by assuming X is normally distributed and use the sample to estimate its mean and standard deviation. Because these are the only parameters of $f(x)$, this is tantamount to estimating $f(x)$ itself. If we use $\bar{x}$ and s as estimates, we get

$$\hat{f}(x) = \frac{1}{s\sqrt{2\pi}}\exp\left[-\frac{1}{2}\left(\frac{x - \bar{x}}{s}\right)^2\right] \tag{17-1}$$

One could evaluate this function to produce a graph or integrate the function to assign probabilities. (Tables of the standard normal could be used for the latter.) Though very easy to apply, the worry with this approach is that if $f(x)$ is far from normal, the graph and associated probabilities would be misleading. Using the cyclone data as an example, the normal curve gives the probability of an extremely heavy year, $P(X > 200)$, of .02. With 3 of the 41 values near or above 200, the normal estimate seems too low.

Another option is to proceed nonparametrically by using the sample to construct a relative frequency histogram as in Figure 2-1. If each histogram value is divided by the interval width, we get a probability density estimate, one for each interval. Though no assumptions have been made about the form of $f(x)$, we do have rather arbitrary decisions about the number and position of class intervals. Depending on the data, small changes can have large effects, both in visual appearance and computed probabilities. As mentioned in Chapter 2, another disadvantage is the roughness of the histogram, with the result that $\hat{f}(x)$ changes abruptly at the class limits.

Kernel Methods

Kernel estimation is a simple approach that avoids the difficulties mentioned above. The method provides smooth estimates without requiring explicit assumptions about the form of $f(x)$. Recall that in the histogram approach, each observation is treated as a point. If the point lies in a probability interval, the entire mass of the point, $1/n$, is added to the probability. Kernel methods, on the other hand, spread the mass of each observation around the observed value. The amount of spread or smearing is governed by a function called the kernel. Normally, the kernel function will be symmetric, so that the point is smeared equally toward higher and lower values. The width of the kernel is adjustable—narrow kernels concentrate mass tightly around central (observed) values, whereas a wide kernel gives more smearing. To estimate probability density, we add the kernel values for all the data points. This operation is shown in Figure 17-2, where the dataset consists of just three points, located at $X = 3$, 4, and 6. Centered over each data point is its kernel function, in this case a standard normal curve. That is, we have placed a Z distribution over each data point; hence, we are using a kernel width of unity. The top curve is the density estimate, obtained by summing the three kernels for 100 different X values. Notice that at intermediate values of X all three kernels contribute substantially to $\hat{f}(x)$, but the tails are determined by the two extreme points. To express the operation mathematically, let $K(\bullet)$

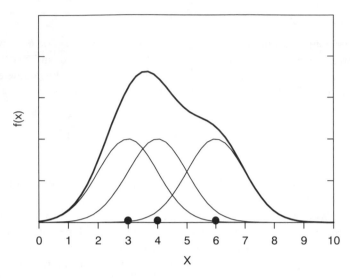

FIGURE 17-2. Kernel probability density.

represent the kernel function and h the degree of smearing, or bandwidth. With this notation, the kernel density estimate is simply

$$\hat{f}(x) = \frac{1}{nh} \sum_{i=1}^{n} K\left(\frac{x - x_i}{h}\right) \tag{17-2}$$

Looking first at the argument to K, we see that the distance between a point x and each data value is expressed as a multiple of h. Because h has the same units as x, the ratio is dimensionless. This scaled distance is used in the kernel function to find the contribution of each observation. With K a decreasing function of distance, the contribution of a point far from x will be small, whereas nearby points will contribute more. The sum of all contributions divided by nh becomes the density estimate. Division by h is required to make $\hat{f}(x)$ a density, with dimensions of probability per unit X.

Several general points can be made about this approach. First, $\hat{f}(x)$ inherits many of its properties from the kernel function. In particular, if K is a probability function, $\hat{f}(x)$ will likewise be a probability function (nonnegative everywhere, with total area equal to unity). Similarly, if K is smooth, $\hat{f}(x)$ will also be smooth. A second general point is that the amount of detail present in $\hat{f}(x)$ varies directly with the bandwidth h. A large value leads to much smearing of each data point, which in turn results in a broad, smooth density function without much fine structure. This behavior is seen in Figure 17-3, which shows kernel estimates for the cyclone data.

Using the wider kernel ($h = 20\%$) results in a very smooth curve without the density minimum near 60% activity. Notice that both curves reflect the positive skew in the data, illustrating the general principle that a symmetric kernel does not force symmetry in $\hat{f}(x)$. In other words, symmetry is one attribute *not* inherited from the kernel. In this example the presence of skewness confirms our suspicion regarding the inadequacy of the normal curve.

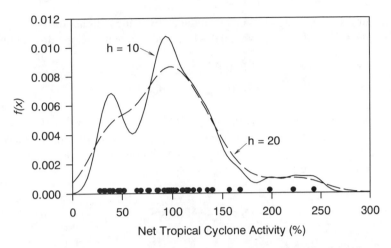

FIGURE 17-3. Kernel probability density estimated using cyclone data.

Using either curve, we get a value of .06 for $P(X > 200)$, about three times larger than the normal estimate. As a final general point about the kernel method, notice that each density estimate requires n evaluations of K. Given that a large number of estimates might be needed to produce a graph or calculate a probability, this method is obviously not suited to hand calculation.

Choice of Kernel and Bandwidth

Having described the method in general terms, several practical matters need to be discussed, foremost of which is the choice of kernel function. There are no hard rules about this, though generally one will choose a symmetric probability distribution. Within the family of symmetric kernels, selection will often be made by balancing computational demands against smoothness. We present three alternatives below, each of which occupies a different point along the tradeoff continuum. Perhaps the simplest choice is a triangular kernel given by

$$K(z) = \begin{cases} 1 - |z|, & |z| \le 1 \\ 0 & , |z| > 1 \end{cases} \qquad (17\text{-}3)$$

where

$$z = \left(\frac{x - \bar{x}}{h} \right)$$

This kernel is just a triangle one unit high and two units wide, and therefore has an area of unity. Its chief advantage is ease of evaluation; only division and subtraction are required, and there is no need to consider data points more than h units away from any x. On the other hand, it does not provide as smooth an estimate as do other kernels. (Though the kernel is continuous, it has breaks in slope at the center and edges. These carry over to slope discontinuities in $\hat{f}$). Roughness can be reduced by using a higher degree function in z. For example, a second degree polynomial will avoid the central slope discontinuity:

$$K(z) = \begin{cases} \frac{3}{4}(1 - z^2), & |z| \leq 1 \\ 0 & , |z| > 1 \end{cases} \tag{17-4}$$

Now K varies smoothly across its midpoint; hence $\hat{f}$ is smoother. One again, K becomes zero a finite distance away from the observations; thus, only nearby observations need be considered. There are, however, slope discontinuities at the edges of this kernel. Still smoother estimates are produced using the normal kernel:

$$K(z) = \frac{1}{\sqrt{2\pi}} \exp\left(-\frac{1}{2}z^2\right) \tag{17-5}$$

This yields a very smooth density function, but at the cost of more arithmetic. There is the added burden of an exponential, and K is technically nonzero everywhere. As a practical matter, however, because the normal curve approaches zero so quickly, an observation more than a few multiples of h distant cannot contribute much to $\hat{f}$. The calculations can therefore be restricted to nearby points if necessary.

Having selected a kernel function, there remains the question of bandwidth, h. The general principles are easy to state: too large a value prevents small-scale variations in f from emerging in $\hat{f}$, whereas too small a value gives estimates that are too variable. For example, looking at the cyclone data, it may be that the local minimum in $\hat{f}$ is fictitious and would not appear in a larger sample. If so, the lower value of h is too small, because it invents a signal where none exists. In that case the 20% bandwidth would be preferred. On the other hand, an alternative interpretation is that the minimum is real, and the larger bandwidth obscures an important feature of cyclone frequencies! The lesson learned from this example is that one should experiment with a range of bandwidths, comparing the resulting density estimates with one another. Results that turn out to be sensitive to h should be reported only with qualification.

This said, how does one choose reasonable values of h for experimentation? Intuition suggests that if the data are highly dispersed in X, a large bandwidth is called for. Intuition likewise suggests that because large samples contain more information, bandwidth should decrease with increasing n, everything else being equal. It so happens that theory and experience confirm these notions. As a first step in choosing the optimum h, Silverman (p. 48) recommends first calculating the sample standard deviation s and the interquartile range R. His suggested bandwidth is then the smaller of the following two values:

$$h_1 = 0.9\,s/n^{1/5} \tag{17-6a}$$

$$h_2 = 0.67\,R/n^{1/5} \tag{17-6b}$$

Notice how much more important dispersion is than sample size. If we think of h as a measure of how much information can be extracted from the sample, we see that a very large n is needed to compensate for widely spaced data points. In the case of the cyclone data, h_1 and h_2 are 21% and 18%.

The equations above provide an easy way to choose h for any dataset and give values that work well for a wide range of density functions. However, they depend only on bulk sample statistics (s, R), not on the details of how the points are arranged. At the expense of more calculations, two refinements are possible that make greater use of sample information. First, it is possible to optimize the choice of bandwidth by selecting a value that minimizes predictive error in $\hat{f}$. The technique is called cross-validation and uses the data themselves to identify the optimum h. The basic idea is that a good choice for h will give a good estimate for $\hat{f}$. Cross-validation uses the data in hand to evaluate how good is a particular h. The value that gives lowest estimated number of errors in $\hat{f}$ is selected. Cross-validation can be used in many other contexts as well and will be described in a later section of this chapter.

A second refinement is to employ a so-called *adaptive* kernel. In this case the bandwidth is allowed to vary with X; small values are used where the data are closely packed, and large values are used where observations are sparse. In this way the kernel adapts to local variations in density, providing high resolution where possible without creating spurious features in low-density regions of X. The approach requires two passes through the data, the first of which uses a fixed kernel, giving preliminary density estimates. These pilot estimates are then used to determine a unique bandwidth for each observation. Though considerably more work, adaptive estimates are especially useful if there are large density variations in the data. Details can be found in the Silverman monograph.

Kernel Methods for Bivariate Data

Thus far, we have been using kernel methods to estimate a function $f(X)$, where the observations lie on the number line. The domain of f is obviously one-dimensional— location within the domain is determined by a single coordinate value. Here we consider generalizing the kernel method for use on bivariate data. The domain is now two-dimensional, and we are after a bivariate function $f(X, Y)$ in which the data are points lying in a plane.

Instead of considering probability density functions, we will first describe how the method works for estimating point density. That is, we are given a set of points in space, and we want to know how the density of points varies throughout the study area. In Chapter 3 we mentioned that Thiessen polygons can be used, but are not very satisfactory as they give density estimates only at the data points. Kernel methods are attractive because they can provide estimates at any location. In addition, if a smooth kernel is used, the resulting density surface will be smooth. Let d_i be the distance between data point i and some location (x, y):

$$d_i = \sqrt{(x - x_i)^2 + (y - y_i)^2}$$

As before, we need to choose a kernel width h and use distances scaled by h in the kernel. For example, the normal kernel is

$$K\left(\frac{d_i}{h}\right) = \frac{1}{2\pi} \exp\left[-\tfrac{1}{2}\left(\frac{d_i}{h}\right)^2\right]$$

With this notation, the density estimate is simply

$$\hat{g}(x, y) = \frac{1}{h^2} \sum_{i=1}^{n} K\left(\frac{d_i}{h}\right)$$

Notice how we divide by h^2 in the equation above. Because h has units of length (miles, kilometers, etc.), the density estimate will have units of points per unit area (e.g., points per square mile, points per square kilometer). The equation above can be evaluated for any x, y location; thus, the method provides a continuous density surface.

EXAMPLE 17-1. Figure 17-4(a) shows a sample of 200 points within a 100 km^2 area. Visual inspection reveals that the northeast is sparsely sampled, but that the southwest has relatively high point density. To supplement this impression, we seek numerical information about the variations in sampling density.

Though we could divide the region into cells and count points within each cell, to do so would involve all the disadvantages associated with histograms. We prefer a method that gives a smooth density estimate, and thus will employ the bivariate kernel method. We choose the bandwith h, to be 1 km, or 10% of the range of the x and y coordinates. As mentioned above, this will constrain the detail present in the resulting density surface. Given the rather large data gaps in the northeastern part of the study area, a smaller value does not seem justfied. Kernel functions are placed over all 200 points and summed at a large number of interior locations. When the resulting values are contoured, Figure 17-4(b) is the result.

The contour map clearly shows the density maximum in the southwest, along with a secondary maximum to the north of the primary maximum. If we had to rely solely on the point data [Figure 17-4(a)], we would probably miss the secondary maximum. Note that with 200 points in a 100-km^2 area, the mean density is 2 points/km^2. Looking at the map, we see that below-average densities are confined to the eastern and extreme northern portions of the study area. Notice that density appears to decrease rapidly near the southern and western edges of the map. This is undoubtedly an edge effect, produced by not sampling outside the map boundaries. Had we done so, those points would contribute to the density estimate within the boundaries, so that estimates near the edges would be larger. Because the normal curve falls off rapidly, there is very little contribution from points farther away than $2h$. In other words, we might have wished for points in a band of width $2h$ surrounding the study area, but no more than that. By the same logic, edge effects extend no more than $2h$ units into the study area. (We do not see an edge effect along the eastern boundary, because densities are so low that missing points have almost nothing to contribute.) As a final point, we remark that the 0.5 contour has small "wiggles" here and there, smaller even than the bandwidth of .1 km. These wiggles are artifacts of the method, an indication that h is too small for these low-density regions. We could use a larger bandwidth, but would lose detail in the high-density areas. A better choice is an adaptive kernel method, which automatically adjusts to variations in density, as described earlier.

We have used g rather than f above to indicate that the equation provides point density, not probability density. The former is appropriate if we are after information about spatial variations in the density of observations. For example, we might be studying earthquake epicenters, and want to map epicenter density. In that case we would very likely

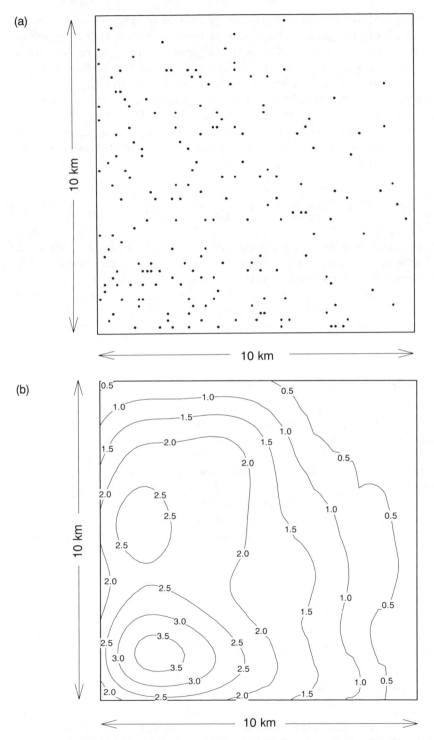

FIGURE 17-4. (a) Sample of 200 points. (b) Density contours of the data in (a).

contour the density function g. As another example, perhaps we are mapping some climatic variable on the basis of values observed at a network of weather stations. The density of stations plays a key role in determining the resolving power of the network; hence, it's likely that we will need a map of station density in order to know how well the network is going to perform. Once again the kernel method can provide an estimate.

A related problem is that of estimating a bivariate probability density function. The domain is still two-dimensional, but this time x and y are not spatial variables. The function we seek, $\hat{f}(x, y)$, is a surface that gives probability density. As discussed in Section 5.5, the volume under the surface is a probability. To convert point density to probability density, we need only divide by the number of data points:

$$\hat{f}(x, y) = \frac{1}{n}\, \hat{g}(x, y) = \frac{1}{nh^2} = \sum_{i=1}^{n} K\left(\frac{d_i}{h}\right)$$

So long as the kernel function is a probability distribution, the estimate $\hat{f}$ will be a probability distribution. If x and y have different units, it will be necessary to standardize them in order for the distance computation to have meaning. A natural choice is to express both as z-scores when computing d. Doing so lets h be interpreted as the width of the kernel in standard deviation units, rather than in raw units like miles or kilometers. In all other regards the bivariate method is a direct extension of the one-dimensional technique.

17.2. Estimating Sampling Error—The Jackknife

Suppose a sample has been used to estimate some characteristic of a population, such as μ, or ρ, or the probability of X exceeding some critical value. As we have seen, hardly ever is the estimate sufficient by itself—it is almost always necessary to provide additional information about its accuracy. This requires that we have some idea about how the sample statistic is distributed, or equivalently, how it varies from sample to sample. The traditional methods covered previously rely on statistical theory to arrive at sampling variability. For example, one can prove that if the standard deviation of X is σ, then $\bar{X}$ has a standard deviation (or standard error) equal to $\sigma/\sqrt{n}$ under random sampling. The standard error of $\bar{X}$ is an uncertainty measure; large values of $\sigma_{\bar{x}}$ imply large doubt about the true value of μ. In actual practice, of course, we are forced to estimate $\sigma_{\bar{x}}$ using sample information. Chapter 7 presented the traditional estimate, $s/\sqrt{n}$, as a good choice.

This section considers a nonparametric method for arriving at such error measures. Called the jackknife, it works when there is no adequate theory that might be used, or when the assumptions required to apply the theory cannot be justified. The jackknife can be used for a variety of uncertainty measures, but we will focus on its use in estimating sampling error as reflected in the variance and/or standard deviation of a statistic.

Jackknife Estimates of Standard Error

The jackknife is most easily understood as a generalization of the traditional method used for the standard error of $\bar{X}$. Recall that the standard error is the square root of the variance of $\bar{X}$. It measures the expected departure between the sample mean and population mean:

$$\sigma_{\bar{x}} = \sqrt{E(\bar{X} - \mu)^2} \tag{17-7}$$

That is, the standard error is the square root of the average squared difference between the estimate and the true value. This is also called the root-mean-square error. We can think of it as the average error, but only in the sense of averaging squared errors. (The average *unsquared* error is bias.) Thus far we have used $s/\sqrt{n}$ as an estimate of $\sigma_{\bar{x}}$:

$$\hat{\sigma}_{\bar{x}} = \frac{s}{\sqrt{n}} = \sqrt{\frac{1}{n(n-1)} \sum_{i=1}^{n}(x_i - \bar{X})^2} \tag{17-8}$$

This equation uses variability in the x_i to arrive at the accuracy of $\bar{X}$ as an estimate of μ. The jackknife takes the more direct approach of using variability in sample means rather than variability in sample observations. To derive the jackknife, we need to rewrite Equation (17-8) in terms of averages taken over subsamples of the original n observations. Imagine n subsamples, each containing $n - 1$ different observations. The subsamples might be arranged as follows:

Subsample No.	Subsample contents	Subsample mean
1	$x_2, x_3, x_4, \ldots, x_n$	$\bar{X}_{(1)}$
2	$x_1, x_3, x_4, \ldots, x_n$	$\bar{X}_{(2)}$
3	$x_1, x_2, x_4, \ldots, x_n$	$\bar{X}_{(3)}$
.	.	.
.	.	.
.	.	.
n	$x_1, x_2, x_3, \ldots, x_n$	$\bar{X}_{(n)}$

Notice that x_1 is missing from the first subsample, x_2 is missing from the second subsample, and so forth. Each subsample has a mean, denoted $\bar{X}_{(1)}$, to indicate that the ith observation is missing (column 3 above). The average of all the means, $\Sigma \bar{X}_{(i)}/n$, is written $\bar{X}_{(\bullet)}$. With this notation and a little algebra, Equation (17-8) can be written

$$\hat{\sigma}_{\bar{x}} = \frac{s}{\sqrt{n}} = \sqrt{\frac{(n-1)}{n} \sum_{i=1}^{n}(\bar{X}_{(i)} - \bar{X}_{(\bullet)})^2} \tag{17-9}$$

We see that the variability of the subsample means around the overall mean is used to arrive at uncertainty in the estimate of μ. Of course, inasmuch as Equations (17-8) and (17-9) give exactly the same value, there is no advantage to the latter, assuming we are interested in the accuracy of $\bar{X}$. However, because it generalizes to other statistics, Equation (17-9) is extremely useful if we want to estimate the standard error of something else. Suppose we are using a sample value $\hat{\theta}$ as an estimate of some population parameter θ. Replacing $\bar{X}$ with $\hat{\theta}$ in Equation (17-9) gives the jackknife estimate of standard error:

$$\hat{\sigma}_J = \sqrt{\frac{(n-1)}{n} \sum_{i=1}^{n}(\hat{\theta}_{(i)} - \bar{\theta}_{(\bullet)})^2} \tag{17-10}$$

This is a complicated expression involving a large number of operations. If we break Equation (17-10) down into its component parts, the jackknife procedure is as follows:

1. Delete a single observation from the sample and compute $\hat{\theta}$ from the remaining $n - 1$ observations. Let this value be $\hat{\theta}_{(i)}$, meaning that θ has been estimated with the ith observation omitted. Repeating for all observations, this gives a total of n distinct $\hat{\theta}_{(i)}$.
2. Average the values from Step 1 to get the jackknife mean

$$\bar{\theta}_{(\bullet)} = \frac{1}{n} \sum_{i=1}^{n} \hat{\theta}_{(i)}$$

3. Calculate the jackknife standard error from

$$\hat{\sigma}_J = \sqrt{\frac{n-1}{n} \sum_{i=1}^{n} (\hat{\theta}_{(i)} - \bar{\theta}_{(\bullet)})^2}$$

We can think of the jackknife as working with multiple "samples" created from the original observations. Of course, each subsample gives rise to a different estimate, and it is precisely this variability that the jackknife uses to determine error in $\hat{\theta}$. If the subsamples give widely diverging estimates, our uncertainty estimate will be large. On the other hand, if all the subsamples give very similar values, we would conclude that $\hat{\theta}$ is likely to be near the true value.

DEFINITION: JACKKNIFE

The jackknife is a resampling procedure in which a sample of size n is used to form n subsamples, each containing $n - 1$ of the original observations. (A different observation is left out of each subsample.) Variation from subsample to subsample is used to estimate the standard error (or variance) of a statistic.

EXAMPLE 17-2. Using the cyclone data as an example, suppose we are interested in the probability of a year being within 50% of normal, i.e., $P(50 < X < 150)$. The kernel method gives .68 as the best guess. At this point we want some idea about how close .68 is to the true probability. Though easy to phrase, this question involves difficult theoretical issues and requires that we have detailed information about $f(x)$, the very function we are estimating! As a result, there is no formula like $s/\sqrt{n}$ to which we can turn. The jackknife provides a simple, but computationally demanding alternative. As shown in Table 17-1, we compute 41 separate estimates of $P(50 < X < 150)$. The variability of these around the mean gives .054 as the jackknife error estimate. We conclude that, on average, the kernel probability will be within 5.4% of the true probability.

Table 17-1 follows the equation in Step 3 above and shows clearly how differences in the $\hat{\theta}_{(i)}$ contribute to $\hat{\sigma}_J$. This is not, however, the most convenient form for calculations. As was the case for the sample standard deviation, it is easier (and computationally more

TABLE 17-1
Calculation of the Jackknife Standard Error

i	x_i	$\hat{\theta}_{(i)}$	$\hat{\theta}_{(i)} - \bar{\theta}_{(\bullet)}$	$(\hat{\theta}_{(i)} - \bar{\theta}_{(\bullet)})^2$
1	28	.694	.014	.000196
2	198	.697	.017	.000289
3	95	.673	−.008	.000049
⋮	⋮	⋮	⋮	⋮
39	65	.678	−.002	.000004
40	104	.673	−.008	.000049
41	38	.691	.010	.000121
		27.899		.002983

$$\bar{\theta}_{(\bullet)} = \frac{27.899}{41} = .680$$

$$\hat{\sigma}_J = \sqrt{40/41(.002983)} = .054$$

accurate) to compute without subtracting the mean from each jackknife. After a little algebra, the equation becomes

$$\hat{\sigma}_J = \sqrt{\frac{(n-1)}{n}\left(\sum_{i=1}^{n}\hat{\theta}_{(i)}^2 - n\bar{\theta}_{(\bullet)}^2\right)} \qquad (17\text{-}11)$$

Discussion

Because the jackknife procedure does not depend on the functional form of $\hat{\theta}$, it can be used to estimate sampling variance for many statistics, including correlation coefficients, regression coefficients, trimmed means, and even mathematical functions of the mean. As an illustration of the functional problem, consider an example from fluvial geomorphology. It can be shown that total flow resistance in a stream varies with the inverse square of mean velocity, or $1/\mu^2$. If the sample mean is transformed to get an estimate of resistance, we need some way of estimating the expected difference between $1/\bar{X}^2$ and $1/\mu^2$. Using $\hat{\theta} = 1/\bar{X}^2$, the jackknife procedure applies directly. Other operations involving logarithms, exponentials, sine functions, etc., can be handled similarly. It should be clear by now that the jackknife is not limited to statistics or functions of statistics. By this, we mean θ need not be a parameter of a probability distribution or a function of a parameter. The requirement is only that $\hat{\theta}$ be a well-defined function of the sample values. Though $\hat{\theta}$ might be a parameter estimate, it might also be something much more complicated, as we saw in the cyclone example. It is this flexibility that allows the jackknife to be used in so many cases where traditional methods fail.

To say that almost anything *can* be jackknifed is not to say everything *should* be jackknifed. There are, in fact, several limitations that deserve mention. First, the theory behind the jackknife calls for $\hat{\theta}$ to be symmetric in the observations. That is, the value of

$\hat{\theta}$ must not depend on the order of the observations in the sample. (All of the measures we have described meet this requirement.) Secondly, if a function of the mean is to be jackknifed, the function must have a continuous derivative. Smooth functions like those cited above are appropriate candidates. Thirdly, because they always give unbiased estimates of variance, linear estimators are preferred for use with the jackknife. (A linear estimator is one that can be written as a linear function of the observations, like $\bar{X}$.) Nonlinear estimators can be used, but caution is needed in some cases. Several classes of nonlinear estimators can be shown to have nonnegative bias, meaning that jackknife errors are possibly biased high, but not low. In these cases the jackknife value can be considered conservative, tending to overstate expected error. For still other highly nonlinear statistics, the jackknife can perform poorly. A good example is the sample median, which should not be jackknifed. (The bootstrap, described later, works much better.)

When jackknifing complicated quantities, it is a good idea to monitor the individual terms of Equation (17-9), as shown in the rightmost column of Table 17-1. The danger is that for some measures, the sum will be dominated by a single term, and the jackknife estimate will not be robust. Another sample, which happens not to include the unusual observation, would give a much different $\hat{\sigma}_J$. This can happen even when the estimate $\hat{\theta}$ is itself relatively insensitive to the outlier. Of course, as we saw with regression analysis, this problem also arises in traditional statistics, and we do not mean to imply that the jackknife is unusually susceptible. Our point, rather, is that while this method has great flexibility and is free from the usual parametric restrictions, it should not be used blindly.

17.3. Estimating Model Error—Cross-Validation

In the chapters on regression analysis, we considered the problem of predicting a dependent variable from one or more independent variables. In the case of both linear and nonlinear models, we used the principle of least squares to arrive at estimates of model coefficients. Our procedure was to define a measure of model error (the sum of squared residuals) and choose coefficients that give the smallest possible error. Recall that as by-product of the fitting procedure, the error sum of squares can be used to estimate predictive error. The estimates differ depending on whether one is predicting the mean Y at some X, or a single Y at some X, but both are based on the scatter around the regression line. Despite its many advantages, the traditional approach can be criticized for the way it uses the same observations for both fitting the model and assessing its predictive power. We may wonder, for example, whether a model with high r^2 would do as well predicting completely new values as it does "predicting" the values used in the fitting procedure. Our worry is that the model might have adjusted itself to the quirks of the' data in hand, with the result that predictive error is underestimated. If we could check the model on another independent dataset, we would be more confident in our assessment of its predictive ability.

Cross-validation offers a simple, nonparametric way of making this type of check on a model. To understand the underlying idea, imagine that we divide our dataset into two distinct parts. One portion is used for fitting, and the other for calculating prediction error. So long as there is no overlap in the two portions, we get an independent test of the model's ability to predict new values. The question, of course, is how to split the dataset? Using the majority of cases for fitting gives better parameter estimates (coefficients), but we are less sure of the error statistics.

On the other hand, if most of the sample is held back for validation, we might get a precise estimate of predictive ability, but the fitting procedure will be less reliable. Cross-validation solves this dilemma by holding back a single observation and fitting the model with the remaining $n - 1$ observations. The model is then used to predict the missing observation, and the resulting error is noted. Next, a different observation is held back, and a new model is estimated. We predict the missing observation and again note the error. This "leave-one-out" procedure is repeated a total of n times, once for each observation. At the end of the process, we will have produced n residuals, all calculated using values not seen by the fitting procedure. We will therefore be able to calculate error statistics from n independent observations, rather than from just the validation subset. At the same time, each model will rest on nearly the entire sample ($n - 1$ observations) and will therefore be more robust than if just a few observations were used.

DEFINITION: CROSS-VALIDATION

Cross-validation refers to testing model predictions on data that are independent of those used to fitted the model. Most often, a model is fitted repeatedly to a dataset with a different observation held back each time. The withheld observations are used as a check on model performance, with error measures calculated across the entire dataset.

We should mention that the term cross-validation sometimes refers to a model-fitting procedure. Data points are successively withheld, and a parameter value is assigned such that prediction of withheld data is maximized. For example, in the cyclone problem, we estimated the probability density function $f(x)$. Doing so required that we choose a value for the kernel bandwidth h. There are several ways to cross-validate for h, all based on the idea of "predicting" withheld data and picking the h that gives the best fit as measured over the entire sample. The use of cross-validation as a fitting procedure is somewhat specialized; thus, we will not consider it further, except to note that it is especially useful in fitting interpolation models (see Wahba, 1990). We will focus instead on cross-validation as a diagnostic technique useful in assessing model performance. It will be easiest to describe if we imagine fitting a model $y = f(x)$, where x and y are the independent and dependent variables, respectively. The procedure is the following:

1. Given n observations, hold back the first and fit the model to the remaining observations. Use the model to predict y_1 from x_1. Calculate the prediction error, ε_1. Repeat for all observations, withholding each in turn. This gives the prediction errors ε_i, $i = 1, 2, \ldots, n$.

2. Compute a summary error measure from the ε_1. Some examples follow:

   Model bias: $\qquad\qquad\qquad\quad \frac{1}{n}\Sigma\varepsilon_i$

   Mean absolute error: $\qquad\quad \frac{1}{n}\Sigma|\varepsilon_i|$

   Root mean square error: $\qquad \sqrt{\frac{1}{n}\Sigma\varepsilon_i^2}$

EXAMPLE 17-3. Cross-Validating a Regression Model. Table 17-2 shows a cross-validation of the household trip data used previously with ordinary least squares regression (Chapter 12). Each row represents a separate cross-validation and shows the intercept and slope estimated from 11 of the 12 observations. Notice how the parameter estimates all differ from one another, as well as from the OLS values of 0.803 and 0.335. Because of this, each model gives rise to a different predicted value and different residual. Not surprisingly, the predictions are worse than those obtained from the traditional approach. Using r^2 as an overall index of performance, we get a cross-validated value of 0.75 versus 0.84 for OLS. In other words, the OLS value overstates the agreement between observed and predicted values by about 12%.

Notice how easily cross-validation yields estimates of prediction error. For example, if mean absolute error is of interest, we simply calculate $1/n \Sigma |\hat{Y}_i - Y_i|$ from the cross-validation data. That is, having fit n separate models, we compute an absolute error from each one and take the average. Table 17-2 shows the corresponding calculation for the root-mean-square error, which is perhaps more appropriate given our use of least squares in fitting the models. Using this index of error, cross-validation gives an average error of .85 trips.

It is more difficult to arrive at an average prediction error for OLS. The standard error of the estimate, $s_{Y \cdot X}$, is easily found, of course. For this dataset the value is .75 trips, smaller

TABLE 17-2
Cross-Validation of the Household Trip Data

Data		Cross-Validation				OLS	
Case No.	Observed Y	Intercept a	Slope b_i	Predicted $\hat{Y}_i$	Residual $\hat{Y}_i - Y$	Predicted $\hat{Y}_i$	Residual $\hat{Y}_{i-}$
1	3	1.434	0.303	4.464	1.464	4.156	1.156
2	5	0.741	0.338	4.798	−0.202	4.826	−0.174
3	5	0.576	0.346	4.387	−0.613	4.491	−0.509
4	4	0.684	0.342	3.758	−0.242	3.82	−0.18
5	6	0.698	0.339	5.444	−0.556	5.497	−0.503
6	7	0.741	0.334	6.091	−0.909	6.167	−0.833
7	9	1.137	0.306	7.573	−1.427	7.843	−1.157
8	7	0.813	0.334	6.819	−0.181	6.838	−0.162
9	8	0.734	0.341	8.233	0.233	8.179	0.179
10	8	0.264	0.376	9.296	1.296	8.849	0.849
11	6	0.807	0.338	6.552	0.552	6.502	0.502
12	5	0.920	0.333	5.910	0.910	5.832	0.832

$\Sigma(Y_i - \hat{Y})^2 = 34.917$ $\Sigma(\hat{Y}_i - Y_i)^2 = 8.687$ $\Sigma(\hat{Y} - Y_i)^2 = 5.666$

$$r^2 = 1 - \frac{8.687}{34.917} = 0.751 \qquad r^2 = 1 - \frac{5.666}{34.917} = 0.838$$

$$RMSE = \sqrt{\frac{8.687}{12}} = 0.851 \qquad s_{y \cdot x} = \sqrt{\frac{5.666}{10}} = 0.753$$

than the cross-validation error estimate. Recall, however, that the $s_{Y \cdot X}$ estimates only the scatter around the regression line, and not uncertainty in the regression line itself. To calculate prediction error, we need both components of variance, as reflected in the standard error of the forecast $s_{\hat{Y}_{new}}$. Unfortunately, this varies with X (Equation 13-11); thus, we would have to calculate a separate error estimate for each observation and then combine them in some way to arrive at an overall index of error. Although this can be done, it is certainly more difficult than using the cross-validation errors. In addition, the error estimates would not be based on independent data, as is the case for cross-validation.

Discussion

Cross-validation is much more flexible than the example above might imply. It can be used with other forms of OLS regression (e.g., multiple regression, polynomial regression) or with altogether different regression methods (e.g., least absolute deviation). In fact, any model fit to data is fair game for cross-validation. Similarly, the choice of error measure used is nearly unlimited. We have focused on mean-square measures, but we could just as easily have worked with other indices. Moreover, there is no requirement that one cross-validate the measure used by the fitting procedure. For example, having fit a model using least absolute deviations, one can cross-validate for relative (percentage) errors. Of course, one would need to make clear which measures are purely diagnostic, and which are used for fitting.

Because both cross-validation and jackknifing employ a "leave-one-out principle" for estimating error, it is worth comparing the two techniques. We have emphasized the jackknife as a way of estimating sampling variation, or (equivalently) the error that arises because one has only a part of the population in hand. Conversely, cross-validation was described as a way of assessing model error, or the inability of a model to perfectly predict some variable of interest. No doubt these model errors are very often a result of sampling, but they arise for other reasons as well, foremost of which is process error. For example, our inability to perfectly predict household trips arises largely from the influence of other variables ignored by the regression equation. Increasing the sample size will not eliminate those influences nor lead to perfect prediction. As a result, we would not expect to see cross-validation error measures shrink with increasing sample size. On the other hand, if we are using a consistent estimator, we would expect $\hat{\theta}$ to approach θ as sample size increases. Jackknife standard errors therefore typically decrease with the addition of more observations. In addition to this fundamental difference of purpose, there are obvious methodological differences as well. In particular, the subsamples used in the jackknife are far from independent of each other, as they share all but two observations.

17.4. Bootstrapping

The bootstrap is a nonparametric method for obtaining information about the distribution of a sample statistic $\hat{\theta}$. Like the jackknife, it is a resampling procedure in which a single sample is used repeatedly to mimic the behavior of multiple samples. Though computationally demanding, the bootstrap is very flexible and can be applied to a wide variety of problems. We will restrict our discussion to its use in (1) estimating the sampling distribution of $\hat{\theta}$, (2) estimating the standard error of $\hat{\theta}$, and (3) obtaining confidence intervals for $\hat{\theta}$.

Bootstrapping is based on the simple idea that if we had a large number of samples, we could obtain information about sampling variability by observing sample-to-sample differences. In actual practice only one sample is available, of course; thus, the bootstrap is forced to create these samples from the original sample. These artificial samples are called *bootstrap samples*, each of which is obtained by sampling with replacement from the original observations. Variability across the bootstrap samples serves as a substitute for variability across truly independent (but unavailable) samples.

DEFINITION: BOOTSTRAP SAMPLE

A bootstrap sample is obtained by sampling uniformly with replacement from an original dataset. Each bootstrap sample consists of n observations, some of which may be identical.

Figure 17-5 shows the bootstrap procedure for a sample of size 4, much smaller than would be used in actual practice. The original sample is resampled B times, giving B bootstrap samples of size n. To construct a bootstrap sample, we generate uniform random numbers between 1 and n inclusive. The random numbers are used to choose which observations appear in the bootstrap sample. Using a uniform random number generator ensures that all observations have an equal probability of appearing in the sample. Because the random numbers are uncorrelated, an observation can appear more than once in the sample and is no less likely to appear again once it has been chosen. Obviously, this procedure amounts to sampling with replacement.

For the example above with four observations, there are only 256 different bootstrap samples that might be drawn. However, the number of different bootstrap samples grows

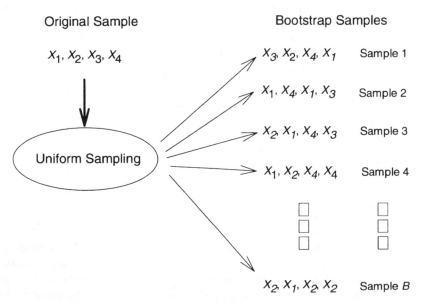

FIGURE 17-5. The bootstrap procedure for a sample of size 4.

rapidly with sample size (see Problem 5); thus for even moderate n the range of outcomes is huge. We calculate $\hat{\theta}$ for each bootstrap sample and use resulting distribution of $\hat{\theta}$ as if it had come from the population, rather than from a single sample. We are pretending that the B bootstrap samples can be used instead of B real samples. The rationale behind this seemingly extreme approach is actually quite straightforward. Obviously, the probability distribution of $\hat{\theta}$ depends on the unknown probability distribution $f(x)$. The bootstrap replaces $f(x)$ with an empirical estimate developed from the sample. In particular, it uses an estimate $\hat{f}(x)$, which assumes all observations are equally likely, as is consistent with random sampling. After all, if every observation has an equal chance of appearing in a bootstrap sample, we are saying that there is nothing special about any particular observation that would cause us to weight it more heavily than another. We emphasize that assuming all *observations* are equally likely is not at all the same as assuming all *values* of X are equally likely. Rather, the distribution of values appearing in a bootstrap sample reflects the distribution of values in the original sample. If the sample is representative of the population, the bootstrap samples will vary one from another much like real samples.

At first thought, the idea of sampling with replacement might seem wrong, as it permits a single observation to appear more than once in a bootstrap sample. However, a particular observation is no more likely than any other to appear multiple times; thus no inherent bias is introduced. Although a single bootstrap sample might contain some observation an unusually large number of times, many bootstrap samples are used, not just one. It would be very unlikely indeed for the same observation to repeatedly appear a disproportionate number of times in the ensemble of bootstrap samples. As a final point about sampling with replacement, notice that if one samples without replacement, only one outcome is possible, and; except for order, all bootstrap samples will be identical to the original sample!

Bootstrap Estimate of a Sampling Distribution

Suppose we have a single sample giving a single estimate $\hat{\theta}$ and would like to know something about the probability distribution of $\hat{\theta}$. For example, we might want to know the chances of obtaining a value twice as large, or half as large, or perhaps we want to know something about the shape of the sampling distribution. This is different than the problem faced in Section 17.1. Here, we are after the probability distribution of $\hat{\theta}$, denoted $f(\hat{\theta})$. Previously the goal was to estimate the $f(x)$, the probability distribution of the random variable underlying $\hat{\theta}$. The procedure to estimate $f(\hat{\theta})$ is simple:

1. Draw a bootstrap sample of size n from the original observations and compute $\hat{\theta}^*_1$, the sample statistic for bootstrap sample number 1. Repeat for a total of B samples, giving the bootstrap values $\hat{\theta}^*_i$, $i = 1, 2, \ldots, B$.
2. Use the relative frequencies of $\hat{\theta}^*_i$ to assign probability density estimates for the sampling distribution $f(\hat{\theta})$.

In step 2, we might calculate the relative frequency over some range of $\hat{\theta}^*_i$, or build a histogram of $\hat{\theta}^*_i$, or perhaps use the kernel method to obtain an estimate $\hat{f}(\hat{\theta})$. The number of bootstrap samples needed (B) depends on how $\hat{f}(\hat{\theta})$ is used. For example, estimating a probability near the tail of the distribution demands more samples than creating a rough histogram. Though there are no hard-and-fast rules, experience shows that the number of samples needed is at most on the order of thousands. If there is doubt about whether or not B is large enough, one can always increase B and note any change in the results.

EXAMPLE 17-4. El-Niño/Southern Oscillation events (atmosphere-ocean changes in the tropical Pacific) are thought to influence climatic conditions at far-away locations. Figure 17-6 shows the relationship between rainfall in sub-Saharan Africa and the Southern Oscillation Index, a measure of surface pressure differences between Darwin, Australia, and Tahiti. The scatter plot indicates that high-index years are associated with higher than normal rainfall, the correlation coefficient (r) being .451. We have a point estimate $\hat{\theta}$ and want to know something about its sampling variability. The classical method is Fisher's Z transformation, but this requires an assumption that both variables are normally distributed. Because the rainfall values are positively skewed, we will use a bootstrap instead of the Z transform.

There are 41 sample pairs x_i, y_i, corresponding to the years 1950–1990. The first step is to construct bootstrap samples of size 41, where each member is an x, y pair for one of the years. In this example a total of 5000 samples were created, each giving a correlation coefficient. The bootstrap correlations range from a low of $-.027$ to a maximum value of .750. Dividing this range into 50 bins gives the histogram shown in Figure 17-7. The histogram shows the frequency of occurrence bin by bin—to assign a probability, the count for a bin would be divided by the number of bootstrap samples B. Notice that the histogram

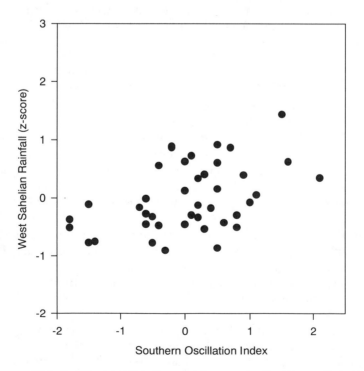

FIGURE 17-6. Scatter plot of Sahelian rainfall versus the Southern oscillation index.

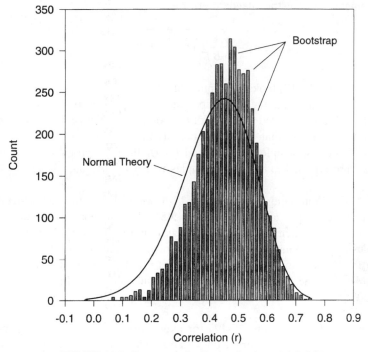

FIGURE 17-7. Bootstrap distribution for the rainfall-SOI.

is not perfectly symmetric, but rather is negatively skewed. This is in qualitative agreement with normal theory using Fisher's transformation, but the amount of skew present in the bootstrap histogram is less. Notice that the bootstrap distribution is more tightly clustered around the sample estimate $\hat{\theta}$, suggesting that the variance in $\hat{\theta}$ is smaller than that indicated by normal theory.

Bootstrap Estimates of Standard Error

As with the jackknife, the goal is to estimate the standard error of a sample statistic $\hat{\theta}$. Once again, multiple samples are created from the original sample, but this time we use random numbers to choose which observations are included in the derived samples. The procedure is as follows:

1. Draw a bootstrap sample of size n from the original n observations and compute $\hat{\theta}_1^*$, the sample statistic for bootstrap sample number 1. Repeat for a total of B samples, giving the bootstrap values $\hat{\theta}_i^*$; $i = 1, 2, 3, \ldots, B$.
2. Average the values from Step 1 to get the mean bootstrap value

$$\bar{\theta}^* = \frac{1}{B} \sum_{i=1}^{B} \hat{\theta}_i^*$$

3. Compute the bootstrap standard error as:

$$\hat{\sigma}_B = \sqrt{\frac{1}{B} \sum_{i=1}^{B} (\hat{\theta}_i^* - \bar{\theta}^*)^2} \qquad (17\text{-}12)$$

Notice that Equation (17-12) looks very much like the standard deviation of $\hat{\theta}^*$. This is no accident, because we are using the bootstrap values $\hat{\theta}_i^*$ as proxies to stand in for values that might be obtained from repeated sampling of the population. Notice that $\hat{\sigma}_B$ depends on all the bootstrap values, rather than just a few near a tail or within a narrow range. Because of this, fewer bootstrap samples are required for $\hat{\sigma}_B$ than for probability estimates. For bootstrap standard errors, 100 bootstrap samples is usually adequate.

EXAMPLE 17-5. Continuing with Example 17-4, we wish to estimate the standard error of r. Recall that the sample gave a value of $r = .451$. Here the goal is to estimate the error in that estimate without resorting to a normality assumption. Table 17-3 shows a few of the calculations using 100 bootstrap samples. The bootstrap standard error for r is .101, which we can interpret as the amount by which we expect r to depart from the true correlation ρ. In other words, .1 is the expected error in the sample value of r. In practice, we would probably go on to compute a confidence interval for ρ, a topic considered in the next section.

We have written Equation (17-12) for expository purposes—once again, a better computing formula is available. For the bootstrap standard error, the computing formula is:

$$\hat{\sigma}_B = \sqrt{\frac{1}{B} \sum_{i=1}^{B} (\hat{\theta}_i^*)^2 - (\bar{\theta}^*)^2} \qquad (17\text{-}13)$$

TABLE 17-3
Calculation of the Bootstrap Standard Error

i	$\hat{\theta}_i^*$	$\hat{\theta}_i^* - \bar{\theta}^*$	$(\hat{\theta}_i^* - \bar{\theta}^*)^2$
1	.426	−.023	.001
2	.647	.198	.039
3	.484	.035	.001
$\vdots$	$\vdots$	$\vdots$	$\vdots$
97	.423	−.026	.001
98	.492	.043	.002
100	.361	−.088	.008
	44.941		1.021

$$\bar{\theta}^* = \frac{44.941}{100} = .449$$

$$\hat{\sigma}_B = \sqrt{\frac{1}{100}(1.021)} = .101$$

Bootstrap Confidence Intervals

A very important use of the bootstrap is to construct confidence intervals for $\hat{\theta}$. When we faced this problem in a parametric context, we relied on a theory regarding the distribution of the sample statistic. If we knew the sampling distribution $f(\hat{\theta})$, we could find an interval likely to contain θ at any desired level of confidence. The bootstrap accomplishes the same thing, but uses the bootstrap distribution instead of a theoretical sampling distribution. It is therefore a nonparametric method based on an *empirical* sampling distribution. We will describe two variants, the percentile and bias-corrected percentile method. Both versions require thousands of samples and should probably not be used with B less than 1000.

Percentile Method

Given a set of bootstrap values, this method is trivially easy to carry out. For example, to construct a 90% confidence interval, we simply use the central 90% of the bootstrap distribution. That is, the lower limit will be the fifth percentile value of $\hat{\theta}^*$, and the upper limit will be the 95th percentile. Similarly, the 95% confidence interval is based on the 2.5th and 97.5th percentiles.

Recall that the 90% confidence level corresponds to an α level of .10, and a 95% confidence level corresponds to an α level of .05. In other words, the confidence *level* is $1 - \alpha$. To find the confidence *interval*, we look for the lower and upper $\alpha/2$ percentiles of $\hat{\theta}^*$. This is easily done by sorting the bootstrap values in ascending order. Given any integer j, the proportion of bootstrap values less than or equal to the jth element in the sorted list is j/B. For example, using 2000 samples, 5% of the values are less than or equal to the 100th element in the list. More generally, the method is:

1. Choose a confidence level $1 - \alpha$.
2. Draw a bootstrap sample of size n from the original observations and compute $\hat{\theta}^*_1$, the sample statistic for bootstrap sample number 1. Repeat for a total of B samples, giving the bootstrap values $\hat{\theta}^*_i$, $i = 1, 2, \ldots, B$.
3. Sort the bootstrap samples in ascending order.
4. Use the lower $\alpha/2$ percentile as the lower limit of the confidence interval θ_{Low}. This can be found as the jth element in the sorted list, where $j = B\alpha/2$.
5. Use the upper $\alpha/2$ percentile as the upper limit of the confidence interval θ_{High}. This can be found from the kth element in the sorted list, where $k = B(1 - \alpha/2)$.

EXAMPLE 17-6. We will use the bootstrap sampling distribution from Example 17-4 to find a 90% confidence interval for the correlation between sub-Saharan rainfall and the southern oscillation index. With 5000 samples, the fifth percentile is at position 250 (5% of 5000). The 95th percentile is at position 4750. After sorting the bootstrap values, we obtain $\theta_{Low} = \hat{\theta}^*_{250} = .262$ and $\theta_{High} = \hat{\theta}^*_{4750} = .610$. The interval given by normal theory is wider, running from .216 to .637.

Bias-Corrected Percentile Method

The percentile method is easy to use but is not always unbiased. At the cost of additional complexity, the method can be improved upon. It can be shown that when the median is being bootstrapped, the bootstrap distribution $f(\hat{\theta}^*)$ is unbiased, or "centered," on the true value. That is, the expected value of the bootstrap median equals the true population median. In the absence of sampling error, we would have the estimate $\hat{\theta}$ equal to the true median and would also have the median of the bootstrap distribution $(\hat{\theta}^*_{B/2})$ equal to the true median. In other words, without sampling error, $\theta = \hat{\theta} = \hat{\theta}^*_{B/2}$. Of course, when we generate a particular bootstrap distribution from a real sample, its 50th percentile will not equal $\hat{\theta}$. Knowing that the difference between $\hat{\theta}$ and $\hat{\theta}^*_{B/2}$ arises from sampling error, we are led to a bias correction based on the difference. The theory behind the correction is beyond an introductory text, but correction itself is easily accomplished.

The adjustment for sample bias does not involve any assumptions of normality, but does require that we use the normal distribution function. Let $\phi(z)$ be the area to the left of z under the normal curve. That is $\phi(z)$ gives $P(Z < z)$, where Z is a standard normal variable. This is recognizable as the cumulative normal curve and should be familiar. Note that column 2 of Table A-3 gives these probabilities; hence it is actually a table of ϕ. The inverse of ϕ is denoted $\phi^{-1}(p)$, and gives a z-value corresponding to a probability value p. The inverse function is like using Table A-3 in reverse; given a probability value in column 2, we find the corresponding from column 1. Notice that $z_\alpha = \phi^{-1}(\alpha)$.

Turning to the bootstrap distribution $\hat{f}(\hat{\theta}^*)$, we can define an analogous cumulative distribution function $\hat{F}(\hat{\theta}^*)$, which gives the proportion of the bootstrap distribution to the left of some value. To put it differently, $\hat{F}$ gives percentiles of the bootstrap distribution. Similarly, there is an inverse function $\hat{F}^{-1}(p)$ which converts a probability into a bootstrap value. With this as background, we can describe the bias-corrected percentile method:

1. Choose a confidence level $1 - \alpha$. Find the lower and upper z-values $z_{\alpha/2} = \phi^{-1}(\alpha/2)$ and $z_{1-\alpha/2} = \phi^{-1}(1 - \alpha/2)$.
2. Draw a bootstrap sample of size from the original observations and compute $\hat{\theta}^*_1$, the sample statistic for bootstrap sample number 1. Repeat for a total of B samples, giving the bootstrap values $\hat{\theta}^*_i$, $i = 1, 2, \ldots, B$.
3. Sort the bootstrap samples in ascending order.
4. Compute $\hat{\theta}$ from the original sample and find its percentile in the bootstrap distribution, $\hat{F}(\hat{\theta}^*)$. Call this $\hat{p}_0$, and find $z_0 = \phi^{-1}(p_0)$.
5. Compute two probabilities, $p_{Low} = \phi(z_{\alpha/2} + 2z_0)$ and $p_{High} = \phi(z_{1-\alpha/2} + 2z_0)$. Use $\theta_{Low} = \hat{F}^{-1}(p_{Low})$ and $\theta_{High} = \hat{F}^{-1}(p_{Low})$ as the lower and upper limits of the confidence interval.

This looks complicated at first glance, but is actually not too different from the percentile method. Before discussion, we present a numerical example. Continuing with the correlation example,

1. We choose a 90% confidence level, giving $\alpha = .1$. The corresponding 5% and 95% z-values are -1.645 and 1.645.
2. We use the 5000 bootstrap samples from Example 17-5.
3. The bootstrap values are sorted, giving the histogram of Figure 17-7.
4. The sample correlation $\hat{\theta}$ is 0.451, giving $p_0 = \hat{F}^{-1}(.541) = .486$. That is, the sample estimate is approximately the 49th percentile of the bootstrap distribution. Converting this to a z-value, yields $z_0 = \phi^{-1}(.486) = -.033$, thus $2z_0 = -.066$.

5. We find the two probabilities

$$p_{\text{Low}} = \phi(-1.645 - .066) = .044 \qquad p_{\text{High}} = \phi(1.645 - .066) = .934$$

and the confidence limits

$$\theta_{\text{Low}} = \hat{F}^{-1}(.044) = .256 \qquad \theta_{\text{High}} = \hat{F}^{-1}(.934) = .606$$

Notice that in the last step p_{Low} and p_{High} call for percentiles slightly lower than the 5th and 95th percentiles of $\hat{\theta}^*$. In this example bias correction reduces θ_{Low} and θ_{High} by almost the same amount, effectively shifting the confidence interval to the left. The change is not large, however, because the bootstrap distribution is so nearly centered on $\hat{\theta}$. In other situations bias correction can lead to a significant change from the simpler percentile method.

Although the bias-corrected method seems much more complicated, in fact z_0 is the only term not present in the simple percentile method. Moreover, the two methods are identical when bias correction is not needed. To see this, suppose that the bootstrap median happens to equal the sample value $\hat{\theta}$. In this case $\hat{\theta}$ is obviously at the 50th percentile of the bootstrap distribution, and p_0 is .5. The corresponding z-value (z_0) is zero. Under these circumstances Step 5 above reduces to

$$\theta_{\text{Low}} = \hat{F}^{-1}(p_{\text{Low}}) = \hat{F}^{-1}(\phi(z_{\alpha/2})) = \hat{F}^{-1}(\alpha/2)$$

and

$$\theta_{\text{High}} = \hat{F}^{-1}(p_{\text{High}}) = \hat{F}^{-1}(\phi(z_{1-\alpha/2})) = \hat{F}^{-1}(1 - \alpha/2)$$

The right-hand members of the equations above call for the lower and upper $\alpha/2$ tails of the bootstrap distribution, which is exactly the same as the percentile method. We see that bias correction only appears when the bootstrap distribution is not centered on $\hat{\theta}$, as is consistent with the motivation for the correction. Because of the way it adjusts to the bias apparent in the data, we think of the bias-corrected method an adaptive technique.

It should be mentioned that both the percentile and bias-corrected percentile methods assume the existence of some function that transforms $\hat{\theta}$ to a normal variable. We do not have to know the form of the function, only that it exists. Actually, reliance on these so-called normal pivotal functions is not unique to bootstrapping. For example, the Fisher Z transformation is a pivotal function for bivariate normal variables. So far as the bootstrap is concerned, we prefer the bias-corrected method over the percentile method because it admits a larger class of pivotal functions, and thus applies in a larger number of situations. There are, however, some instances for which no appropriate pivotal function exists, and neither the percentile nor bias-corrected intervals will be correct. Unfortunately, these situations are likely to be complicated, offering little hope for verifying the assumption regarding a pivotal quantity. This is a little like having to assume bivariate normality, but not having data available to check on the assumption. However, in the case of bootstrap confidence intervals, the assumption we make is much less restrictive than bivariate normality. Our point, therefore, is not that bootstrap confidence intervals are unreliable, but rather that they are not foolproof.

Discussion

It should be clear that the bootstrap is a very general method and can be used in many contexts other than those described here. Common to all applications is the requirement that the values being bootstrapped form a random sample. It is worth repeating that this means the observations must be identically distributed and uncorrelated. This restriction does rule out a number of situations where one might be tempted to bootstrap. For example, if the data exhibit systematic temporal variation, one cannot bootstrap a time series directly. If values in one year are related to those of nearby years, we want bootstrap samples that preserve that temporal correlation. Simple random sampling of the original data will not suffice, as it generates uncorrelated sequences. Whereas the real world might not admit a very high value followed by a low value, the bootstrap sample would contain no similar restriction and would therefore be a poor surrogate for repeated sampling of the real world.

Note that this requirement for a random sample is not unusual, and in fact random sampling is nearly always assumed. Bootstrapping is therefore no more restricted by this requirement than are the traditional techniques we have described in other chapters. Given its simplicity and wide applicability, it's not surprising that the bootstrap is becoming ever more common in geographic research. It is particularly useful in modeling spatial processes. One has a single expression of the process (perhaps observed at multiple points in time) and will want to know if the observations are consistent with some hypothesized theoretical model. That is, could the observations have arisen from the hypothesized model? In the same way that classical methods allow one to test simple hypotheses about parameters, the bootstrap provides a way to examine more complicated hypotheses.

Further Reading

P. Diaconis and B. Efron, "Computer-Intensive Methods in Statistics," *Scientific American* 248, (1983) 110–130.

B. Efron and R. Tibshirani, "Bootstrap Methods for Standard Errors, Confidence Intervals, and Other Measures of Statistical Accuracy," *Statistical Science* 1, (1986) 54–77.

B. Efron, *The Jackknife, the Bootstrap, and Other Resampling Plans* (Philadelphia: Society for Industrial and Applied Mathematics, 1982).

W. M., Gray, C. W. Landsea, P. W. Mielke, J. J. Berry, "Predicting Atlantic Seasonal Hurricane Activity 6–11 Months in Advance," Weather and Forecasting 7 (1992) 440–455.

C. Z. Mooney and R. D. Duv, *Bootstrapping: A Nonparametric Approach to Statistical Inference*. (Beverly Hills: Sage Publications, 1994).

B. W. Silverman, *Density Estimation for Statistics and Data Analysis* (New York: Chapman and Hall, 1986).

G. Wahba, *Spline Models for Observational Data* (Philadelphia: Society for Industrial and Applied Mathematics, 1990).

Problems

1. Explain the workings the following methods:
   - a. Kernel probability density estimate
   - b. Cross-validation
   - c. Jackknife
   - d. Bootstrap

2. How does the choice of kernel bandwidth affect the estimate $\hat{f}(x)$?

3. Why is the normal distribution a good kernel function?

4. Use the kernel method to estimate the probability distribution for the DO data of Table 2-1. Graph the resulting function and compare with Figure 2-1.

5. Given a sample of size 12, how many different bootstrap samples are possible?

6. Give three examples of statistics well suited for jackknifing.

7. Are there any statistics that should not be jackknifed?

8. Suppose one uses least squares to fit a linear regression model. Is it possible for the residual sum of squares to be larger than the corresponding value obtained by cross-validation?

9. What is the fundamental assumption behind bootstrapping?

10. Explain the percentile method for constructing bootstrap confidence intervals.

APPENDIX:
STATISTICAL
TABLES

TABLE A-1
Binomial Distribution

Entry is probability $P(X = x) = \binom{n}{x} p^x (1-p)^{n-x}$

n x	.01	.02	.03	.04	.05	.06	.07	.08	.09	
2 0	0.9801	0.9604	0.9409	0.9216	0.9025	0.8836	0.8649	0.8464	0.8281	2
1	0.0198	0.0392	0.0582	0.0768	0.0950	0.1128	0.1302	0.1472	0.1638	1
2	0.0001	0.0004	0.0009	0.0016	0.0025	0.0036	0.0049	0.0064	0.0081	0 2
3 0	0.9703	0.9412	0.9127	0.8847	0.8574	0.8306	0.8044	0.7787	0.7536	3
1	0.0294	0.0576	0.0847	0.1106	0.1354	0.1590	0.1816	0.2031	0.2236	2
2	0.0003	0.0012	0.0026	0.0046	0.0071	0.0102	0.0137	0.0177	0.0221	1
3	0.0000	0.0000	0.0000	0.0001	0.0001	0.0002	0.0003	0.0005	0.0007	0 3
4 0	0.9606	0.9224	0.8853	0.8493	0.8145	0.7807	0.7481	0.7164	0.6857	4
1	0.0388	0.0753	0.1095	0.1416	0.1715	0.1993	0.2252	0.2492	0.2713	3
2	0.0006	0.0023	0.0051	0.0088	0.0135	0.0191	0.0254	0.0325	0.0402	2
3	0.0000	0.0000	0.0001	0.0002	0.0005	0.0008	0.0013	0.0019	0.0027	1
4	0.0000	0.0000	0.0000	0.0000	0.0000	0.0000	0.0000	0.0000	0.0001	0 1
5 0	0.9510	0.9039	0.8587	0.8154	0.7738	0.7339	0.6957	0.6591	0.6240	5
1	0.0480	0.0922	0.1328	0.1699	0.2036	0.2342	0.2618	0.2866	0.3086	4
2	0.0010	0.0038	0.0082	0.0142	0.0214	0.0299	0.0394	0.0498	0.0610	3
3	0.0000	0.0001	0.0003	0.0006	0.0011	0.0019	0.0030	0.0043	0.0060	2
4	0.0000	0.0000	0.0000	0.0000	0.0000	0.0001	0.0001	0.0002	0.0003	1
5	0.0000	0.0000	0.0000	0.0000	0.0000	0.0000	0.0000	0.0000	0.0000	0 5
6 0	0.9415	0.8858	0.8330	0.7828	0.7351	0.6899	0.6470	0.6064	0.5679	6
1	0.0571	0.1085	0.1546	0.1957	0.2321	0.2642	0.2922	0.3164	0.3370	5
2	0.0014	0.0055	0.0120	0.0204	0.0305	0.0422	0.0550	0.0688	0.0833	4
3	0.0000	0.0002	0.0005	0.0011	0.0021	0.0036	0.0055	0.0080	0.0110	3
4	0.0000	0.0000	0.0000	0.0000	0.0001	0.0002	0.0003	0.0005	0.0008	2
5	0.0000	0.0000	0.0000	0.0000	0.0000	0.0000	0.0000	0.0000	0.0000	1
6	0.0000	0.0000	0.0000	0.0000	0.0000	0.0000	0.0000	0.0000	0.0000	0 6
7 0	0.9321	0.8681	0.8080	0.7514	0.6983	0.6485	0.6017	0.5578	0.5168	7
1	0.0659	0.1240	0.1749	0.2192	0.2573	0.2897	0.3170	0.3396	0.3578	6
2	0.0020	0.0076	0.0162	0.0274	0.0406	0.0555	0.0716	0.0886	0.1061	5
3	0.0000	0.0003	0.0008	0.0019	0.0036	0.0059	0.0090	0.0128	0.0175	4
4	0.0000	0.0000	0.0000	0.0001	0.0002	0.0004	0.0007	0.0011	0.0017	3
5	0.0000	0.0000	0.0000	0.0000	0.0000	0.0000	0.0000	0.0001	0.0001	2
6	0.0000	0.0000	0.0000	0.0000	0.0000	0.0000	0.0000	0.0000	0.0000	1
7	0.0000	0.0000	0.0000	0.0000	0.0000	0.0000	0.0000	0.0000	0.0000	0 7
8 0	0.9227	0.8508	0.7837	0.7214	0.6634	0.6096	0.5596	0.5132	0.4703	8
1	0.0746	0.1389	0.1939	0.2405	0.2793	0.3113	0.3370	0.3570	0.3721	7
2	0.0026	0.0099	0.0210	0.0351	0.0515	0.0695	0.0888	0.1087	0.1288	6
3	0.0001	0.0004	0.0013	0.0029	0.0054	0.0089	0.0134	0.0189	0.0255	5
4	0.0000	0.0000	0.0001	0.0002	0.0004	0.0007	0.0013	0.0021	0.0031	4
5	0.0000	0.0000	0.0000	0.0000	0.0000	0.0000	0.0001	0.0001	0.0002	3
6	0.0000	0.0000	0.0000	0.0000	0.0000	0.0000	0.0000	0.0000	0.0000	2
7	0.0000	0.0000	0.0000	0.0000	0.0000	0.0000	0.0000	0.0000	0.0000	1
8	0.0000	0.0000	0.0000	0.0000	0.0000	0.0000	0.0000	0.0000	0.0000	0 8
9 0	0.9135	0.8337	0.7602	0.6925	0.6302	0.5730	0.5204	0.4722	0.4279	9
1	0.0830	0.1531	0.2116	0.2597	0.2985	0.3292	0.3525	0.3695	0.3809	8
2	0.0034	0.0125	0.1262	0.0433	0.0629	0.0840	0.1061	0.1285	0.1507	7
3	0.0001	0.0006	0.0019	0.0042	0.0077	0.0125	0.0186	0.0261	0.0348	6
4	0.0000	0.0000	0.0001	0.0003	0.0006	0.0012	0.0021	0.0034	0.0052	5
5	0.0000	0.0000	0.0000	0.0000	0.0000	0.0001	0.0002	0.0003	0.0005	4
6	0.0000	0.0000	0.0000	0.0000	0.0000	0.0000	0.0000	0.0000	0.0000	3
7	0.0000	0.0000	0.0000	0.0000	0.0000	0.0000	0.0000	0.0000	0.0000	2
8	0.0000	0.0000	0.0000	0.0000	0.0000	0.0000	0.0000	0.0000	0.0000	1
9	0.0000	0.0000	0.0000	0.0000	0.0000	0.0000	0.0000	0.0000	0.0000	0 9
	.99	.98	.97	.96	.95	.94	.93	.92	.91	x n

p

(continued)

TABLE A-1 (continued)

n x	.10	.15	.20	.25	.30	.35	.40	.45	.50	
					p					
2 0	0.8100	0.7225	0.6400	0.5625	0.4900	0.4225	0.3600	0.3025	0.2500	2
1	0.1800	0.2550	0.3200	0.3750	0.4200	0.4550	0.4800	0.4950	0.5000	1
2	0.0100	0.0225	0.0400	0.0625	0.0900	0.1225	0.1600	0.2025	0.2500	0 2
3 0	0.7290	0.6141	0.5120	0.4219	0.3430	0.2746	0.2160	0.1664	0.1250	3
1	0.2430	0.3251	0.3840	0.4219	0.4410	0.4436	0.4320	0.4084	0.3750	2
2	0.0270	0.0574	0.0960	0.1406	0.1890	0.2389	0.2880	0.3341	0.3750	1
3	0.0010	0.0034	0.0080	0.0156	0.0270	0.0429	0.0640	0.0911	0.1250	0 3
4 0	0.6561	0.5220	0.4096	0.3164	0.2401	0.1785	0.1296	0.0915	0.0625	4
1	0.2916	0.3685	0.4096	0.4219	0.4116	0.3845	0.3456	0.2995	0.2500	3
2	0.0486	0.0975	0.1536	0.2109	0.2646	0.3105	0.3456	0.3675	0.3750	2
3	0.0036	0.0115	0.0256	0.0469	0.0756	0.1115	0.1536	0.2005	0.2500	1
4	0.0001	0.0005	0.0016	0.0039	0.0081	0.0150	0.0256	0.0410	0.0625	0 4
5 0	0.5905	0.4437	0.3277	0.2373	0.1681	0.1160	0.0778	0.0503	0.0312	5
1	0.3280	0.3915	0.4096	0.3955	0.3601	0.3124	0.2592	0.2059	0.1562	4
2	0.0729	0.1382	0.2048	0.2637	0.3087	0.3364	0.3456	0.3369	0.3125	3
3	0.0081	0.0244	0.0512	0.0879	0.1323	0.1811	0.2304	0.2757	0.3125	2
4	0.0004	0.0022	0.0064	0.0146	0.0283	0.0488	0.0768	0.1128	0.1562	1
5	0.0000	0.0001	0.0003	0.0010	0.0024	0.0053	0.0102	0.0185	0.0312	0 5
6 0	0.5314	0.3771	0.2621	0.1780	0.1176	0.0754	0.0467	0.0277	0.0156	6
1	0.3543	0.3993	0.3932	0.3560	0.3025	0.2437	0.1866	0.1359	0.0938	5
2	0.0984	0.1762	0.2458	0.2966	0.3241	0.3280	0.3110	0.2780	0.2344	4
3	0.0146	0.0415	0.0819	0.1318	0.1852	0.2355	0.2765	0.3032	0.3125	3
4	0.0012	0.0055	0.0154	0.0330	0.0595	0.0951	0.1382	0.1861	0.2344	2
5	0.0001	0.0004	0.0015	0.0044	0.0102	0.0205	0.0369	0.0609	0.0938	1
6	0.0000	0.0000	0.0001	0.0002	0.0007	0.0018	0.0041	0.0083	0.0156	0 6
7 0	0.4783	0.3206	0.2097	0.1335	0.0824	0.0490	0.0280	0.0152	0.0078	7
1	0.3720	0.3960	0.3670	0.3115	0.2471	0.1848	0.1306	0.0872	0.0547	6
2	0.1240	0.2097	0.2753	0.3115	0.3177	0.2985	0.2613	0.2140	0.1641	5
3	0.0230	0.0617	0.1147	0.1730	0.2269	0.2679	0.2903	0.2918	0.2734	4
4	0.0026	0.0109	0.0287	0.0577	0.0972	0.1442	0.1935	0.2388	0.2734	3
5	0.0002	0.0012	0.0043	0.0115	0.0250	0.0466	0.0774	0.1172	0.1641	2
6	0.0000	0.0001	0.0004	0.0013	0.0036	0.0084	0.0172	0.0320	0.0547	1
7	0.0000	0.0000	0.0000	0.0001	0.0002	0.0006	0.0016	0.0037	0.0078	0 7
8 0	0.4305	0.2725	0.1678	0.1001	0.0576	0.0319	0.0168	0.0084	0.0039	8
1	0.3826	0.3847	0.3355	0.2670	0.1977	0.1373	0.0896	0.0548	0.0312	7
2	0.1488	0.2376	0.2936	0.3115	0.2965	0.2587	0.2090	0.1569	0.1094	6
3	0.0331	0.0839	0.1468	0.2076	0.2541	0.2786	0.2787	0.2568	0.2188	5
4	0.0046	0.0185	0.0459	0.0865	0.1361	0.1875	0.2322	0.2627	0.2734	4
5	0.0004	0.0026	0.0092	0.0231	0.0467	0.0808	0.1239	0.1719	0.2188	3
6	0.0000	0.0002	0.0011	0.0038	0.0100	0.0217	0.0413	0.0703	0.1094	2
7	0.0000	0.0000	0.0001	0.0004	0.0012	0.0033	0.0079	0.0164	0.0312	1
8	0.0000	0.0000	0.0000	0.0000	0.0001	0.0002	0.0007	0.0017	0.0039	0 8
9 0	0.3874	0.2316	0.1342	0.0751	0.0404	0.0207	0.0101	0.0046	0.0020	9
1	0.3874	0.3679	0.3020	0.2253	0.1556	0.1004	0.0605	0.0339	0.0176	8
2	0.1722	0.2597	0.3020	0.3003	0.2668	0.2162	0.1612	0.1110	0.0703	7
3	0.0446	0.1069	0.1762	0.2336	0.2668	0.2716	0.2508	0.2119	0.1641	6
4	0.0074	0.0283	0.0661	0.1168	0.1715	0.2194	0.2508	0.2600	0.2461	5
5	0.0008	0.0050	0.0165	0.0389	0.0735	0.1181	0.1672	0.2128	0.2461	4
6	0.0001	0.0006	0.0028	0.0087	0.0210	0.0424	0.0743	0.1160	0.1641	3
7	0.0000	0.0000	0.0003	0.0012	0.0039	0.0098	0.0212	0.0407	0.0703	2
8	0.0000	0.0000	0.0000	0.0001	0.0004	0.0013	0.0035	0.0083	0.0176	1
9	0.0000	0.0000	0.0000	0.0000	0.0000	0.0001	0.0003	0.0008	0.0020	0 9
	.90	.85	.80	.75	.70	.65	.60	.55	.50	x n
					p					

TABLE A-1 (continued)

n	x	.01	.02	.03	.04	.05	.06	.07	.08	.09		
							p					
10	0	0.9044	0.8171	0.7374	0.6648	0.5987	0.5386	0.4840	0.4344	0.3894	10	
	1	0.0914	0.1667	0.2281	0.2770	0.3151	0.3438	0.3643	0.3777	0.3851	9	
	2	0.0042	0.0153	0.0317	0.0519	0.0746	0.0988	0.1234	0.1478	0.1714	8	
	3	0.0001	0.0008	0.0026	0.0058	0.0105	0.0168	0.0248	0.0343	0.0452	7	
	4	0.0000	0.0000	0.0001	0.0004	0.0010	0.0019	0.0033	0.0052	0.0078	6	
	5	0.0000	0.0000	0.0000	0.0000	0.0001	0.0001	0.0003	0.0005	0.0009	5	
	6	0.0000	0.0000	0.0000	0.0000	0.0000	0.0000	0.0000	0.0000	0.0001	4	
	7	0.0000	0.0000	0.0000	0.0000	0.0000	0.0000	0.0000	0.0000	0.0000	3	
	8	0.0000	0.0000	0.0000	0.0000	0.0000	0.0000	0.0000	0.0000	0.0000	2	
	9	0.0000	0.0000	0.0000	0.0000	0.0000	0.0000	0.0000	0.0000	0.0000	1	
	10	0.0000	0.0000	0.0000	0.0000	0.0000	0.0000	0.0000	0.0000	0.0000	0	10
12	0	0.8864	0.7847	0.6938	0.6127	0.5404	0.4759	0.4186	0.3677	0.3225	12	
	1	0.1074	0.1922	0.2575	0.3064	0.3413	0.3645	0.3781	0.3837	0.3827	11	
	2	0.0060	0.0216	0.0438	0.0702	0.0988	0.1280	0.1565	0.1835	0.2082	10	
	3	0.0002	0.0015	0.0045	0.0098	0.0173	0.0272	0.0393	0.0532	0.0686	9	
	4	0.0000	0.0001	0.0003	0.0009	0.0021	0.0039	0.0067	0.0104	0.0153	8	
	5	0.0000	0.0000	0.0000	0.0001	0.0002	0.0004	0.0008	0.0014	0.0024	7	
	6	0.0000	0.0000	0.0000	0.0000	0.0000	0.0000	0.0001	0.0001	0.0003	6	
	7	0.0000	0.0000	0.0000	0.0000	0.0000	0.0000	0.0000	0.0000	0.0000	5	
	8	0.0000	0.0000	0.0000	0.0000	0.0000	0.0000	0.0000	0.0000	0.0000	4	
	9	0.0000	0.0000	0.0000	0.0000	0.0000	0.0000	0.0000	0.0000	0.0000	3	
	10	0.0000	0.0000	0.0000	0.0000	0.0000	0.0000	0.0000	0.0000	0.0000	2	
	11	0.0000	0.0000	0.0000	0.0000	0.0000	0.0000	0.0000	0.0000	0.0000	1	
	12	0.0000	0.0000	0.0000	0.0000	0.0000	0.0000	0.0000	0.0000	0.0000	0	12
15	0	0.8601	0.7386	0.6333	0.5421	0.4633	0.3953	0.3367	0.2863	0.2430	15	
	1	0.1303	0.2261	0.2938	0.3388	0.3658	0.3785	0.3801	0.3734	0.3605	14	
	2	0.0092	0.0323	0.0636	0.0988	0.1348	0.1691	0.2003	0.2273	0.2496	13	
	3	0.0004	0.0029	0.0085	0.0178	0.0307	0.0468	0.0653	0.0857	0.1070	12	
	4	0.0000	0.0002	0.0008	0.0022	0.0049	0.0090	0.0148	0.0223	0.0317	11	
	5	0.0000	0.0000	0.0001	0.0002	0.0006	0.0013	0.0024	0.0043	0.0069	10	
	6	0.0000	0.0000	0.0000	0.0000	0.0000	0.0001	0.0003	0.0006	0.0011	9	
	7	0.0000	0.0000	0.0000	0.0000	0.0000	0.0000	0.0000	0.0001	0.0001	8	
	8	0.0000	0.0000	0.0000	0.0000	0.0000	0.0000	0.0000	0.0000	0.0000	7	
	9	0.0000	0.0000	0.0000	0.0000	0.0000	0.0000	0.0000	0.0000	0.0000	6	
	10	0.0000	0.0000	0.0000	0.0000	0.0000	0.0000	0.0000	0.0000	0.0000	5	
	11	0.0000	0.0000	0.0000	0.0000	0.0000	0.0000	0.0000	0.0000	0.0000	4	
	12	0.0000	0.0000	0.0000	0.0000	0.0000	0.0000	0.0000	0.0000	0.0000	3	
	13	0.0000	0.0000	0.0000	0.0000	0.0000	0.0000	0.0000	0.0000	0.0000	2	
	14	0.0000	0.0000	0.0000	0.0000	0.0000	0.0000	0.0000	0.0000	0.0000	1	
	15	0.0000	0.0000	0.0000	0.0000	0.0000	0.0000	0.0000	0.0000	0.0000	0	15
20	0	0.8179	0.6676	0.5438	0.4420	0.3585	0.2901	0.2342	0.1887	0.1516	20	
	1	0.1652	0.2725	0.3364	0.3683	0.3774	0.3703	0.3526	0.3282	0.3000	19	
	2	0.0159	0.0528	0.0988	0.1458	0.1887	0.2246	0.2521	0.2711	0.2818	18	
	3	0.0010	0.0065	0.0183	0.0364	0.0596	0.0860	0.1139	0.1414	0.1672	17	
	4	0.0000	0.0006	0.0024	0.0065	0.0133	0.0233	0.0364	0.0523	0.0703	16	
	5	0.0000	0.0000	0.0002	0.0009	0.0022	0.0048	0.0088	0.0145	0.0222	15	
	6	0.0000	0.0000	0.0000	0.0001	0.0003	0.0008	0.0017	0.0032	0.0055	14	
	7	0.0000	0.0000	0.0000	0.0000	0.0000	0.0001	0.0002	0.0005	0.0011	13	
	8	0.0000	0.0000	0.0000	0.0000	0.0000	0.0000	0.0000	0.0001	0.0002	12	
	9	0.0000	0.0000	0.0000	0.0000	0.0000	0.0000	0.0000	0.0000	0.0000	11	
	10	0.0000	0.0000	0.0000	0.0000	0.0000	0.0000	0.0000	0.0000	0.0000	10	
	11	0.0000	0.0000	0.0000	0.0000	0.0000	0.0000	0.0000	0.0000	0.0000	9	
	12	0.0000	0.0000	0.0000	0.0000	0.0000	0.0000	0.0000	0.0000	0.0000	8	
	13	0.0000	0.0000	0.0000	0.0000	0.0000	0.0000	0.0000	0.0000	0.0000	7	
	14	0.0000	0.0000	0.0000	0.0000	0.0000	0.0000	0.0000	0.0000	0.0000	6	
	15	0.0000	0.0000	0.0000	0.0000	0.0000	0.0000	0.0000	0.0000	0.0000	5	
	16	0.0000	0.0000	0.0000	0.0000	0.0000	0.0000	0.0000	0.0000	0.0000	4	
	17	0.0000	0.0000	0.0000	0.0000	0.0000	0.0000	0.0000	0.0000	0.0000	3	
	18	0.0000	0.0000	0.0000	0.0000	0.0000	0.0000	0.0000	0.0000	0.0000	2	
	19	0.0000	0.0000	0.0000	0.0000	0.0000	0.0000	0.0000	0.0000	0.0000	1	
	20	0.0000	0.0000	0.0000	0.0000	0.0000	0.0000	0.0000	0.0000	0.0000	0	20
		.99	.98	.97	.96	.95	.94	.93	.92	.91	x	n
							p					

(continued)

TABLE A-1 (continued)

						p						
n	x	.10	.15	.20	.25	.30	.35	.40	.45	.50		
10	0	0.3487	0.1969	0.1074	0.0563	0.0282	0.0135	0.0060	0.0025	0.0010	10	
	1	0.3874	0.3474	0.2684	0.1877	0.1211	0.0725	0.0403	0.0207	0.0098	9	
	2	0.1937	0.2759	0.3020	0.2816	0.2335	0.1757	0.1209	0.0763	0.0439	8	
	3	0.0574	0.1298	0.2013	0.2503	0.2668	0.2522	0.2150	0.1665	0.1172	7	
	4	0.0112	0.0401	0.0881	0.1460	0.2001	0.2377	0.2508	0.2384	0.2051	6	
	5	0.0015	0.0085	0.0264	0.0584	0.1029	0.1536	0.2007	0.2340	0.2461	5	
	6	0.0001	0.0012	0.0055	0.0162	0.0368	0.0689	0.1115	0.1596	0.2051	4	
	7	0.0000	0.0001	0.0008	0.0031	0.0090	0.0212	0.0425	0.0746	0.1172	3	
	8	0.0000	0.0000	0.0001	0.0004	0.0014	0.0043	0.0106	0.0229	0.0439	2	
	9	0.0000	0.0000	0.0000	0.0000	0.0001	0.0005	0.0016	0.0042	0.0098	1	
	10	0.0000	0.0000	0.0000	0.0000	0.0000	0.0000	0.0001	0.0003	0.0010	0	10
12	0	0.2824	0.1422	0.0687	0.0317	0.0138	0.0057	0.0022	0.0008	0.0002	12	
	1	0.3766	0.3012	0.2062	0.1267	0.0712	0.0368	0.0174	0.0075	0.0029	11	
	2	0.2301	0.2924	0.2835	0.2323	0.1678	0.1088	0.0639	0.0339	0.0161	10	
	3	0.0852	0.1720	0.2362	0.2581	0.2397	0.1954	0.1419	0.0923	0.0537	9	
	4	0.0213	0.0683	0.1329	0.1936	0.2311	0.2367	0.2128	0.1700	0.1208	8	
	5	0.0038	0.0193	0.0532	0.1032	0.1585	0.2039	0.2270	0.2225	0.1934	7	
	6	0.0005	0.0040	0.0155	0.0401	0.0792	0.1281	0.1766	0.2124	0.2256	6	
	7	0.0000	0.0006	0.0033	0.0115	0.0291	0.0591	0.1009	0.1489	0.1934	5	
	8	0.0000	0.0001	0.0005	0.0024	0.0078	0.0199	0.0420	0.0762	0.1208	4	
	9	0.0000	0.0000	0.0001	0.0004	0.0015	0.0048	0.0125	0.0277	0.0537	3	
	10	0.0000	0.0000	0.0000	0.0000	0.0002	0.0008	0.0025	0.0068	0.0161	2	
	11	0.0000	0.0000	0.0000	0.0000	0.0000	0.0001	0.0003	0.0010	0.0029	1	
	12	0.0000	0.0000	0.0000	0.0000	0.0000	0.0000	0.0000	0.0001	0.0002	0	12
15	0	0.2059	0.0874	0.0352	0.0134	0.0047	0.0016	0.0005	0.0001	0.0000	15	
	1	0.3432	0.2312	0.1319	0.0668	0.0305	0.0126	0.0047	0.0016	0.0005	14	
	2	0.2669	0.2856	0.2309	0.1559	0.0916	0.0476	0.0219	0.0090	0.0032	13	
	3	0.1285	0.2184	0.2501	0.2252	0.1700	0.1110	0.0634	0.0318	0.0139	12	
	4	0.0428	0.1156	0.1876	0.2252	0.2186	0.1792	0.1268	0.0780	0.0417	11	
	5	0.0105	0.0449	0.1032	0.1651	0.2061	0.2123	0.1859	0.1404	0.0916	10	
	6	0.0019	0.0132	0.0430	0.0917	0.1472	0.1906	0.2066	0.1914	0.1527	9	
	7	0.0003	0.0030	0.0138	0.0393	0.0811	0.1319	0.1771	0.2013	0.1964	8	
	8	0.0000	0.0005	0.0035	0.0131	0.0348	0.0710	0.1181	0.1647	0.1964	7	
	9	0.0000	0.0001	0.0007	0.0034	0.0116	0.0298	0.0612	0.1048	0.1527	6	
	10	0.0000	0.0000	0.0001	0.0007	0.0030	0.0096	0.0245	0.0515	0.0916	5	
	11	0.0000	0.0000	0.0000	0.0001	0.0006	0.0024	0.0074	0.0191	0.0417	4	
	12	0.0000	0.0000	0.0000	0.0000	0.0001	0.0004	0.0016	0.0052	0.0139	3	
	13	0.0000	0.0000	0.0000	0.0000	0.0000	0.0001	0.0003	0.0010	0.0032	2	
	14	0.0000	0.0000	0.0000	0.0000	0.0000	0.0000	0.0000	0.0001	0.0005	1	
	15	0.0000	0.0000	0.0000	0.0000	0.0000	0.0000	0.0000	0.0000	0.0000	0	15
20	0	0.1216	0.0388	0.0115	0.0032	0.0008	0.0002	0.0000	0.0000	0.0000	20	
	1	0.2702	0.1368	0.0576	0.0211	0.0068	0.0020	0.0005	0.0001	0.0000	19	
	2	0.2852	0.2293	0.1369	0.0669	0.0278	0.0100	0.0031	0.0008	0.0002	18	
	3	0.1901	0.2428	0.2054	0.1339	0.0716	0.0323	0.0123	0.0040	0.0011	17	
	4	0.0898	0.1821	0.2182	0.1897	0.1304	0.0738	0.0350	0.0139	0.0046	16	
	5	0.0319	0.1028	0.1746	0.2023	0.1789	0.1272	0.0746	0.0365	0.0148	15	
	6	0.0089	0.0454	0.1091	0.1686	0.1916	0.1712	0.1244	0.0746	0.0370	14	
	7	0.0020	0.0160	0.0545	0.1124	0.1643	0.1844	0.1659	0.1221	0.0739	13	
	8	0.0004	0.0046	0.0222	0.0609	0.1144	0.1614	0.1797	0.1623	0.1201	12	
	9	0.0001	0.0011	0.0074	0.0271	0.0654	0.1158	0.1597	0.1771	0.1602	11	
	10	0.0000	0.0002	0.0020	0.0099	0.0308	0.0686	0.1171	0.1593	0.1762	10	
	11	0.0000	0.0000	0.0005	0.0030	0.0120	0.0336	0.0710	0.1185	0.1602	9	
	12	0.0000	0.0000	0.0001	0.0008	0.0039	0.0136	0.0355	0.0727	0.1201	8	
	13	0.0000	0.0000	0.0000	0.0002	0.0010	0.0045	0.0146	0.0366	0.0739	7	
	14	0.0000	0.0000	0.0000	0.0000	0.0002	0.0012	0.0049	0.0150	0.0370	6	
	15	0.0000	0.0000	0.0000	0.0000	0.0000	0.0003	0.0013	0.0049	0.0148	5	
	16	0.0000	0.0000	0.0000	0.0000	0.0000	0.0000	0.0003	0.0013	0.0046	4	
	17	0.0000	0.0000	0.0000	0.0000	0.0000	0.0000	0.0000	0.0002	0.0011	3	
	18	0.0000	0.0000	0.0000	0.0000	0.0000	0.0000	0.0000	0.0000	0.0002	2	
	19	0.0000	0.0000	0.0000	0.0000	0.0000	0.0000	0.0000	0.0000	0.0000	1	
	20	0.0000	0.0000	0.0000	0.0000	0.0000	0.0000	0.0000	0.0000	0.0000	0	20
		.90	.85	.80	.75	.70	.65	.60	.55	.50	x	n
						p						

TABLE A-2
Poisson Distribution

Poisson probabilities

					λ					
X	0.1	0.2	0.3	0.4	0.5	0.6	0.7	0.8	0.9	1.0
0	.9048	.8187	.7408	.6703	.6065	.5488	.4966	.4493	.4066	.3679
1	.0905	.1637	.2222	.2681	.3033	.3293	.3476	.3595	.3659	.3679
2	.0045	.0164	.0333	.0536	.0758	.0988	.1217	.1438	.1647	.1839
3	.0002	.0011	.0033	.0072	.0126	.0198	.0284	.0383	.0494	.0613
4		.0001	.0002	.0007	.0016	.0030	.0050	.0077	.0111	.0153
5				.0001	.0002	.0004	.0007	.0012	.0020	.0031
6							.0001	.0002	.0003	.0005
7										.0001

					λ					
X	1.5	2.0	2.5	3.0	3.5	4.0	4.5	5.0	6.0	7.0
0	.2231	.1353	.0821	.0498	.0302	.0183	.0111	.0067	.0025	.0009
1	.3347	.2707	.2052	.1494	.1057	.0733	.0500	.0337	.0149	.0064
2	.2510	.2707	.2565	.2240	.1850	.1465	.1125	.0842	.0446	.0223
3	.1255	.1804	.2138	.2240	.2158	.1954	.1687	.1404	.0892	.0521
4	.0471	.0902	.1336	.1680	.1888	.1954	.1898	.1755	.1339	.0912
5	.0141	.0361	.0668	.1008	.1322	.1563	.1708	.1755	.1606	.1277
6	.0035	.0120	.0278	.0504	.0771	.1042	.1281	.1462	.1606	.1490
7	.0008	.0034	.0099	.0216	.0385	.0595	.0824	.1044	.1377	.1490
8	.0001	.0009	.0031	.0081	.0169	.0298	.0463	.0653	.1033	.1304
9		.0002	.0009	.0027	.0066	.0132	.0232	.0363	.0688	.1014
10			.0002	.0008	.0023	.0053	.0104	.0181	.0413	.0710
11				.0002	.0007	.0019	.0043	.0082	.0225	.0452
12				.0001	.0002	.0006	.0016	.0034	.0113	.0264
13					.0001	.0002	.0006	.0013	.0052	.0142
14						.0001	.0002	.0005	.0022	.0071
15							.0001	.0002	.0009	.0033
16									.0003	.0014
17									.0001	.0006
18										.0002
19										.0001

Table entries are $P(X = x/\lambda)$.

TABLE A-3
Standard Normal Probabilities

z	$P(-z < Z < z)$	$P(|Z| > z)$	$P(Z > z)$	$P(Z < -z)$	$P(Z < z)$	$P(Z > -z)$
0.50	0.383	0.617	0.309	0.309	0.691	0.691
0.60	0.451	0.549	0.274	0.274	0.726	0.726
0.70	0.516	0.484	0.242	0.242	0.760	0.758
0.80	0.576	0.424	0.212	0.212	0.788	0.788
0.90	0.632	0.368	0.184	0.184	0.816	0.816
1.00	0.683	0.317	0.159	0.159	0.841	0.841
1.28	**0.800**	**0.200**	**0.100**	**0.100**	**0.900**	**0.900**
1.50	0.866	0.134	0.067	0.067	0.933	0.933
1.60	0.890	0.110	0.055	0.055	0.945	0.945
1.65	**0.900**	**0.100**	**0.050**	**0.050**	**0.950**	**0.950**
1.70	0.911	0.089	0.045	0.045	0.955	0.955
1.80	0.928	0.072	0.036	0.036	0.964	0.964
1.90	0.943	0.057	0.029	0.029	0.971	0.971
1.96	**0.950**	**0.050**	**0.025**	**0.025**	**0.975**	**0.975**
2.00	0.954	0.046	0.023	0.023	0.977	0.977
2.10	0.964	0.036	0.018	0.018	0.982	0.982
2.20	0.972	0.028	0.014	0.014	0.986	0.986
2.30	0.979	0.021	0.011	0.011	0.989	0.989
2.40	0.984	0.016	0.008	0.008	0.992	0.992
2.50	0.988	0.012	0.006	0.006	0.994	0.994
2.58	**0.990**	**0.010**	**0.005**	**0.005**	**0.995**	**0.995**
2.60	0.991	0.009	0.005	0.005	0.995	0.995
2.70	0.993	0.007	0.003	0.003	0.997	0.997
2.80	0.995	0.005	0.003	0.003	0.997	0.997
2.90	0.996	0.004	0.002	0.002	0.998	0.998
3.00	0.997	0.003	0.001	0.001	0.999	0.999
3.10	0.998	0.002	0.001	0.001	0.999	0.999
3.20	0.999	0.001	0.001	0.001	0.999	0.999
3.30	0.999	0.001	0.000	0.000	1.000	1.000
3.40	0.999	0.001	0.000	0.000	1.000	1.000
3.50	1.000	0.000	0.000	0.000	1.000	1.000

TABLE A-4
t Distribution

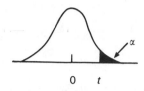

df \ α	.10	.05	.025	.01	.005
1	3.078	6.314	12.706	31.821	63.657
2	1.886	2.920	4.303	6.965	9.925
3	1.638	2.353	3.182	4.541	5.841
4	1.533	2.132	2.776	3.747	4.604
5	1.476	2.015	2.571	3.365	4.032
6	1.440	1.943	2.447	3.143	3.707
7	1.415	1.895	2.365	2.998	3.499
8	1.397	1.860	2.306	2.896	3.355
9	1.383	1.833	2.262	2.821	3.250
10	1.372	1.812	2.228	2.764	3.169
11	1.363	1.796	2.201	2.718	3.106
12	1.356	1.782	2.179	2.681	3.055
13	1.350	1.771	2.160	2.650	3.012
14	1.345	1.761	2.145	2.624	2.977
15	1.341	1.753	2.131	2.602	2.947
16	1.337	1.746	2.120	2.583	2.921
17	1.333	1.740	2.110	2.567	2.898
18	1.330	1.734	2.101	2.552	2.878
19	1.328	1.729	2.093	2.539	2.861
20	1.325	1.725	2.086	2.528	2.845
21	1.323	1.721	2.080	2.518	2.831
22	1.321	1.717	2.074	2.508	2.819
23	1.319	1.714	2.069	2.500	2.807
24	1.318	1.711	2.064	2.492	2.797
25	1.316	1.708	2.060	2.485	2.787
26	1.315	1.706	2.056	2.479	2.779
27	1.314	1.703	2.052	2.473	2.771
28	1.313	1.701	2.048	2.467	2.763
29	1.311	1.699	2.045	2.462	2.756
30	1.310	1.697	2.042	2.457	2.750
40	1.303	1.684	2.021	2.423	2.704
60	1.296	1.671	2.000	2.390	2.660
120	1.289	1.658	1.980	2.358	2.617
∞	1.282	1.645	1.960	2.326	2.576

TABLE A-5
One-Tailed Probabilities of *T*

	Degrees of freedom							
t-value	1	2	3	4	5	6	7	8
0.5	0.352	0.333	0.326	0.322	0.319	0.317	0.316	0.315
0.6	0.328	0.305	0.295	0.290	0.287	0.285	0.284	0.283
0.7	0.306	0.278	0.267	0.261	0.258	0.255	0.253	0.252
0.8	0.285	0.254	0.241	0.234	0.230	0.227	0.225	0.223
0.9	0.267	0.232	0.217	0.210	0.205	0.201	0.199	0.197
1.0	0.250	0.211	0.196	0.187	0.182	0.178	0.175	0.173
1.1	0.235	0.193	0.176	0.167	0.161	0.157	0.154	0.152
1.2	0.221	0.177	0.158	0.148	0.142	0.138	0.135	0.132
1.3	0.209	0.162	0.142	0.132	0.125	0.121	0.117	0.115
1.4	0.197	0.148	0.128	0.117	0.110	0.106	0.102	0.100
1.5	0.187	0.136	0.115	0.104	0.097	0.092	0.089	0.086
1.6	0.178	0.125	0.104	0.092	0.085	0.080	0.077	0.074
1.7	0.169	0.116	0.094	0.082	0.075	0.070	0.066	0.064
1.8	0.161	0.107	0.085	0.073	0.066	0.061	0.057	0.055
1.9	0.154	0.099	0.077	0.065	0.058	0.053	0.050	0.047
2.0	0.148	0.092	0.070	0.058	0.051	0.046	0.043	0.040
2.1	0.141	0.085	0.063	0.052	0.045	0.040	0.037	0.034
2.2	0.136	0.079	0.058	0.046	0.040	0.035	0.032	0.029
2.3	0.131	0.074	0.052	0.041	0.035	0.031	0.027	0.025
2.4	0.126	0.069	0.048	0.037	0.031	0.027	0.024	0.022
2.5	0.121	0.065	0.044	0.033	0.027	0.023	0.020	0.018
2.6	0.117	0.061	0.040	0.030	0.024	0.020	0.018	0.016
2.7	0.113	0.057	0.037	0.027	0.021	0.018	0.015	0.014
2.8	0.109	0.054	0.034	0.024	0.019	0.016	0.013	0.012
2.9	0.106	0.051	0.031	0.022	0.017	0.014	0.011	0.010
3.0	0.102	0.048	0.029	0.020	0.015	0.012	0.010	0.009
3.1	0.099	0.045	0.027	0.018	0.013	0.011	0.009	0.007
3.2	0.096	0.043	0.025	0.016	0.012	0.009	0.008	0.006
3.3	0.094	0.040	0.023	0.015	0.011	0.008	0.007	0.005
3.4	0.091	0.038	0.021	0.014	0.010	0.007	0.006	0.005
3.5	0.089	0.036	0.020	0.012	0.009	0.006	0.005	0.004
4.0	0.078	0.029	0.014	0.008	0.005	0.004	0.003	0.002
4.5	0.070	0.023	0.010	0.005	0.003	0.002	0.001	0.001
5.0	0.063	0.019	0.008	0.004	0.002	0.001	0.001	0.001

(continued)

TABLE A-5 (continued)

t-value	Degrees of freedom							
	9	10	15	20	30	50	100	Infinity
0.5	0.315	0.314	0.312	0.311	0.310	0.310	0.309	0.309
0.6	0.282	0.281	0.279	0.278	0.277	0.276	0.275	0.274
0.7	0.251	0.250	0.247	0.246	0.245	0.244	0.243	0.242
0.8	0.222	0.221	0.218	0.217	0.215	0.214	0.213	0.212
0.9	0.196	0.195	0.191	0.189	0.188	0.186	0.395	0.184
1.0	0.172	0.170	0.167	0.165	0.163	0.161	0.160	0.159
1.1	0.150	0.149	0.144	0.142	0.140	0.138	0.137	0.136
1.2	0.130	0.129	0.124	0.122	0.120	0.118	0.116	0.115
1.3	0.113	0.111	0.107	0.104	0.102	0.100	0.098	0.097
1.4	0.098	0.096	0.091	0.088	0.086	0.084	0.082	0.081
1.5	0.084	0.082	0.077	0.075	0.072	0.070	0.068	0.067
1.6	0.072	0.070	0.065	0.063	0.060	0.058	0.056	0.055
1.7	0.062	0.060	0.055	0.052	0.050	0.048	0.046	0.045
1.8	0.053	0.051	0.046	0.043	0.041	0.039	0.037	0.036
1.9	0.045	0.043	0.038	0.036	0.034	0.032	0.030	0.029
2.0	0.038	0.037	0.032	0.030	0.027	0.025	0.024	0.023
2.1	0.033	0.031	0.027	0.024	0.022	0.020	0.019	0.018
2.2	0.028	0.026	0.022	0.020	0.018	0.016	0.015	0.014
2.3	0.023	0.022	0.018	0.016	0.014	0.013	0.012	0.011
2.4	0.020	0.019	0.015	0.013	0.011	0.010	0.009	0.008
2.5	0.017	0.016	0.012	0.011	0.009	0.008	0.007	0.006
2.6	0.014	0.013	0.010	0.009	0.007	0.006	0.005	0.005
2.7	0.012	0.011	0.008	0.007	0.006	0.005	0.004	0.003
2.8	0.010	0.009	0.007	0.006	0.004	0.004	0.003	0.003
2.9	0.009	0.008	0.005	0.004	0.003	0.003	0.002	0.002
3.0	0.007	0.007	0.004	0.004	0.003	0.002	0.002	0.001
3.1	0.006	0.006	0.004	0.003	0.002	0.002	0.001	0.001
3.2	0.005	0.005	0.003	0.002	0.002	0.001	0.001	0.001
3.3	0.005	0.004	0.002	0.002	0.001	0.001	0.001	0.000
3.4	0.004	0.003	0.002	0.001	0.001	0.001	0.000	0.000
3.5	0.003	0.003	0.002	0.001	0.001	0.000	0.000	0.000
4.0	0.002	0.001	0.001	0.000	0.000	0.000	0.000	0.000
4.5	0.001	0.001	0.000	0.000	0.000	0.000	0.000	0.000
5.0	0.000	0.000	0.000	0.000	0.000	0.000	0.000	0.000

TABLE A-6
Two-Tailed Probabilities of T

	Degrees of freedom							
t-value	1	2	3	4	5	6	7	8
0.5	0.705	0.667	0.651	0.643	0.638	0.635	0.632	0.631
0.6	0.656	0.609	0.591	0.581	0.575	0.570	0.567	0.565
0.7	0.611	0.556	0.534	0.523	0.515	0.510	0.507	0.504
0.8	0.570	0.508	0.482	0.469	0.460	0.454	0.450	0.447
0.9	0.533	0.463	0.434	0.419	0.409	0.403	0.398	0.394
1.0	0.500	0.423	0.391	0.374	0.363	0.356	0.351	0.347
1.1	0.470	0.386	0.352	0.333	0.321	0.313	0.308	0.303
1.2	0.442	0.353	0.316	0.296	0.284	0.275	0.269	0.264
1.3	0.417	0.323	0.284	0.263	0.250	0.241	0.235	0.230
1.4	0.395	0.296	0.256	0.234	0.220	0.211	0.204	0.199
1.5	0.374	0.272	0.231	0.208	0.194	0.184	0.177	0.172
1.6	0.356	0.251	0.208	0.185	0.170	0.161	0.154	0.148
1.7	0.339	0.231	0.188	0.164	0.150	0.140	0.133	0.128
1.8	0.323	0.214	0.170	0.146	0.132	0.122	0.115	0.110
1.9	0.308	0.198	0.154	0.130	0.116	0.106	0.099	0.094
2.0	0.295	0.184	0.139	0.116	0.102	0.092	0.086	0.081
2.1	0.283	0.171	0.127	0.104	0.090	0.080	0.074	0.069
2.2	0.272	0.159	0.115	0.093	0.079	0.070	0.064	0.059
2.3	0.261	0.148	0.105	0.083	0.070	0.061	0.055	0.050
2.4	0.251	0.138	0.096	0.074	0.062	0.053	0.047	0.043
2.5	0.242	0.130	0.088	0.067	0.054	0.047	0.041	0.037
2.6	0.234	0.122	0.080	0.060	0.048	0.041	0.035	0.032
2.7	0.226	0.114	0.074	0.054	0.043	0.036	0.031	0.027
2.8	0.218	0.107	0.068	0.049	0.038	0.031	0.027	0.023
2.9	0.211	0.101	0.063	0.044	0.034	0.027	0.023	0.020
3.0	0.205	0.095	0.058	0.040	0.030	0.024	0.020	0.017
3.1	0.199	0.090	0.053	0.036	0.027	0.021	0.017	0.015
3.2	0.193	0.085	0.049	0.033	0.024	0.019	0.015	0.013
3.3	0.187	0.081	0.046	0.030	0.021	0.016	0.013	0.011
3.4	0.182	0.077	0.042	0.027	0.019	0.014	0.011	0.009
3.5	0.177	0.073	0.039	0.025	0.017	0.013	0.010	0.008
4.0	0.156	0.057	0.028	0.016	0.010	0.007	0.005	0.004
4.5	0.139	0.046	0.020	0.011	0.006	0.004	0.003	0.002
5.0	0.126	0.038	0.015	0.007	0.004	0.002	0.002	0.001

(continued)

TABLE A-6 (continued)

	Degrees of freedom							
t-value	9	10	15	20	30	50	100	Infinity
0.5	0.629	0.628	0.624	0.623	0.621	0.619	0.618	0.617
0.6	0.563	0.562	0.557	0.555	0.553	0.551	0.550	0.549
0.7	0.502	0.500	0.495	0.492	0.489	0.487	0.486	0.484
0.8	0.444	0.442	0.436	0.433	0.430	0.427	0.426	0.424
0.9	0.392	0.389	0.382	0.379	0.375	0.372	0.370	0.368
1.0	0.343	0.341	0.333	0.329	0.325	0.322	0.320	0.317
1.1	0.300	0.297	0.289	0.284	0.280	0.277	0.274	0.271
1.2	0.261	0.258	0.249	0.244	0.240	0.236	0.233	0.230
1.3	0.226	0.223	0.213	0.208	0.204	0.200	0.197	0.194
1.4	0.195	0.192	0.182	0.177	0.172	0.168	0.165	0.162
1.5	0.168	0.165	0.154	0.149	0.144	0.140	0.137	0.134
1.6	0.144	0.141	0.130	0.125	0.120	0.116	0.113	0.110
1.7	0.123	0.120	0.110	0.105	0.099	0.095	0.092	0.089
1.8	0.105	0.102	0.092	0.087	0.082	0.078	0.075	0.072
1.9	0.090	0.087	0.077	0.072	0.067	0.063	0.060	0.057
2.0	0.077	0.073	0.064	0.059	0.055	0.051	0.048	0.046
2.1	0.065	0.062	0.053	0.049	0.044	0.041	0.038	0.036
2.2	0.055	0.052	0.044	0.040	0.036	0.032	0.030	0.028
2.3	0.047	0.044	0.036	0.032	0.029	0.026	0.024	0.021
2.4	0.040	0.037	0.030	0.026	0.023	0.020	0.018	0.016
2.5	0.034	0.031	0.025	0.021	0.018	0.016	0.014	0.012
2.6	0.029	0.026	0.020	0.017	0.014	0.012	0.011	0.009
2.7	0.024	0.022	0.016	0.014	0.011	0.009	0.008	0.007
2.8	0.021	0.019	0.013	0.011	0.009	0.007	0.006	0.005
2.9	0.018	0.016	0.011	0.009	0.007	0.006	0.005	0.004
3.0	0.015	0.013	0.009	0.007	0.005	0.004	0.003	0.003
3.1	0.013	0.011	0.007	0.006	0.004	0.003	0.003	0.002
3.2	0.011	0.009	0.006	0.004	0.003	0.002	0.002	0.001
3.3	0.009	0.008	0.005	0.004	0.002	0.002	0.001	0.001
3.4	0.008	0.007	0.004	0.003	0.002	0.001	0.001	0.001
3.5	0.007	0.006	0.003	0.002	0.001	0.001	0.001	0.000
4.0	0.003	0.003	0.001	0.001	0.000	0.000	0.000	0.000
4.5	0.001	0.001	0.000	0.000	0.000	0.000	0.000	0.000
5.0	0.001	0.001	0.000	0.000	0.000	0.000	0.000	0.000

Two-Tailed F-Table

Table A-7 can be used in testing hypotheses about the variances of two populations. Most often, the null hypothesis will be that the population variances are equal, with the alternate hypothesis being they are unequal. The theory behind the test requires that the underlying variables are normally distritubed. If so, and if the population variances are equal, the ratio of the sample variances, s_1^2/s_2^2, follows an F distribution. In a two-sided test, the null hypothesis $\sigma_1^2 = \sigma_2^2$ is rejected if the ratio of sample variances is far from unity in either direction. To reject H_0 at some level α, the probability of landing in *either* tail must be less than or equal to α. Alternatively, the *PROB-VALUE* of the same test is $P(F > s_1^2/s_2^2) + P(F < s_2^2/s_1^2)$. Table A-7 gives these two-tailed probabilities in the column labeled α. To use the table, *take the larger of the two sample variances as s_1^2*. (The observed ratio will never be less than unity.) Find the observed F ratio in the body of the table using the appropriate degrees of freedom:

$$r_1 = n_1 - 1 \quad \text{and} \quad r_2 = n_2 - 1$$

The probabilities are two-tailed probabilities, meaning that each tail contains $\alpha/2$, as shown in the diagram below.

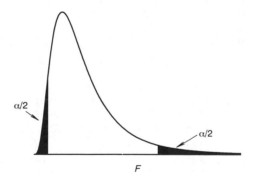

Example:

$H_0: \sigma_1^2 = \sigma_2^2$ $H_A: \sigma_1^2 \neq \sigma_2^2$

$n_1 = 25, s_1 = 75.5$ $n_2 = 13, s_2 = 25$

Observed f: $f = \dfrac{75.5}{25} = 3.02$ Degrees of freedom: $r_1 = 24, r_2 = 12$

PROB-VALUE: .05 (see page 617)

TABLE A-7
Two-Tailed F Table

r_1	α	1	2	3	4	5	6	7	8	9
						$r1$				
1	.2	39.9	49.5	53.59	55.83	57.24	58.2	58.91	59.44	59.86
	.1	161.4	199.5	215.71	224.58	230.16	233.99	236.77	238.88	240.54
	.05	647.8	799.5	864.16	899.58	921.85	937.11	948.22	956.66	963.28
	.02	4,052.2	4,999.5	5,403.35	5,624.58	5,763.65	5,858.99	5,928.36	5,981.07	6,022.47
	.01	16,210.7	19,999.5	21,614.74	22,499.58	23,055.8	23,437.11	23,714.57	23,925.41	24,091
	.001	1,621,138	2,000,000	2,161,518	2,250,000	2,305,620	2,343,750	2,371,494	2,392,578	2,409,138
2	.2	8.53	9.00	9.16	9.24	9.29	9.33	9.35	9.37	9.38
	.1	18.51	19.00	19.16	19.25	19.30	19.33	19.35	19.37	19.38
	.05	38.51	39.00	39.17	39.25	39.30	39.33	39.36	39.37	39.39
	.02	98.50	99.00	99.17	99.25	99.30	99.33	99.36	99.37	99.39
	.01	198.50	199.00	199.17	199.25	199.30	199.33	199.36	199.37	199.39
	.001	1,998.50	1,999.00	1,999.17	1,999.25	1,999.30	1,999.33	1,999.36	1,999.38	1,999.39
3	.2	5.54	5.46	5.39	5.34	5.31	5.28	5.27	5.25	5.24
	.1	10.13	9.55	9.28	9.12	9.01	8.94	8.89	8.85	8.81
	.05	17.44	16.04	15.44	15.10	14.88	14.73	14.62	14.54	14.47
	.02	34.12	30.82	29.46	28.71	28.24	27.91	27.67	27.49	27.35
	.01	55.55	49.80	47.47	46.19	45.39	44.84	44.43	44.13	43.88
	.001	266.55	236.61	224.70	218.25	214.20	211.41	209.38	207.83	206.61
4	.2	4.54	4.32	4.19	4.11	4.05	4.01	3.98	3.95	3.94
	.1	7.71	6.94	6.59	6.39	6.26	6.16	6.09	6.04	6.00
	.05	12.22	10.65	9.98	9.60	9.36	9.20	9.07	8.98	8.90
	.02	21.20	18.00	16.69	15.98	15.52	15.21	14.98	14.80	14.66
	.01	31.33	26.28	24.26	23.15	22.46	21.97	21.62	21.35	21.14
	.001	106.22	87.44	80.09	76.12	73.63	71.92	70.66	69.71	68.95
5	.2	4.06	3.78	3.62	3.52	3.45	3.40	3.37	3.34	3.32
	.1	6.61	5.79	5.41	5.19	5.05	4.95	4.88	4.82	4.77
	.05	10.01	8.43	7.76	7.39	7.15	6.98	6.85	6.76	6.68
	.02	16.26	13.27	12.06	11.39	10.97	10.67	10.46	10.29	10.16
	.01	22.78	18.31	16.53	15.56	14.94	14.51	14.20	13.96	13.77
	.001	63.61	49.78	44.42	41.53	39.72	38.47	37.56	36.86	36.31
6	.2	3.78	3.46	3.29	3.18	3.11	3.05	3.01	2.98	2.96
	.1	5.99	5.14	4.76	4.53	4.39	4.28	4.21	4.15	4.10
	.05	8.81	7.26	6.60	6.23	5.99	5.82	5.70	5.60	5.52
	.02	13.75	10.92	9.78	9.15	8.75	8.47	8.26	8.10	7.98
	.01	18.63	14.54	12.92	12.03	11.46	11.07	10.79	10.57	10.39
	.001	46.08	34.80	30.45	28.12	26.65	25.63	24.89	24.33	23.88

TABLE A-7 (continued)

r_2	α	10	12	15	20	24	30	60	120	∞
						r_1				
1	.2	60.19	60.71	61.22	61.74	62	62.26	62.79	63.06	63.33
	.1	241.88	243.91	245.95	248.01	249.05	250.1	252.2	253.25	254.31
	.05	968.63	976.71	984.87	993.1	997.25	1,001.41	1,009.8	1,014.02	1,018.24
	.02	6,055.85	6,106.32	6,157.28	6,208.73	6,234.63	6,260.65	6,313.03	6,339.39	6,365.76
	.01	24,224.49	24,426.37	24,630.21	24,835.97	24,939.56	25,043.63	25,253.14	25,358.57	25,464.03
	.001	2,422,485	2,442,672	2,463,056	2,483,632	2,493,990	2,504,396	2,525,348	2,535,890	2,546,436
2	.2	9.39	9.41	9.42	9.44	9.45	9.46	9.47	9.48	9.49
	.1	19.4	19.41	19.43	19.45	19.45	19.46	19.48	19.49	19.5
	.05	39.4	39.41	39.43	39.45	39.46	39.46	39.48	39.49	39.5
	.02	99.4	99.42	99.43	99.45	99.46	99.47	99.48	99.49	99.5
	.01	199.4	199.42	199.43	199.45	199.46	199.47	199.48	199.49	199.5
	.001	1,999.4	1,999.42	1,999.43	1,999.45	1,999.46	1,999.47	1,999.48	1,999.49	1,999.5
3	.2	5.23	5.22	5.2	5.18	5.18	5.17	5.15	5.14	5.13
	.1	8.79	8.74	8.7	8.66	8.64	8.62	8.57	8.55	8.53
	.05	14.42	14.34	14.25	14.17	14.12	14.08	13.99	13.95	13.9
	.02	27.23	27.05	26.87	26.69	26.6	26.5	26.32	26.22	26.13
	.01	43.69	43.39	43.08	42.78	42.62	42.47	42.15	41.99	41.83
	.001	205.63	204.13	202.62	201.08	200.31	199.53	197.95	197.15	196.35
4	.2	3.92	3.9	3.87	3.84	3.83	3.82	3.79	3.78	3.76
	.1	5.96	5.91	5.86	5.8	5.77	5.75	5.69	5.66	5.63
	.05	8.84	8.75	8.66	8.56	8.51	8.46	8.36	8.31	8.26
	.02	14.55	14.37	14.2	14.02	13.93	13.84	13.65	13.56	13.46
	.01	20.97	20.7	20.44	20.17	20.03	19.89	19.61	19.47	19.33
	.001	68.35	67.42	66.48	65.53	65.05	64.56	63.58	63.08	62.58
5	.2	3.3	3.27	3.24	3.21	3.19	3.17	3.14	3.12	3.11
	.1	4.74	4.68	4.62	4.56	4.53	4.5	4.43	4.4	4.37
	.05	6.62	6.52	6.43	6.33	6.28	6.23	6.12	6.07	6.02
	.02	10.05	9.89	9.72	9.55	9.47	9.38	9.2	9.11	9.02
	.01	13.62	13.38	13.15	12.9	12.78	12.66	12.4	12.27	12.14
	.001	35.86	35.19	34.5	33.8	33.44	33.09	32.36	31.99	31.62
6	.2	2.94	2.9	2.87	2.84	2.82	2.8	2.76	2.74	2.72
	.1	4.06	4	3.94	3.87	3.84	3.81	3.74	3.7	3.67
	.05	5.46	5.37	5.27	5.17	5.12	5.07	4.96	4.9	4.85
	.02	7.87	7.72	7.56	7.4	7.31	7.23	7.06	6.97	6.88
	.01	10.25	10.03	9.81	9.59	9.47	9.36	9.12	9	8.88
	.001	23.52	22.96	22.4	21.83	21.54	21.25	20.65	20.35	20.04

(*continued*)

TABLE A-7 (continued)

r_2	α	r_1								
		1	2	3	4	5	6	7	8	9
7	.2	3.59	3.26	3.07	2.96	2.88	2.83	2.78	2.75	2.72
	.1	5.59	4.74	4.35	4.12	3.97	3.87	3.79	3.73	3.68
	.05	8.07	6.54	5.89	5.52	5.29	5.12	4.99	4.90	4.82
	.02	12.25	9.55	8.45	7.85	7.46	7.19	6.99	6.84	6.72
	.01	16.24	12.40	10.88	10.05	9.52	9.16	8.89	8.68	8.51
	.001	36.99	27.21	23.46	21.44	20.17	19.30	18.66	18.17	17.78
8	.2	3.46	3.11	2.92	2.81	2.73	2.67	2.62	2.59	2.56
	.1	5.32	4.46	4.07	3.84	3.69	3.58	3.5	3.44	3.39
	.05	7.57	6.06	5.42	5.05	4.82	4.65	4.53	4.43	4.36
	.02	11.26	8.65	7.59	7.01	6.63	6.37	6.18	6.03	5.91
	.01	14.69	11.04	9.6	8.81	8.3	7.95	7.69	7.5	7.34
	.001	31.56	22.75	19.39	17.58	16.44	15.66	15.08	14.64	14.29
9	.2	3.36	3.01	2.81	2.69	2.61	2.55	2.51	2.47	2.44
	.1	5.12	4.26	3.86	3.63	3.48	3.37	3.29	3.23	3.18
	.05	7.21	5.71	5.08	4.72	4.48	4.32	4.2	4.1	4.03
	.02	10.56	8.02	6.99	6.42	6.06	5.8	5.61	5.47	5.35
	.01	13.61	10.11	8.72	7.96	7.47	7.13	6.88	6.69	6.54
	.001	27.99	19.87	16.77	15.11	14.06	13.34	12.8	12.4	12.08
10	.2	3.29	2.92	2.73	2.61	2.52	2.46	2.41	2.38	2.35
	.1	4.96	4.1	3.71	3.48	3.33	3.22	3.14	3.07	3.02
	.05	6.94	5.46	4.83	4.47	4.24	4.07	3.95	3.85	3.78
	.02	10.04	7.56	6.55	5.99	5.64	5.39	5.2	5.06	4.94
	.01	12.83	9.43	8.08	7.34	6.87	6.54	6.3	6.12	5.97
	.001	25.49	17.87	14.97	13.41	12.43	11.75	11.25	10.87	10.56
12	.2	3.18	2.81	2.61	2.48	2.39	2.33	2.28	2.24	2.21
	.1	4.75	3.89	3.49	3.26	3.11	3	2.91	2.85	2.8
	.05	6.55	5.1	4.47	4.12	3.89	3.73	3.61	3.51	3.44
	.02	9.33	6.93	5.95	5.41	5.06	4.82	4.64	4.5	4.39
	.01	11.75	8.51	7.23	6.52	6.07	5.76	5.52	5.35	5.2
	.001	22.24	15.3	12.66	11.25	10.35	9.74	9.28	8.94	8.66
15	.2	3.07	2.7	2.49	2.36	2.27	2.21	2.16	2.12	2.09
	.1	4.54	3.68	3.29	3.06	2.9	2.79	2.71	2.64	2.59
	.05	6.2	4.77	4.15	3.8	3.58	3.41	3.29	3.2	3.12
	.02	8.68	6.36	5.42	4.89	4.56	4.32	4.14	4	3.89
	.01	10.8	7.7	6.48	5.8	5.37	5.07	4.85	4.67	4.54
	.001	19.51	13.16	10.76	9.48	8.66	8.1	7.68	7.37	7.11

TABLE A-7 (continued)

r_2	α	r_1								
		10	12	15	20	24	30	60	120	∞
7	.2	2.7	2.67	2.63	2.59	2.58	2.56	2.51	2.49	2.47
	.1	3.64	3.57	3.51	3.44	3.41	3.38	3.3	3.27	3.23
	.05	4.76	4.67	4.57	4.47	4.41	4.36	4.25	4.2	4.14
	.02	6.62	6.47	6.31	6.16	6.07	5.99	5.82	5.74	5.65
	.01	8.38	8.18	7.97	7.75	7.64	7.53	7.31	7.19	7.08
	.001	17.47	16.99	16.5	16	15.75	15.49	14.97	14.71	14.44
8	.2	2.54	2.5	2.46	2.42	2.4	2.38	2.34	2.32	2.29
	.1	3.35	3.28	3.22	3.15	3.12	3.08	3.01	2.97	2.93
	.05	4.3	4.2	4.1	4	3.95	3.89	3.78	3.73	3.67
	.02	5.81	5.67	5.52	5.36	5.28	5.2	5.03	4.95	4.86
	.01	7.21	7.01	6.81	6.61	6.5	6.4	6.18	6.06	5.95
	.001	14.01	13.58	13.14	12.69	12.46	12.22	11.75	11.51	11.26
9	.2	2.42	2.38	2.34	2.3	2.28	2.25	2.21	2.18	2.16
	.1	3.14	3.07	3.01	2.94	2.9	2.86	2.79	2.75	2.71
	.05	3.96	3.87	3.77	3.67	3.61	3.56	3.45	3.39	3.33
	.02	5.26	5.11	4.96	4.81	4.73	4.65	4.48	4.4	4.31
	.01	6.42	6.23	6.03	5.83	5.73	5.62	5.41	5.3	5.19
	.001	11.81	11.42	11.01	10.59	10.38	10.16	9.72	9.49	9.26
10	.2	2.32	2.28	2.24	2.2	2.18	2.16	2.11	2.08	2.06
	.1	2.98	2.91	2.85	2.77	2.74	2.7	2.62	2.58	2.54
	.05	3.72	3.62	3.52	3.42	3.37	3.31	3.2	3.14	3.08
	.02	4.85	4.71	4.56	4.41	4.33	4.25	4.08	4	3.91
	.01	5.85	5.66	5.47	5.27	5.17	5.07	4.86	4.75	4.64
	.001	10.32	9.94	9.56	9.17	8.96	8.76	8.34	8.12	7.91
12	.2	2.19	2.15	2.1	2.06	2.04	2.01	1.96	1.93	1.9
	.1	2.75	2.69	2.62	2.54	2.51	2.47	2.38	2.34	2.3
	.05	3.37	3.28	3.18	3.07	3.02	2.96	2.85	2.79	2.73
	.02	4.3	4.16	4.01	3.86	3.78	3.7	3.54	3.45	3.36
	.01	5.09	4.91	4.72	4.53	4.43	4.33	4.12	4.01	3.9
	.001	8.43	8.09	7.74	7.37	7.19	7	6.61	6.41	6.2
15	.2	2.06	2.02	1.97	1.92	1.9	1.87	1.82	1.79	1.76
	.1	2.54	2.48	2.4	2.33	2.29	2.25	2.16	2.11	2.07
	.05	3.06	2.96	2.86	2.76	2.7	2.64	2.52	2.46	2.4
	.02	3.8	3.67	3.52	3.37	3.29	3.21	3.05	2.96	2.87
	.01	4.42	4.25	4.07	3.88	3.79	3.69	3.48	3.37	3.26
	.001	6.91	6.59	6.26	5.93	5.75	5.58	5.21	5.02	4.83

(*continued*)

TABLE A-7 (continued)

r_2	α	r_1 1	2	3	4	5	6	7	8	9
20	.2	2.97	2.59	2.38	2.25	2.16	2.09	2.04	2	1.96
	.1	4.35	3.49	3.1	2.87	2.71	2.6	2.51	2.45	2.39
	.05	5.87	4.46	3.86	3.51	3.29	3.13	3.01	2.91	2.84
	.02	8.1	5.85	4.94	4.43	4.1	3.87	3.7	3.56	3.46
	.01	9.94	6.99	5.82	5.17	4.76	4.47	4.26	4.09	3.96
	.001	17.19	11.38	9.2	8.02	7.27	6.76	6.38	6.09	5.85
24	.2	2.93	2.54	2.33	2.19	2.1	2.04	1.98	1.94	1.91
	.1	4.26	3.4	3.01	2.78	2.62	2.51	2.42	2.36	2.3
	.05	5.72	4.32	3.72	3.38	3.15	2.99	2.87	2.78	2.7
	.02	7.82	5.61	4.72	4.22	3.9	3.67	3.5	3.36	3.26
	.01	9.55	6.66	5.52	4.89	4.49	4.2	3.99	3.83	3.69
	.001	16.17	10.61	8.51	7.39	6.68	6.18	5.82	5.54	5.31
30	.2	2.88	2.49	2.28	2.14	2.05	1.98	1.93	1.88	1.85
	.1	4.17	3.32	2.92	2.69	2.53	2.42	2.33	2.27	2.21
	.05	5.57	4.18	3.59	3.25	3.03	2.87	2.75	2.65	2.57
	.02	7.56	5.39	4.51	4.02	3.7	3.47	3.3	3.17	3.07
	.01	9.18	6.35	5.24	4.62	4.23	3.95	3.74	3.58	3.45
	.001	15.22	9.9	7.89	6.82	6.13	5.66	5.31	5.04	4.82
60	.2	2.79	2.39	2.18	2.04	1.95	1.87	1.82	1.77	1.74
	.1	4	3.15	2.76	2.53	2.37	2.25	2.17	2.1	2.04
	.05	5.29	3.93	3.34	3.01	2.79	2.63	2.51	2.41	2.33
	.02	7.08	4.98	4.13	3.65	3.34	3.12	2.95	2.82	2.72
	.01	8.49	5.79	4.73	4.14	3.76	3.49	3.29	3.13	3.01
	.001	13.55	8.65	6.81	5.82	5.2	4.76	4.44	4.19	3.98
120	.2	2.75	2.35	2.13	1.99	1.9	1.82	1.77	1.72	1.68
	.1	3.92	3.07	2.68	2.45	2.29	2.18	2.09	2.02	1.96
	.05	5.15	3.8	3.23	2.89	2.67	2.52	2.39	2.3	2.22
	.02	6.85	4.79	3.95	3.48	3.17	2.96	2.79	2.66	2.56
	.01	8.18	5.54	4.5	3.92	3.55	3.28	3.09	2.93	2.81
	.001	12.8	8.1	6.34	5.39	4.79	4.37	4.06	3.82	3.62
∞	.2	2.71	2.3	2.08	1.95	1.85	1.77	1.72	1.67	1.63
	.1	3.84	3	2.61	2.37	2.21	2.1	2.01	1.94	1.88
	.05	5.02	3.69	3.12	2.79	2.57	2.41	2.29	2.19	2.11
	.02	6.64	4.61	3.78	3.32	3.02	2.8	2.64	2.51	2.41
	.01	7.88	5.3	4.28	3.72	3.35	3.09	2.9	2.75	2.62
	.001	12.12	7.6	5.91	5	4.42	4.02	3.72	3.48	3.3

TABLE A-7 (continued)

r_2	α	r_1								
		10	12	15	20	24	30	60	120	∞
20	.2	1.94	1.89	1.84	1.79	1.77	1.74	1.68	1.64	1.61
	.1	2.35	2.28	2.2	2.12	2.08	2.04	1.95	1.9	1.84
	.05	2.77	2.68	2.57	2.46	2.41	2.35	2.22	2.16	2.09
	.02	3.37	3.23	3.09	2.94	2.86	2.78	2.61	2.52	2.42
	.01	3.85	3.68	3.5	3.32	3.22	3.12	2.92	2.81	2.69
	.001	5.66	5.37	5.07	4.75	4.59	4.42	4.08	3.9	3.71
24	.2	1.88	1.83	1.78	1.73	1.7	1.67	1.61	1.57	1.53
	.1	2.25	2.18	2.11	2.03	1.98	1.94	1.84	1.79	1.73
	.05	2.64	2.54	2.44	2.33	2.27	2.21	2.08	2.01	1.94
	.02	3.17	3.03	2.89	2.74	2.66	2.58	2.4	2.31	2.21
	.01	3.59	3.42	3.25	3.06	2.97	2.87	2.66	2.55	2.43
	.001	5.13	4.85	4.56	4.25	4.09	3.93	3.59	3.41	3.22
30	.2	1.82	1.77	1.72	1.67	1.64	1.61	1.54	1.5	1.46
	.1	2.16	2.09	2.01	1.93	1.89	1.84	1.74	1.68	1.62
	.05	2.51	2.41	2.31	2.2	2.14	2.07	1.94	1.87	1.79
	.02	2.98	2.84	2.7	2.55	2.47	2.39	2.21	2.11	2.01
	.01	3.34	3.18	3.01	2.82	2.73	2.63	2.42	2.3	2.18
	.001	4.65	4.38	4.09	3.8	3.64	3.49	3.15	2.97	2.78
60	.2	1.71	1.66	1.6	1.54	1.51	1.48	1.4	1.35	1.29
	.1	1.99	1.92	1.84	1.75	1.7	1.65	1.53	1.47	1.39
	.05	2.27	2.17	2.06	1.94	1.88	1.82	1.67	1.58	1.48
	.02	2.63	2.5	2.35	2.2	2.12	2.03	1.84	1.73	1.6
	.01	2.9	2.74	2.57	2.39	2.29	2.19	1.96	1.83	1.69
	.001	3.82	3.57	3.3	3.02	2.87	2.71	2.38	2.19	1.98
120	.2	1.65	1.6	1.55	1.48	1.45	1.41	1.32	1.26	1.19
	.1	1.91	1.83	1.75	1.66	1.61	1.55	1.43	1.35	1.25
	.05	2.16	2.05	1.94	1.82	1.76	1.69	1.53	1.43	1.31
	.02	2.47	2.34	2.19	2.03	1.95	1.86	1.66	1.53	1.38
	.01	2.71	2.54	2.37	2.19	2.09	1.98	1.75	1.61	1.43
	.001	3.46	3.22	2.96	2.68	2.53	2.38	2.04	1.83	1.59
∞	.2	1.6	1.55	1.49	1.42	1.38	1.34	1.24	1.17	1.01
	.1	1.83	1.75	1.67	1.57	1.52	1.46	1.32	1.22	1.02
	.05	2.05	1.95	1.83	1.71	1.64	1.57	1.39	1.27	1.02
	.02	2.32	2.19	2.04	1.88	1.79	1.7	1.47	1.33	1.03
	.01	2.52	2.36	2.19	2	1.9	1.79	1.53	1.36	1.03
	.001	3.14	2.9	2.65	2.38	2.23	2.07	1.71	1.48	1.04

TABLE A-8
χ^2 Distribution

Entry is $\chi^2(a; v)$ where $P[\chi^2(v) \leq \chi^2(a; v)] = a$

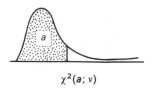

$\chi^2(a; v)$

df						a				
v	.005	.010	.025	.050	.100	.900	.950	.975	.990	.995
1	0.0^4393	0.0^3157	0.0^3982	0.0^2393	0.0158	2.71	3.84	5.02	6.63	7.88
2	0.0100	0.0201	0.0506	0.103	0.211	4.61	5.99	7.38	9.21	10.60
3	0.072	0.115	0.216	0.352	0.584	6.25	7.81	9.35	11.34	12.84
4	0.207	0.297	0.484	0.711	1.064	7.78	9.49	11.14	13.28	14.86
5	0.412	0.554	0.831	1.145	1.61	9.24	11.07	12.83	15.09	16.75
6	0.676	0.872	1.24	1.64	2.20	10.64	12.59	14.45	16.81	18.55
7	0.989	1.24	1.69	2.17	2.83	12.02	14.07	16.01	18.48	20.28
8	1.34	1.65	2.18	2.73	3.49	13.36	15.51	17.53	20.09	21.96
9	1.73	2.09	2.70	3.33	4.17	14.68	16.92	19.02	21.67	23.59
10	2.16	2.56	3.25	3.94	4.87	15.99	18.31	20.84	23.21	25.19
11	2.60	3.05	3.82	4.57	5.58	17.28	19.68	21.92	24.73	26.76
12	3.07	3.57	4.40	5.23	6.30	18.55	21.03	23.34	26.22	28.30
13	3.57	4.11	5.01	5.89	7.04	19.81	22.36	24.74	27.69	29.82
14	4.07	4.66	5.63	6.57	7.79	21.06	23.68	26.12	29.14	31.32
15	4.60	5.23	6.26	7.26	8.55	22.31	25.00	27.49	30.58	32.80
16	5.14	5.81	6.91	7.96	9.31	23.54	26.30	28.85	32.00	34.27
17	5.70	6.41	7.56	8.67	10.09	24.77	27.59	30.19	33.41	35.72
18	6.26	7.01	8.23	9.39	10.86	25.99	28.87	31.53	34.81	37.16
19	6.84	7.63	8.91	10.12	11.65	27.20	30.14	32.85	36.19	38.58
20	7.43	8.26	9.59	10.85	12.44	28.41	31.41	34.17	37.57	40.00
21	8.03	8.90	10.28	11.59	13.24	29.62	32.67	35.48	38.93	41.40
22	8.64	9.54	10.98	12.34	14.04	30.81	33.92	36.78	40.29	42.80
23	9.26	10.20	11.69	13.09	14.85	32.01	35.17	38.08	41.64	44.18
24	9.89	10.86	12.40	13.85	15.66	33.20	36.42	39.36	42.98	45.56
25	10.52	11.52	13.12	14.61	16.47	34.38	37.65	40.65	44.31	46.93
26	11.16	12.20	13.84	15.38	17.29	35.56	38.89	41.92	45.64	48.29
27	11.81	12.88	14.57	16.15	18.11	36.74	40.11	43.19	46.96	49.64
28	12.46	13.56	15.31	16.93	18.94	37.92	41.34	44.46	48.28	50.99
29	13.12	14.26	16.05	17.71	19.77	39.09	42.56	45.72	49.59	52.34
30	13.79	14.95	16.79	18.49	20.60	40.26	43.77	46.98	50.89	53.67
40	20.71	22.16	24.43	26.51	29.05	51.81	55.76	59.34	63.69	66.77
50	27.99	29.71	32.36	34.76	37.69	63.17	67.50	71.42	76.15	79.49
60	35.53	37.48	40.48	43.19	46.46	74.40	79.08	83.30	88.38	91.95
70	43.28	45.44	48.76	51.74	55.33	85.53	90.53	95.02	100.4	104.2
80	51.17	53.54	57.15	60.39	64.28	96.58	101.9	106.6	112.3	116.3
90	59.20	61.75	65.65	69.13	73.29	107.6	113.1	118.1	124.1	128.3
100	67.33	70.06	74.22	77.93	82.36	118.5	124.3	129.6	135.8	140.2

TABLE A-9
Critical Values of Kolmogorov-Smirnov Test Statistic D

n	.20	.10	.05	.02	.01	n	.20	.10	.05	.02	.01
1	.900	.950	.975	.990	.995	21	.226	.259	.287	.321	.344
2	.684	.776	.842	.900	.929	22	.221	.253	.281	.314	.337
3	.565	.636	.708	.785	.829	23	.216	.247	.275	.307	.330
4	.493	.565	.624	.689	.734	24	.212	.242	.269	.301	.323
5	.447	.509	.563	.627	.669	25	.208	.238	.264	.295	.317
6	.410	.468	.519	.577	.617	26	.204	.233	.259	.290	.311
7	.381	.436	.483	.538	.576	27	.200	.229	.254	.284	.305
8	.358	.410	.454	.507	.542	28	.197	.225	.250	.279	.300
9	.339	.387	.430	.480	.513	29	.193	.221	.246	.275	.295
10	.323	.369	.409	.457	.489	30	.190	.218	.242	.270	.290
11	.308	.352	.391	.437	.468	31	.187	.214	.238	.266	.285
12	.296	.338	.375	.419	.449	32	.184	.211	.234	.262	.281
13	.285	.325	.361	.404	.432	33	.182	.208	.231	.258	.277
14	.275	.314	.349	.390	.418	34	.179	.205	.227	.254	.273
15	.266	.304	.338	.377	.404	35	.177	.202	.224	.251	.269
16	.258	.295	.327	.366	.392	36	.174	.199	.221	.247	.265
17	.250	.286	.318	.355	.381	37	.172	.196	.218	.244	.262
18	.244	.279	.309	.346	.371	38	.170	.194	.215	.241	.258
19	.237	.271	.301	.337	.361	39	.168	.191	.213	.238	.255
20	.232	.265	.294	.329	.352	40	.165	.189	.210	.235	.252

Approximation for $n > 40$: $\dfrac{1.07}{\sqrt{n}}$ $\dfrac{1.22}{\sqrt{n}}$ $\dfrac{1.36}{\sqrt{n}}$ $\dfrac{1.52}{\sqrt{n}}$ $\dfrac{1.63}{\sqrt{n}}$

Answers to Selected Exercises

Chapter 2

3. Household income. The existence of extreme values tend to make the median a better measure of central tendency in this case. Note that many government agencies report median rather than average household income in their studies.

4. a) 2.4
   b) 3
   c) Property 1: $(-8 - 2.4) + (-2 - 2.4) + (3 - 2.4) + (5 - 2.4) + (14 - 2.4) = 0$
      Property 2: $(-8 - 2.4)^2 + (-2 - 2.4)^2 + (3 - 2.4)^2 + (5 - 2.4)^2 + (14 - 2.4)^2 = 269.20$
      Property 3: $|-8 - 3| + |-2 - 3| + |3 - 3| + |5 - 3| + |14 - 3| = 29$

5. c) Mean = 204.33, Variance = 552.88, Median = 203.5, Mode = 200, Skewness = .11

6. c) Mean = 74.95, Variance = 281.30, Median = 78, Mode = 90, Skewness = −.18

7. The coefficients of variation are .44, .46, and .48 for regions A, B, and C, respectively. So, household incomes are most even in Region A and least in region C.

8. Using Pearson's measure of skewness, we calculate .38 for A, .42 for B, and .25 for C. Therefore income is least skewed in region C and most in B.

Chapter 4

3. b) (i) (48.33, 60.83); (ii) (51.75, 67.5)
   c) 22.47

4. a) Location Quotients

	Primary	Manufacturing	Services
North	1.75	0.14	0.69
South	0.38	1.78	1.19
East	0.40	1.68	1.26
West	0.29	1.78	1.34

b) The coefficients of localization are 0.34, 0.39 and 0. 14 respectively for the three sectors.

5. b) (i) (58.33,56.67); (ii) (51.72,45:86)
c) 38.55

9. Minimum is 0, maximum ∞.

Chapter 5

5. Since we are interested in rankings, we are interested in the order of the neighborhoods, and we must use permutations. Since a complete ranking will include all 5 of the neighborhoods, we must calculate $P(5, 5) = 5!/(5 - 5)! = 5! = 120$. 24 of the permutations begin with each of the letters A to E. For each sequence beginning with each of the five letters there are 4 groups, one for each of the remaining four letters. For example, for the letter A there are 4 groups AB, AC, AD, and AE. For each of these four groups there are 6 permutations of the remaining 3 letters. For example, for orderings beginning with AB these 6 sequences are $ABCDE, ABCED, ABDCE, ABDEC, ABECD$, and $ABEDC$.

6. This is an application of the hypergeometric rule in which one route must be selected from those available between the towns. The total number of combinations is

$$C_1^4 \cdot C_1^3 \cdot C_1^5 \cdot C_1^2 = \frac{4!}{1!(4 - 1)!} \cdot \frac{3!}{1!(3 - 1)!} \cdot \frac{5!}{1!(5 - 1)!} \cdot \frac{2!}{1!(2 - 1)!} = 4 \cdot 3 \cdot 5 \cdot 2 = 120$$

8. a) $P(\text{female}) = 1 - P(\text{male}) = 1 - .55 = .45$
b) $P(\text{married}) = 1 - P(\text{umarried}) = 1 - .80 = .20$.
c) Within a contingency table, using given information and answers to a) and b) above, we have

	Married	Umarried	
Male	0.05		0.05
Female			0.45
	0.20	0.80	1.00

We can determine other cells by subtraction. Therefore, for example $P(\text{male} \cap \text{unmarried}) = P(\text{male}) + P(\text{unmarried}) - P(\text{male} \cup \text{unmarried})$. The last term is simply $.55 - .05 = .50$, so the required probability is $.55 + .80 - .50 = .85$.
d) Using multiplication theorem, $P(\text{female} \cup \text{unmarried}) = P(\text{female} \mid \text{unmarried}) P(\text{unmarried})$ where $P(\text{female} \mid \text{unmarried}) = P(\text{female} \cup \text{unmarried})/P(\text{unmarried})$

= .30/.80 = .375. Therefore, P(female $\cup$ unmarried) = .375(.80) = .30. This can be more easily determined by simple subtraction within the contingency table above.

9. Each of the digits is independent and we can use the hypergeometric rule:

$$C_1^3 \cdot C_1^4 \cdot C_1^9 \cdot C_1^9 \cdot C_1^9 \cdot C_1^9 = 3 \cdot 4 \cdot 9 \cdot 9 \cdot 9 \cdot 9 = 78732.$$

10. In unsimplified form, $C_1^{26} \cdot C_1^{26} \cdot C_1^{26} \cdot C_1^{26} \cdot C_1^{26} \cdot C_1^{26}$, (b) $C_1^{10} \cdot C_1^{10} \cdot C_1^{10} \cdot C_1^{10} \cdot C_1^{10} \cdot C_1^{10}$, and (c) $C_1^{26} \cdot C_1^{26} \cdot C_1^{26} \cdot C_1^{10} \cdot C_1^{10} \cdot C_1^{10}$.

11. a) $P(A \cap B) = P(A) + P(B) - P(A \cup B) = 1/3 + 2/3 - 1/3$
    b) $P(B|A) = P(B \cup A)/P(A)$ or equivalently $P(A \cup B)/P(B) = 1/3 - 2/3 = 1/2$
    d) No, since $P(B|A) \neq P(B)$; $2/3 \neq 1/2$

13. a) $(1 - 0.1)(1 - 0.1)(1 - 0.1) = .9^3 = .729$
    b) $(0.1)(1 - 0.1)(1 - 0.1) = .081$
    c) No, day to day weather cannot be considered independent. If it is clear today there is a higher probability that tomorrow will be clear than rainy.

14. a) $90/200 = 0.45$
    b) $25/200 = 0.125$
    c) $10/200 = .05$
    d) $50/100 = .50$

15. a) 0.50
    b) 0.40

Chapter 6

4. a) In order to be a valid probability model, all probabilities must be nonnegative and they must add to 1. By inspection, we see that all are probabilities are positive valued. Second, note .30 +.40 + .15 + .15 = 1.0. It is thus a valid distribution.
    b) $E(X) = .30(0) + .40(1) +.15(2) +.15(3) = 0 + .4 +.3 +.45 = 1.15$
    c) Using the tabular format presented in Section 6.1:

x_i	$P(x_i)$	x_i^2	$x_i^2 P(x_i)$
0	0.40	0	0.00
1	0.30	1·	0.30
2	0.15	4	0.60
3	0.15	9	1.35
			2.25

and $V(X) = 2.25 - (1.15)^2 = 2.25 - 1.3225 = 0.9275.$
    d) 1

5. c) $E(X) = .13(0) + .14(1) + .21(2) + .18(3) + .10(4) + .08(5) + .07(6) + .05(7)$
$+ .03(8) + .01(9)$
$= 0 + .14 + .42 + .54 + .40 + .40 + .42 + .35 + .24 + .09$
$= 3.00$

for $V(X)$

x_i	$P(x_i)$	x_i^2	$x_i^2 P(x_i)$
0	0.13	0	0.00
1	0.14	1	0.30
2	0.21	4	0.60
3	0.18	9	1.35
4	0.10	16	1.60
5	0.08	25	2.00
6	0.07	36	2.52
7	0.05	49	2.45
8	0.03	64	1.92
9	0.01	81	0.81
			13.55

and $V(X) = 13.55 - 3.00^2 = 4.55$.

d) First, we calculate the standard deviation as the square root of $V(X)$ or $\sqrt{4.55} = 2.133$. Two standard deviations around the mean $E(X)$ defines the range $[3 - 2(2.133), 3 + 2(2.133)]$ or $[0, 7.266]$. From the table we add $P(x)$ for values 0 to 7 inclusive, yielding .96.

7. a) A uniform random variable takes on the values 1 to 9. Each has a probability of 1/9, so $P(3) = P(7) = 1/9$.
b) This includes the five x values 1, 2, 3, 4, and 5, so $P(X < 6) = 5/9$.
c) This includes the four x values 6, 7, 8, and 9, so $P(X > 6) = 4/9$.
d) $(9 + 1)/2 = 5$.
e) $(9^2 - 1)/12 = 80/12 = 6.67$.

8. a) Using the formula for a Poisson variable

$$P(3) = \frac{e^{-3}3^3}{3!} = .224 \text{ where, } e = 2.71828.$$

b) Using the tables of the Poisson we must add the values for 0, 1, 2, 3, 4, and 5, $P(X < 6) = .0498 + .1494 + .2240 + .2240 + .1680 + .1008 = .916$.
c) Using the table, we calculate $.0504 + .0216 + .0081 + .0027 + .0008 + .0002 + .0001$ .839. Note that this contains more values than problem 7 above.
d) 3
e) 3

9. a) For a binomial with $n = 7$ and $\pi = .3$, $P(2) = C(7,2)(0.3)^2(0.7)^5 = .3177$.
b) $P(5) + P(6) + P(7) = .0250 + .0036 + .0002 = .2582$.

c) $1 - P(0) = 1 - .0824 = .9176.$
d) $P(7) = .0002$
e) No. Because of the movement of weather systems, weather on successive days cannot be considered independent events.

10. a)

Number of points	Number of quadrats
0	9
1	9
2	5
3	2

b) 1
c)

Number of points	Number of quadrats
0	9.20
1	9.20
2	4.60
3	1.53
4+	0.47

d) yes

11. a)

Number of points	Number of quadrats
0	75
1	20
2	5

b) 0.3
c)

Number of points	Number of quadrats
0	74.1
1	22.2
2	3.3
3+	0.3

d) yes

16. b) Random, clustered, dispersed, dispersed, dispersed

Chapter 7

10. a) $\mu = 6$, $\sigma = 2.619$

10. b) The possible samples and corresponding means are:

AA 4.0	BB 3.0	CC 5.0	DD 7.0	EE 4.0	FF 8.0	GG 11
AB 3.5	BC 4.0	CD 6.0	DE 5.5	EF 6.0	FG 9.5	
AC 4.5	BD 5.0	CE 4.5	DF 7.5	EG 7.5		
AD 5.5	BE 3.5	CF 6.5	DG 9.0			
AE 4.0	BF 5.5	CG 8.0				
AF 6.0	BG 7.0					
AG 7.5						

10. c) All samples are equally likely, so appear with probability 1/28. The average sample mean is $\mu_{\bar{x}} = 6$, and its standard deviation is $\sigma_{\bar{x}} = 1.852$.

Chapter 8

4. 99% Confidence Interval: [29.01, 38.81]

6. 95% Confidence Interval [.642, .778]

8. $n = 384$

Chapter 9

7.

$$T = \frac{52 - 40}{32/5} = 1.85$$

a) At $\alpha = 0.05$ level, the critical value for T is 2.064 for a two-tailed test. We cannot reject H_0.
b) PROB-VALUE $\approx .08$

8.

$$T = \frac{32 - 2.6}{108/4} = 2.77$$

a) At the $\alpha = .05$ level, the critical value for T is 2.064. We reject H_0.
b) PROB-VALUE $\approx .01$.

10. The decision rule is given, thus the critical region is pre-determined. The upper tail of the critical region is to the right of

$$Z = \frac{625 - 600}{100/\sqrt{75}} = 2.17$$

From the normal table (A-3), $\alpha/2 = .05$, thus $\alpha = .03$

Chapter 10

2. After forming a difference column, we find a mean difference of 10.88, with a standard deviation of 1.46. Adopting a null hypothesis of no difference, this gives

$$T = \frac{10.88 - 0}{1.46/\sqrt{8}} = 7.46$$

a) The two-tailed PROB-VALUE is approximately .0001.
b) The one-tailed PROB-VALUE is approximately .00005.

4. Assuming the variances are equal, we form the pooled variance estimate

$$S_p^2 = \frac{(60 - 1)(4.2^2 + 3.7^2)}{(60 - 1) + (60 - 1)} = 31.33$$

This gives $\hat{\sigma}_{\bar{x}_1 - \bar{x}_2} = \sqrt{31.33} \ \sqrt{1/60 + 1/60} = 1.02$
a) The 95% confidence interval is therefore

$$7.3 - 6.8 \pm 2(1.02) = [-1.54, 2.54]$$

b) To test the equity of variances, we find $F = 4.2^2/3.4^2 = 1.53$. The PROB-VALUE is only 0.1, thus there is good reason to doubt the hypothesis of equal variances.

Chapter 11

4) With $H_0; p = .5$, $H_A: p > .5$, $n = 12$, the binomial table can be used to find the probability of obtaining 8 or more students favoring Professor Fontanez, when there is no preference in the population at large. We get

$$P(X \geq 8) = .1208 + .0537 + .0161 + .0029 + .0002 = .1937.$$

If we reject the null hypothesis in favor of the alternative, the risk of error is about 20%. We cannot conclude Fontanez is preferred.

6) Though only the sum of ranks is needed, note that the average rank is much lower for the control:

	Sum of Ranks	n	Mean Rank
Control	36	8	4.5
XLC	100	8	12.5
Total	136	16	8.5

The exact probability associated with S = 136 is .0001, thus we reject the null hypothesis of no difference. If we were to use the normal approximation (note that sample sizes are too small for reliable results), we would obtain Z = 3.3, and a PROB-VALUE of .0009.

8) The sum of negative ranks is −36, the sum of positive ranks is 9. Using the normal approximation gives Z = 1.54. The associated two-tailed PROB-VALUE is .062. The null hypothesis of no difference cannot be rejected at the α = .05 level.

11) The expected frequencies are:

	On Floodplain	Off Floodplain
Purchased Insurance	39	21
No Insurance	26	14

The null hypothesis is no relationship between insurance coverage and location. The associated observed Chi-Square statistic is

$$\chi^2 = \frac{5^2}{39} + \frac{5^2}{16} + \frac{5^2}{14} + \frac{5^2}{21} = 4.58$$

For α = .05, the critical value of χ^2 is 3.84. We reject the null hypothesis.

13) The median value is 86.5. We find 15 runs in the sample. The expected number of runs is 16.0, and $P(X \le 15)$ = .424. If we reject the null hypothesis of no difference in favor of a one-sided alternative, we run a large risk of error. We do not reject the null hypothesis.

14) The median is 3.43. We find 8 runs in the sample, as compared to the expected value of 8.0. The exact probability $P(X \le 8)$ is .617. We cannot reject the null hypothesis of no difference.

Chapter 12

6b) $r = .967$

6c) $r_s = .988$

8) $\tau = .956$

Chapter 13

5. c) Intercept $a = 79.97$; regression coefficient $b = .04$
   d) Standard error of estimate $= 4.90$; coefficient of determination $= .80$
   f) Intercept $a = 79.97$; regression coefficient $= .4$
   h) Observation 9; observation 10

6. c) Intercept $a = .53.47$; regression coefficient $b = .0014$
   d) Standard error of estimate $= 13.20$; coefficient of determination $= .06$

Chapter 15

7) The sample autocorrelation and partial autocorrelations are:

Lag	Autocorrelation	Partial Autocorrelation
1	0.792	0.792
2	0.549	0.211
3	0.408	0.137
4	0.250	−0.212
5	0.199	0.275
6	0.236	0.046
7	0.253	0.066
8	0.261	−0.007
9	0.200	−0.164
10	0.046	−0.182
11	−0.099	−0.094
12	−0.211	−0.100
13	−0.304	−0.117
14	−0.336	−0.078
15	−0.287	0.075

Chapter 17

5) There are 8,916,100,448,256 distinct samples.

Index